Bioceramics

Volume 7

Edited by

Örjan H. Andersson
Åbo Akademi University, Turku, Finland

Risto-Pekka Happonen
Antti Yli-Urpo
University of Turku, Finland

Pergamon

Butterworth-Heinemann Ltd
Linacre House, Jordan Hill, Oxford OX2 8DP

A member of the Reed Elsevier plc group

OXFORD LONDON BOSTON
MUNICH NEW DELHI SINGAPORE SYDNEY
TOKYO TORONTO WELLINGTON

First published 1994

British Library Cataloguing in Publication Data
A catalogue record for this book is available from the British Library

Library of Congress Cataloguing in Publication Data
A catalogue record for this book is available from the Library of Congress

ISBN 0 7506 21753

Cover picture

The cover shows a scanning electron microscopy image, enhanced with artificial colour, showing crystals formed during firing of bioactive glass coating. Field width: 40 μm. (by Örjan H. Andersson and Nikita Beliaev)

Printed in Great Britain at The University Press, Cambridge

Bioceramics

Volume 7

Proceedings of the 7th International Symposium on Ceramics in Medicine

Turku, Finland, July 1994

Edited by
Ö.H. Andersson
R-P. Happonen
A. Yli-Urpo

Organizing Committee

President	Kaj H. Karlsson, Åbo Akademi University
Chairman	Antti Yli-Urpo, University of Turku
Vice Chairman	Risto-Pekka Happonen, University of Turku
Secretary General	Örjan H. Andersson, Åbo Akademi University
Member	Allan Aho, Turku University Central Hospital

Preface

The Seventh International Symposium on Ceramics in Medicine was held in Turku/Åbo, Finland, on 28–30 July 1994. The meeting was organized by the Biomaterials Research Group at the Åbo Akademi University and University of Turku.

The Proceedings, *Bioceramics 7*, provides an overview of the latest developments in this field of research, as presented at the Symposium. The topics included vary from preparation of materials to cellular response and clinical applications, thus reflecting the truly interdisciplinary nature of the field. The articles demonstrate advances in our fundamental knowledge and innovative approaches to the development of new bioactive materials.

The International Advisory Board contributed significantly to the scientific quality of the symposium and *Bioceramics 7*. Each of nearly 100 abstracts submitted to the organizers was reviewed by three members of the Board. The following experts comprised the International Advisory Board: W. Bonfield (London), W.T. Butler (Houston), J.E. Davies (Toronto, Canada), K. de Groot (The Netherlands), P. Ducheyne (Philadelphia, USA), L.-E. Eriksson (Gothenburg, Sweden), U.M. Gross (Berlin, Germany), G.W. Hastings (London, UK), L.L. Hench (Gainesville, USA), T. Kokubo (Kyoto, Japan) and T. Yamamuro (Kyoto, Japan).

We do hope that the book will further stimulate and promote the communication between the disciplines involved in bioceramics research. The benefit of the patient remains the driving force behind our efforts.

Örjan Andersson   Risto-Pekka Happonen   Antti Yli-Urpo

CONTENTS

Review

Preparation of Bioceramics

Reaction Kinetics and Mechanisms of Bioactive Ceramics

Bioactive Ceramics and Tissue Reactions

Cells and Materials

Bone Induction

Bioinert and Wear-Resistant Materials

Clinical Results

Review

Bioceramics, Volume 7 Edited by Ö.H. Andersson and A. Yli-Urpo
(*Proceedings of the 7th International Symposium on Ceramics in Medicine, Turku, Finland, July 1994*)

Bioactive Ceramics: Theory and Clinical Applications

L.L. Hench
Advanced Materials Research Center, University of Florida, One Progress Blvd., #14, Alachua, FL 32615, USA

ABSTRACT

Twenty five years of development of bioactive ceramics are reviewed with emphasis on: 1) the sequence of interfacial reactions between implant and tissue, 2) relation of clinical applications to the composition, structure and physical form of the implant (bulk, powders, coatings, composites). Two classes of bioactivity are proposed: Class A bioactive implants are osteoproductive due to both intracellular and extracellular factors and Class B bioactive implants that are osteoconductive due to extracellular factors. Future directions of the field are proposed including: 1) use of biomolecular design of bioactive surfaces to achieve regeneration of tissues, and 2) biomechanical design of bioactive composites to create optimized interfacial gradients in strength, toughness and elastic properties.

INTRODUCTION

Ideally, the physical and chemical properties of an implant material should match the tissue being replaced. This is seldom possible. Nearly all implants are a compromise in form and function and in biochemistry and biomechanics. The challenge in the theory, design and application of biomaterials is to minimize this inevitable compromise. The inevitability lies in the fact that natural tissues are alive. A critical feature of living tissues is their capacity to regenerate and repair themselves. Man-made, prosthetic materials are required only when the natural repair process is insufficient due to disease, damage, or aging. The combination of any two of these factors in a patient; i.e., disease and damage or damage and aging, is especially debilitating. Thus, fracture of an arthritic hip in a 75 year old patient usually requires a total hip replacement to restore function and mobility. However, poor bone quality leads to a poor prognosis for full recovery.

Implants cannot repair themselves. Small cracks irreversibly grow larger and eventually lead to fracture. Small movements at an implant interface inevitably become larger and result in pain and loosening of the prosthesis. Implants cannot modify their shape or structure in response to stress, as do living tissues. Bones remodel their cross sectional area to optimize the distribution of an applied load. Implants only change in response to their environment by a deterioration in mechanical properties due to corrosion or crack growth.

No prosthetic material implanted in living tissues is inert. All materials elicit a response from the tissues at their interface. There are four general types of implant-tissue reactions which lead to different means of attaching prostheses to the musculo-skeletal system.[1] The first implant material used in surgery, and often still the material of choice, is the patient's own bone, termed an autologous implant. It is a resorbable implant and is eventually remodeled as new bone forms. The disadvantages are: low strength, second site surgery is often required, and supply is limited. Freeze dried cadaver bone, *homograft*, is an important alternative but viral infections are a concern.

Historically, the first prosthetic implant materials were biologically inactive and were attached to tissues by mechanical interlock. This means of attachment is termed morphological fixation. The self-curing polymer "bone cement" (PMMA) is extensively used to provide a stable interlock between bone and total joint prostheses. There are millions of implants in people with this type of interface. The success rate has been high, >85% after 5 years, and large numbers of people have benefited from the restoration of function of hips and knees by total joint prostheses.

The Problem: Implant Failure Rate in an Aging Population

The rapidly growing use of implants combined with the increasing life expectancy of patients leads to the following problem: an increasing fraction of patients will outlive the expected lifetime of their implants. Revision surgeries have an even greater probability of failure. Thus, conditions exist for an autocatalytic growth in implant failures and revision surgery. A new generation of implant materials and prostheses designs are needed with double the lifetime of current devices (20-30 years) in order to solve this problem. This is the current challenge in the molecular design of prosthetic materials. The long range challenge is to provide bioactive implants that will simulate repair and regeneration of tissues.

The Source of the Problem: Interfacial Failure

For most implants, failure originates at the interface between the biomaterial and its host tissue. Two factors contribute to this interfacial failure. They are biomechanical and biochemical in nature and depend on the type of tissue being replaced. It is difficult to prevent motion at the interface of devices used to replace parts of the musculo-skeletal system because cyclic mechanical loads are transferred through those tissues. Joint replacements are subjected to interfacial cyclic shear stresses. Interfacial bonds between dissimilar materials are generally weakest in shear. Thus, it is difficult to achieve long term interfacial stability between bones and implants because of the shear stresses at the interface (Table 1). Extensive wear can also occur in the articulating surfaces of prosthetic joints. The wear of ultrahigh molecular weight polyethylene used in joint prostheses changes local stress distributions and accelerates interfacial failure. Polyethylene wear debris migrates into tissues and causes additional damage to bone as a consequence of chronic inflammation. The combination of wear and deterioration of the bone-implant

Table 1

THE PROBLEM	THE SOLUTION
STABILITY OF THE BONE-IMPLANT INTERFACE STRESS SHIELDING OF BONE	A NEW GENERATION OF IMPLANTS DESIGNED TO: - ACHIEVE INTERFACIAL STABILITY BY BONDING TO BONE AND SOFT CONNECTIVE TISSUES - MINIMIZE STRESS SHIELDING BY HAVING AN ELASTIC MODULUS SIMILAR TO BONE - MINIMIZE FRACTURE AND FATIGUE BY USE OF TOUGHENING TECHNOLOGY - ASSIST IN THE AUGMENTATION AND REGENERATION OF TISSUES INSTEAD OF REPLACING TISSUES
LEADS TO WEAR DEBRIS IMPLANT LOOSENING DETERIORATION OF BONE FRACTURE PAIN REVISION SURGERY	

interface can be catastrophic and is the cause of the largest proportion of failed orthopedic implants.

Figures 1 and 2 illustrate the extent of the biomechanical problem at the interface between bone and orthopaedic implants. Figure 1 shows the deterioration in trabecular bone volume as a function of age of women.[2] By the time women reach 60 years of age many have lost nearly half of their bone mass. Figure 1 shows the consequence of losing bone mass; the strength of long bones and vertebrae are reduced with age which increases the probability of fracture. On Fig. 2 is plotted the range of modulus of elasticity of cortical bone and cancellous bone.[2] Values for cortical bone are between 7 to 25 GPa, depending upon age and type of bone, and between 0.05 GPa to 0.5 GPa for cancellous bone. The composite structure of cortical bone results in a high fracture toughness, K_{1C}=2 to 12 MPa $m^{-1/2}$, and a high strain to failure, 1 to 3%, which is important with respect to implant interfacial behavior.

Cancellous bone, also called trabecular or spongy bone, is less dense than cortical bone and consequently has a lower modulus of elasticity and higher strain to failure, 5 to 7%. The difference in stiffness between the two types of natural bone tissues ensures a gradient in mechanical load across a bone. Many types of implants for repair of the skeletal system are in contact with both trabecular and cortical bone. It is impossible for most prosthetic materials to produce similar gradients of stiffness between an implant and its host tissues. Aging has its most severe effect on the amount and strength of trabecular bone. Women are especially susceptible to loss of trabecular bone (Fig. 1).

The left side of Fig. 2 shows the range of moduli of elasticity of various "bioinert" metallic and ceramic prostheses. There are large differences between the moduli of bone and those of current orthopaedic biomaterials. For example, medical grade

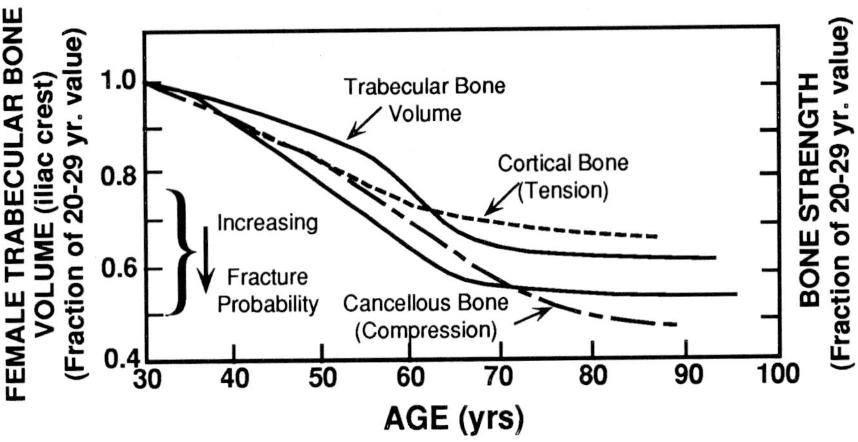

Figure 1. Effect of age on female trabecular bone volume of the Iliac crest and strength of cortical and cancellous bone. Data plotted from Coupron et al., 1975, Melson, 1978 and Yamada, 1970.

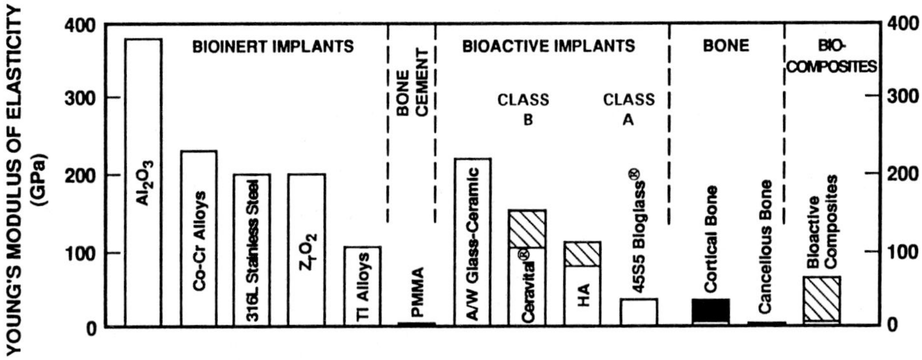

Figure 2. Modulus of elasticity (GPa) for prosthetic materials compared with bone (Hench and Wilson, 1993).

alumina (>99% Al_2O_3) has a Young's modulus of 380 GPa. The modulus mismatch between an alumina implant and its cortical bone interface is in the range of 15-55X. The modulus mismatch with cancellous bone is enormous; e.g., 760X to 7600X.

The problem with a mismatch in elastic modulus across an interface is that the higher modulus implant will carry most of the load. The bone will be "stress shielded". This is undesirable because living bone must be under some tensile load to remain healthy. Bone that is unloaded or is loaded in compression will undergo a biological change that leads to resorbtion and weakening. The interface between a stress shielded bone and an implant will deteriorate as the bone structure is weakened. Loosening and/or fracture of the bone, the interface, or the implant can result. Wear debris at the interface will accelerate weakening of stress shielded bone due to increased cellular activity involved in removing the foreign material. Use of PMMA bone cement also results in death of bone at the interface due to an exothermic reaction during polymerization. The dead bone is less elastic than living bone and is easier to fracture. Cellular remodeling of the dead bone also occurs which leads to additional weakening of the implant-bone interface.

Bioactive Ceramics as a Solution to the Problem

Certain compositions of glasses, glass-ceramics and ceramics composed primarily of SiO_2-CaO-Na_2O-P_2O_5 develop a layer of hydroxy-carbonate-apatite (HCA) on their surface during exposure to body fluids. Collagen fibrils are incorporated within the growing inorganic mineral layer which produces a mechanically strong bond between the implant and the host tissue. This type of tissue attachment is termed *bioactive fixation* and is described in numerous reviews and books.[1-6]

The rate of bonding of bioactive implants to bone depends on composition of the material, as shown in Fig. 3. Compositions with 42-53 mol% SiO_2 form a bond to bone very rapidly, within days and also form an adherent bond with soft tissues.[6] Compositions with 54-60 mol% SiO_2 require two to four weeks to form a bonded bone interface and do not form a bond with soft tissues. Compositions with >60 mol% SiO_2 do not bond with bone or soft tissues and are nearly bioinert materials.

Limitations of Bioactive Ceramics

Twenty five years of research on bioactive materials make it possible to compare *in vivo* and *in vitro* results of the SiO_2-Na_2O-CaO-P_2O_5 glasses with various glass-ceramics (Ceravital® and A/W glass-ceramics), SiO_2-CaO-P_2O_5 glasses, porous

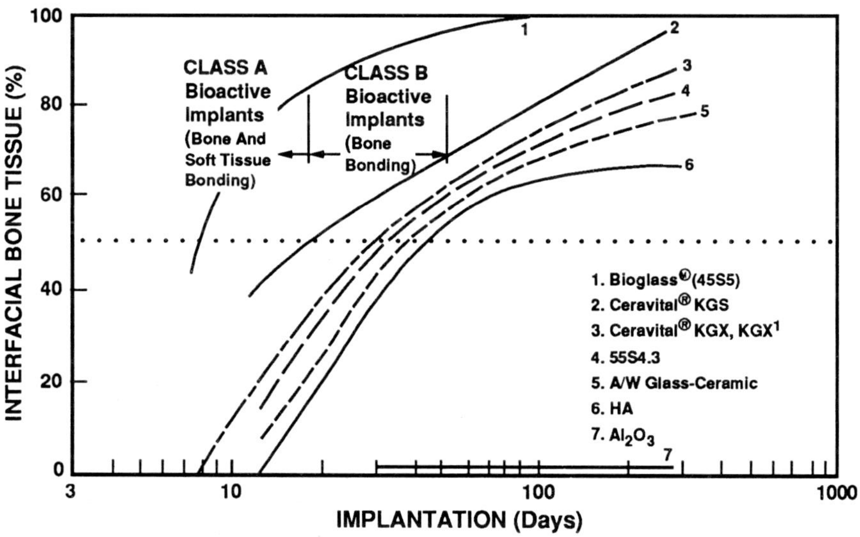

Figure 3. Difference in rate of interfacial bone bonding (rat or rabbit tibia) of Class A vs Class B bioactive implants.

SiO_2 gel-glasses, many multicomponent bioactive glasses and synthetic hydroxyapatite (HA). All of the above materials form a mechanically strong interfacial bond with bone. The strength of the bond is generally equivalent to or greater than the strength of the host bone, depending upon test conditions. With the exception of A/W glass-ceramic, most bioactive ceramics have a flexural strength, strain-to-failure, and fracture toughness that is less than bone. Also, the elastic moduli of the stronger and tougher bioactive glass-ceramics are greater than both cortical and cancellous bone (Fig. 2). The long term consequence of a large mismatch of elastic modulus across the interface of a load bearing bioactive ceramic bonded to bone is not known. A large gradient in elastic modulus may lead to extensive stress shielding of the bone and eventual fracture of the bone distal or proximal to the implant. Because of these biomechanical limitations, clinical use of bioactive ceramics has been restricted to several areas:[1-5] 1) as coatings on fracture metallic alloys for load bearing applications; 2) as vertebral prostheses and iliac crest replacement where the loads are primarily compressive; 3) as relatively non-load bearing devices, such as dental implants to maintain the alveolar ridge and middle ear prostheses; 4) for bone augmentation; and 5) for bone restoration, such as required in periodontal defects. See refs. 1-3 for a summary of clinical applications.

Another limitation of bioactive ceramics is that the materials with high fracture toughness and high flexural strength also have a high elastic modulus and a low bioactive index (defined as $I_B = 100/t_{0.5bb}$; e.g. the inverse of the time required for more than 50% of the implant interface to be bonded to bone) (Fig. 3). Many compositions bond to bone but very few bond rapidly.[8] Thus, there is a compromise between the mechanical behavior of bioactive implants and their bioactivity. Clinical applications have evolved such as to minimize this compromise. The clinical uses of A/W glass-ceramic are based upon its excellent mechanical properties and rate of bonding is not critical. In contrast, use of 45S5 Bioglass® particulate in repair of periodontal defects takes advantage of its high bioactivity index and minimizes mechanical property requirements.

Theory of Bioactivity (Class A and Class B Bioactive Materials)

The large difference in rate of bone bonding to bioactive implants indicates that different biochemical factors are occurring at the implant-tissue interface. This difference in rate of tissue proliferation is amplified when the bioactive materials are compared in the form of particulates. Wilson and Low[9] compared 45S5 Bioglass® particulate with HA and other calcium phosphate granules in the treatment of surgically created periodontal defects. Use of the most bioactive powders (45S5 Bioglass®) led to a more rapid and complete repair of surgically created periodontal defects in *Patus* monkeys than was observed for the lower bioactivity HA or resorbable tricalcium phosphate (TCP) powders. Bioactive glass powders with high bioactivity also resulted in reattachment of the periodontal ligament and little or no epithelial downgrowth. The highly bioactive glass powders (I_B = 8 to 10) showed both osteoconduction *and* osteoproduction of bone, whereas the powders with lower I_B values (I_B = 3 to 4) exhibited only osteoconduction. *Osteoproduction* has been defined by Wilson as *"The process whereby a bioactive surface is colonized by osteogenic stem cells free in the defect environment as a result of surgical intervention."* This differs from *osteoconduction*, where the implant simply provides a biocompatible interface along which bone migrates. Recall that *osteoinduction* is considered to be the production of bone in an ectopic site, i.e., one in which bone is not normally found, such as muscle or skin. No alloplastic implant is osteoinductive regardless of level of bioactivity.

Wilson et al.[10] have also compared the effectiveness of 45S5 Bioglass® with autologous bone in the augmentation of canine ribs. The bioactive glass led to marginally more bone growth than autologous bone in this model. Mixtures of bioactive glass and autologous bone where even more effective, with a 50/50 mix resulting in twice the quantity of new bone in the 6 week experiment than autologous bone alone. The surface of the bioactive glass particles were rapidly colonized by osteoblasts which produced substantial quantities of new bone as early as two weeks. These findings have been extended in a rabbit tibial model by Oonishi et al.[11] where 45S5 Bioglass® particulate has been shown to enhance bone growth many times faster than HA particulate.

The *in vivo* studies indicate that there are two classes of bioactive materials; Class A bioactivity leads to osteoproduction and Class B bioactivity leads to osteoconduction. The key question is *"Why the different tissue response?"* I propose the following hypothesis as the answer, *"Class A, osteoproductive, bioactivity occurs when a material elicits both an intracellular and an extracellular response at its interface. Class B, osteoconductive, bioactivity occurs when a material elicits only an extracellular response at its interface."*

Figure 4 summarizes the *in vivo* and *in vitro* data that support the above hypothesis. Class A bioactive materials release soluble silicon in the form of silicic acid due to ion exchange with H^+ and H_3O^+ (Fig. 4A). The release occurs immediately upon exposure of the glass to an *in vitro* solution or to body fluids. The concentration of silicon in solution continues until a solubility limit is reached which depends greatly upon pH and the concentration of other species in solution which can lead to precipitation of complex silicate phases. Network dissolution increases the rate of silicon release. Class A bioactive glasses release soluble silicon by both ion exchange and network dissolution. Class B bioactive materials have low or zero rates of ion exchange, low rates of network dissolution and therefore release small, or zero amounts of soluble silicon (Fig. 4A).

The importance of soluble silicon on bone growth was first demonstrated by Carlisle and coworkers[12] in chicks and confirmed by Schwarz and Milne in rat.[13] Carlisle's studies indicate that silicon serves both a *"specific metabolic role in connective tissue at a cellular level"* and a *"structural role"* where silicon is chemically combined in the glycosaminoglycan-protein complexes of ground substance that surround collagen, elastic fibers and cells.

Studies by Keeting et al.[14] on human osteoblast-like (hOB) cells have identified a potential metabolic function served by soluble silicon. They show that the dissolution

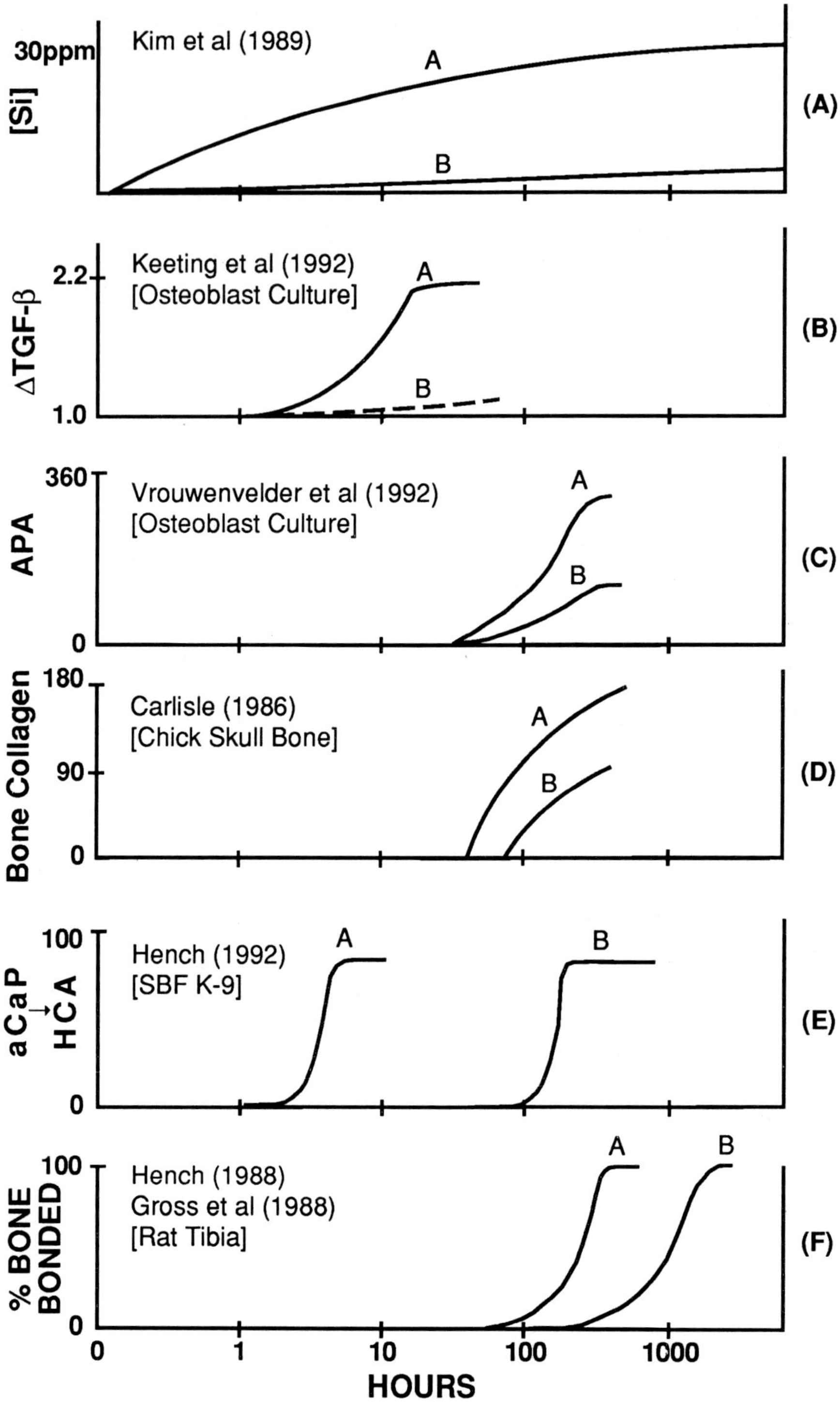

Figure 4. Time dependence of Class A vs Class B bioactive materials.

product of zeolite-A, which is high in soluble silicon, *"was a potent mitogen for hOB cells."* The 48 hour inorganic treatment induced a dose-dependent increase in DNA synthesis in the cultures and increased by 3X the proportion of cells in mitosis. Alkaline phosphatase activity (APA) and osteocalcin release was enhanced but collagen production per individual cell was not altered. Induction of genetically controlled autocrine factors appear to be responsible for the mitogenic response. In particular, the steady-state mRNA levels of transforming growth factor $ß_1$ (TGF-$ß_1$), which is a potent mitogen for osteoblasts, were increased. The inorganic solution induced the release of the latent form of TGF-ß protein into the conditioned medium within 6 hours (Fig 4B). Keeting et al. conclude, *"ZA induces the proliferation and differentiation of cells of the osteoblast lineage."* Figure 4B compares the time dependent release of TGF-$ß_1$ from cells exposed to soluble silicon (Class A environment) with cells not exposed to soluble silicon (Class B environment).

Studies by Vrouwenvelder et al.[15] of primary osteoblast-like cultures grown on Class A (45S5 Bioglass®) and B substrates (synthetic HA) (Fig. 4C) demonstrate the longer term consequences of the intracellular mitogenic effect of soluble silicon depicted in Fig. 4B. Figure 4C shows that by 6 days there was an enhanced quantity of APA from the cultures grown on the Class A substrates, by 8 days the APA concentration had doubled. DNA content of the cultures was also greater on the Class A substrates indicating a mitogenic enhancement.

The enhanced mitosis of osteoblasts, through the stimulation of formation and release of TGF-$ß_1$, explains the concentration effects of soluble silicon on the growth and mineralization of the skull bones of chicks reported by Carlisle.[12] Bone grew rapidly in the presence of silicon (Class A environment) whereas a silicon deficient environment (Class B) retarded bone growth.

It is proposed that Class A implants provide both an intracellular effect on bone proliferation due to release of soluble silicon and also an extracellular effect by chemisorbtion of bone growth proteins, such as TGF-$ß_1$, on the implant surface. The extracellular effect is due to the high surface area and nanometer scale porosity of the hydrated silica gel and hydroxycarbonate apatite (HCA) layer which forms rapidly on Class A bioactive materials.[1] Figure 4E summarizes the surface reaction studies of various bioactive glasses and glass-ceramics.[16] Class A materials form a silica-gel layer (not shown) within minutes of exposure to body fluids. Damen and Ten Cate[17] have shown that soluble silicon accelerates the precipitation of an amorphous calcium phosphate (a-CaP) phase from solution. The a-CaP phase forms within the pores of the silica gel layer where the porosity and silanols provide a heterogeneous nucleation mechanism for (HCA) crystallization.[18] Thus, the crystalline HCA layer develops within a few hours on Class A bioactive substrates whereas it takes many days, or even weeks, to form on Class B materials. The negatively charged silica gel and defect HCA crystals provide sites for chemisorbtion of TGF-$ß_1$ and other growth factors released from proliferating osteoblasts. The adsorbed growth factors enhance the differentiation and mitosis of stem cells at the implant interface, thereby creating an autocatalytic growth of bone on Class A implants.

The *in vivo* consequence of the autocatalytic process of cell proliferation is illustrated in Fig. 4F, based upon rat tibial implant data compiled by Hench and Clark[19] and Gross et al.[4] Class A materials begin to develop an interfacial bone bond within a few days and by 20-30 days there is 100% interfacial bonding. In contrast, Class B implants require several months to develop a large fraction of interfacial bonded bone. The comparative study of Class A (45S5 Bioglass®) versus Class B (HA) particulate in a rabbit tibial model by Oonishi et al.[11] shows results equivalent to Fig. 4F, as do the canine and monkey studies of Wilson et al.[9,10]

The Future: Class A Bioactive Composites

Molecularly designed composite materials that mimic the structure and properties of bone offer the greatest potential for solving the problems of interfacial stability and stress shielding of load-bearing prostheses, as described by Bonfield.[20] This is because the modulus of elasticity and fracture toughness of a composite is a function of many variables which can be controlled during processing, including modulus of the matrix material and modulus, volume fraction, aspect ratio and orientation of the dispersed phase. Various combinations of a low-elastic-modulus matrix, such as polyethylene, and a higher elastic-modulus dispersed phase, such as bioactive glass, can yield composites with properties that are similar to those of bone. This type of composite should eliminate stress shielding of bone because there is no mismatch in elastic modulus across the implant-bone interface. Use of a dispersed phase that has Class A bioactivity can ensure that rapid bone bonding occurs so that the stresses are transferred across the interface. Orienting the dispersed phase, in the form of fibers or whiskers, can provide anisotropic mechanical properties, a characteristic of natural bones.

A new direction of research shows promise for tailoring Class A bioactivity at a molecular level, e.g., the discovery that alkoxide-based sol-gel derived SiO_2-CaO-P_2O_5 glasses have an expanded compositional range of bioactivity over glasses or ceramics made by traditional processes.[21] Figure 5 compares melt-derived bioactive

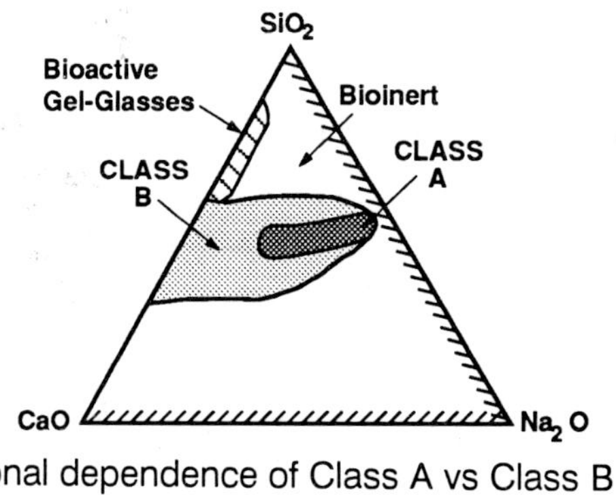

Figure 5. Compositional dependence of Class A vs Class B bioactive glasses and gel-glasses.

glasses and glass-ceramics with this new compositional range of gel-glasses. Gel-derived glasses with even 100% SiO_2 develop HCA layers in SBF whereas the limit for melt-glasses is 60%. Gel-glasses heterogeneously nucleate HCA crystals on their surface within minutes of exposure to physiological solutions. The origin of these important differences is the large volume fraction of porosity, the large concentration of silanols and three-membered and four-membered siloxane rings on the surface of gel-glasses.

Thus, Class A bioactive gel-glasses offer several potential advantages over current bioactive ceramics:

1) Potential for controlling intracellular mitogenic factors by regulating rate of soluble silicon release and solution pH without network dissolution.

2) Enhancement of the extracellular adsorption of growth factors by rapid nucleation of HCA by surface pores.

3) Potential for controlled time release of growth factors or hormones, such as parathyroid hormone from pores, for therapeutic reversal of osteoporosis or osteopenia at an implant interface.

4) Enhanced binding of a polymeric matrix to produce a toughened composite that has Class A bioactivity.

This new direction of research requires understanding the interaction of organic and inorganic species at a molecular level. Recently the author and Jon West have shown, using quantum mechanical molecular orbital (MO) calculations, that surface silanol sites can bind with specific charge sites on protein molecules.[22] This MO method offers the potential for predicting the various biological interactions that are occurring with Class A bioactive materials. Thus, there is hope that in the future bioceramics will be used to enhance the regeneration of tissues instead of the replacement of tissues.

ACKNOWLEDGEMENT

The authors gratefully acknowledge the financial support of the Air Force Office of Scientific Research, Grant No. F49620-92-J-0351.

REFERENCES

1. Hench, L.L. J. Am. Ceram. Soc. 1991, 74[7], 1487-1510.

2. Hench, L.L. and Wilson, J. In An Introduction to Bioceramics, World Scientific Publishing Co., London, 1993, 1-24.

3. Yamamuro, T., Hench, L.L., and Wilson, J. In Handbook on Bioactive Ceramics, Vols. I and II, CRC Press, Boca Raton, Florida, 1990.

4. Gross, U., et al. CRC Critical Reviews in Biocomp. 1988, 4, 2.

5. Davis, J.E. In The Bone-Biomaterial Interface, University of Toronto Press, 1991.

6. Wilson, J. and Nolletti, D. In Handbook of Bioactive Ceramics, Vol. I, CRC Press, Boca Raton, Florida, 1990, 283-302.

7. Wilson, J., et al., "Clinical Applications of Bioglass® Implants," this proceedings, 1994.

8. Hench, L.L. In The Bone-Biomaterial Interface, 1991, University of Toronto Press, 33-44.

9. Wilson, J. and Low, S.B. J. Appl. Biomaterials 1992, 3, 123-129.

10. Wilson, J., Yu, L.T., and Beale, B.S. In Bioceramics 5, Kobonshi Kankokai, Inc., Kyoto, Japan, 1992, 139-146.

11. Oonishi, H., et al., "Bone Growth into Spaces Between 45S5 Bioglass® Granules," this proceedings, 1995.

12. Carlisle, E. In Silicon Biochemistry, Wiley, New York, 1986, 123-136.

13. Schwarz, K. and Milne, D.B. Nature 1972, 239, 333-334.

14. Keeting, P.E., Oursler, M.J., Wiegand, K.E., Bonde, S.K., Spelsberg, T.C., Riggs, B.L. J. Bone and Mineral Res. 1992, 7[11], 1281-1289.

15. Vrouwenvelder, W.C.A., Groot, C.G., and de Groot, K. J. Biomed. Mater. Res. 1993, 27, 465-475.

16. Hench, L.L. and LaTorre, G.P. In Bioceramics 5, Kobonshi Kankokai, Inc., Kyoto, Japan, 1992, 67-74.

17. Damen, J.J.M. and Ten Cate, J.M. J. Dent. Res 1989, 68(9), 1355-1359.

18. Pereira, M.M., Clark, A.E. and Hench, L.L., "Effect of Texture on the Rate of Hydroxyapatite Formation on Silica Gel Surface," J. Am. Ceram. Soc., 1994.

19. Hench, L.L. and Clark, A.E. In Biocompatibility of Orthopaedic Implants, Vol II, CRC Press. Boca Raton, Florida, 1982, Chapter 6.

20. Bonfield, W. Ann. NY Acad. Sci. 1988, 523, 173.

21. Li, R. Clark, A.E., Hench, L.L. J. Appl. Biomaterials. 1991, 2, 231-239.

22. Hench, L.L. and West, J.K. In Bioceramics 6, Butterworth-Heinemann Ltd., Oxford, England, 1993, 35-40.

Preparation of Bioceramics

Bioceramics, Volume 7 Edited by Ö.H. Andersson and A. Yli-Urpo
(*Proceedings of the 7th International Symposium on Ceramics in Medicine, Turku, Finland, July 1994*)

Fracture Toughness Evaluation of Sintered Hydroxyapatite

J.A. Planell,[1] M. Vallet-Regí,[2] E. Fernández,[1] L.M. Rodríguez,[2] A. Salinas,[2] O. Bermúdez,[3] B. Baraduc,[4] F.J. Gil[1] and F.C.M. Driessens[1]

[1] Dpt. de Ciència dels Materials i Enginyeria Metal.lúrgica-UPC. Avda. Diagonal 647, E-08028-Barcelona-Spain.
[2] Dpto. de Química Inorgánica y Bioinorgánica – Facultad de Farmacia – Universidad Complutense. E-28040-Madrid-Spain.
[3] Dpto. de Física. Universidad del Cauca – Popayán-Colombia.
[4] Ecole Nationale d'Ingenieurs de TARBES (ENIT). Avenue d'Azereix-B.P. 1629. 65016-Tarbes-Cedex – France.

ABSTRACT

Indentation tests are widely used for the evaluation of fracture toughness of dense hydroxyapatite. However, a wide variety of equations for the calculation of fracture toughness of ceramic materials by means of a Vickers indentation are available. These equations are meant to reproduce the values obtained when conventional mechanical testing methods are used. In the present work six different equations widely used in ceramic studies are considered and the results obtained are compared with those obtained when SENB specimens are tested in four point bending. The role played by the Youngs' modulus and the microstructure of the material are considered.

INTRODUCTION

The interest in the use of calcium phosphates, and more specifically hydroxyapatite (HA), as bone substitutive material has grown during the last two decades and consistent research has yielded an abundant and highly relevant scientific literature. This is the case for the preparation, microstructure and mechanical properties of dense polycrystalline sintered HA [1-8].

The starting HA powders may be either prepared in the laboratory [1,5,7] or commercially supplied [2-4,6,8]. They have been thoroughly characterized by means of different techniques such as XRD, IR, TGA, photocolourimetric analysis, AAS, TDA, surface area analysis, particle size analysis, SEM, EDS, TEM, ED, dilatometry and OM. The sintered materials have been similarly characterized and important parameters like porosity and grain size have been evaluated respectively by mercury porosimetry and SEM, OM and replica specimens observed by TEM.

Given the structural function that a bone substitute should perform in service, the mechanical properties are always taken into account, mainly in terms of strength. When fracture toughness is considered [3-6,8], it is understood as K_{IC} , being the reported values never much higher than 1 MPa.m$^{1/2}$. In most cases, K_{IC} is evaluated by means of indentation methods. One of the main advantages of the indentation methods is that only a small amount of material is required. However, the revision of the literature on indentation K_{IC} testing of ceramic materials reveals that a great number of equations is available [9]. Such equations are meant to adjust the values obtained with conventional specimens such as SENB, CVNB and DCB.

The present work compares the K_{IC} results obtained using six different K_{IC} indentation equations [10-15] with those obtained with SENB specimens tested in 4 point bending. The influence of the Young's modulus (E) of the material is also analyzed.

EXPERIMENTAL METHODS

The initial HA powder has been synthesized in the laboratory and characterized by means of XRD (Siemens), TGA (Seiko) and IR (Perkin Elmer).

The HA powder was synthetized by reacting $Ca(NO_3)_2$ (1 M) with KH_2PO_4 (0.6 M) dissolved in H_2O free of CO_2. By adding NaOH (2 M), a pH = 10 is reached. After 16 hours agitation and 24 hours rest, the precipitate was centrifuged and aged during a week in a 0.1 M calcium nitrate solution. The aged precipitate was washed 4 times in boiling water during 1.5 hours and centrifuged again. Finally it was dried during 48 hours in an oven at 104 °C. This is the initial powder or material 1. The TGA showed a continuous loss of weight of about 6% between 20 °C and 300 °C. Up to 600 °C the weight does not change significantly, but between 700 °C and 850 °C the weight drops again steeply by about 4%. Then, up to 1200 °C, weight is lost at an approximate constant rate. The overall weight loss between 20 °C and 1200 °C amounts about 13%. In order to reduce the weight loss when sintering, the initial powder was annealed at 850 °C during 1 hour, and this powder is known as material 2. The TGA of material 2 shows that the weight stays constant up to 520 °C and it decreases at an approximately constant rate by about 3% up to 1200 °C.

Beam specimens of 58 mm x 8 mm x 3.5 mm were moulded with material 2 in a die where a pressure of 250 MPa was maintained during 6 minutes. The samples were then sintered at 1000 °C during 1 hour (material 3) and 22 hours (material 4) respectively, and at 1200 °C during 1 hour (material 5) and 22 hours (material 6) respectively. The IR spectra of these materials are very similar, although the intensity of the OH bands is higher in material 2 than in materials 3 - 6. The absence of absorption bands at 1150 cm^{-1} shows that there is no amorphous calcium phosphate. Carbonate is not detected. XRD reveals minor amounts of CaO.

The density of the specimens was measured using Archimedes' principle. The percentage of the theoretical density (3.156 g/cm^3) and porosity are shown in Table 1. 5 specimens (SENB) of each material were notched, whilst the rest were kept unnotched for E evaluation. The tests were performed in 4 point bending up to fracture in an Instron electromechanic machine equipped with a 10 KN capacity load cell and working at a cross-head speed of 0.5 mm/min. E was calculated as:

$$E = \frac{K \cdot slope \cdot L^3}{W \cdot B^3} \quad (1)$$

where K = 0.17, corresponding to a span ratio of 2, L is the length of the sample, W its width and B its thickness. Shrinkage from the initial size was observed when the specimens were fired, as shown in Table 1.

Table 1.- Percentages of shrinkage, theoretical density and porosity, Young's moduli and fracture toughness (SENB specimens) of the 4 materials.

Mat	L (%)	W (%)	B (%)	% D_{th}	% Por	E (GPa)	K_{IC} (MPa.m$^{1/2}$)
3	3	3 ± 1	24 ± 2	75	25	3 ± 1	0.34 ± 0.01
4	3	6 ± 1	19 ± 3	77	23	13 ± 1	0.38 ± 0.05
5	10	13 ± 1	24 ± 1	83	17	23 ± 7	0.90 ± 0.04
6	12	15 ± 1	27 ± 2	95	5	25 ± 9	0.95 ± 0.11

The fracture toughness is evaluated in a 4 point bending test according to

$$K_{IC} = Y S \sqrt{a} \qquad S = \frac{3 P l}{B W^2}$$

$$Y = 1.99 - 2.47 \left(\frac{a}{W}\right) + 12.97 \left(\frac{a}{W}\right)^2 - 23.17 \left(\frac{a}{W}\right)^3 + 24.8 \left(\frac{a}{W}\right)^4 \quad (2)$$

The notch a was machined in such a way that a/W ≈ 0.5, and l corresponds to the span of the upper rollers of the testing jig (10 mm). The indentation tests were performed in a Matzuzawa microhardness tester at 4.9 and 9.8 N loads, with Vickers and Knoop indenters. A SEM JEOL 6400 was used for the fractographic studies.

RESULTS AND DISCUSSION

The elastic moduli and fracture toughness values obtained in 4 point bending tests for SENB and 4 point bending specimens are shown in Table 1.

From the fractographic analysis it was observed that for materials 3, 4 and 5 the crack propagation is intergranular, the porosity is high and the grain size is not larger than 1 μm. However, for material 6, the average grain size is about 4 μm, the porosity is low, and surprisingly, the crack propagation is transgranular in the denser regions, whilst it is intergranular in the less dense regions, as shown in Figure 1. A specimen with a compactation defect, a crack parallel to the long axis, showed that the sintering process is correct and grains grow homogeneously being porosity low, as shown in Figure 2. In the compact regions, near the defect, transgranular fracture is observed.

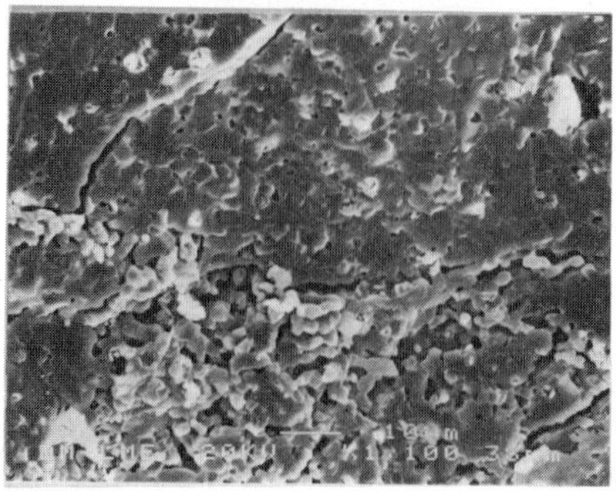

Figure 1. Compacted and porous regions where transgranular and intergranular fracture can be respectively observed.

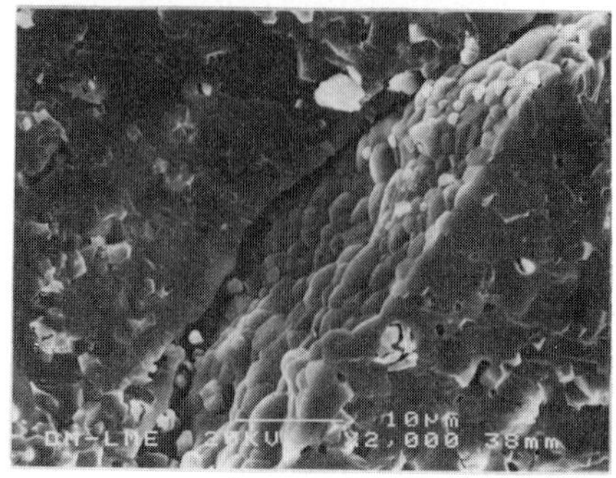

Figure 2. Compactation defect, which allows to see the sintered grains. Transgranular fracture is observed around the defect.

The equations used to calculate K_{IC} by means of the indentation method are shown in Table 2 in terms of Vickers hardness. P is the indenting load, a is the half diagonal of the Vickers indentation, and c is the length of the crack emanating from the indentation. In equation 4, the parameter 1.51 has been calculated in terms of the Poisson ratio of HA, taken as 0.27. Being the aim of these equations to adjust the values obtained in conventional mechanical tests, their parameters may not have a clear physical meaning. Equations 1, 2, 5 and 6 consider that the emanating radial cracks are median half-penny type, equations 3 and 6 model Palmquist radial cracks and finally equation 4 is a universal equation adjusting any kind of radial crack.

Table 2. Selected equations for fracture toughness measurement of hydroxyapatite ceramics with Vickers hardness.

Number	Equation	Reference
1	$0.0154*(E/H_v)^{1/2}*(P/c^{3/2})$	[10]
2	$0.0095*(E/H_v)^{2/3}*(P/c^{3/2})$	[11]
3	$0.0143*(E/H_v)^{2/3}*(P/c^{3/2})*[(c/a) - 1]^{-1/2}$	[12]
4	$0.0495*(H_v a^{1/2})*(E/H_v)^{2/5}*(c/a)^{[(c/18a) - 1.51]}$	[13]
5	$0.0303*(H_v a^{1/2})*(E/H_v)^{2/5}*\log(8.4\ a/c)$	[14]
6	$0.0782*(H_v a^{1/2})*(E/H_v)^{2/5}*(c/a)^{-1.56}$	[15]

The hardness measurements in materials 3,4 and 5 were difficult to carry out and large errors were introduced due to their porosity. This is the reason why the indentation studies were only considered for material 6. The Vickers hardness measured with 1 Kg load during 15 s is 353 ± 28, in Kg/mm^2.

In all equations, K_{IC} depends on the ratio E/H_v. This means that K_{IC} can be evaluated either directly from the E values obtained in the 4 point bending test and the H_v values obtained by a Vickers indentation, or by measuring the elastic recovery of a Knoop indentation according to the relationship of Marshall, Noma and Evans [16]:

$$\frac{b'}{a'} = \frac{b}{a} - \alpha\left(\frac{H}{E}\right) \tag{3}$$

where b' and a' are the elastically recovered short and long diagonals of a Knoop indentation, b and a are the short and long diagonals of the Knoop indenter, E is the Youngs' modulus, H the hardness measured with a Vickers indenter and α is a correlation parameter. This relationship yields a value E = 78 ± 22 GPa, about a factor of 3 higher than the value obtained in 4 point bending. b'/a' from a Knoop indentation gives directly the ratio E/H which is transformed into E/H_v and introduced into the different equations. Figures 3 and 4 compare the ranges obtained with each equation with that obtained in 4 point bending tests using SENB specimens.

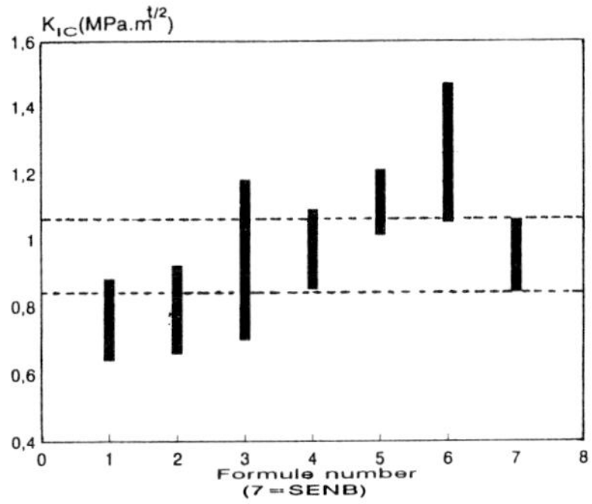

Figure 3. Comparison of K_{IC} obtained from the 6 equations considered and from SENB tests (E obtained from a Knoop indentation).

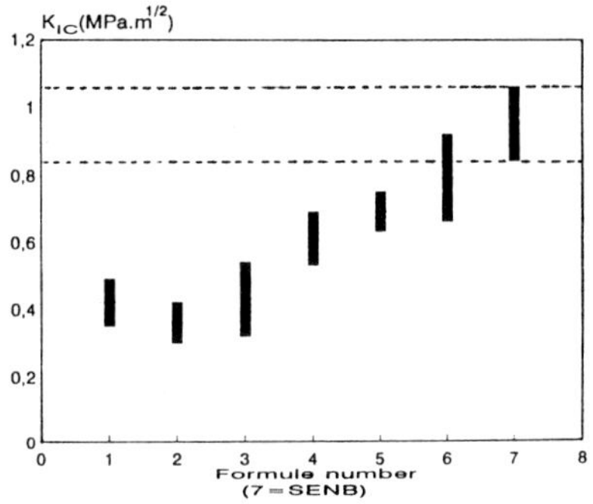

Figure 4. Comparison of K_{IC} obtained from the 6 equations considered and from SENB tests (E obtained from 4 point bending tests).

It can be noticed that K_{IC} increases consistently from equation 1 to equation 6, with the exception of equation 2 when E is taken from a conventional bending test. This behaviour can be explained if the parameters of the different equations are carefully analyzed. Equations 3 and 4 give values closer to those obtained with an SENB specimen in 4 point bending when E/H is obtained by means of a Knoop indentation. When statistical deviations are taken into account, equation 4 is the most accurate. At variance, when E obtained from a 4 point bending test is introduced into the equations, none gives a value similar to those obtained with SENB specimens.

The present calculations reveal that the value of E which is introduced into the equations plays a very relevant role. It is therefore very important the way in which the elastic modulus is measured. The main causes which may explain the differences between the values obtained by indentation and by 4 point bending are that porosity and compactation defects reduce the elastic modulus values in conventional tests, and that the Knoop microindentation tests are carried out on well compacted surfaces of the specimen. It is worth mentioning that equation 4, which is formulated as universal for all kinds of ceramics, seems to adjust rather well to the present results, mainly, when the modulus is measured by means of a Knoop indentation.

ACKNOWLEDGEMENTS

The authors acknowledge the financial support given by the Spanish CICYT through the projects MAT93-1259-E and MAT93-0207.

REFERENCES

1. M Jarcho, C Bolen, M Thomas, J Bobick, J Kay and R Doremus, J. Mat. Sci., 11, (1976), 2027-2035.

2. J Peelen, B Rejda and K de Groot, Ceramurgia International, 4, 2, (1978) 71-74.

3. S Best, W Bonfield and C Doyle, Bioceramics 1, (1988), 68-73.

4. S Best, W Bonfield and C Doyle, Bioceramics 2, (1989), 57-64.

5. K Ioku, M Yoshimura and S Somiya, Biomaterials, 11, (1990), 57-61.

6. J Santos, S Morrey, G Hastings and F Monteiro, Bioceramics 4, (1991), 71-78.

7. P Wang and T Chaki, J. Mat. Sci.: Materials in Medicine, 4, (1993), 150-158.

8. J Santos, J Knowles, R Reis, F Monteiro and G Hastings, Biomaterials, 15, 1, (1994), 5-10.

9. C Ponton and R Rawlings, Mater. Sci. and Technology, 5, (1989), 865-872.

10. G Anstis, P Chantikul, B Lawn and D Marshall,J. Am. Ceram. Soc. 64,9,(1981), 533-538.

11. M Laugier, J. Mater. Sci. Lett., 4, (1985), 1539-1541.

12. M Laugier, J. Mater. Sci. Lett., 6, (1987), 355-356.

13. K Liang, G Orange and G Fantozzi, J. Mater. Sci. Lett., 25, (1990), 207-214.

14. J Blendell, Ph.D. Thesis, MIT Press, Cambridge, MA (1979).

15. J Lankford, J. Mater. Sci. Lett., 1, (1982), 493-496.

16. D Marshall, T Noma and A Evans, J. Am. Ceram. Soc., 65, (1982), 175-176.

Bioceramics, Volume 7 Edited by Ö.H. Andersson and A. Yli-Urpo
(Proceedings of the 7th International Symposium on Ceramics in Medicine, Turku, Finland, July 1994)

Rietveld Analysis on the Effect of Sintering Conditions on the Structure of Hydroxyapatite

J.C. Knowles,[1] J.A. Horton[2] and W. Bonfield[1]

[1] IRC in Biomedical Materials, [2] Department of Materials,
Queen Mary and Westfield College, Mile End Road, London E1 4NS, UK

ABSTRACT

Two types of hydroxyapatite (HA) were used in this study, a commercially available material produced by hydrothermal methods and an experimental material, prepared by the authors from solution. Discs were pressed from both types of HA and samples from both groups were subjected to different firing temperatures, to compare the effect of increasing firing temperature on crystal structure. For the basic unit cell parameters, the major changes were seen along the 'c' axis in the change in unit cell volume with temperature. However, there were significant differences between the two HA types. For the wet prepared HA, the unit cell showed a decrease in size with increasing temperature. The commercial type remained constant until the higher firing temperatures were reached, when an increase in volume was seen. The major changes seen on the atomic level were in the Ca(2)-Ca(2) bonds and also Ca(2)-OH bonds. Again significant differences were seen between the two types of HA

INTRODUCTION

HA may be synthesised by a variety of different routes. Two major methods used are a hydrothermal method (1), where lime is treated in a flowing gaseous atmosphere at elevated temperatures and a second method, where HA is precipitated from solution (2) when two liquid reagents are mixed. The first method has been patented and is used to produce commercial quantities of HA. The second method is used within the IRC for producing HA under controlled conditions. These two differing methods will produce HA with differing crystal structure and furthermore, variation in processing environment, such as temperature, may affect the crystal structure. Having produced the HA, in order to construct devices for clinical use, a thermal processing method must be undertaken and this heat treatment process can significantly affect the structure of the HA and thus affect its long term clinical survival.

The crystal structure of apatite $Ca_5(PO_4)_3F_2$ was determined by De Jong (3) some sixty years ago. HA possesses an hexagonal structure with space group P63/m and is built up from a three dimensional framework of calcium oxide and phosphate polyhedra. Within the structure, two calcium sites occur, the first: Ca(1) is located in an octahedral site and the second: Ca(2) is seven coordinate. The second calcium type is arranged such that it forms channels running parallel to the *c* axis. Within these channels are located the hydroxy ions, with disorder along the *c* axis, being shifted away from the 2a site (0,0,0.25 normally occupied in the fluorapatites, to half occupy the 4e site (0,0,0.19). The phosphate tetrahedra in HA are almost regular.

MATERIALS AND METHODS

Sample Processing

A commercially available HA was used (Plasma Biotal, UK, batch P81). The powder was sieved to produce a free running powder and then approximately 4g of the powder was pressed at $288MNm^{-2}$ to form 30mm diameter discs. The discs were then dried at 70°C for a minimum of 24 hours. The dried discs were placed on an alumina tile and then heated at 4°C min^{-1} up to the sintering temperature (within the range of 1000-1350°C). The discs were held at the sintering temperature for one hour before furnace cooling to room temperature.

The second HA was prepared using ANALAR grade chemicals and a precipitation type reaction, following the basic principles established by Barralet et al (2). Once precipitated, the HA was washed several times to remove any residual unreacted chemicals and traces of ammonia and the powder was then dried. Having obtained a dry powder, exactly the same processing conditions as described above were used.

X-Ray Diffraction

X-Ray powder diffraction data were collected for each sample on a Siemens D5000 diffractometer in flat plate geometry using Cu-Kα (λ=1.5418Å) and a graphite reflected beam monochromator. Data was acquired from 5 to 110° 2θ with a step size of 0.02° and with a count time of 12 seconds per step. The crystal structure for each HA sample was refined by the Rietveld method, using the program GSAS (3). The starting model used in refinement was based on data from a single crystal determination (4). A pseudo-Voigt model was used to model the peak shapes and an asymmetry parameter was refined. Scattering factors for neutral atoms were assumed. In each case, five background parameters, a scale factor, four peak shape variables, cell parameters and a zero point correction were refined before variation of the structural parameters. For all atoms, isotropic thermal parameters were refined and a site occupancy parameter refined for the hydroxy oxygen.

RESULTS

Examining the unit cell parameters, both samples showed similar trends for change in 'a' axis length with temperature. However, the change in 'c' axis with temperature showed disparities between the two materials. The wet prepared HA exhibited a small and approximately linear shrinkage with increasing temperature. The P81 HA showed an increase in 'c' axis length and this increased significantly at the higher temperatures. This increase reflected an overall unit cell volume decrease for the wet prepared HA and increase at the higher temperatures for the P81.

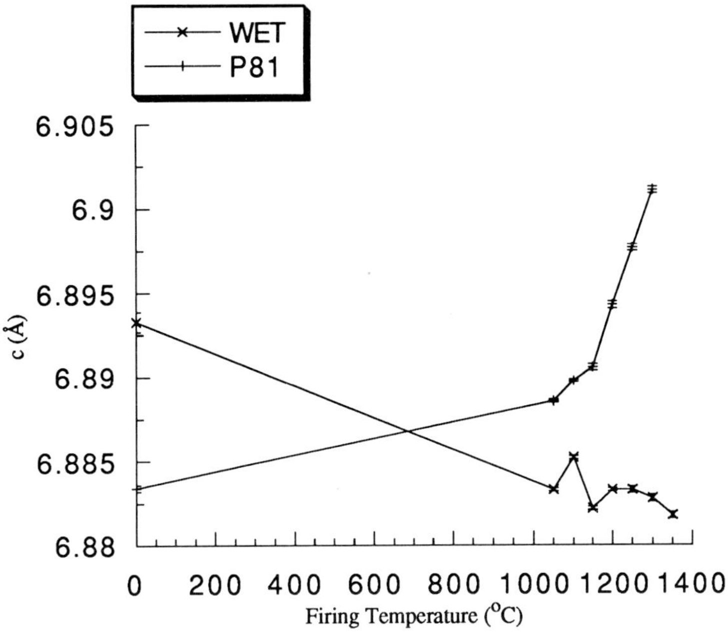

Figure 1. Effect of Firing Temperature on c axis length.

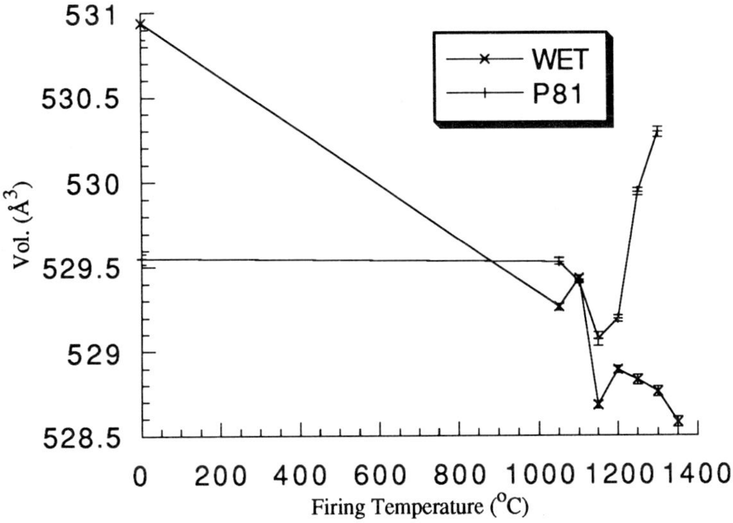

Figure 2. Effect of Firing Temperature on overall unit cell volume.

A more detailed examination of the structural changes revealed that the most significant changes occurred around the hydroxyl channels. These are channels running parallel to the 'c' axis. Figure 3 shows the change in the Ca(2)-Ca(2) bond length. For the P81, little change was seen in this bond length with temperature. The wet prepared HA however, showed a marked decrease at the highest firing temperature of 1350°C. These bonds are closely associated with the hydroxyl channel size and by using the following formula:

$$Rc = \frac{d[Ca(2)\text{-}Ca(2)]}{2} \tan \frac{\pi}{6}$$

a channel radius (Rc) may be calculated, which has been plotted in figure 4, for both types of HA against firing temperature. The Rc values for each HA type, mirror the values seen for the Ca(2)-Ca(2) bond length, as expected.

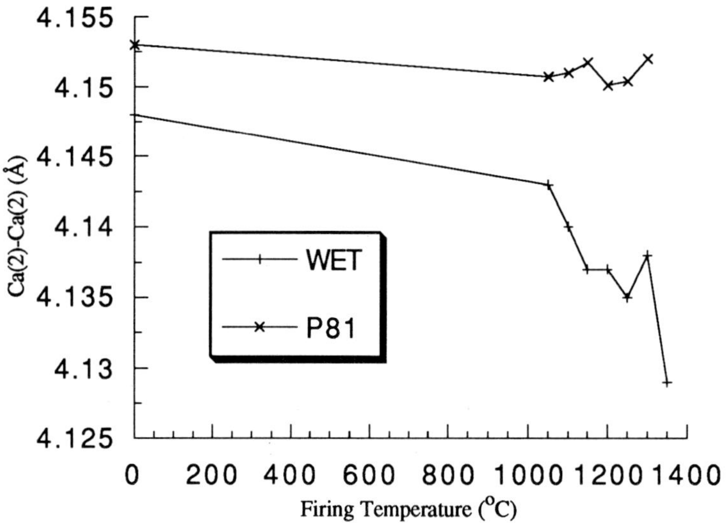

Figure 3. Effect of firing temperature on Ca(2)-Ca(2) bond length.

A further measurement that may be made is the distortion index, calculated using the bond angles with the PO_4 tetrahedra. Using the following formula:

$$\text{Dind} = \frac{\sum_{i=1}^{i=6} (\theta - 109.17)^2}{6} (\theta = angle O - P - O)$$

From Figure 5, major differences in the distortion index for the starting powder may be seen. The wet produced powder has a very large distortion index and the P81 raw powder has a much lower index, indicating a more relaxed structure. Following firing, to the first temperature point of 1050°C, both powders assume approximately similar values. The distortion indices remain at approximately similar values until both increase at the higher firing temperatures.

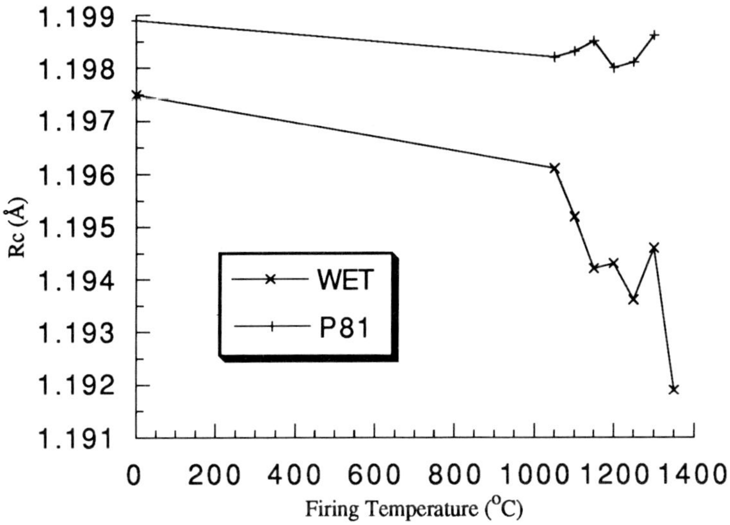

Figure 4. Effect of firing temperature on hydroxyl channel radius.

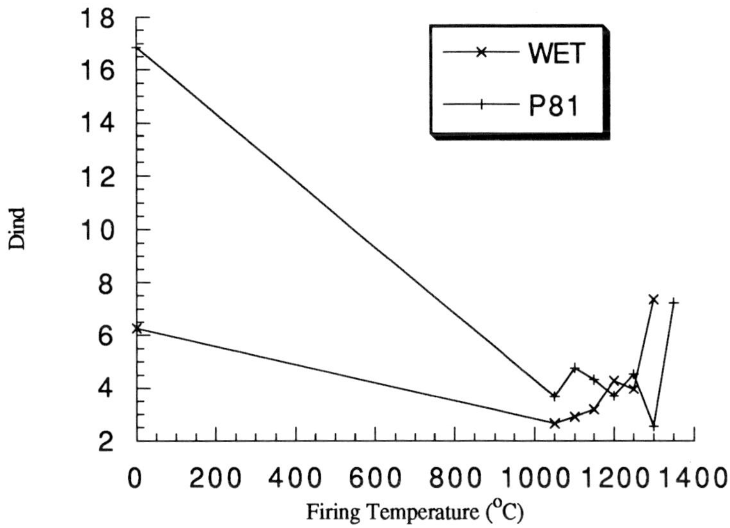

Figure 5. Effect of firing temperature on distortion index.

DISCUSSION AND CONCLUSIONS

When examining the results, it should be borne in mind, that the changes seen, whilst being significant (note SD bars on figures 1,2 and 3) are small. For example, the change in 'c' axis length for the P81 varies from 6.883Å to approximately 6.913Å, a change of only 0.03Å. Changes seen within the basic unit cell parameters come from atomic level changes within the structure and for the wet prepared HA, the majority of changes are occurring in the bonds associated with the hydroxyl channels. The wet prepared HA shows trends consistent with those previously reported (6) which were accounted for as being due to loss of CO_2 from the lattice structure allowing contraction of the hydroxyl channels. The commercial powder, however, does not exhibit any major changes in the hydroxyl channels and associated bonds, yet major changes must be occurring to give a significant unit cell volume increase with increasing temperature. The O(H) site occupancy shows a reduction, to an expected value of around 0.5.The Ca(2)-OH and Ca(2)-Ca(2) both show a reduction in size, indicating as expected, that carbonate loss is occurring. The average P-O bond length for the PO_4 tetrahedra has also decreased significantly. This data is summarised in table 1.

Firing Temperature (^oC)	Average Ca(1)-O (Å)	Average Ca(2)-O (Å)	Average P-O (Å)	Ca(2)-OH (Å)
pre-sintered powder	2.552	2.464	1.537	2.400(2)
1050	2.554	2.462	1.540	2.406(2)
1100	2.553	2.462	1.540	2.402(2)
1150	2.557	2.459	1.538	2.403(3)
1200	2.557	2.462	1.534	2.396(2)
1250	2.557	2.464	1.535	2.392(3)
1300	2.568	2.466	1.522	2.384(4)

Table 1. Variation of bond lengths with temperature.

In conclusion, significant structural variations may be detected between powders, introduced during production and these may give rise to significant structural changes when the ceramic is subsequently thermally processed.

ACKNOWLEDGEMENTS

The continuing support of the SERC for the IRC is gratefully acknowledged.

REFERENCES

(1) British Patent 1,586,915 (1993)
(2) Barralet, J.E., Best, S.M. and Bonfield, W. (1993) Bioceramics Vol.6, 179.
(3) DeJong, W.S. Rec. Trav. Chim. 1926, 45, 445
(4) Larson, A., Von Dreele, R.B. and Lujan Jr., M. GSAS-Generalised Crystal Structure Analysis System Neutron Scattering Centre, Los Alamos National Laboratory, California, 1990.
(5) Kay, M.I., Young, R.A. and Posner, A.S. Nature (London), 1964, 204 p1050.
(6) Knowles, J.C. and Abrahams, I. J. Mat. Chem. 1994, 4, 185.

Bioceramics, Volume 7 Edited by Ö.H. Andersson and A. Yli-Urpo
(*Proceedings of the 7th International Symposium on Ceramics in Medicine, Turku, Finland, July 1994*)

Emulsion Conditioning: A Route to Stronger and More Reliable Hydroxyapatite Ceramics

M.G.S. Murray,[§] J. Wang,[*] C.B. Ponton[*§] and P.M. Marquis[*§†]

[§] School of Metallurgy and Materials, [*] IRC in Materials for High Performance Applications,
[†] School of Dentistry,
The University of Birmingham, Edgbaston, Birmingham, B15 2TT, UK

Abstract

The production of high strength hydroxyapatite ceramics has to date been hindered by the formation of agglomerates during the powder processing stages. In this paper it is shown how the introduction of an emulsion conditioning stage has modified the nature of the agglomerates to produce a "soft" deformable powder leading to improvements in both strength and consistency of sintered HA. On sintering at 1100°C for 2 hours, the conventionally processed and the emulsion refined material produced fracture strengths of 70.4 MPa and 201.1 MPa, respectively. The effect of powder agglomeration on densification during both the initial and intermediate/late stages of sintering is discussed.

Introduction

Hydroxyapatite (HA) is the major inorganic constituent of natural teeth and bone. It exhibits excellent tissue responses and so is regarded as a potentially important implant material [1]. Many research groups have worked on producing and developing high strength synthetic HA [2-4]. The HA powders produced were agglomerated to some extent, although their crystallite sizes were fine. The agglomerate sizes and strengths depend on the powder processing parameters chosen [5]. Hard agglomerates lead to differential sintering and hence a reduction in the mechanical properties of the sintered ceramic [6]. This restricts the potential applications of HA. Thus, its principal current clinical uses are as bone substitutes in non-load bearing regions and as a coating on metal prostheses.

A potential means of producing stronger HA is to modify certain established conventional processing routes. The present author introduced a novel alternative route at Bioceramics 5 in Kyoto, Japan [7], which used emulsion technology to fabricate dense HA with a significant reduction in the critical defect sizes. Using emulsions as a means to prepare fine, agglomerate-free ceramic powders is a relatively new idea [8-12]. The present work discusses the subjecting of an HA powder precursor to an emulsion conditioning stage, in terms of its effect on the HA powder precursor characteristics and densification behaviour, and hence, the microstructure and mechanical properties of the sintered powder compact.

Experimental Procedure

Powder precursors for sintered HA ceramics were fabricated via a conventional processing route (CP) and an emulsion conditioning route (EC). The conventional method involves the initial formation of a suspension of 0.5 moles of $Ca(OH)_2$ powder$ in 1litre of distilled water. This suspension was stirred vigorously at 2000 r.p.m whilst adding 1 litre of 0.3 M H_3PO_4 (aq) acid* at a drip rate of 1 to 2 drops per second whilst controlling the pH (pH >7) and temperature (25°C) to form a gelatinous precipitate [13]. Once the precipitate had been stirred and aged for 120 hours, it was dried at 50°C for 12 hours. The as-dried powder was ground in a mortar and pestle to eliminate large powder lumps.

For the EC route, an oil-in-water (O/W) emulsion system was prepared by mixing 228 ml of distilled water with 60 ml of ethanol$ as a cosurfactant, 12 ml of non-ionic surfactant (Triton X-100$), and 300 ml of a pure vegetable oil. The mixture was further homogenised in an ultrasonic bath for 40 minutes, giving a stable emulsion. Excess oil was removed using a separating funnel, leaving only the O/W emulsion. A second batch of the gelatinous calcium phosphate was prepared as detailed above. The O/W emulsion was then added to the gelatinous solution in a weight ratio of 3:4 at a drip rate of 1 to 2 drops per second whilst being ultrasonically agitated and stirred for one hour. The sol was then centrifuged for 30 minutes at 2500 r.p.m it remove any excess emulsion phases and to form a HA gel. This was then dried at 50°C for 12 hours. In order to ensure the complete removal of any oil, the dried solid was subjected to solvent exchange by washing in 1,1,1 trichloroethane for 5 minutes.

Two types of powder compacts were prepared by single-ended uniaxial cold pressing; 2.5 g powder in a 7.5 mm diameter die and 1.15 g powder in a 24 mm diameter die, pressed at 5 MPa and 75 MPa, respectively. The former discs, which were used for the dilatometry study, exhibited thicknesses of 7.2 mm and 6.1 mm for the CP and the EC HA powder, respectively. The dilatometry study was carried out at a heating rate of 5°C per minute up to 1100°C in air in a tube furnace. The 24 mm diameter powder compacts were sintered in an electric furnace at 1100°C. Each sample was heated from room temperature to 800°C at 2°C per minute, held for 3 hours, then heated at 2°C per minute up to 1100°C. This temperature was held for two hours before the samples were cooled to room temperature at a rate of 5°C per minute.

The sintered pellets were characterised for sintering shrinkage and sintered density. Green density measurements were made on the basis of specimen mass and dimensions. For those pellets sintered at 1100°C, where the open porosity level was low, the density was measured using Archimedes' method in water. Microstructural characterisation using SEM was performed on fracture surfaces of the sintered HA. Average fracture strengths were determined using a standard Shell test [16]; at least 15 tests were made for each material.

Simultaneous thermogravimetry and differential thermal analysis (DTGA) at the rate of 10°C per minute in air were carried out for both processing routes using a Stanton Redcroft STA 781 apparatus.

$ B.D.H. Chemicals Ltd., Analar grade * Aldrich. Chemicals Ltd., Analar grade

Experimental Results and Discussion

Agglomeration is the major source of detrimental flaws in the conventional production (CP) of HA. Agglomerates are formed during the precipitation stage of the amorphous calcium phosphate. Vigorous stirring during the ageing process does reduce the average particle size from 12 μm to 4.7 μm. However, whilst this is a substantial reduction in particle size, agglomerates of up to 20 μm in size are still present. These agglomerates persist through the subsequent ceramic powder processing stages to produce large inherent pores and flaws. Improvements in the mechanical properties of HA will be achieved only if control over the quality and nature of the HA powder particle precursors is obtained.

An emulsion conditioning (EC) stage was developed in order to prevent the formation of hard agglomerates [7]. The aim of this stage was to modify the surface energy and chemical characteristics of the oil so that it would encapsulate the HA powder particles originating from the agglomerates being broken up by ultrasonic agitation and mechanical shear mixing. Encapsulation of the HA particles by the oil would form a physical barrier against particle reagglomeration. Centrifuging produced a HA gel containing particles which had been conditioned to a sterically dispersed state by being coated with a thin film of surfactant and oil.

The emulsion refinement step proved to be effective in "softening" particle agglomerates present in the CP powder, and hence a more deformable powder was produced. Optimisation of the weight ratio addition of emulsion to HA has further enhanced the properties of the EC material. This was confirmed by the measured green and sintered densities, and the respective percentage open porosities, which are given in Table 1 together with the mean biaxial fracture strength, σ_f, sample standard deviation, S and the Weibull modulus, m in Table 1.

By using EC, the green density increased from 1.12 to 1.49 gcm^{-3}, and the sintered density accordingly increased from 2.92 (92.1% theoretical (TD)) to 3.07 gcm^{-3} (97.2% TD) with an associated reduction in the percentage open porosity from 9.4 to 2.3%. By comparison, the calculated sintering shrinkages give sintered densities of 78% TD and 96% TD for the CP and EC materials, respectively. The percentage open porosity results for the CP material are higher than the upper limit of 4%, defined for accurate Archimedes' density measurements. The higher sintered density and lower percentage open porosity of the EC material is attributed to the production of more deformable soft agglomerates that can be broken down by conventional powder pressing methods.

Comparing the sintered CP and EC material microstructures in Figures 1a and 1b respectively, supports this view. Figure 1a shows the remnant now-densified large agglomerates surrounded by void space; this indicative of agglomeration and subsequent differential densification. Figure 1b shows a more uniform, reduced porosity microstructure with no sign of agglomerate-induced defects. The porosity characteristics due to the two processes also differ [15].

The increase in the packing density coupled with the control over the formation of agglomerates has produced significant improvements in both the fracture strength and reliability of the emulsion conditioned material; see Table 1.

Table 1: Comparison of properties obtained from the different processing routes.

Processing method :	Green Density (g/cm^3) *	Density (g/cm^3)	% Open Porosity	σ_f (MPa)	S (MPa)	m
Conventional (CP)	1.12	2.92	9.4	70.4	15.4	5.7
Emulsion conditioned (EC)	1.49	3.07	2.3	201.1	13.1	10.0

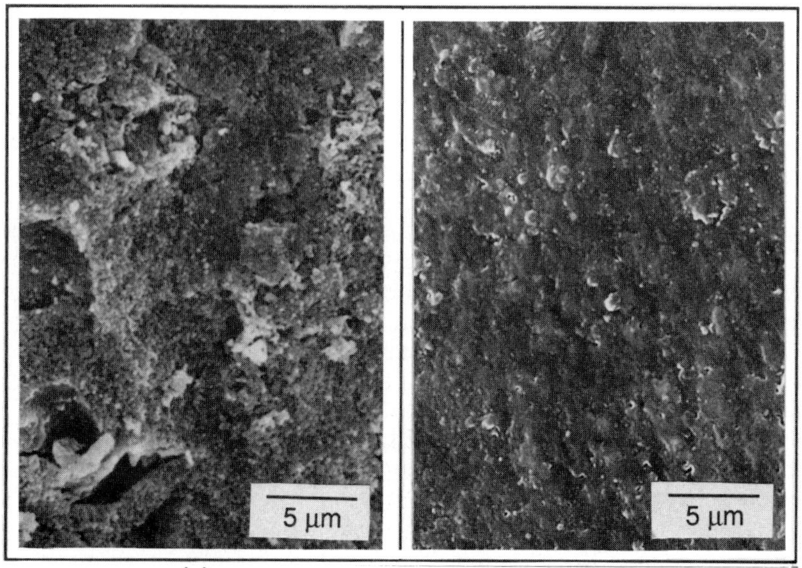

(a) (b)

Figure 1: Scanning electron micrograph illustrating (a) the agglomerated microstructure of the conventionally produced HA and (b) the homogeneous microstructure of the emulsion conditioned HA (magnification X 3500).

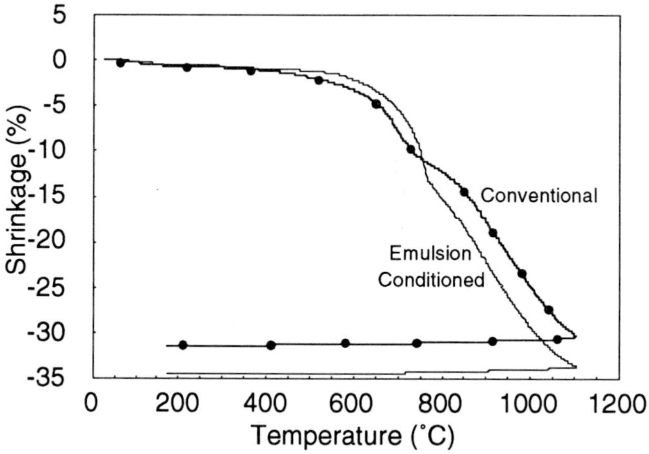

Figure 2: Dilatometry for the conventionally processed and emulsion conditioned HA from 25˚C to 1100˚C.(Heating rate of 5˚C/minute)

The fracture strength has increased from 70.4 MPa for the conventional route to 201.1 MPa for the EC material. These results compare very favourably with the highest values recorded by Jarcho [2] and they were achieved without the need for elaborate finishing methods. The fracture strength increase is accompanied by a narrowing of its distribution; hence, the increase in the Weibull Modulus from 6 to 10 and a decrease in the sample standard deviation from 15.4 to 13.1 MPa, leading to greater reliability. These property and reliability improvements show clearly that the critical defect size has been reduced and that the distribution of defect characteristics is more uniform. This indicates that the large defects, which normally originate from agglomerates, have been eliminated.

The densification behaviour was investigated in terms of sintering shrinkage in a dilatometry study. The powder compacts exhibited a green density of 28% and 34% theoretical density, for the CP and EC respectively. Figure 2 shows the sintering shrinkage as a function of sintering temperature. Little shrinkage was observed at temperatures below 400°C for both materials. The onset temperature for noticeable shrinkage is 400°C for the CP powder compact compared with 540°C for the EC. From 500°C to 630°C, the former exhibits a shrinkage rate which is almost doubles that of the latter. However, over the temperature range of 720 to 860°C, the shrinkage rate of the CP powder compact slows down markedly, whereas that of the EC powder compact increases significantly. The decrease in the sintering rate of the CP material is attributed to the change from intra-agglomerate to inter-agglomerate pore densification. This is not apparent in the EC material owing to the elimination of the agglomerates during processing. The latter exhibited a total sintering shrinkage of 34.1% on heating up to 1100°C, compared with 29.6% for the former. A theoretical model describing the effects of powder agglomeration has been developed [15].

The phase development and mass loss with increasing temperature was monitored using simultaneous DTGA, see Figures 3 (a,b). From room temperature up to 250°C, both powder precursors demonstrated a steady mass decrease with increasing temperature due to the loss of free water and volatile organic residuals. There follows a sharp mass loss by the EC powder precursor, indicating that the burn-off of organic residuals occurs over the temperature range of 300 to 350°C. A well established endothermic peak is associated with the organic burnout. Both powder precursors exhibit a second mass loss and exothermic peaks indicating the formation of crystalline HA between 600-900˚Cfor the CP and 810-890˚C for the EC powders.

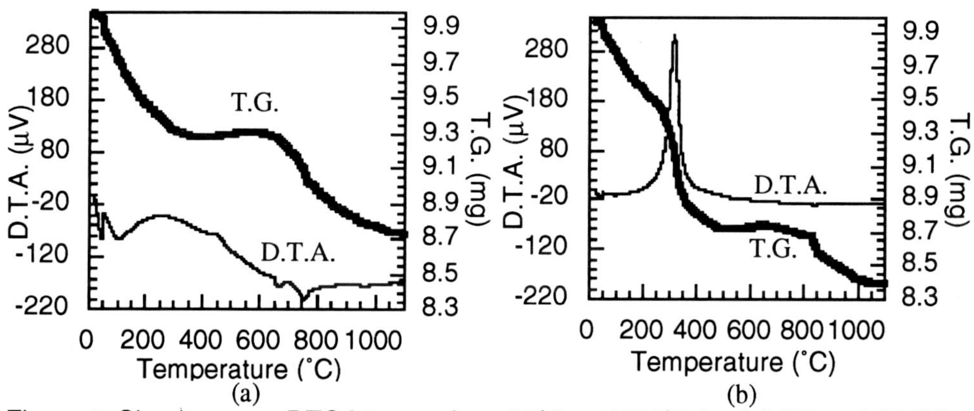

Figure 3: Simultaneous DTGA traces from 25˚C to 1100˚C for (a) CP and (b) EC

Thus, the EC process apparently increases the crystallisation temperature of amorphous calcium phosphate.

The data also show that the CP material loses mass gradually whereas the EC material exhibits a sudden drop of mass over a narrower temperature range; Figures 3 (a, b). The escape rate of volatile species, e.g. water or organic residuals, is related to the number of interconnected diffusion channels (pores). This rate is reduced in agglomerated powders due to packing density differences, but gives a sharpdrying temperature for a constant heating rate in non-agglomerated powders as seen in the EC powder.

Conclusions

The results show that it is possible to generate a dense, high strength monolithic HA ceramic using simple uniaxial powder pressing simply by introducing an emulsion conditioning stage to a conventional HA synthesis process. This stage beneficially eliminates large non-deformable agglomerates from the powder. It has been shown that agglomerate size has a marked effect on both the sinterability of the powder compact and the microstructure of the sintered ceramic. Optimisation of the weight ratio addition of emulsion to HA has further improved the properties of the material, and the improved Weibull modulus and greater fracture strength should lead to a substantial increase in the survival probability of the emulsion conditioned material at low applied stresses. Thus, it is now exhibits comparable mechanical performance to the apatite glass-ceramics currently attracting interest for a number of clinical applications.

Acknowledgements

The authors would like to thank the U.K. Science and Engineering Research Council for the provision of a Ph.D. research studentship for M.G.S. Murray.

References

1. Hench, L.L., J. Am. Ceram. Soc., **74**, (1991), 1487-1510.
2. Jarcho, M., Bolen, C.H., Thomas, M.B., Bobick, J., Kay, J.K. and Doremus, R.H., J. Mat. Sci., **11**, (1976), 2027-2035.
3. Quinton, A., Nass, R. and Schmidt, H., 2nd ECerS proc., Augsburg, Germany, (1991), 209-214.
4. Hattori, T. and Iwadate, Y., J. Am. Ceram. Soc., **73**, (1990), 1803-1805.
5. Lange, F.F., J. Am. Ceram. Soc., **72**, (1989), 3-15.
6. Kellett, B. and Lange, F.F., J. Am. Ceram. Soc., **67**, (1984), 369-371.
7. Murray, M.G.S., Ponton, C.B. and Marquis, P.M., 5th Int. Symp. Ceramics in Medicine, Ed. by Yamamuro, T., Kokudo, T. and Nakamura, T., Kyoto, Japan, Nov., (1992), 15-22.
8. Reyen, P., Bastius, H., and Fielder, M., Ceramic powders, Ed. Vincenzine, P., Elsevier, Amsterdam, Netherlands, (1983), 499.
9. Celikkaya, A. and Akinc, M., Ceramic Transactions, Vol 1A, Ceramic powder science II, Edited by Messing, G.L., Fuller, E.R. and Hausner, Am. Ceram Soc., Westerville, Ohio, (1988), 110-117.
10. Kanai, T., Rhine W.E. and Bowen, H.K., ***ibid***, 119-127.
11. Ramamurthi, S.D., Xu, Z. and Payne, D.A., J. Am. Ceram. Soc., **73**, (1990), 2760-2763.
12. Maher, G.H., Hutchins, C.E. and Ross, S.D., Ceram. Bull., **72,** No. 5, (1993), 72-76.
13. Akao, M., Aoki, H. and Kato, K., J. Mat. Sci., **16**, (1981), 809-812.
14. Timoshenko, S. and Wainwsky-kreigers, 'Plates and Shells', Mcgraw Hill, (1959), 71.
15. Wang, J., Murray, M.G.S., Ponton, C.B. and Marquis, P.M., 3rd ECerS proc., Vol.1, Madrid (1993), 715-720.

Bioceramics, Volume 7 Edited by Ö.H. Andersson and A. Yli-Urpo
(*Proceedings of the 7th International Symposium on Ceramics in Medicine, Turku, Finland, July 1994*)

Preparation of P Ion-implanted Silica Glass for Radiotherapy

M. Kawashita, T. Yao, F. Miyaji, T. Kokubo, G.H. Takaoka* and I. Yamada*
Division of Material Chemistry, Faculty of Engineering, Kyoto University, Sakyo-ku, Kyoto 606-01, Japan
* Ion Beam Engineering Experimental Laboratory, Faculty of Engineering, Kyoto University, Sakyo-ku, Kyoto 606-01, Japan

ABSTRACT

P-31 can be activated to ß-emitter P-32 with 14.3 d half-life by neutron bombardment. A chemically durable glass microsphere containing a large amount of phosphorus is believed to be useful for *in situ* irradiation of cancers. In the present study, a chemically durable silica glass was ion-implanted with phosphorus and subjected to heat treatments under various conditions. The phosphorus ion was saturated at a dose of about 5×10^{16} cm^{-2} under 20 keV. The as-implanted sample completely lost the phosphorus by soaking in water at 95°C, since the surface of the glass was damaged by the ion implantation. When the implanted sample was heat-treated at 400°C in H_2, the phosphorus colloids grew and then when it was heat-treated at 900°C in O_2, the phosphorus colloids were oxidized at the surface and encapsulated by the resulting SiO_2-P_2O_5 glass layer. Thus treated glass showed little dissolution of both the silicon and phosphorus even in the hot water and is expected to be useful for the *in situ* radiotherapy.

INTRODUCTION

Radiotherapy is one of the effective treatment of cancers. External irradiations, however, often cause damages to healthy tissues. Recently, Ehrhardt et al. showed that a glass microsphere of 20 to 30 μm in diameter of the composition Y_2O_3 40, Al_2O_3 20, SiO_2 40 wt% is useful for *in situ* irradiation of cancers (1). Subjected to neutron bombardment, Y-89 in the glass is activated to ß-emitter Y-90 with a half-life of 64.1 h. The glass is insoluble in body fluids and non-toxic. Injected into the neighborhood of a liver tumor, the glass microspheres give large local irradiation of the short-ranged highly ionizing ß-ray to the tumor.

The short half-life of 64.1 h for Y-90, however, may result in the substantial decay before the treatment. P-31 with 100 % natural abundance similarly to Y-89 can be activated to ß-emitter P-32 with half-life 14.3 d by neutron bombardment. The biological effectiveness of P-32 is about four times as large as that of Y-90. But highly phosphorus-containing glass prepared by the conventional melting method are usually less chemically durable.

A silica glass has a high chemical durability, and neither silicon nor oxygen is activated by neutron bombardment. It is expected that phosphorus-containing glasses with high chemical durability could be obtained by the phosphorus ion implantation into a silica glass. In the present study, a silica glass was ion-implanted with phosphorus then heat-treated under various conditions. Chemical durabilities of thus treated glasses were examined in hot water. The results were discussed in terms of the surface structure of the glasses.

EXPERIMENTAL PROCEDURE

Preparation

Highly pure silica glass (metallic impurities<0.5 ppm, OH<100 ppm) made by the vapor-phase axial deposition (SUMIQUARTZ SK-1300, Sumitomo Metal Industries Ltd., Tokyo, Japan) was cut into a rectangular specimen of 10 mm×10 mm×1 mm in size, annealed at 1100°C for 1 h. Phosphorus ions were implanted at a dose of 1×10^{16}, 5×10^{16}, 1×10^{17} or 2×10^{17} cm^{-2} at 20 keV. After the ion implantation, every sample was preserved in a nitrogen atmosphere in order to prevent the oxidation of phosphorus. Ion implanted samples were subjected to heat treatments under various conditions.

Soaking test

The samples thus treated were soaked in 20 ml of distilled water at 95°C for 7 d in a polypropylene bottle, shaken at 3 cm stroke length and 120 min^{-1} frequency. The amounts of phosphorus and silicon leached out from the samples were measured by an inductively coupled plasma atomic emission spectrometer (SPS 1500, Seiko Instruments Inc., Tokyo, Japan).

Analysis

The distributions of phosphorus in the samples were analyzed by the Rutherford backscattering spectrometry (RBS) using 2 MeV $^4He^+$ ions with 170° incident angle. Surface structures of the samples were analyzed by Fourier transform infrared (FT-IR) spectrometer (SR-5M, Japan Spectroscopic Co. Ltd., Tokyo, Japan) with the reflection angle of 30°. The P_{2p} binding energies were measured by X-ray photoelectron spectroscopy (XPS) (ESCA MT5500, ULVAC-phi Co. Ltd., Chigasaki, Japan) using MgKα X-ray as the source. Measured binding energies were corrected by referring the binding energy of C_{1s} (284.6 eV) of the hydrocarbon layer adsorbed on the sample surfaces.

RESULTS AND DISCUSSION

Figure 1 shows the peak area of phosphorus in the RBS spectra of silica glass implanted with phosphorus ions at 20 keV as a function of dose. The peak area increased with increasing dose, but showed a almost constant values for doses more than 5×10^{16} cm^{-2}. This means that the phosphorus implanted into the glass is saturated in its amount at the dose of about 5×10^{16} cm^{-2}.

P_{2p} XPS spectra of the sample as-implanted with P^+ion at the dose of 5×10^{16} cm^{-2} at 20 keV are shown in the left hand side column of Figure 2 as a function of depth from the surface. Two peaks, assigned to elemental phosphorus at about 130 eV and assigned to oxides at about 134 eV, are observed at the surface. The peak intensity of the oxides decreased with the increasing depth. Only the peak of elemental phosphorus appeared for the deeper regions than 1 nm. It is considered that implanted phosphorus exists as elemental phosphorus and that a part of the phosphorus at the surface is oxidized by the atmospheric oxygen. After the ion implantation, the sample was colored brown, which indicates the formation of phosphorus colloids (2).

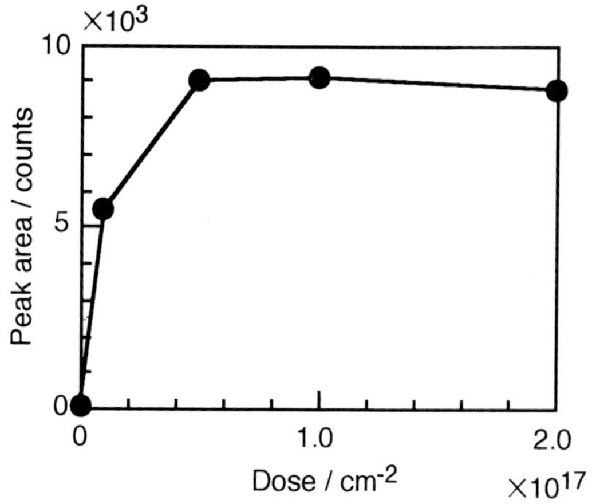

Figure 1 Peak areas of phosphorus in RBS spectra of P^+-implanted samples as a function of dose.

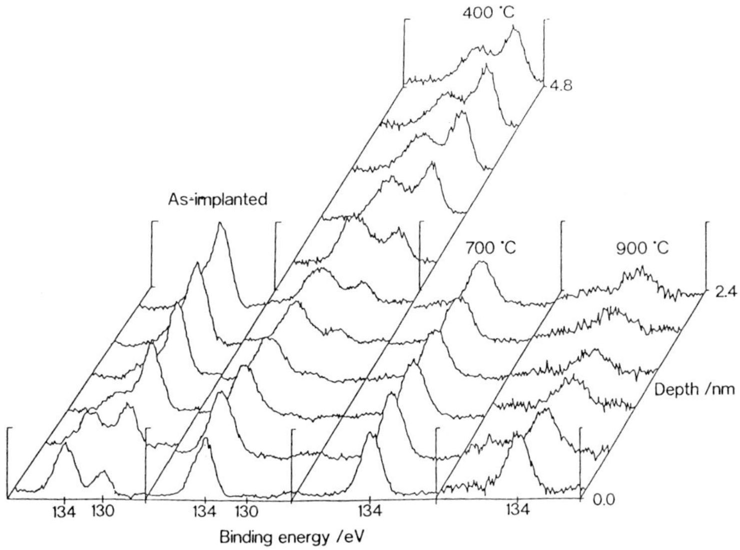

Figure 2 P2p XPS spectra of the sample implanted with P^+ ion and heat-treated at various temperatures in air for 1 h.

The concentrations of the phosphorus and silicon leached out from the as-implanted samples into the hot water for 7 d are shown in Table 1 (sample 2). It can be seen from Table 1 that the original silica glass has a high chemical durability, but that as-implanted with P^+ion dissolves appreciable amount of the phosphorus as well as the silicon. The amount of the silicon dissolved from the glass, i.e. 1.238 ppm, indicates that a surface layer more than 45 nm thick of the silica glass was dissolved. Since the mean and the standard deviation of the projected range of the phosphorus ions implanted at 20 keV is estimated at about 20 nm and 8.4 nm, respectively (3), the phosphorus implanted into the glass completely dissolved into the water within 7 d. In fact, the RBS spectrum of the sample after soaking for 7 d showed no peak of the phosphorus.

Figure 3 shows the FT-IR reflection spectrum of the as-implanted sample in comparison with that of the not-implanted original sample. The as-implanted sample showed slightly larger reflectance in a region around 1000 cm^{-1} assigned to non-bridging oxygen bonded to silicon atom (4). This indicates that silica network in the surface layer was broken by the ion implantation. This structural damage of the surface layer might be responsible for the decrease in the chemical durability of the silica glass due to the ion-implantation. Therefore, the implanted sample was heat-treated at various temperatures in the air for 1 h in order to heal the damage.

Figure 4 shows the peak areas of phosphorus in the RBS spectra of the P^+-implanted samples as a function of heat-treatment temperature. It can be seen from Fig. 4 that the phosphorus implanted to the sample decreases with increasing heat-treatment temperature, especially above 700°C. This might be attributed to vaporization of the phosphorus through the glass, since the red phosphorus has sublimation point at 416°C (5).

Table 1 Concentrations of P and Si leached out from the samples implanted with P^+ion and heat-treated under various conditions into water at 95°C for 7 d

Sample No.	Treatment	Concentration / ppm	
		P	Si
1	original glass	not detected	0.058
2	as-implanted glass	0.265	1.238
3	heated at 400°C in air for 1 h	0.199	0.324
4	heated at 700°C in air for 1 h	0.182	0.448
5	heated at 900°C in air for 1 h	0.155	0.403
6	heated at 400°C in H_2 for 1 h then at 900°C in O_2 for 1 h	0.122	0.190
7	heated at 400°C in H_2 for 2 h then at 900°C in O_2 for 2 h	0.038	0.060

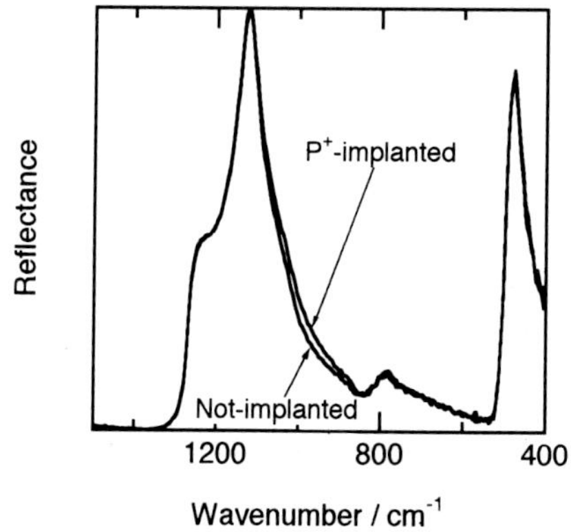

Figure 3 FT-IR reflection spectra of the as-implanted sample and the not-implanted one.

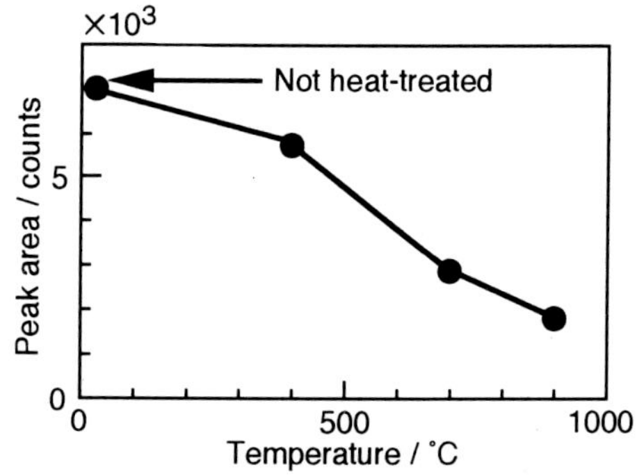

Figure 4 Peak areas of phosphorus in RBS spectra of the P^+-implanted samples as a function of heat-treatment temperature.

P_{2p} XPS spectra of the samples implanted with P^+ion and heat-treated at 400, 700 and 900°C are shown in Fig. 2. The peak assigned to oxidized phosphorus increased in its relative intensity to that of the elemental phosphorus and reached to deeper region for 400°C-treated sample. Above 700°C, only the oxide peak was observed, and its absolute intensity decreased with increasing heat treatment temperature. This indicates that the implanted phosphorus tend to be oxidized and escaped from the glass by the heat treatments. Consequently, the P^+-implanted sample became colorless by the heat treatments.

Figure 5 shows FT-IR reflection spectrum of the P^+-implanted sample heat-treated at 900°C, in comparison with that of the not-implanted original glass. The former is identical to the latter. This indicates that the structural damage caused by the ion implantation at the surface of the glass was healed by the heat treatment.

The concentration of the phosphorus and silicon leached out from the P^+-implanted samples heat-treated at various temperatures into the hot water for 7 d are shown in Table 1 (samples 3-5). It can be seen from Table 1 that the dissolution of both the phosphorus and silicon from the P^+-implanted sample considerably decreased by the heat treatments but not remarkably. This decrease in the dissolution might be partly attributed to the healing of the structural damage at the surface, but also attributed to the decrease in the amount of the phosphorus present in the sample. Dissolution of appreciable amount of the phosphorus and silicon even from heat-treated sample might be attributed to the formation of a chemically not-durable SiO_2-P_2O_5 glass near the surface of the sample. RBS spectrum of the P^+-implanted sample heat-treated at 900°C showed no peak of the phosphorus after soaking in the hot water for 7 d, as shown in Fig. 6. In order to heal the structural damage of the P^+-implanted sample without oxidizing and losing the phosphorus, two-step heat treatment was attempted. First, the P^+-implanted sample was heat-treated at 400°C, below the sublimation point of red phosphorus, in H_2 atmosphere for 1 or 2 h, in order to make the phosphorus colloids grow larger without oxidation. Second, the sample was heat-treated at 900°C in O_2 atmosphere for 1 or 2 h, in order to heal the structural damage at the surface and also to oxidize the phosphorus colloids at their surfaces and encapsulate them by a thin SiO_2-P_2O_5 layer. Figure 7 schematically shows the process of such capsulation of phosphorus colloids in SiO_2-P_2O_5 glass.

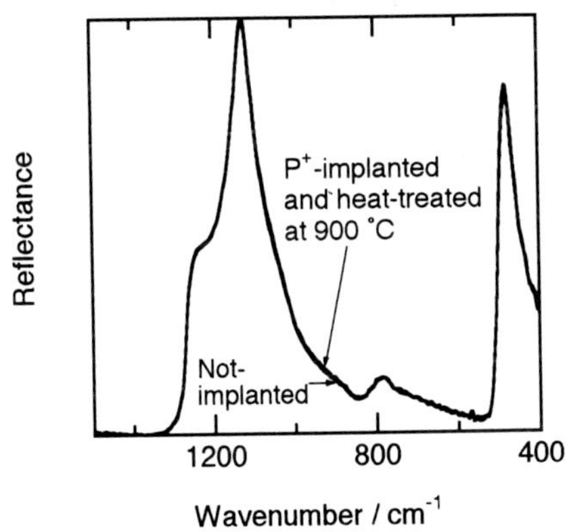

Figure 5 FT-IR reflection spectra of the P^+-implanted sample heat-treated at 900 °C, in comparison with that of the not-implanted sample.

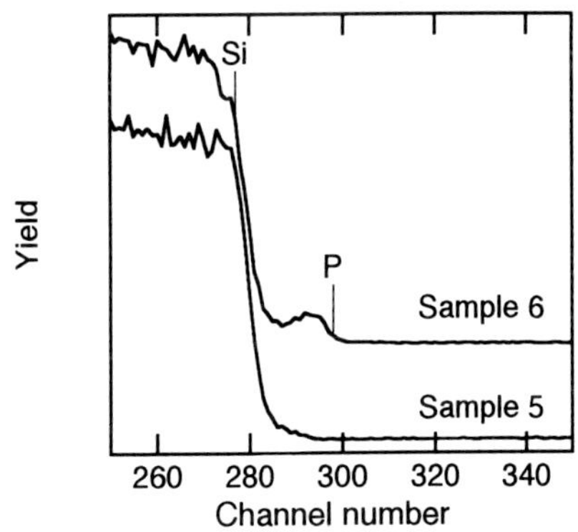

Figure 6 RBS spectra of sample 5 after soaking for 7 d and sample 6 after soaking for 14 d.

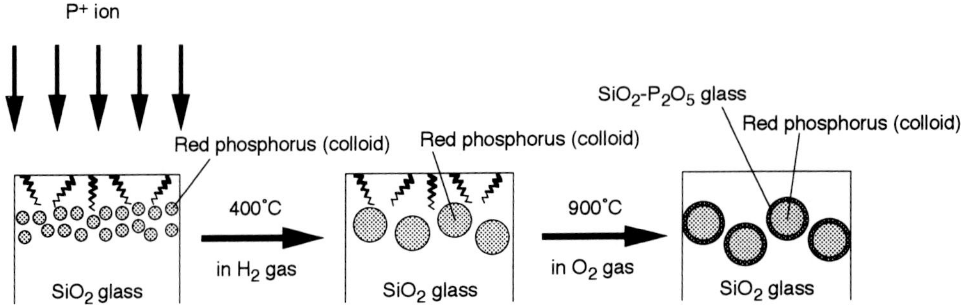

Figure 7 Schematic diagram of capsulation of phosphorus.

The concentrations of the phosphorus and silicon leached out from the P^+-implanted sample subjected to the two-step heat treatment are shown in Table 1 (samples 6, 7). It can be seen from Table 1 that the dissolution of both the phosphorus and silicon from the P^+-implanted sample is significantly reduced by such two-step heat treatment, especially by the treatment for 2 h. The samples remained to show brown color due to the phosphorus colloids even after the two-step heat treatment. The sample subjected to the treatment for 1 h (sample 6) showed the phosphorus peak in RBS spectrum even after soaking in the hot water for 14 d, as shown in Figure 6. It is believed that thus prepared chemically durable silica glass containing large amount of the phosphorus ion is useful for the *in situ* radiotherapy of cancers.

ACKNOWLEDGMENT

This work was supported by a Grant-in-Aid for Scientific Research, The Ministry of Education, Science and Culture, Japan.

REFERENCES

1. Ehrhardt, G.J. and Day, D.E. Nucl. Med. Biol. (1987) 14 233-242
2. Hosono, H. Suzuki, Y. Abe, Y. Oyoshi, K. and Tanaka, S. J. Non. Cryst. Solids (1992) 142 287-290
3. Wilson, R.G. and Brewer, G.R. In: Ion Beams with Applications to Ion Implantation A Willey-Interscience Publication, U.S.A. (1973) 62 120
4. Filgueiras, M.R. Torre, L.T. Hench, L.L. J. Biomed. Mat. Res. (1993) 27 445-453
5. Corbridge, D.E.C. In: Phosphorus Elsevier, The Netherlands, (1990) 120

Bioceramics, Volume 7 Edited by Ö.H. Andersson and A. Yli-Urpo
(*Proceedings of the 7th International Symposium on Ceramics in Medicine, Turku, Finland, July 1994*)

Preparation of Hydroxyapatite Fiber by using Alginate and its Application to Bioceramics

J. Hayashizaki, S. Ban, N. Arimoto, H. Kataoka, K. Miyazawa* and J. Hasegawa
Department of Dental Material Science, Department of Orthodontics*, School of Dentistry, Aichi-Gakuin University, Chikusa-ku, Nagoya, 464 Japan

ABSTRACT

Hydroxyapatite fibers were prepared by a reaction of ion exchange of potassium alginate. To investigate the biocompatibility of the hydroxyapatite fibers, surface changes of these fibers were investigated *in vitro* and *in vivo*. *In vitro* experiment: the fibers were immersed in distilled water and a simulated body fluid (SBF) at 37°C for 9 weeks. After immersion, many voids were observed on the surface of the fibers in distilled water, whereas, the small flake-like particles precipitated on the surface of the fiber and the micro-pores were partially formed around the grain boundary in the SBF. *In vivo* experiment: diffusion chambers filled with these fibers were implanted in subcutaneous tissue of rats. At 3 weeks after implantation, small pores and granular precipitates were observed on the surface of the fibers covered with fascia. These results suggest that the hydroxyapatite fibers are useful biocompatible ceramics which are especially suitable for filling materials.

INTRODUCTION

Bioceramics used in dental implants and orthopaedic prosthesis have to indicate excellent biocompatibility and osteoconductivity resulting in direct apposition with bone tissue. Hydroxyapatite [$Ca_{10}(PO_4)_6(OH)_2$] is one of the bioceramics which have been most frequently applied in medicine and dentistry in the last two decades. Various preparation methods for hydroxyapatite were reported by many investigators. It is well known that clinical success requires the simultaneous achievement of a stable interface with connective tissue and a match of the mechanical behavior of the implant with the tissue to be replaced. It has been showed that the surface dissolution and the bone formation of porous hydroxyapatite occurred initially on the surface of the ceramic and progressed toward the center of the pore. These properties are of great importance in biomaterials because tissue ingrowth, the stimulation of new bone formation, and the biological fixation of the implanted material all depend upon the nature of the pore-microstructure configuration [1]. Thus, hydroxyapatite fibers having a large surface area undoubtedly strongly and rapidly connect to bone, and indicate excellent biocompatibility. In this study, the hydroxyapatite fibers were prepared by using a reaction of ion exchange of potassium alginate (K-Alg), and its biocompatibility was investigated by immersion tests *in vitro* and *in vivo*.

MATERIALS AND METHOD

Hydroxyapatite fibers were prepared as follows. A mixture sol of 5% K-Alg and 2.8 % $Na_4P_2O_7 \cdot 10H_2O$ was used as starting material, and the spinning solution was prepared by dissolving 500 ml of 0.10 M $Ca(OH)_2$ and 500 ml of 0.15 M $Ca(CH_3COO)_2$ and adjusted to pH 7 by adding of CH_3COOH. The sol was pushed out from the syringe with 0.40 mm - diameter nozzle into the spinning solution. The sol was immediately gelled by a reaction of ion exchange and the gel-fibers were aged to complete the gelation in the spinning solution. The gel-fibers were rinsed with distilled water to remove the spinning solution covering the fibers and were dried at 37°C. Finally, hydroxyapatite fibers were prepared by sintering of the dry-fibers at 900°C for 3 hours in air.

In vitro, the hydroxyapatite fibers (about 10 mg) were immersed in 20 ml of distilled water and the SBF at 37°C for 8 hr, 12 hr, 1 day, 1 week, 3 weeks, and 9 weeks. The SBF was prepared by dissolving given amounts of reagent-grade chemicals of NaCl, $NaHCO_3$, KCl, K_2HPO_4, $MgCl_2 \cdot 6H_2O$, and $CaCl_2 \cdot 2H_2O$ into distilled water. The solution was buffered at a pH of 7.2 with 50 mM tryshydroxyaminomethane [$(CH_2OH)_3CNH_2$] and an adequate amount of hydrochloric acid [HCl]. The ion concentrations in the SBF were similar to those of human blood plasma [2]. After immersion, the specimens were rinsed with distilled water and dried at 37°C in air. *In vivo*, diffusion chambers, made from lucite rings bounded by 0.45 μm membrane filters, were respectively filled with 10 mg of the hydroxyapatite fibers. The chambers were implanted in subcutaneous tissue of 6-week-old wister rats for 1 week, 3 weeks, and 9 weeks. After implantation, the rats were sacrificed and the chambers were recovered from the bodies. After drying at 37°C, the specimens were disassembled from the chambers. After drying, the surface of the specimens immersed *in vitro* and *in vivo were* examined by field-emission scanning electron microscopy (FE-SEM) and Fourier transform infrared (FT-IR) micro-spectroscopy.

RESULTS AND DISCUSSION

FE-SEM of the fiber before immersion test demonstrated the close aggregation of crystals separated by grain boundaries (Figure 1). And micro-pores were observed around the grain boundary on the surface of the fiber. The diameter of the fibers varied in the range of 40 - 50 μm which depended on the nozzle size used for spinning and the dilution of alginate [3]. FT-IR spectrum of the fiber showed characteristic absorption bands of hydroxyapatite. X-ray diffraction study revealed that the fibers were identified to be hydroxyapatite with high crystallinity and contained small amounts of CaO. It seems that the formation of CaO was due to the presence of excess Ca^{2+} ion in the dry-fibers and the amounts was less than 6 wt%. Iwasaki et al. reported that hydroxyapatite fibers were prepared by using sodium alginate and $CaCl_2$, $Ca(CH_3COO)_2$, and CH_3COOH, and CaO was removed by using 1 % NH_4Cl at 60°C [4]. In our previous study, it was found that amounts of CaO depend on the concentration of $Na_4P_2O_7 \cdot 10H_2O$, aging time in the spinning solution, and rinsing time with distilled water [3]. Furthermore, chrole-apatite was formed when the gel-fibers were insufficiently rinsed [5]. In the present study, K-Alg and $Ca(OH)_2$ were used instead of sodium alginate and

Figure 1 FE-SEM micrographs of the surface of the fiber sintered at 900°C. Right photo is higher magnification of left one.

Figure 2 FE-SEM micrographs of the surface of the hydroxyapatite fibers immersed in distilled water for 1 week and 9 weeks. Right photos are higher magnification of left ones.

$CaCl_2$, respectivery. It was found that no chloride was formed even when the gel-fibers were not rinsed, because no chloride ion was used in this preparation system. It was also found that the amounts of CaO was small and the aging time to complete the gelation was short, since the purity and the ion exchange rate of K-Alg were higher than those of sodium alginate. Therefore, it suggests that the preparation method used in the present study is suitable for obtaining hydroxy-

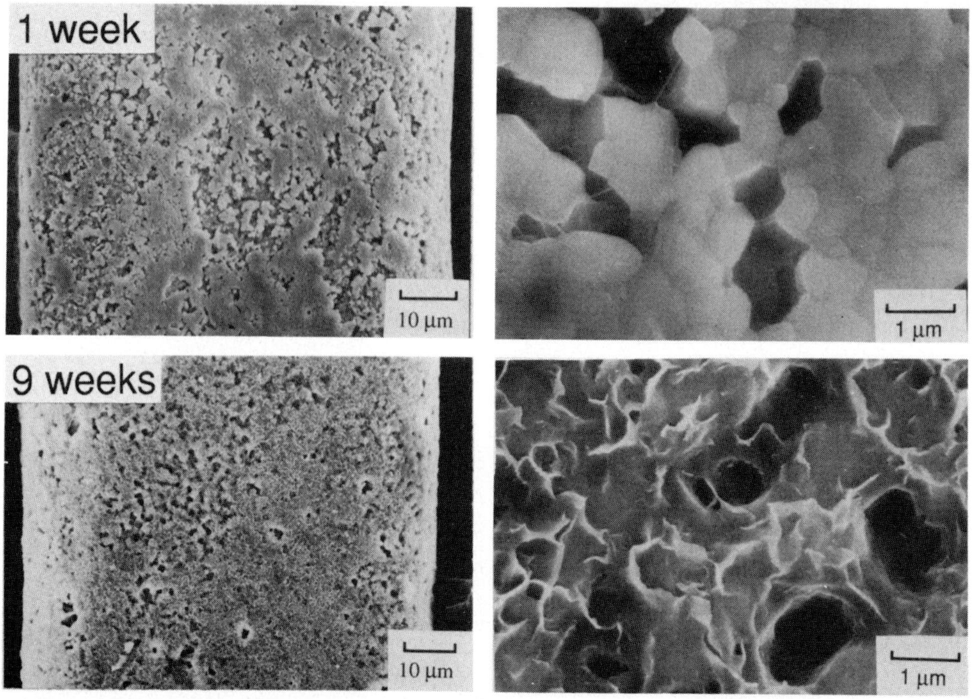

Figure 3 FE-SEM micrographs of the surface of the hydroxyapatite fibers immersed in the SBF for 1 week and 9 weeks. Right photos are higher magnification of left ones.

-apatite fibers with high purity.

Figure 2 shows FE-SEM micrographs of the surface of the hydrxyapatite fibers immersed in distilled water for 1 week and 9 weeks. In comparison to the surface before immersion as shown in Figure 1, it was found that the surface of the fiber was roughened by the dissolution of the crystal grains, and the edge of the crystal grains became sharp. It was also found that the pores on the surface of the fiber immersed for 9 weeks were larger and deeper than those for 1 week.

Figure 3 shows FE-SEM micrographs of the surface of the hydroxyapatite fibers immersed in SBF for 1 week and 9 weeks. In comparison to the surface before immersion as shown in Figure 1, it was found that the crystal grains of hydroxyapatite fibers partially dissolved at 1 week after implantation and the flake-like crystals deposited on the surface of the grains at 9 weeks after implantation.

Figure 4 shows FE-SEM micrographs of the surface of the hydroxyapatite fiber implanted in the rats in the diffusion chambers for 1 week and 3 weeks. At 1 week after implantation, hydroxyapatite grains were covered with fascia and small spherical precipitates were observed on the surface. At 3 weeks after implantation, small pores and granular precipitates were observed on the surface of the fibers covered with fascia. It was also found that the amounts of the precipitates at 3 weeks were significantly larger than that at 1 week.

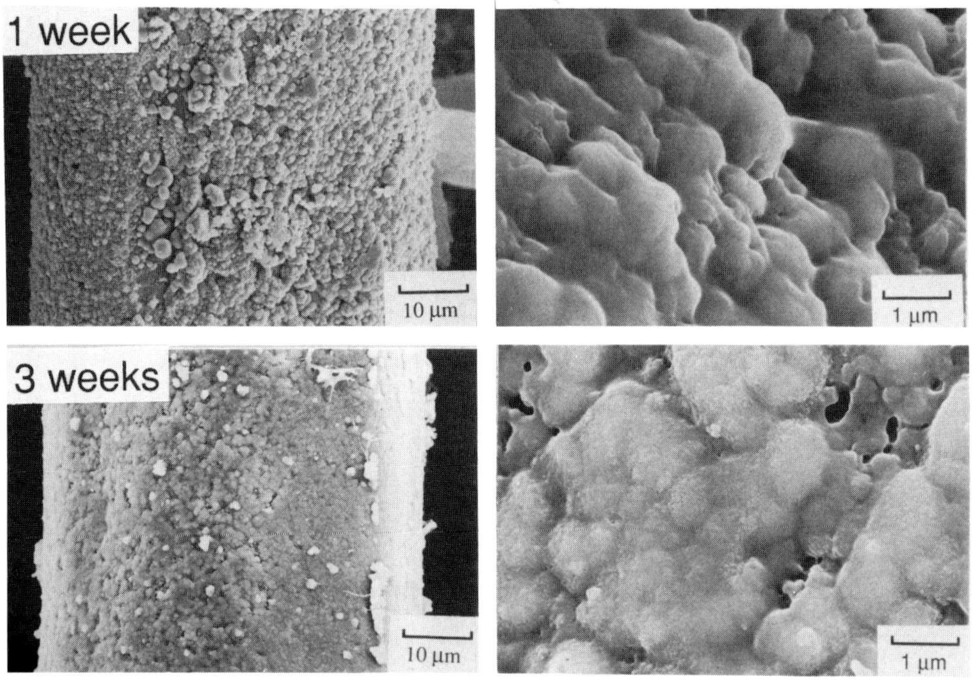

Figure 4 FE-SEM micrographs of the surface of the hydroxyapatite fibers implanted in the rats for 1 week and 3 weeks. Right photos are higher magnification of left ones.

Figure 5 shows FT-IR micro-spectra of the surface of the hydroxyapatite fibers before and after immersion *in vitro* for 9 weeks, and *in vivo* for 3 weeks. The spectrum of the fiber before immersion had the OH absorption band around 3570 cm^{-1} due to structural hydroxyl ions of hydroxyapatite crystal, the OH bands around 3640 cm^{-1} and 875cm^{-1} due to $Ca(OH)_2$ hydrated from CaO, and the C-O bands around 1550 cm^{-1}, 1460 cm^{-1} and 1420 cm^{-1} indicating the presence of carbonate. In the spectra of the fibers immersion *in vitro* and *in vivo*, these sharp OH bands due to $Ca(OH)_2$ were not observed. Therefore, it seems that the pores observed in the fibers after the immersion tests (Figure 2 - 4) were formed by the dissolution of CaO. In the spectrum of the fiber after immersion in the SBF and after implantation in the rats, the OH band around 3570 cm^{-1} due to hydroxyapatite crystal was relatively small compared to that before immersion, and the OH band around 870cm^{-1} due to the HPO_4 was observed. Thus, these results demonstrate that these precipitates in SBF and in the rats are carbonate apatite with low crystallinity and are similar to those of biological apatite. Daculsi et al. demonstrated that needle-like precipitations of apatite crystals were observed on the surface of hydroxyapatite after 3 weeks implantation in mice and appeared perpendicular to the hydroxyapatite crystal surface. And the new crystals were also similar to the biological apatite [6]. Ban et al. showed that the needle-like precipitates *in vitro* are Ca-deficient carbonate apatite with low crystallinity and are similar to those of biological apatite observed in a natural bone of dogs [7]. It is considered that the calcium phosphate layer formed on the surface of bioactive materials implanted in the body plays an essential role in

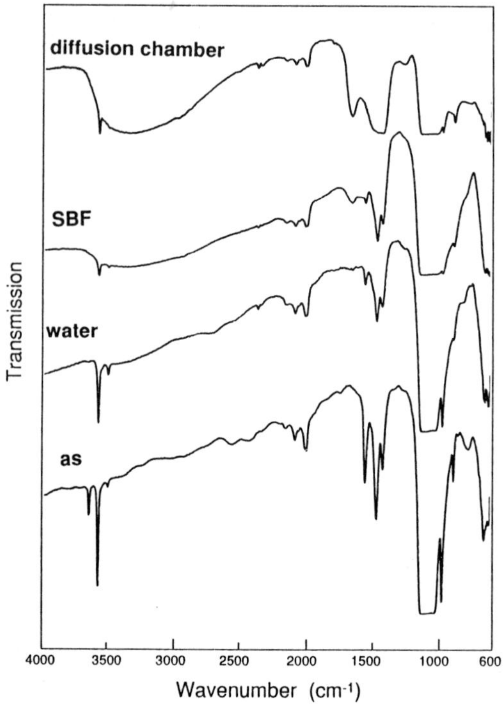

Figure 5 Transmission FT-IR micro-spectra of hydroxyapatite fibers before and after immersion tests *in vitro* for 9 weeks and *in vivo* for 3 weeks.

forming the chemical bond of the bioactive material to the living bone. Furthermore, the spectrum of the specimen implanted in the rats showed the broad H_2O band around 3000 - 3600 cm^{-1} and the amide I (C=O) band around 1655 cm^{-1} indicating the presence of protein such as fascia. These results confirm that these hydroxyapatite fibers have excellent biocompatibility and can be applied as bioceramics such as filling materials.

REFERENCES

1. Salyer K., Holmes R. and Johns D. J. Dent. Res. (1977) 56 B 173
2. Ban S., Matsuura M., Arimoto N., Hayashizaki J., Itoh Y. and Hasegawa J. Dent. Mater. J. (1993) 12-2 106-117
3. Hayashizaki J., Ban S. and Hasegawa J. J. J. Dent. Mater. (1993) 12 46-47
4. Iwasaki H. Ceramics (1989) 24 295-299
5. Hayashizaki J., Ban S. and Hasegawa J. Trans. 9th Symp. Apatite (1993) Gifu Japan 3-4
6. Daculsi G., LeGeros R. Z., Heughebaert M. and Barbienx I. Calcif. Tissue Int. (1990) 46 20-27
7. Ban S., Maruno S., Iwata H. and Itoh H. J. Biomed. Mat. Res. (1994) 28 65-71

Reaction Kinetics and Mechanisms of Bioactive Ceramics

Bioceramics, Volume 7 Edited by Ö.H. Andersson and A. Yli-Urpo
(*Proceedings of the 7th International Symposium on Ceramics in Medicine, Turku, Finland, July 1994*)

Dependence of Bone-like Apatite Formation on Structure of Silica Gel

T. Kokubo, S.B. Cho, K. Nakanishi, N. Soga, C. Ohtsuki,* T. Kitsugi,** T. Yamamuro and T. Nakamura***

Division of Material Chemistry, Faculty of Engineering, Kyoto University, Sakyo-ku, Kyoto 606-01, Japan
* Faculty of Engineering, Okayama University, Okayama 700 Japan
** Faculty of Medicine, Kyoto University, Sakyo-ku, Kyoto 606-01 Japan
*** Research Center for Biomedical Engineering, Kyoto University, Sakyo-ku, Kyoto 606-01 Japan

ABSTRACT

It has been shown that the prerequisite for glasses and glass-ceramics to bond to living bone is the formation of biologically active bone-like apatite on their surfaces, and that a hydrated silica on their surfaces plays an important role in nucleating the apatite. In the present study, structure of the hydrated silica responsible for the apatite nucleation was investigated. Silica gels were prepared by hydrolysis and polycondensation of tetraethoxysilane(TEOS) in aqueous solution containing no polymer, polyacrylic acid or polyethylene glycol. All these gels contained large volume of nanometer-size pore, silanol groups and D2 defect and showed high dissolution rate of the silica. Despite it, only the gel prepared in the solution containing polyethylene glycol formed the apatite on its surface in the simulated body fluid. The same gel did not form it when the gel lost its nanometer-size pores, silanol groups and high dissolution rate of the silica by heat treatment at 1000 ℃. The heat-treated gel recovered its apatite-forming ability by a H_2O gas plasma treatment. This indicates that only a certain type of structural unit of the silanol group is responsible for the apatite nucleation.

INTRODUCTION

The prerequisite for glasses and glass-ceramics to bond to living bone is the formation of a layer of biologically active bone-like apatite on their surfaces in the body [1,2]. It has been proposed that hydrated silica which is formed on their surfaces in the body plays an important role in nucleating the apatite[3-6]. Recently, it was confirmed experimentally that a pure silica gel prepared by a sol-gel method induces the formation of the apatite on its surface in a simulated body fluid with ion concentrations nearly equal to those of the human blood plasma a little increased in its pH or the calcium concentrations[7-10]. A pure silica glass synthesized by vertical axial deposition and quartz single crystal synthesized by hydrothermal method, however, do not induce it [7]. These results suggest that the silanol group abundant on the surface of the silica gel is responsible for the apatite nucleation. On the other hand, Hench et al recently proposed on the basis of molecular orbital calculation that not all the kind of the silanol group but only the silanol group forming trigonal siloxane can induce the apatite nucleation [11]. In the present study, in order to investigate the structure of the silica gel responsible for the apatite nucleation, apatite forming ability of various kinds of silica gel prepared by different methods and subjected to different treatments was investigated in the simulated body fluid and its result was discussed in terms of the structure of the gels.

EXPERIMENTAL PROCEDURE

Gel preparation

Silica gels were prepared by the hydrolysis and polycondensation of tetraethoxysilane (TEOS) in aqueous solution containing no polymer, polyacrylic acid(HPAA, M.W 30.000) or polyethylene glycol(PEG, M.W 10.000) . First, ion-exchanged distilled water in 8 g, was added with no polymer, 0.6 g HPAA or 0.7 g PEG and with 0.61 g concentrated nitric acid (62 wt%). Then 7 ml TEOS was added to the above solutions under vigorous stirring. After stirring for 5 min, the solution was transferred into a plastic petri dish with its top tightly sealed, and was kept at 40 °C in air-circulating oven for gelation. After it was aged for 18 h, the obtained wet gel was immersed in 1 mol/dm^3 nitric acid for 6 h to remove the polymers and dried at 40 °C for 6 days. The nitric acid was renewed every 2 h.

Heat treatment of gel

After the obtained wet gel was dried, it was heated up to various temperatures in the range from 600 to 1000 °C at a rate of 100 °C/h, and held at the respective temperatures for 2 h, then allowed to cool to room temperature.

Plasma treatment of gel

Rectangular specimens of 6 x 6 x 1.5 mm^3 in size of the gel heat-treated at 1000 °C was subjected to the plasma treatment in H_2O gas for 30 or 60 sec. The plasma treatment was carried out at a pressure of 30 Pa and 50 mW power, under an electric field of 13.56MHz by using Plasma Polymerization System (BP-2, SAMCO Ltd., Kyoto, Japan).

Characterization of gel

The volume of pore smaller than 200 nm in diameter were measured with an automatic nitrogen adsorption pore size analyzer (ASAP-2000, Micromeritics, Tokyo, Japan).The microstructures of the silica gels were observed under a scanning electron microscope (SEM: S-2500CX, Hitachi Ltd., Tokyo, Japan). Laser Raman spectroscopy(T64000, Jobin-Yvon, France) was employed to examine the D1 and D2 defects and silanol group of the gels. All Raman spectra were obtained using the 514.5 nm line of an Ar^+ laser in the 180° reflection configuration.

Soaking gel in a simulated body fluid

Rectangular specimens of 6 x 6 x 1.5 mm^3 in size, at least three for each soaking time, were cut from the gels. The specimens were immersed in 15 ml of a simulated body fluid (SBF) which had almost equal ion concentrations to those of the human blood plasma (Na^+ 142.0, K^+ 5.0, Mg^{2+}1.5, Ca^{2+} 2.5, Cl^- 147.8, HCO_3^- 4.2, HPO_4^{2-} 1.0, SO_4^{2-} 0.5 mM). In the case of the plasma treatment, the gel was soaked in a SBF immediately after the treatment. It has been confirmed that this fluid can reproduce fairly precisely apatite formation on the surfaces of various kinds of bioactive glasses and glass-ceramics in vivo [12].

Surface analysis

After the specimens were soaked in SBF for various periods, they were removed from the fluid and immersed into 10 ml of ion-exchanged distilled water for 5 min. The specimens were, then, dried at room temperature. Their surfaces were analyzed using thin-film X-ray diffraction(2651A, Rigaku Co., Tokyo, Japan) and SEM observation.

Measurement of element concentrations

The concentration of the silicon dissolved from the gels into the simulated body fluid were measured with inductively coupled plasma (ICP) atomic emission spectroscopy (SPS1500VR, Seiko Instruments Inc., Tokyo, Japan). For ICP atomic emission spectroscopy, one milliliter of the fluid was drawn from the bottle and added to 10 ml of ion-exchanged distilled water.

RESULTS AND DISCUSSION

The results of the measurements of volume of nanometer size pores and SEM observation of micrometer size pore for the gels which were prepared in three kinds of different media and heat-treated at 600 °C are shown in Table 1. All these gels contained large volume of mesopore, ranging 1.7 to 200 nm. Among them, two kinds of the gels prepared in the media containing polymers also had macropore in micrometer size. The gel prepared in the medium containing PEG had also micropore less than 1.7 nm. Figure 1 shows Raman spectra of these three kinds of gels. It can be seen from Fig. 1 that all these gels contain almost same amounts of silanol groups as well as tetrasiloxane D1 defect and trisiloxane D2 defect. Figure 2 shows concentration of the silicon dissolved from these three kinds of gels into SBF as a function of soaking time. The solubility of the silica in the pure water is almost 2 mM. It can be seen from Fig. 2 that all these gels dissolve the silica up to the solubility at almost same rates. These results indicate that all these gels have almost same surface structure. Figure 3 shows thin film X-ray diffraction patterns of the surfaces of the gels prepared in

Table 1. Structure and properties of the silica gels

Sample	Medium for preparing gel	Heat treatment (°C)	Volume of nanometer pore (cm^3/g)[1] micro-	macro-	Size of macropore (μm)	SiOH[2]	D1[2]	D2[2]	Dissolution[2]
PA-6	TEOS+H_2O+HPAA	600	0.001	0.384	1	++++	---	****	XXXX
P0-6	TEOS+H_2O	600	0	0.571	0	++++	---	****	XXXX
PE-6	TEOS+H_2O+PEG	600	0.106	0.275	2	++++	---	****	XXXX
PE-8	TEOS+H_2O+PEG	800	0.032	0.176	2	++	---	***	XXXX
PE-10	TEOS+H_2O+PEG	1000	0	0.004	2		---	*	X
PE-10-PT[3]	TEOS+H_2O+PEG	1000	0	0.004	2		---	*	XX

1) The nanometer-size pore is classified into micropore less than 1.7 nm and mesopore ranging 1.7 nm- 200 nm.
2)The number of symbols represents relative content.
3)PE-10-PT is the gel heat-treated at 1000 °C and then subjected to plasma treatment for 30 or 60 second.

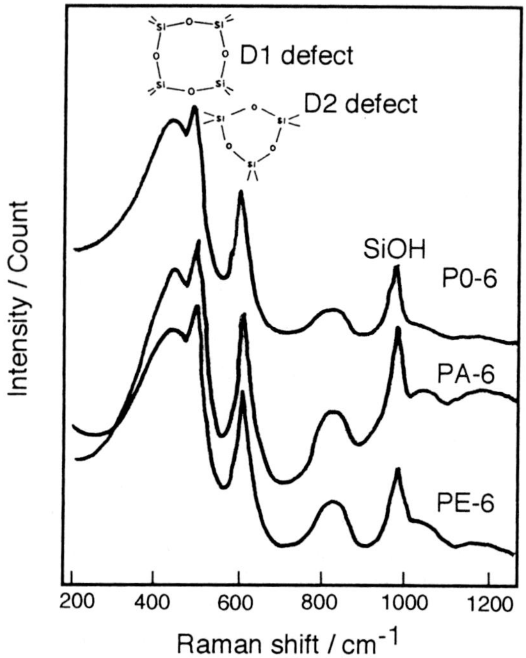

Figure 1. Raman spectra of the silica gels prepared in different media and heat-treated at 600 °C.

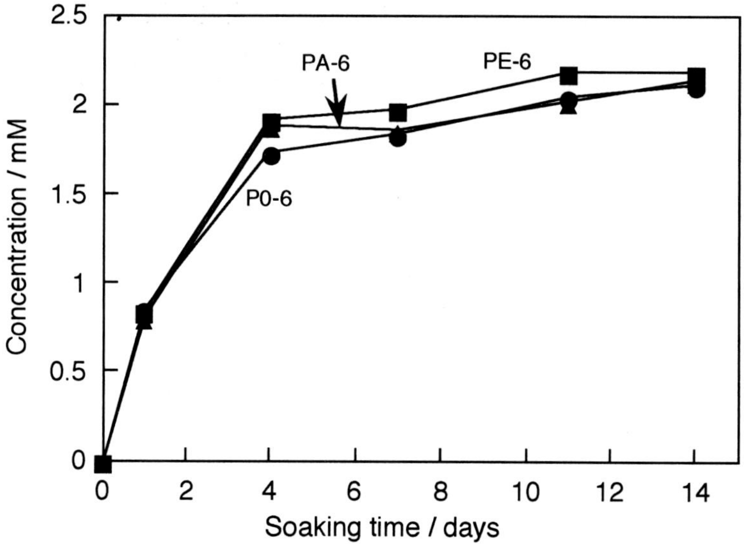

Figure 2. Concentration of silicon dissolved from the gels prepared in different media and heat-treated at 600°C as a function of soaking time in SBF

different media, heat-treated at 600 °C and then soaked in SBF for 2 weeks. Only the gel prepared in the PEG-containing medium formed the apatite on it. This indicates that not all the silanol group or D1 and D2 defects but only a certain type of the silanol group or defect characteristic to the gel prepared in PEG-containing medium is effective for the apatite nucleation.

When the gel prepared in the PEG-containing medium was heat-treated at 800 °C, nanometer size pores as well as the silanol group and D2 defect considerably decreased but dissolution rate of the silica did not decrease, as shown in Table 1. When the same gel was heat-treated at 1000 °C, nanometer size pores and the silanol group were almost completely removed. The D2 defect as well as dissolution rate of the silica remarkably decreased, as shown in Table 1. Figure 4 shows thin film X-ray diffraction patterns of these heat-treated gels after soaking in SBF for 2 weeks. The gel heat-treated at 800 °C formed the apatite on it similar to that heat-treated at 600 °C, but the gel heat-treated at 1000 °C did not form it. This indicates that the surface structure closely related to dissolution of the silica is responsible for the apatite nucleation.

When the gel heat-treated at 1000 °C(PE-10) was subjected to the H_2O plasma treatment, no appreciable change was observed in its pore structure as well as in the amounts of the silanol group and D1 and D2 defects, but the dissolution rate of the silica increased to some extent, as shown in Table 1. Figure 5 shows thin film X-ray diffraction patterns of the plasma-treated gels after soaking in SBF for 3 weeks. It can be seen from Fig. 5 that the gel heat-treated at 1000 °C recovers its apatite-forming ability by the H_2O plasma treatment. This indicates that the silanol group which was formed on the surface of the heat-treated gel by the H_2O plasma treatment and closely related to the dissolution of silica is responsible for the apatite nucleation, although it can not be detected by the Raman spectroscopy. From these results, it can be concluded that only a certain type of the silanol group, that is formed by hydrolysis of a special type of siloxane group, could be responsible for the apatite nucleation. In order to identify it, further study is required.

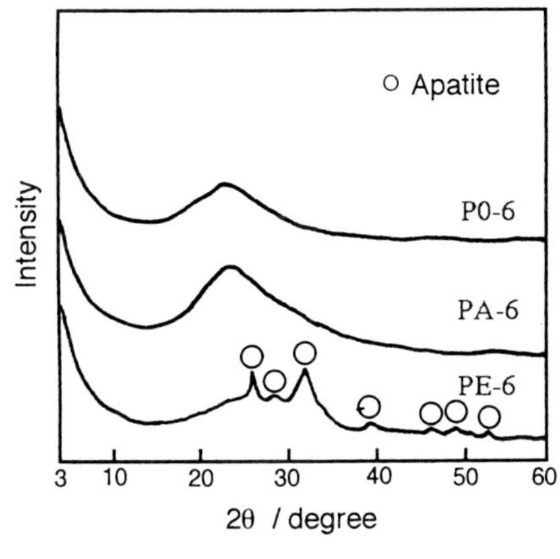

Figure 3. Thin film X-ray diffraction patterns of the silica gel prepared in different media , heat-treated at 600 °C and soaked in the SBFfor 2 weeks.

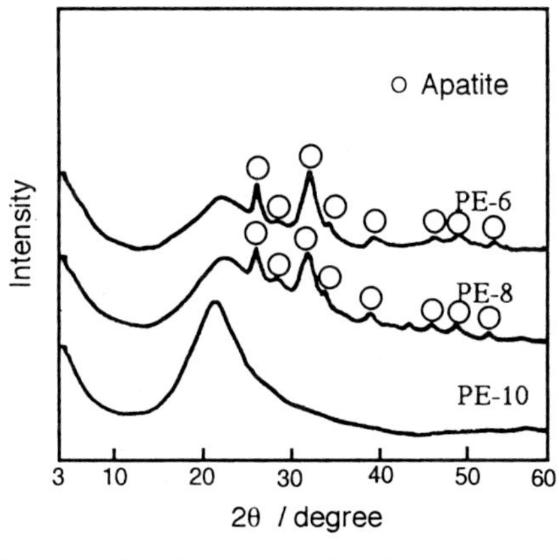

Figure 4. Thin film X-ray diffraction patterns of the silica gel prepared in the PEG-containing medium, heat-treated at various temperatures and then soaked in the SBF for 2 weeks.

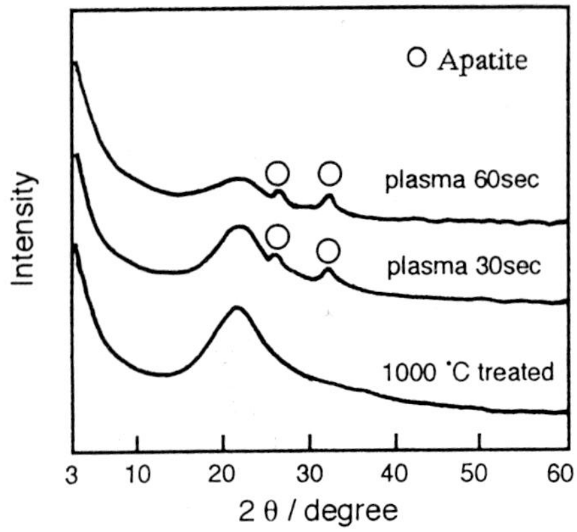

Figure 5. Thin film X-ray diffraction patterns of the silica gel plasma-treated and then soaked in the SBF for 3 weeks.

ACKNOWLEDGMENT

This study was supported by a Grant-in Aid for Scientific Research of the Asahi Glass foundation.

REFERENCE

1. T. Kokubo, Biomaterials, 1991, 12, 155-163
2 L. L. Hench, J. Am. Ceram. Soc., 1991, 74, 1487-1510
3 L. L. Hench, In Bioceramics: Material Characteristics Versus in vivo Behavior, Annual of New York Academy of Science Vol. 523, New York, 1988, 54-71
4 K. H. Karlsson, K. Froberg and T. Ringbom, J. Non-Cryst. Solids, 1989, 112, 69-72
5 C. Ohtsuki, T. Kokubo, K. Takatsuka and T. Yamamuro, J. Ceram. Soc. Japan, 1991, 99, 1-6
6 C. Ohtsuki, T. Kokubo and T. Yamamuro, J. Non-Cryst. Solids, 1992, 143,84-92
7 P. Li, C. Ohtsuki, T. Kokubo, K. Nakanishi, N. Soga, T. Nakamura, T. Yamamuro, J. Am. Ceram. Soc., 1992, 75, 2094-2097
8 P. Li, C. Ohtsuki, T. Kokubo, K. Nakanishi, N. Soga, T. Nakamura, T. Yamamuro, J. Appli. Biomater., 1993, 4, 221-229
9 P. Li, C. Ohtsuki, T. Kokubo, K. Nakanishi, N. Soga, T. Nakamura, T. Yamamuro, J. Mater. Sci.: Medicine, 1993, 4, 127-131
10 P. Li, T. Kokubo, K. Nakanishi, K. de Groot, Biomaterials, 1993, 14, 963-968
11 J. H. West, L. L. Hench, In Bioceramics Vol. 5, Kobunnhi Kankokai, Kyoto, 1991, 75-86
12 T. Kokubo, H. Kushitani, S. Sakka, T. Kitsugi and T. Yamamuro, J. Biomed. Mater. Res., 1990, 24, 721-734

Bioceramics, Volume 7 Edited by Ö.H. Andersson and A. Yli-Urpo
(*Proceedings of the 7th International Symposium on Ceramics in Medicine, Turku, Finland, July 1994*)

Effect of Surface Area to Volume Ratio on *In Vitro* Surface Reactions of Bioactive Glass Particulates

D.C. Greenspan,[1] J.P. Zhong[2] and G.P. LaTorre[2]
[1] U.S. Biomaterials Corporation, One Progress Blvd., #23, Alachua, FL 32615, USA
[2] Advanced Materials Research Center, University of Florida, One Progress Blvd., #14, Alachua, FL 32615, USA

ABSTRACT

The *in vitro* surface reactions of bioactive glass particulates with the composition 45% SiO_2, 24.5% Na_2O, 24.5% CaO and 6% P_2O_5 (45S5 Bioglass®) were studied as a function of surface area to volume ratios in a tris (hydroxyaminomethane) buffer. Formation of the calcium phosphate layer characteristic of this bioactive glass was measured using Fourier Transform infrared spectroscopy. Particle sizes of the bioactive glass studied were 710-500 μm, 710-90 μm and 90-38 μm. The changes in pH and ion concentration of the reaction solutions were also measured.

INTRODUCTION

The bonding of bioactive glasses to bone has been shown to be related to the formation of a hydroxycarbonate apatite layer which is developed on the surface of the glass both *in vitro* and *in vivo*.[1,2] The *in vitro* studies have established the kinetics of a series of surface reactions which lead to this apatite formation on bulk materials.[3,4] Recent clinical interest in bioactive glasses has been extended to include particulates for bone grafting applications.[5] These applications require that these surface reactions be established and understood for the various particle size ranges. The objective of this work is to determine the effects of surface area of particles to volume ratio of solution (SA/V) on the rate of hydroxyapatite formation *in vitro*. The effects of both solution pH and ion concentration were also investigated to determine their role in this formation.

MATERIALS AND METHODS

The Bioglass® powders used for this study were made from 45S5 frit (44.9% SiO_2, 24.49% CaO, 24.55% Na_2O and 5.99% P_2O_5). The frit was ball milled for one hour and then mechanically separated through stainless steel sieves to produce the following particle size ranges: 710-500 μm, 710-90 μm and 90-38 μm. After sieving the powders were analyzed using a Coulter model LS100 particle size analyzer to determine the actual particle size distributions. To determine the surface area of each powder size range nitrogen or krypton adsorption was used.

A dynamic *in vitro* test model was used to evaluate the bioactivity of each particle size range as a function of changes in SA/V. Previous studies[6,7] comparing static versus dynamic *in vitro* testing of Bioglass® powders have established the use of dynamic testing for a fixed SA/V ratio. Samples were placed into 500 ml flasks containing Tris buffer which was preheated to 37°C in a Labline orbital shaker at 175 rpm. The sample weights were adjusted to obtain the desired SA/V ratios for 200 ml of solution. The samples were reacted for 10, 30, 60, 180, 360 and 1200 minutes. The solutions were then vacuum filtered into clean, dry flasks and the powders were rinsed with acetone. The powders and solutions were saved for FTIR and ICP analysis respectively. The FTIR spectra were collected using a Nicolet 20SXB FTIR spectrometer with a diffuse reflectance stage. The ICP data was collected using a Perkin Elmer P40 plasma emission spectrometer. The pH change as a function of time for each test condition was determined using a Fisher Scientific model accumet® 910 pH meter with a combination electrode. The test solutions were placed in a water bath at 37°C under constant stirring during these measurements.

The selection of SA/V ratios for each particle size range was based on a series of *in vitro* tests. The criteria for this *in vitro* QA test requires that the 20 hr. FTIR reflectance spectra is free of any spectral response from the bulk glass and that the resultant spectra is solely the result of a fully developed HA layer. In these tests, for each of the three particle sizes selected, an SA/V ratio which yielded the best test conditions for a twenty hour quality assurance test for bulk glasses[6] was established. Based on these results SA/V ratios above and below this initial SA/V were investigated to establish the boundaries at which the *in vitro* behavior would deviate from the original (middle) SA/V selected.

RESULTS

The results for the pH measurements are seen in Fig. 1 for the selected particle size ranges and SA/V ratios. The pH data show a rapid increase in pH during the first two hours of reaction time and then a more gradual increase of pH up to the twenty hour period for all cases shown. In general the higher SA/V ratios had a faster rate of increase in pH and a higher final pH when compared to the lower SA/V ratios.

The results of the ICP analysis for elemental phosphorus are seen in Fig. 2 for the selected test conditions. The results are shown only for phosphorus for the sake of brevity and do not include the other alkaline species which also contribute to the pH behavior discussed previously. The ICP results indicate that the higher SA/V ratios reached a maximum concentration for phosphorus within the first hour of exposure to the buffered solution and then a rapid decrease in concentration between one and two hours reaction time for all three particle size ranges. After this rapid decrease the phosphorus concentration remains very low up to the twenty hour time period. In contrast the lower SA/V ratios achieved a maximum

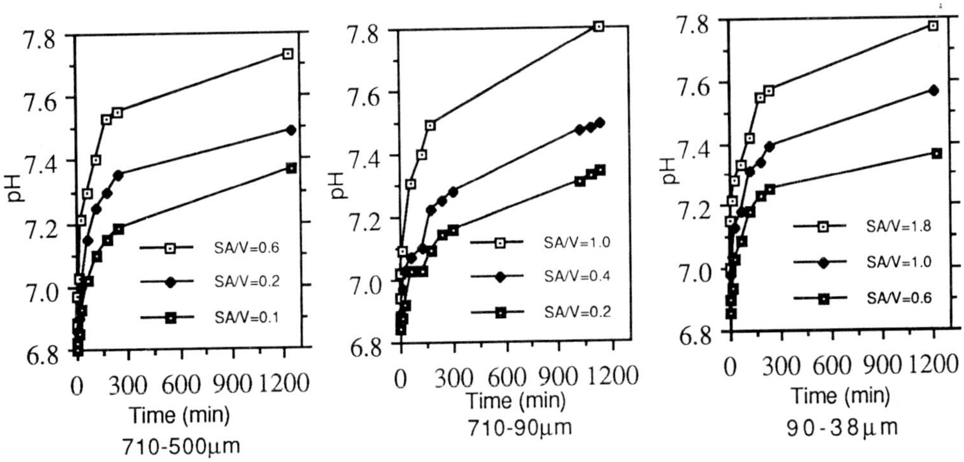

Figure 1. pH of reaction solutions vs. times for 710-500 μm, 710-90 μm and 90-38 μm in tris buffer at the specified SA/V ratios.

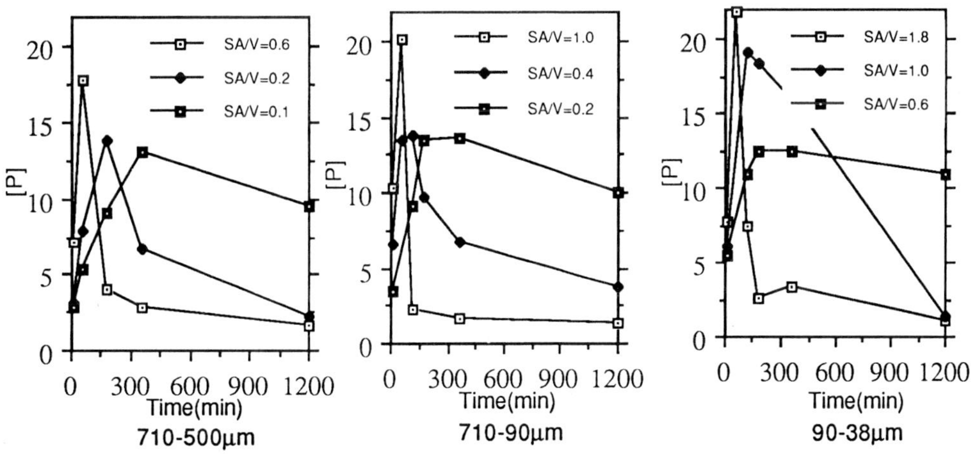

Figure 2. Phosphate concentration of reaction solutions vs. times for 710-500 μm, 710-90 μm and 90-38 μm in tris buffer at the specified SA/V ratios.

concentration within two to three hours, for the middle and lowest SA/V respectively, and then a much slower decrease in concentration relative to the higher SA/V ratio. For these lower SA/V ratios the phosphorus concentration remains higher relative to the high SA/V ratios to twenty hours.

The FTIR results for the selected test conditions are seen in Figs. 3a, b and c. The peaks at 608 cm^{-1} and 566 cm^{-1} are assigned to P-O bending vibrations from the hydroxycarbonate apatite (HCA) layer formed on the glass powder surface after precipitation of calcium and phosphorus from solution. These peaks are seen for each test condition at 20 hrs. reaction time. The peak at 470 cm^{-1} is assigned to the Si-O-Si stretching vibration from the underlying silica rich layer produced after ion exchange with the solution. This layer is clearly seen at 3 hrs. on all spectra, but becomes masked by the development of the HCA layer by 20 hrs. reaction time especially at lower SA/V ratios.

DISCUSSION

The establishment of the reaction kinetics of bulk bioactive glasses has been demonstrated within a fairly limited range of SA/V ratio.[6] The FTIR spectra in Fig. 3a, b and c show the same trends as found in bulk glass. It was noted that the precipitation of amorphous apatite and its crystallization into HCA was slower in current tests when compared to static tests run on bulk 45S5.[4]

After 20 hours in tris buffer the FTIR spectra from the highest SA/V ratios for all three particle sizes studied show a crystalline calcium phosphate layer (P-O bending peaks at 560 cm^{-1} and 610 cm^{-1}) and a significant residual Si-O-Si bending peak at 470 cm^{-1} in Fig. 3. This indicates that the HCA which is formed is very thin, allowing the characteristic vibrations of the underlying silica rich layer to be detected. By comparison, the lower SA/V ratios for the 710-500 μm and 710-90 μm particles show strong HCA peaks and very little underlying Si-O-Si peaks indicating a much thicker calcium phosphate layer.

The difference in the rate of HCA formation may be attributed to the elevated pH as seen in Fig. 1 and phosphorus concentrations seen in Fig. 2 during the early reaction times for the highest SA/V ratios. The rapid increase in pH accompanied by a high concentration of phosphorous can be responsible for the rapid precipitation of hydroxyapatite from solution.[8,9] However, the early formation of calcium phosphate layer and the elevated pH may be inhibiting continued ion exchange. Since the glass is the only source of Ca and P ions to create the calcium phosphate layer in this test system, any interruption of ion exchange could be responsible for the thin calcium phosphate layer formed at 20 hrs at the highest SA/V ratios. It is also possible that the rapid increase in pH is altering the solubilization and subsequent repolymerization of the silica rich layer which is necessary in these glasses in order to form the crystallized calcium phosphate layer.

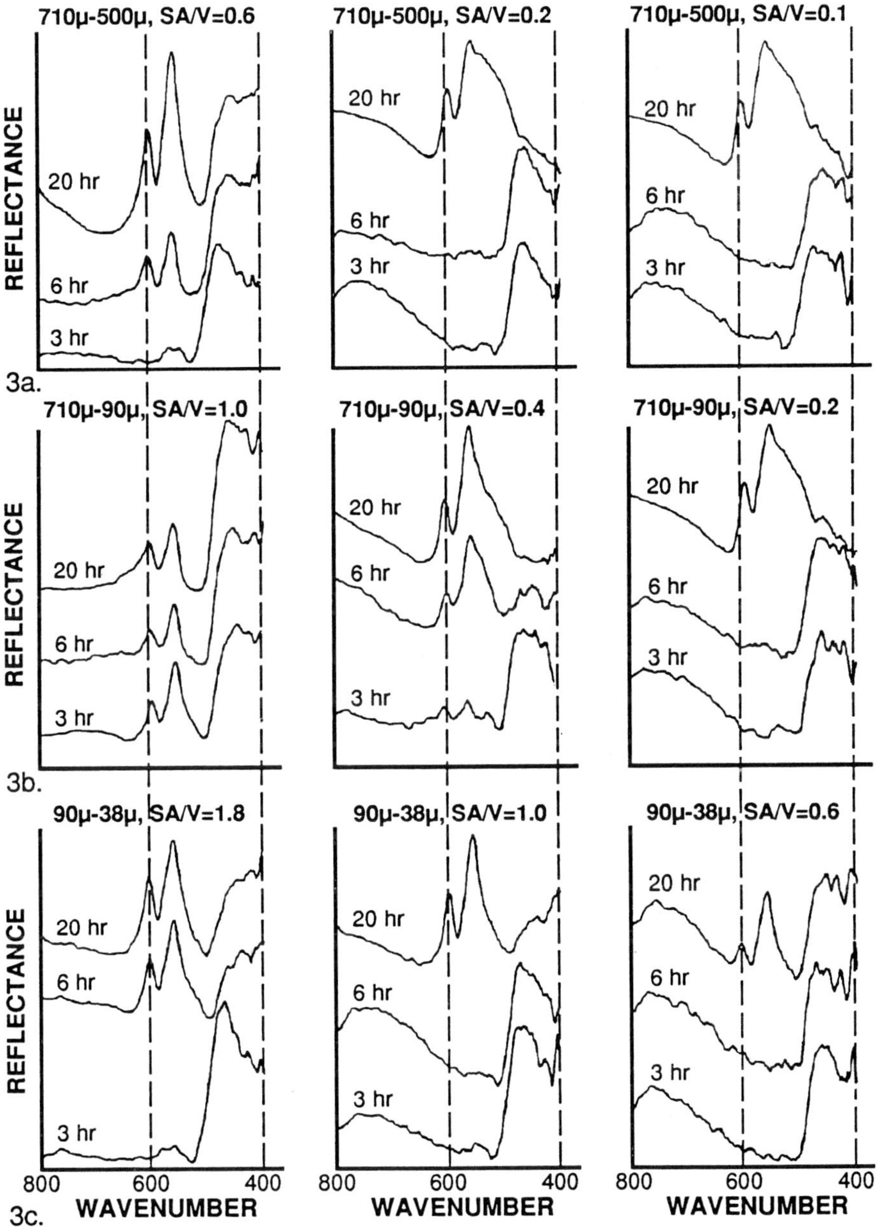

Figure 3. FTIR spectra for 710-500 μm, 710-90 μm and 90-38 μm reacted in tris buffer for 3 hrs, 6 hrs, and 20 hrs at the specified SA/V ratios.

The *in vitro* FTIR data also suggests that particles of different sizes behave differently under the same SA/V test conditions. When comparing the 710-500 μm and 90-38 μm particles at SA/V of 0.6 cm^{-1} it is apparent that the rate of HCA formation is much slower for the smaller particles. This may be attributed to physical differences such as radius of curvature of small versus large particles or to other *in vitro* variable such as mass to volume ratios of the powder and test solution.

CONCLUSIONS

This study has shown that the *in vitro* development of the silica rich reaction layer and calcium phosphate layer of Bioglass® particles is dependant on the SA/V ratio of the test. At high SA/V ratios, there is a rapid increase on pH and ion release from the glass which results in rapid formation of an initial calcium phosphate layer. This layer remains thin over time, with a residual silica rich layer present. At lower SA/V ratios, there is a less pronounced change in pH, a more gradual development initially of the calcium phosphate layer, but a more well developed calcium phosphate layer over the 20 hour duration of the test.

Clearly, the limitations of the test conditions make it difficult to project these result to *in vivo* systems. The correlation between the *in vitro* response of bulk Bioglass® and its *in vivo* behavior is well documented. These results suggest that there may be very different responses *in vivo* to particulates of very different sizes.

ACKNOWLEDGEMENTS

The authors would like to thank Mike Robinson, Keith Lobel and Geri Greenspan for the technical assistance during this study. Two of the authors (JPZ and GPL) would like to acknowledge the financial support of the Air Force office of Scientific Research under Grant No. F49620-92-J-0351.

REFERENCE

1. Hench, L.L. J. Am. Ceram. Soc., 1991, 74[7], 1487-510.
2. Andersson, O.A. and Karlsson, J. Non-Cryst. Solids, 1990, 119, 290-296.
3. Hench, L.L. In Bioceramics 3, Joy E. Hulbert and Samuel F. Hulbert, eds., Rose-Hulman Institute of Technology, Terre Haute, Indiana, 1992, 43-45.
4. Hench, L.L., LaTorre, G.P. In Bioceramics 5, T. Yamamuro, T. Kokubo and T. Nakamura, eds., Kobonshi Kankokai, Inc., Kyoto, Japan, 1992, 67-74.
5. Wilson, J., and Low, S.B. J. Appl. Biomat., 1992, 3, 123-129.
6. Warren, L.D., Clark, A.E. and Hench, L.L. J. Biomed. Maters. Res. l989, 23[A2], 201-209.
7. Li, R., Clark, A.E. and Hench, L.L. J. Appl. Biomat., 1992, 2, 231-239.
8. Brown, W.E. and Chow, L.C. J. Crystal Growth, 1981, 53, 31-41.
9. Li, P., Ohtsuki, C., Kokubo, T., et al. J. Appl. Biomat. 1993, 4, 221-229.

Bioceramics, Volume 7 Edited by Ö.H. Andersson and A. Yli-Urpo
(*Proceedings of the 7th International Symposium on Ceramics in Medicine, Turku, Finland, July 1994*)

The Kinetics of Bioactive Ceramics Part VII: Binding of Collagen to Hydroxyapatite and Bioactive Glass

J.P. Zhong, G.P. LaTorre and L.L. Hench
Advanced Materials Research Center, University of Florida, One Progress Blvd., #14, Alachua, FL 32615 USA

ABSTRACT

The *in vitro* reaction of Type I insoluble collagen in a simulated body fluid (SBF) is analyzed for growth of hydroxyapatite (HA) crystals using Fourier transform IR spectroscopy and SEM-EDX. The negative surface potential of collagen leads to precipitation and growth of HA within the fibers. Exposure of collagen to a bioactive glass (45S5 Bioglass®) and SBF results in collagen becoming embedded within a growing HA layer and thereby creates a strongly bonded collagen-glass interface. The *in-vitro* results correspond with the *in vivo* behavior of 45S5 Bioglass® which bonds to both soft connective tissues and bone.

INTRODUCTION

Bone forms by deposition and crystallization of hydroxy carbonate apatite (HCA) onto a framework of collagen fibrils with osteoblasts embedded in the organic-inorganic matrix. The site of initial crystallization of HCA within collagen fibrils and reasons why Ca and P concentrate at the site are still uncertain.[1] Bonding of bioactive implant materials with bone also involves the precipitation and crystallization of HCA and the bonding with collagen and cells,[2] as discussed in Parts I-VI of this series of papers. Bioactive bonding occurs when collagen is incorporated within the surface HCA layer.[2]

Materials with negative surface potential can induce the precipitation and crystallization of calcium phosphate[3] and stimulate bone formation.[4] Materials considered to be bioactive; i.e., synthetic HA, bioactive glasses, porous silica gel, Vycor glass, SiO_2 and TiO_2 powders, and sephadex beads,[4] have a negative surface potential in their electric double layer in aqueous solution of pH 7.25. The negative surface potential appears to be important in the precipitation and crystallization of HCA. Collagen also has a negative surface potential in a solution of pH 7.25.[5] Thus, precipitation and crystallization of HCA is expected to occur on collagen fibers *in vitro*. This paper compares HCA formation in a collagen + SBF solution with that on a bioactive glass + collagen + SBF solution.

®Trademark University of Florida, Gainesville, FL 32610

MATERIALS AND METHODS

Type I insoluble collagen (bovine Achilles tendon, Sigma Chemical Company, St. Louis, MO, USA) was used. For the bonding study, 45S5 Bioglass® discs (10 x 1.5 mm) were used.[2] A simulated body fluid (SBF-9) was prepared with the ion concentration nearly equal to that of human blood plasma[6] and buffered to pH 7.25 at 37°C. The 0.1g collagen fiber was tested statically in a flask with 200 ml SBF at 37°C for 2 weeks. Bonding of collagen with the bioactive glass surface was observed using SEM. The Bioglass® discs were exposed to SBF with 0.05g collagen using 25 ml SBF (SA/V = 0.1 cm^{-1}) at 37°C for 2 weeks. Surface characteristics of the samples were evaluated by FTIR spectra, SEM and EDX.

RESULTS

Figure 1 shows the FTIR transmission spectrum of the collagen in the SBF solution after two weeks, compared with unreacted collagen. There are three more peaks for the reacted collagen sample which correspond to the apatite crystal P-O bending vibrations at 1030 cm^{-1}, 602 cm^{-1} and 565 cm^{-1}.[3] Figure 2 shows an SEM micrograph of the collagen fibers reacted in SBF. A calcium phosphate phase has grown on the fiber surfaces as confirmed by EDX. Figure 3 shows the SEM of the bioactive glass surface after reacting with collagen and SBF. Collagen fibers are embedded within the glass surface reaction layer. The calcium phosphate crystals, confirmed by EDX in Fig. 3, have grown on the collagen fibers and incorporated the fibers within the glass surface.

DISCUSSION

Amino acids within collagen have functional groups of COOH and COH that can dissociate to COO^- and CO^-, which give rise to the formation of an electric double layer.[7] The zeta potential in the electric double layer for collagen is about -10 mV in an aqueous solution of pH 7.25.[5] The activities of ions can be different between the bulk solution and negative potential region. From the well-known Bolzmann distribution, the difference, as previously used by Li[3], is:

$$a_{ie}/a_{is} = \exp(-Z_i F \varnothing_o / RT) \qquad (1)$$

where, a_{is} and a_{ie} are ion activities in the solution and inside the electric potential region, respectively. $\varnothing_o$ is the electric potential, -10 mV for collagen; Z_i is the ion valency, 2 for Ca^{2+}; F is the Faraday constant, 96489 Coul/mole; R is the Universal Gas constant and T is the solution temperature, 310°K. Thus, for Ca^{2+}: $a_{ie} = 2.12a_{is}$. The greater activity implies that Ca^{2+} will accumulate in the electric double layer first, then allowing the attraction of phosphate in the form of HPO_4^{2-} and $H_2PO_4^-$ which may lead to super-saturation, precipitation and crystallization of HCA within collagen (Fig. 2) filling the "holes" in the collagen bands.[1,5] Since a carboxyl-terminal end with high negative potential is present in the "holes", it is likely that precipitation and crystallization of HCA initially take place there.

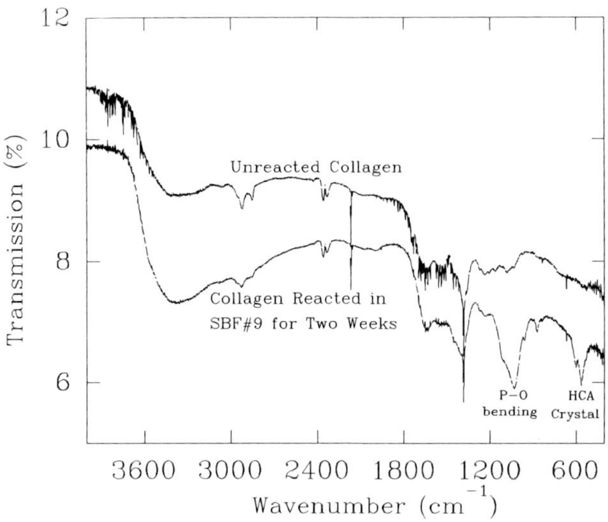

Figure 1. FTIR transmission of unreacted collagen and the collagen reacted in SBF #9 for two weeks.

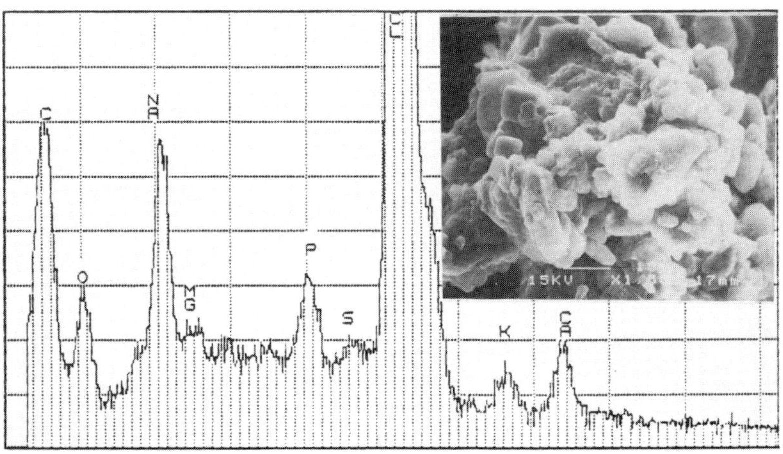

Figure 2. SEM micrograph of flake-like apatite crystals growing on collagen fiber surfaces and EDX of the area with the crystals.

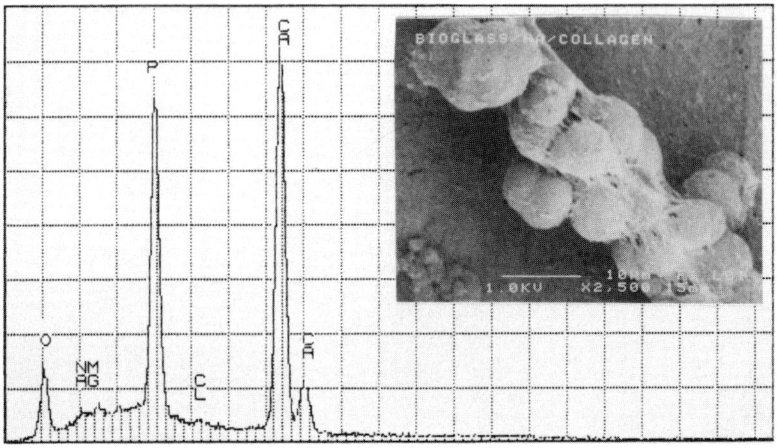

Figure 3. SEM micrograph of the integration of apatite crystals and collagen fibers on Bioglass® surface and EDX of the grape-like crystals.

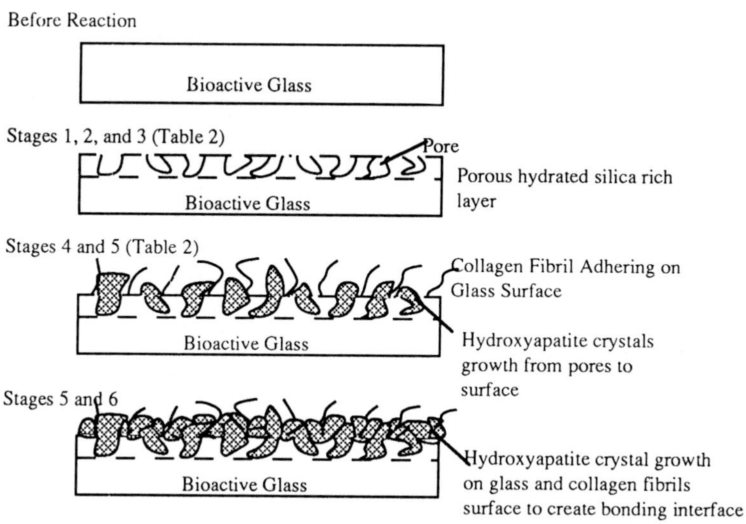

Figure 4. Process of bonding of bioactive glass with connective tissue.

A list of materials tested for bioactivity is in Table 1. All the bioactive materials have a negative surface potential while bio-inert materials have a positive surface potential. Based on intensive studies on bioactive glasses, Hench has proposed a model in which five interfacial stages occur on the glass surface to develop the apatite layer which leads to bonding[2] (Table 2). Calcium phosphate crystals grow and bind on the collagen fiber surfaces (Fig. 2). When collagen fibers are adsorbed on a bioactive glass surface, integration of the crystal layer and the fibers will occur since the crystals grow on both glass and fiber surfaces.

Hench et al.[8] and Iler[9] reviewed the types of chemical bonds which could be present in the bonding of collagen with SiO_2-based glasses. The two most likely bonds are electrostatic, ionic bonds and hydrogen bonds. In the tropocollagen molecule the carboxyl-terminal end is highly negative while the amino-terminal end is highly positive. This positive end can be chemisorbed by the negatively charged silanol-rich glass surface, as modelled by West and Hench.[10] Thus, the process of bonding of 45S5 Bioglass® with connective tissue can be summarized as in Fig. 4.

CONCLUSION

Collagen, a connective tissue with negative surface potential, can induce the precipitation and crystallization of calcium phosphate on its surface in a simulated body fluid. The negative surface potential increases the activities of cations, which result in precipitation and crystallization.

Table 1. Bioactive Materials and Surface Zeta Potential

Materials	Surface Zeta Potential at pH 7.25	Bioactive Description and Implications
Hydroxylapatite (Ca/P: 1.58-1.64)	negative	Main inorganic phase of bones and teeth Bonding with living bone as a biomaterial
Bioglass® (45S5)	negative	Bonding with living bone
Porous silica gel	≈ -20 mV	Induces crystallization of calcium phosphate
Silica particles	≈ -20 mV*	Induces precipitation of calcium phosphate
Titania	negative	Induces precipitation of calcium phosphate
Vycor® glass	≈ -20 mV	Induces precipitation of calcium phosphate
Type I Collagen	≈ -10 mV*	Organic phase of living bone, induces crystallization of calcium phosphate
Sephadex beads	negative	Bonds with living bone
Sephadex beads	positive	No bonding with bone
Alumina	≈ +15 mV†	Bio-inert

*From Model 501 Laser Zee Meter, Pen Kem Inc., Bedford Hills, NY. †From Matec ESA 8000 System.

Table 2. Reaction Stages

Stage (1)	Rapid exchange of Na^+ or K^+ with H^+ or H_3O^+ from solution, $$\text{Si-O-Na}^+ + H^+ + OH^- \rightarrow \text{Si-OH} + Na^+_{(solution)} + OH^-$$
Stage (2)	Loss of soluble silica in form of $Si(OH)_4$ to the solution resulting from breakage of Si-O-Si bonds and formation of Si-OH (silanols) at the glass solution interface. $$2(\text{Si-O-Si}) + H_2O = \text{Si-OH} + \text{OH-Si}$$
Stage (3)	Condensation and repolymerization of a SiO_2-rich layer on the surface depleted in alkalis and alkaline earth cations. $$\text{Si-OH} + \text{OH-Si} = \text{-Si-O-Si-} + H_2O$$
Stage (4)	Migration of Ca^{2+} and PO_4^{3-} groups to the surface through the SiO_2-rich layer forming a $CaO\text{-}P_2O_5$-rich film on top of the SiO_2-rich layer, followed by growth of the amorphous $CaO\text{-}P_2O_5$-rich film by incorporation of soluble calcium and phosphates.
Stage (5)	Crystallization of the amorphous $CaO\text{-}P_2O_5$ film by incorporation of OH^-, CO_3^{2-}, or F anions from solution to form a mixed hydroxyl-carbonate, fluorapatite layer.

Collagen fibrils are chemisorbed on the SiO_2 gel layer of the bioactive glass surface during reaction in a physiological environment which is followed by precipitation and crystallization of hydroxy carbonate apatite on both the glass and collagen fiber surfaces. This leads to a firm bonding at the inorganic-organic interface and is important in the bonding of Bioglass® and other bioactive materials to living tissue.

ACKNOWLEDGEMENT

The authors gratefully acknowledge financial support of the Air Force Office of Scientific Research under Grant No. F49620-92-J-0351.

REFERENCES

1. Boskey, A.L. In The Chemistry and Biology of Mineralized Tissues, W.T. Butler, ed., Ebsco Media, Inc., Birmingham, AL, 1985.
2. Hench, L.L. J. Am. Ceram. Soc., 1991, 74[8], 1487-1510.
3. Li, P.J. and Zhang, F.P. J. Non-Cryst. Solids, 1990, 119, 112-118.
4. Krukowski, M.,, Shively, R.A., Osdoby, P., and Eppley, B.L. J. Oral. Maxillofac. Surg., 1990, 48, 468-475.
5. Gustavson, K.H. The Chemistry and Reactivity of Collagen, Academic Press, NY, 1956.
6. Li, P.J., Ohtsuki, C., Kokubo, T., Nakanishi, K. and Soga, N. J. Am. Ceram. Soc., 1992, 75[8], 2094-2097.
7. Stryer, L. Biochemistry, W. H. Freeman, NY, 1981.
8. Hench, L.L., Splinter, R.J., Allen, W.C., and Greenlee, T.K., Jr. J. Biomed. Maters. Res., 1972, 2[1], 117-141.
8. Iler, R.K. The Chemistry of Silica, John Wiley & Sons, NY, 1979.
10. West, J.K. and Hench, L.L. "AM1 Molecular Orbital Calculations of Silica-Alanine-Nitrogen Interaction," J. Biomed. Maters. Res., in press.

Bioceramics, Volume 7 Edited by Ö.H. Andersson and A. Yli-Urpo
(*Proceedings of the 7th International Symposium on Ceramics in Medicine, Turku, Finland, July 1994*)

Short-term Reaction Kinetics of Bioactive Glass in Simulated Body Fluid and in Subcutaneous Tissue

Ö.H. Andersson,[1] K. Vähätalo,[2] A. Yli-Urpo,[2] R.-P. Happonen[2] and K.H. Karlsson[1]

[1] Dept. Chem. Eng. Åbo Akademi University, FIN-20500 ÅBO/TURKU, Finland
[2] Inst. of Dentistry, University of Turku, FIN-20520 TURKU, Finland

ABSTRACT

Reaction kinetics of a bioactive glass (S53P4) were compared for up to 72 hours in simulated body fluid (SBF) and in subcutaneous tissue. The samples were studied by scanning electron microscopy and energy dispersive X-ray analysis. Glass tested subcutaneously or in SBF at SA/V of 0.4 cm^{-1} induced calcium phosphate formation more rapidly than glass tested in SBF at a SA/V of 0.1 cm^{-1}. However, a pure apatite surface layer developed more rapidly at the low than at the high SA/V ratio or subcutaneously. The calcium phosphate accumulation started later but continued more rapidly at the low than at the high SA/V ratio. Attachment of fibers, most likely collagen, and their mineralisation at the glass surface indicates ability of the glass to bond to soft tissue.

INTRODUCTION

Bioactive glasses and glass-ceramics can be tested in tris buffer solution or simulated body fluid (SBF) for investigation of various aspects such as bioactivity, durability, quality assurance etc. The SBF contains inorganic ions in concentrations close to those in blood plasma and is buffered with tris. Testing in pure tris solution makes it possible to study dissolution and precipitation reactions in detail by solution analysis (1). This is important in fundamental studies on reaction mechanisms and kinetics of bioactive glasses. A limitation of the use of tris is that only materials that release a sufficient amount of calcium and phosphate for apatite formation can be tested.

There is often a need for a rapid and simple *in vitro* test that can be used to predict *in vivo* behaviour with respect to bioactivity. It is generally agreed that for a glass to be bioactive it must have the ability to induce calcium phosphate formation at its surface after implantation. Even if this is not a sufficient criterion, as demonstrated for Al-containing glasses (2), formation of apatite at the surface of a material *in vitro* is a strong indication for bioactivity. There seems to be little controversy regarding the use of SBF as the first qualitative test for bioactivity. However, the knowledge about the correlation between reactions *in vivo* and *in vitro* appears insufficient. Kokubo et al. reported formation of apatite on the A/W glass-ceramic in 7 days in SBF (3) compared with 5 to 10 days in rabbit tibia (4).

One parameter that affects the reaction kinetics of bioactive glass in SBF is the surface area to solution volume ratio (SA/V). It is not possible to determine the SA/V ratio in the implantation site. However, by comparing samples tested *in vivo* with samples tested under known SA/V ratios in SBF it is in principle possible to indirectly estimate the effective SA/V ratio in the implantation site. The present work compares short-term reaction kinetics in subcutaneous tissue with kinetics in SBF at two different SA/V ratios.

MATERIALS AND METHODS

Plates, 10 mm x 2 mm x 1 mm, were cut from a block of bioactive glass, S53P4, of the composition SiO_2 53.0; CaO 20.0; Na_2O 23.0; P_2O_5, 4.0 weight-%. The plates were ground in ethanol on 600 grit SiC paper and washed ultrasonically in ethanol. *In vitro* testing was done in SBF at SA/V of 0.1 cm^{-1} and 0.4 cm^{-1}. Testing of glass plates in subcutaneous tissue was done as previously described (5). Two samples were implanted in each rat and three rats were used at each test period. From each rat one sample was fixed in formalin for 24 hours and the other in ethanol for 30 seconds. Test periods *in vitro* and *in vivo* were 8, 24 and 72 hours. The surface reaction layer was partly removed by scratching with a scalpel, thus producing a fracture cross-section of the reaction layer. After coating with carbon the samples were studied by scanning electron microscopy (SEM) and energy dispersive X-ray analysis (EDXA).

RESULTS

Regardless of testing time, all surfaces of the subcutaneous samples fixed in formalin were covered with a layer of newly precipitated apatite. This was not the case for the ethanol rinsed specimens which, thus, were analysed by EDXA.

At 8 hours the surface tested at SA/V of 0.1 cm^{-1} consisted essentially of Si-gel. On the contrary, the calcium phosphate accumulation had started on the surface of the specimens tested at SA/V of 0.4 cm^{-1} or subcutaneously (Fig. 1). In SEM the surfaces of all specimens showed a smooth appearance with no apatite crystals.

At 24 hours, a layer of dendritic calcium phosphate crystals forming sphere-like aggregates, about 1.5 μm in size, had precipitated on the surface of the specimens tested at SA/V of 0.1 cm^{-1} (Fig. 2 a,b). EDXA indicated extensive calcium phosphate accumulation (Fig. 1). For specimens tested at SA/V of 0.4 cm^{-1} the surface consisted of a smooth layer of small spheres, 100-150 nm in diameter (Fig. 2 c,d). The surface layer was thin and the spheres did not form aggregates. The calcium phosphate content was lower than for specimens tested at SA/V of 0.1 cm^{-1} (Fig. 1). The pH of the SBF was 8.0 and 8.3 for SA/V of 0.1 cm^{-1} and SA/V of 0.4 cm^{-1}, respectively. For the subcutaneously tested specimens a thin smooth surface layer on top of the subsurface Si-rich layer was observed (Fig. 2 e,f). No crystals were seen on the surface even if EDXA showed presence of calcium phosphate. The subsurface layer was in all samples nearly pure silica according to EDXA. This layer had a smooth appearance for the subcutaneously tested specimens, whereas it was coarse for the samples tested in SBF (Fig. 2 a, c, e).

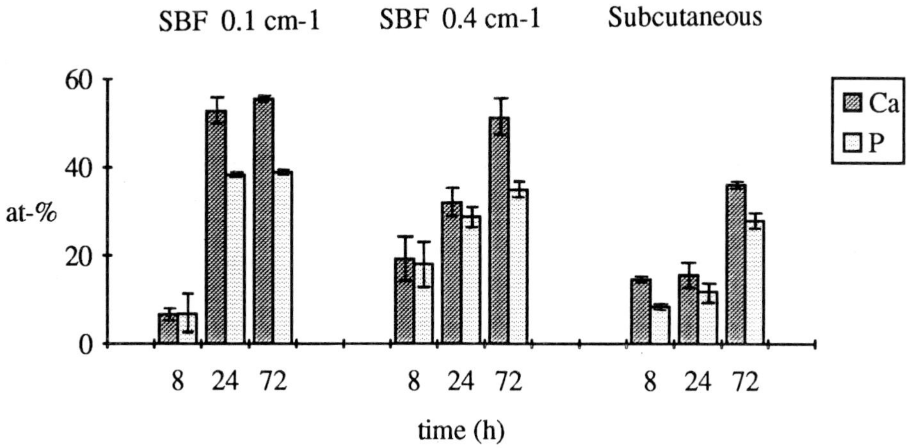

Figure 1. Calcium and phosphorus content at the surface of glass S53P4 as a function of time in SBF and subcutaneously. Average values of EDXA of three samples.

Samples tested at both SA/V-ratios were covered by apatite crystals at 72 hours as observed by SEM and EDXA. The calcium phosphate content at the surface of samples tested subcutaneously was lower than that of samples tested in SBF (Fig. 1).

Organic material was found on the surface of subcutaneously tested specimens. The formalin fixed samples showed more organic material than the ethanol fixed ones. Fibers of several hundred micrometer length were seen on the glass surfaces after 24 hours. The fibers appeared highly mineralised (Fig. 3).

DISCUSSION

The expected gradual increase in calcium phosphate content was observed at the surface of ethanol fixed specimens. However, precipitated calcium phosphate crystals were seen on the surface of formalin fixed specimens at all time points. This shows that the reactions continue during fixing in formalin as reported recently (6). Thus, figure 3 shows a situation after 24 hours subcutaneously followed by 24 hours under continued reactions *in vitro*. However, corresponding mineralised fibers were seen also on ethanol fixed specimens after 24 hours.

El-Ghannam et al. (7) found similar fibers on glass 45S5 after 7 days in an osteoblast culture. Hench and Wilson reported mineralisation of collagen attached to the surface of glass 45S5 *in vitro* (8). The presence of the mineralised fibers, assumed to be collagen, on glass S53P4 indicates that soft tissue bonds to the glass. Glass S53P4 contains 53 weight-% SiO_2, which is just above the highest content for glasses showing soft tissue bonding in the work by Wilson and Noletti (9).

Only ethanol fixed specimens were EDX analysed. In SBF, at SA/V of 0.1 cm^{-1} the Si-gel surface contained little calcium and phosphate at 8 hours. At SA/V of 0.4 cm^{-1} and in subcutaneous tissue the calcium phosphate formation had clearly started already at this time point. Thus, the calcium phosphate formation starts earlier at higher SA/V ratios.

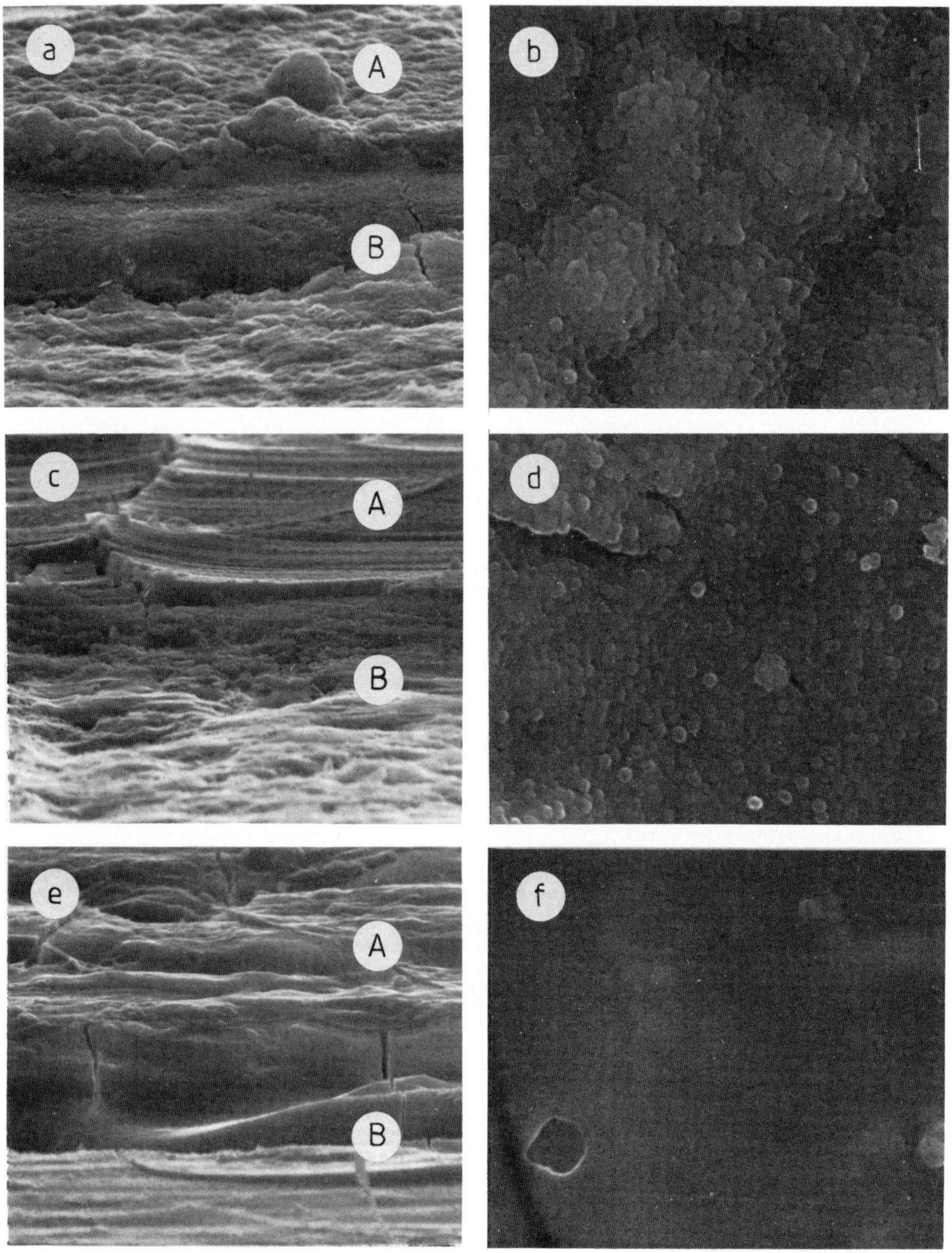

Figure 2. SEM images of specimens tested 24 h in SBF at SA/V of 0.1 cm^{-1} (a,b), SBF at SA/V of 0.4 cm^{-1} (c,d), and subcutaneously (e,f). Specimens scratched with scalpel and viewed along the surface (a,c,e). Surface layer (A) detached by fracture within the subsurface Si-rich layer (B). Field width 25 μm. Corresponding images of surface layer (A) taken perpendicularly to the surface (b, d, f). Field width 5 μm.

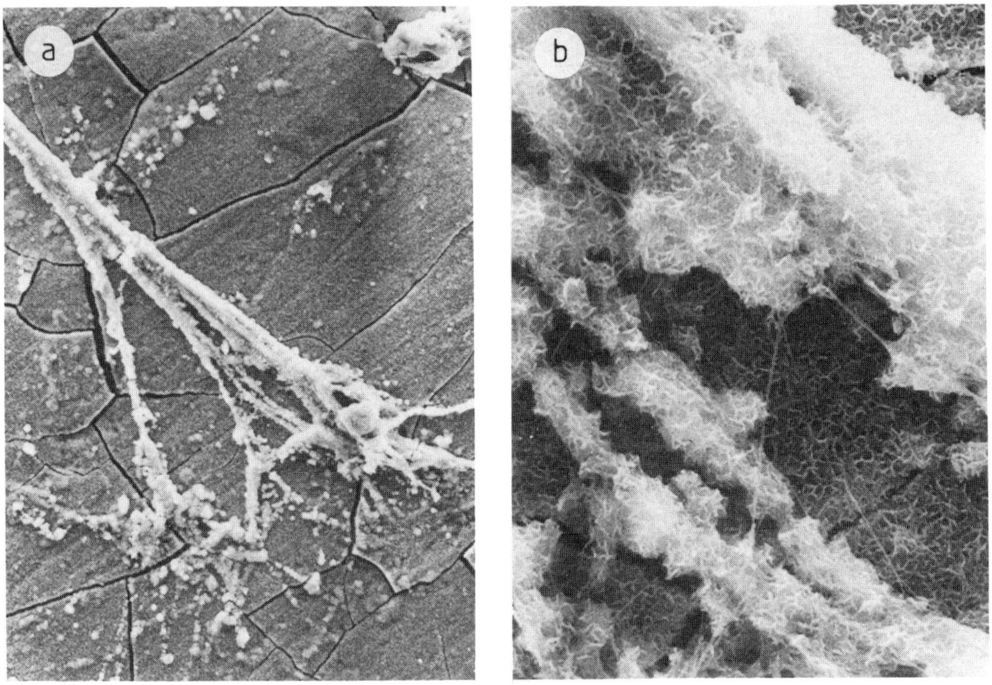

Figure 3. Fibers attached to the surface of glass S53P4 after 24 hours in subcutaneous tissue. Calcium phosphate crystals cover the glass surface as well as the fibers. a) Field width 240 μm. b) Field width 30 μm. Formalin fixed sample.

At 24 hours, a calcium phosphate surface layer completely covered the surface of the glass tested in SBF at SA/V of 0.1 cm^{-1}. Silicon was still detected at the surface of the specimens tested at SA/V of 0.4 cm^{-1}. The subcutaneously tested specimens showed a considerable amount of silicon in the EDX analysis. By SEM, dendritic crystals forming sphere-like aggregates, about 1.5 μm in size, could be seen on the surface of specimens tested in SBF at SA/V of 0.1 cm^{-1}. At the surface of samples tested at SA/V of 0.4 cm^{-1} spheres 100 - 150 nm in diameter were seen. These may be precursors for the larger calcium phosphate aggregates appearing later (c.f. Fig. 2 a,b). However, the spheres may also be Si-gel which contains calcium phosphate. Subcutaneously tested samples showed smooth surface morphology.

After the rapid calcium phosphate formation on specimens tested subcutaneously or in SBF at SA/V of 0.4 cm^{-1} the apatite growth is slower than on the SA/V 0.1 cm^{-1} specimens. This may be due to a more rapid increase in pH. At 24 hours the pH was 8.0 for SA/V of 0.1 cm^{-1} and 8.3 for SA/V of 0.4 cm^{-1}. A higher pH may be expected to decrease the rate of leaching of calcium and phosphate. However, further studies are required.

On specimens tested in SBF the first precipitated calcium phosphate has a low Ca/P ratio. At the time of appearance of spherulitic calcium phosphate aggregates the Ca/P-ratio is higher and comparable to that on specimens tested *in vivo*. It seems that *in vivo* there may be some mechanism which regulates the precipitation and/or the concentration of calcium and phosphate at the glass surface preventing the formation of the calcium phosphate with the low Ca/P ratio.

It is not possible to determine the effective SA/V ratio *in vivo* due to factors such as fluid content of surrounding tissue, blood circulation, clot formation, cell activity etc. The present results suggest that the short-term conditions in the subcutaneous implantation site correspond to a situation with a high SA/V-ratio. However, due to the gradual retardation in ion release and the dynamic nature of the environment the situation may change with time to a situation corresponding to a low SA/V-ratio.

CONCLUSIONS

Bioactive glass tested subcutaneously or in SBF at SA/V of 0.4 cm^{-1} induce calcium phosphate formation more rapidly than glass tested in SBF at a SA/V of 0.1 cm^{-1}. However, a pure apatite surface layer develops more rapidly at the low than at the high SA/V ratio or subcutaneously. Thus, the calcium phosphate accumulation starts later but continues more rapidly at the low SA/V ratio. The results suggest that the subcutaneous implantation site provides an environment which initially corresponds to a high SA/V ratio.

Organic fibers assumed to be collagen rapidly attach to the surface of glass S53P4 in subcutaneous tissue. The fibers mineralise and become embedded in the calcium phosphate surface layer. This indicates that not only bone but also soft tissue bonds to the glass.

REFERENCES

1. Andersson, Ö.H., Rosenqvist, J., Karlsson, K.H. J. Biomed. Mater. Res (1993) 27 941-948
2. Andersson, Ö.H., Liu, G., Kangasniemi, K., Juhanoja, J. J. Mater. Sci.: Materials in Medicine (1992) 3 145-150
3. Kokubo, T., Kushitani, H., Sakka, S., Kitsugi, T., Yamamuro, T., J. Biomed. Mater. Res. (1990) 24 721-734
4. Kitsugi, T., Nakamura, T., Yamamuro, T., Kokubo, T., Shibuya, T., Takagi, M. J. Biomed. Mater. Res. (1987) 21 1255-1271
5. Happonen, R.-P., Vähätalo, K., Yli-Urpo, A., Andersson, Ö.H. In: Bioceramics Volume 5, Eds. T. Yamamuro, T. Kokubo, T. Nakamura, Kobunshi Kankokai, Inc., Kyoto, Japan, (1992) 185-190
6. Kieswetter, K., Bauer, T.W., Brown, S.A., Van Lente, F., Merrit, K. J. Biomed. Mater. Res. (1994) 28 281-287
7. El-Ghannam, A., Ducheyne, P., Shapiro, I., In: Bioceramics Volume 6 , Eds. P. Ducheyne and D. Christiansen, Butterworth-Heinemann Ltd., UK (1993) 143-149
8. Hench, L.L, Wilson, J., In: Silicon Biochemistry, Wiley, Chester (1986) 231-246
9. Wilson, J., Noletti, D., In: CRC Handbook of Bioactive Ceramics, Vol. 1, Eds T. Yamamuro, L.L. Hench, June Wilson, CRC Press Inc., Boca Raton, FL (1988) 283-302

Bioceramics, Volume 7 Edited by Ö.H. Andersson and A. Yli-Urpo
(*Proceedings of the 7th International Symposium on Ceramics in Medicine, Turku, Finland, July 1994*)

Effects of Al_2O_3 and TiO_2 on Bioactivity of CaO-SiO_2 Glasses

C. Ohtsuki, A. Osaka and T. Kokubo*
Department of Bioengineering Science, Faculty of Engineering, Okayama University, Tsushima, Okayama 700, Japan
* Division of Material Chemistry, Faculty of Engineering, Kyoto University, Sakyo-Ku, Kyoto 606-01, Japan

ABSTRACT

In order to investigate the effects of Al_2O_3 and TiO_2 on the bioactivity of glasses, compositional dependence of apatite formation on the glasses in the systems CaO-SiO_2-Al_2O_3 and CaO-SiO_2-TiO_2 were studied *in vitro*. The addition of Al_2O_3 and TiO_2 to CaO,SiO_2-based glasses suppressed the dissolution of Ca(II) and Si(IV) and hence reduced the increasing rate of supersaturation of the simulated body fluid with respect to apatite. Those additives was also suggested to decrease apatite nucleation ability of the hydrated silica.

INTRODUCTION

Since the discovery of Bioglass® in the early 1970s (1), various types of glasses and glass-ceramics have been found to bond to living bone, and some of them are already clinically used. It is already confirmed that the essential condition for glasses and glass-ceramics to bond to living bone is formation of a biologically active apatite layer on their surfaces in the body (2). The ability of the formation of the apatite layer is significantly dependent on composition of the glasses. Addition of Al_2O_3 and/or TiO_2 to the bioactive glasses and glass-ceramics remarkably suppresses their ability of apatite formation (3,4). It has not, however, been revealed fundamentally how Al_2O_3 and TiO_2 effectively suppress the ability of apatite formation of bioactive glasses and glass-ceramics. In the present study, in order to understand this issue, the compositional dependence of the apatite formation on glasses in the systems CaO-SiO_2-Al_2O_3 and CaO-SiO_2-TiO_2 was studied *in vitro*. Binary CaO-SiO_2 system gives the simplest composition of glasses showing bioactivity (5), and hence are suitable as the basic composition in this type of fundamental study.

EXPERIMENTAL

Glasses of various compositions in the systems CaO-SiO_2-Al_2O_3 and CaO-SiO_2-TiO_2 were prepared by the conventional melt-quenching technique. Rectangular specimens 15x10x1 mm^3 of the obtained glasses were polished with 3-4 $\mu m\phi$ diamond paste, and washed with acetone and distilled water in an ultrasonic cleaner. They were immersed in 35 ml of an acellular simulated body fluid with ion concentrations nearly equal to those of the human blood plasma (Na^+ 142.0, K^+ 5.0, Mg^{2+} 1.5, Ca^{2+} 2.5, Cl^- 147.8, HCO_3^- 4.2, HPO_4^{2-} 1.0 and SO_4^{2-} 0.5 mM). It was buffered at pH 7.25 with 50 mM trishydroxymethyl-aminomethane ($(CH_2OH)_3CNH_2$) and 45 mM hydrochloric acid (HCl), and its temperature was kept at 36.5 ^oC. It had been confirmed that this fluid can reproduce fairly precisely apatite formation on the surfaces of various types of bioactive glasses and glass-ceramics *in vivo* (4). After soaked for predetermined intervals, the specimens were removed from the fluid and gently washed with acetone. Apatite formation on the surface of the specimens was examined with thin-film X-ray diffraction and Fourier transform infrared (FT-IR) reflection spectroscopy. These two techniques enable detection of a layer only about 1 μm thick at the surface. Changes in concentrations of calcium, phosphorus, silicon, aluminum and titanium in the simulated body fluid with the immersion of the glasses were measured by inductively coupled plasma emission (ICP) spectroscopy. Proton concentration was measured by a pH meter.

RESULTS AND DISCUSSION

Figure 1 shows the compositional dependence of the apatite formation on the surface of glasses in the systems CaO-SiO_2-Al_2O_3 and CaO-SiO_2-TiO_2 soaked in the simulated body fluid for 30 days. In CaO-SiO_2-Al_2O_3 system, the glasses containing Al_2O_3 less than 1.5 mol% formed the apatite layer on their surfaces in the simulated body fluid by 30 days as well as Al_2O_3-free CaO-SiO_2 glasses, but those containing Al_2O_3 more than 1.7 mol% did not form even after 30 days. $CaO{\cdot}SiO_2$ glass containing as little as 0.3 mol% of TiO_2 did not form the apatite layer. These results indicate that only small amounts of Al_2O_3 and TiO_2 remarkably suppress the apatite formation of glasses.

Figure 2 shows FT-IR reflection spectra of the surfaces of some of the glasses before and after being soaked in the simulated body fluid for various periods. Their compositions are given in Table 1. Increased reflection peaks of Si-O stretching vibration in Fig. 2 for the glasses soaked in the fluid is attributed to formation of silica hydrogel layer on their surfaces. It can been seen from Fig. 2 that CS glass forms a silica hydrogel and amorphous calcium phosphate on its surface, prior to the formation of crystalline apatite. The rate of the surface structural changes of CaO·SiO_2 glass is reduced by addition of Al_2O_3 and TiO_2. Addition of more than 10% Al_2O_3 or TiO_2 makes CaO·SiO_2 glass inert against the simulated body fluid.

Figure 3 shows changes in element concentrations and pH in the fluid due to the immersion of CS and CSA2 glasses, as an example of the results. CS glass showed appreciable increases in the calcium and silicon concentrations, accompanied with an distinct decrease in the phosphorus concentration. The rate of such changes in element concentrations decreased with increasing amount of additional Al_2O_3 or TiO_2 to CaO·SiO_2 composition. The CaO·SiO_2 glasses containing Al_2O_3 or TiO_2 more than 10 mol% did not show any changes in element concentration of the fluid. Aluminum and titanium were not detected in all

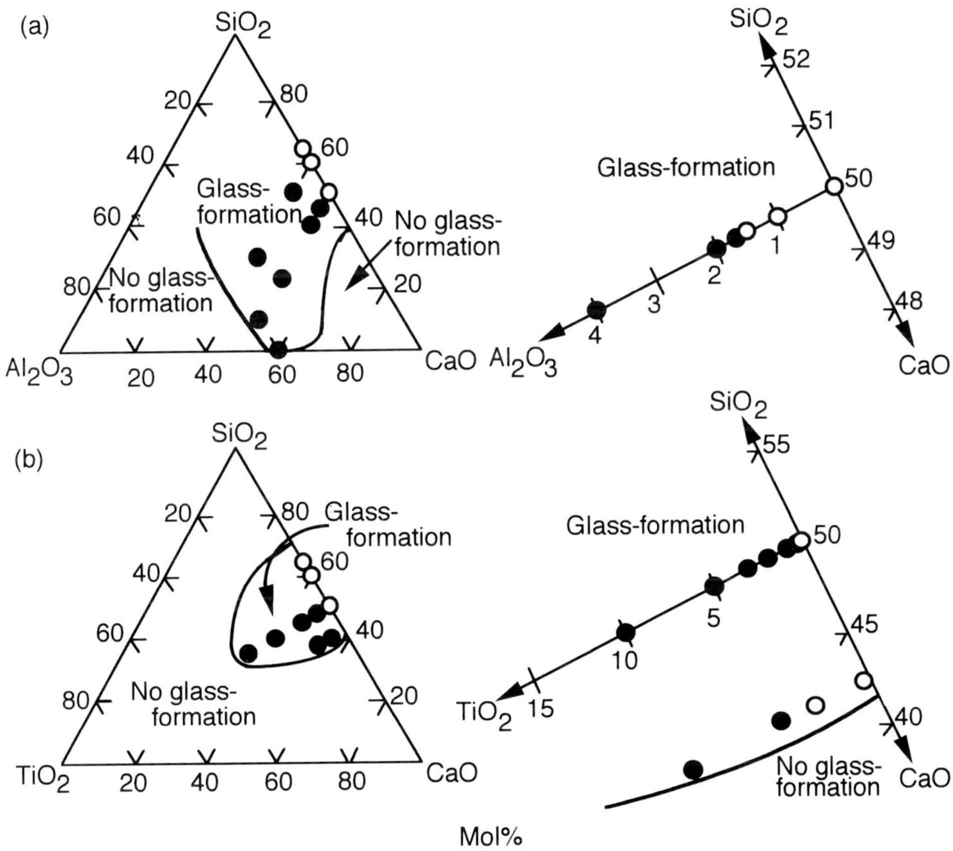

Figure 1 Compositional dependence of apatite formation on the surface of the glasses in the systems (a)CaO-SiO_2-Al_2O_3 and (b)CaO-SiO_2-TiO_2 after being soaked in the simulated body fluid for 30 days.
○: apatite formation, ●: no apatite formation.

the measured fluid. Ionic activity products of apatite in the fluids were calculated from the measured element concentrations and pH, and were plotted in Fig. 4. Solubility product of apatite in aqueous solution is 5.5×10^{-118} at 37 ^{o}C. The simulated body fluid is, therefore, already supersaturated with respect to the apatite even before the immersion of the glasses.

Our previous report has shown that the apatite formation on CaO,SiO_2-based glasses is caused by the increase in degree of supersaturation of surrounding body fluid with respect to apatite, and by a specific catalytic effect of hydrated silica

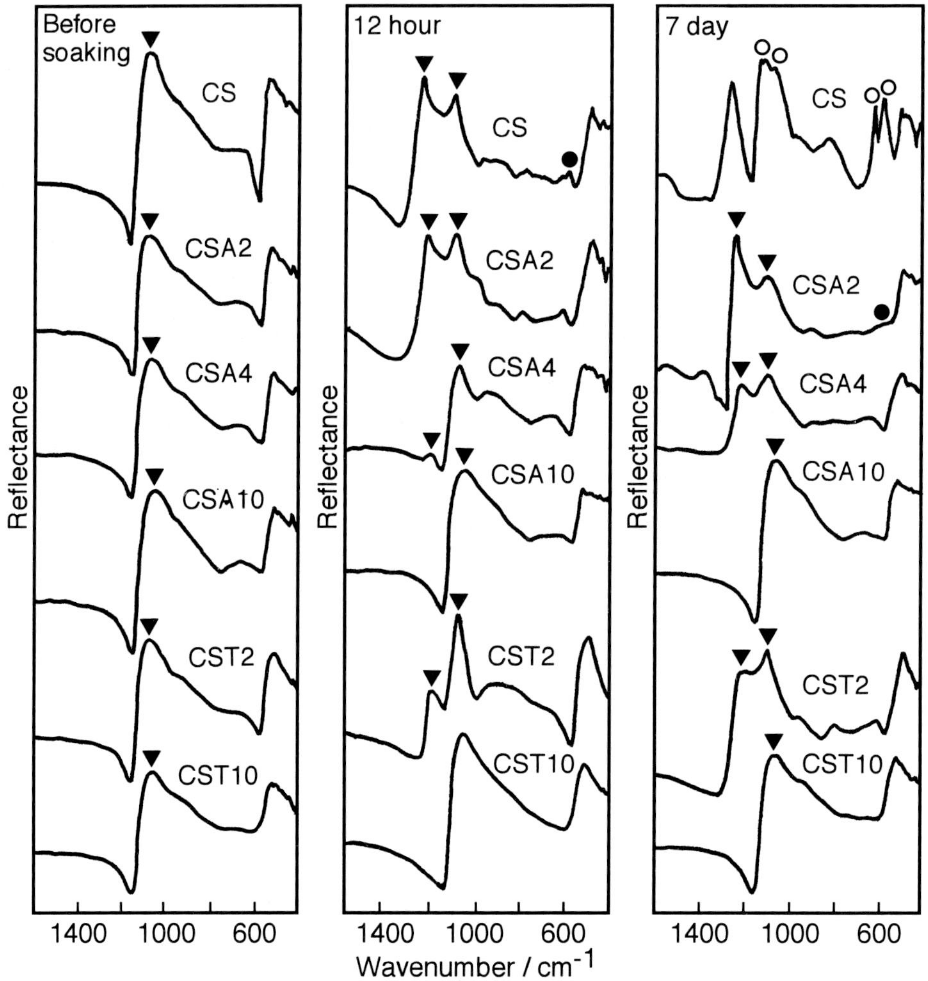

Figure 2 FT-IR reflection spectra of the surfaces of some of the glasses before and after being soaked in the simulated body fluid for various periods.
▼: Si-O stretching vibration, ●: amorphous calcium phosphate,
○: crystalline apatite.

Table 1 Compositions of some of the examined glasses.

Name	Composition (mol%)			
	CaO	SiO_2	Al_2O_3	TiO_2
CS	50.0	50.0	0.0	0.0
CSA2	49.0	49.0	2.0	0.0
CSA4	48.0	48.0	4.0	0.0
CSA10	45.0	45.0	10.0	0.0
CST2	49.0	49.0	0.0	2.0
CST10	45.0	45.0	0.0	10.0

formed on the surfaces of the glasses for apatite nucleation (5,6). Figure 4 indicates that increasing rate of the degree of supersaturation of the simulated body fluid, i.e., the difference between the ionic activity product and the solubility product of the apatite, decreased with increasing the amounts of added Al_2O_3 and TiO_2. Furthermore, addition of these components retards the formation of silica hydrogel on the surface of the glasses, as shown on Fig. 2. Al_2O_3 and TiO_2 are generally known as components which can improve the chemical durability of glasses. The reduction of apatite formation is, therefore, interpreted in terms of the suppressing effect of Al_2O_3 and TiO_2 on the dissolution of Ca(II) and Si(IV) ions from the glass. There is another possible factor to reduce the apatite formation. In the case of the glasses containing small amounts of Al_2O_3 or TiO_2, the degree of supersaturation of the surrounding fluid gradually continues to increase with increasing soaking time, and exceeds that due to the immersion of $CaO{\cdot}SiO_2$ glass after 7 days immersion. Nevertheless, the former glasses did not form the

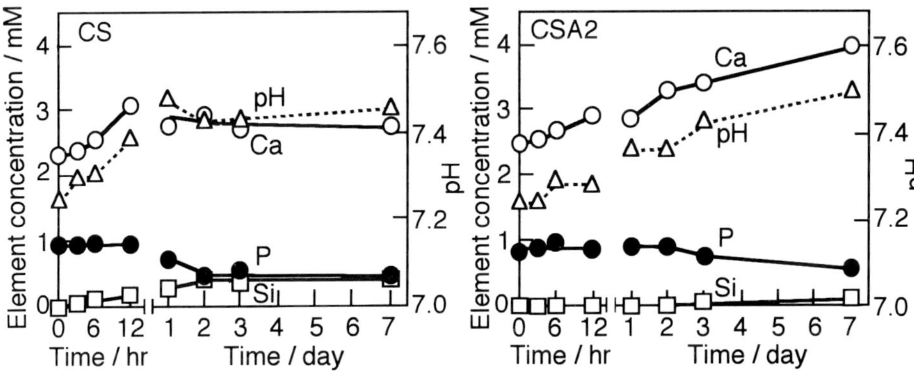

Figure 3 Changes in element concentrations and pH of the simulated body fluid due to immersion of CS and CSA2 glasses.

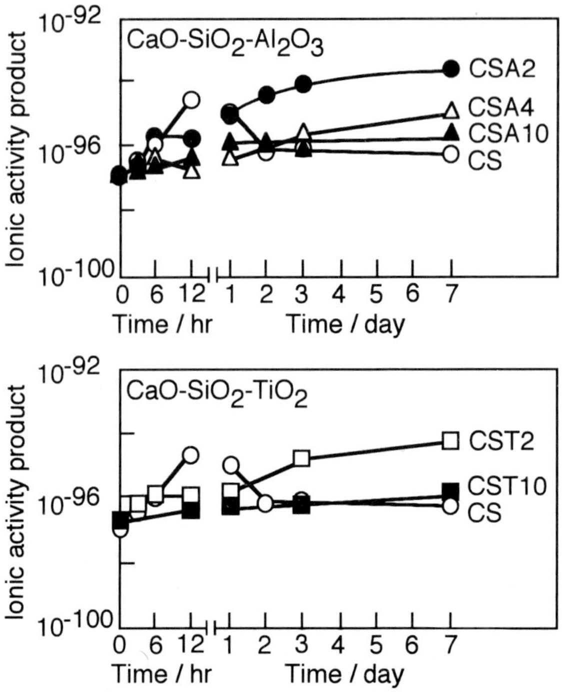

Figure 4 Changes in ionic activity product of the simulated body fluid due to immersion of the glasses.

apatite layer on their surfaces, whereas the latter one did. This suggests that apatite nucleation ability of the hydrated silica is decreased by addition of Al_2O_3 or TiO_2 .

REFERENCES

1. Hench, L.L., Splinter, R.J., Allen, W.C. and Greenlee, T.K. J. Biomed. Mater. Res. Symp. (1972) 2, 117-141.
2. Kokubo, T. J. Non-Cryst. Solids (1990) 120, 138-151.
3. Andersson, Ö.H., Liu, G., Niemi, K.H., Niemi, L., Miettinen, J. and Juhanoja, J. J. Mater. Sci.: Mater. Med. (1990) 1, 219-227.
4. Kokubo, T., Ito, S., Huang, Z.T., Hayashi, T., Sakka, S., Kitsugi, T. and Yamamuro, T. J. Biomed. Mater. Res. (1990) 24, 331-343.
5. Ohtsuki, C., Kokubo T. and Yamamuro, T. J. Non-Cryst. Solids (1992) 143, 84-92.
6. Li, P., Ohtsuki, C., Kokubo, T., Nakanishi, K., Soga, N., Nakamura, T. and Yamamuro, T. J. Am. Ceram. Soc. (1992) 75, 2094-2097.

Bioceramics, Volume 7 Edited by Ö.H. Andersson and A. Yli-Urpo
(*Proceedings of the 7th International Symposium on Ceramics in Medicine, Turku, Finland, July 1994*)

FT-Raman Spectroscopic Analysis of Natural Bones and their Comparison with Bioactive Glasses and Hydroxyapatite

I. Rehman,* R. Smith,* L.L. Hench† and W. Bonfield*
* IRC in Biomedical Materials, QMW, University of London, London E1 4NS, UK
† Advanced Material Research Center, University of Florida, USA

ABSTRACT:
Fourier transform Raman (FT-Raman) spectroscopy is used to compare bone samples with the HCA layer developed on a bioactive glass (45S5 Bioglass®). FT-Raman measurements were made on whole bone samples using a spectrometer equipped with a Nd-YVO4 laser which eliminates fluorescence from the organic phase. The spectroscopic features are similar for all species of bone tested. The vibrational frequencies of CO_3^{2-} and PO_4^{3-} groups in bone mineral, obtained from both whole bone and deproteinated bone, are equivalent to the frequencies measured for the HCA layer that develops on a bioactive glass, when reacted in a simulated physiological solution (K-9).

INTRODUCTION:
Fourier Transform Raman (FTR) spectroscopy has been used in the past to study both wet and dry bone and Bioglass® samples and it provides spectral details of the hydroxy carbonate apatite (HCA) mineral phase and the organic phase[1]. The C-O and P-O vibrational modes appear on the bioactive glass surface as early as 2 hours following immersion in K-9 solution. Initially, the C-O and P-O vibrations are characteristic of an amorphous calcium phosphate phase. The intensity of the C-O and P-O peaks increases with immersion time as crystallisation of hydroxyl carbonate apatite phase progresses, resulting in a spectrum that is equivalent to the FT-Raman spectrum of bone mineral. These surface spectra of bioactive glasses are very different to those obtained from commercial calcium phosphate ceramics, which do not match the FT-Raman spectra of natural bones.

FT-Raman spectroscopy is a powerful analytical technique for biomedical applications which provides molecular level information on functional groups of both inorganic and organic phases[1-4]. Sample preparation is minimal, the measurement is non-destructive and small amounts of material (micrograms to nanograms) can be analysed. Peak assignments in the FT-Raman work (Table 1) are based upon a study of non-biological, synthetic carbonated apatites using chemical analysis, Infrared and Raman spectroscopy[5]. A conventional Raman study of human femur cortical bone showed phosphate peaks that were similar to those from HA samples and agreed closely with the results reported by Walters et. al.[2] and Nelson[5].

MATERIALS SELECTION:

Cortical bone obtained from the tibial mid-shaft region of (where such bone is found in suitable quantities) of normal human tissue (mean age 59 years, male) has been studied as the specific material relevent to weight bearing orthopaedic applications. The spectra obtained were compared with that of Bioglass® and HA (implant materials), baboon and bovine bone of two years age samples . All the bone samples were obtained in a frozen state, at 0°C.

EXPERIMENTAL PROCEDURE:

Details of selection, preparation, hydrazine deproteination and FT-Raman analysis of human cortical bone tissue have been described elsewhere[1]. All bone samples were obtained from the mid-shaft region of the femur and maintained frozen until analysis following 30 min of thawing. Only the human bone sample was subject to deproteination, while the baboon and bovine bone samples were studied fresh. Details of 45S5 glass preparation are described elsewhere[6,7]. The simulated body fluid (SBF) solution used was equivalent to that of Kokubo et.al.[8] and is designated SBF K-9 by Filgueiras et.al.[7] in which SBF solution effects on the kinetics of 45S5 Bioglass® were described.

RESULTS:

The FT-Raman spectra of human bone samples are given in Figure 1, incorporation with those from bovine and baboon. The most striking feature of the spectrum of whole bone tissue samples obtained from FT-Raman is the absence of fluorescence (Fig. 1), and well-defined peaks from the organic phase, at 2935-2882 cm^{-1} (Table 1). Peaks at 1062, 1036, 952 & 584cm^{-1} are present in all the spectra of bone samples (Table 1). The FT-Raman spectrum of 45S5 Bioglass® samples prior to exposure to a SBF solution and after 2, 6,and 24 hours are shown in Figure 2A, 2B,2C and 2D respectively. The glass spectra are dominated by phosphate vibrational modes (Table 1), which differ from the peaks of HA. The carbonate modes in the glass appear at 1062 $cm^{-1}$and 1036 cm^{-1} overlap, yielding a fairly broad, but intense, peak in the 100C-1100 cm^{-1} region. The phosphate(stretch) mode at 952 cm^{-1} is sharp and intense. The phosphate(bend) mode at 580 cm^{-1} is intense but somewhat broadened. The 422 cm^{-1} phosphate(bend) is not very well resolved. The large intensities for the Raman shifts of the P-O vibrations are important considering that the glass contains only 6 weight % P_2O_5. (The other components are: 45 % SiO_2, 24.5% CaO and 24.5% Na_2O). The intensity of Si-O vibrational modes is very weak. The peak at 625-650 cm^{-1} is due to mixed stretching and bending vibrations of Si-O-Si bonds which overlap the bend mode of the phosphate vibrations. The large intensity of the phosphate modes dominate the FT-Raman spectrum of the glass, enhancing analysis of the HCA layer. The FT-Raman spectra of deproteinated human bone, synthetic hydroxyapaptite and Bioglass® after 24 hours of immersion in SBF solution are compared in Figure 3 and their peak assignments in Table 1.

Crystallisation of the α-CaP layer to form HCA occurs after 6 hr in the SBF-K9 solution[8]. This time is three times longer than required for the formation of crystalline HCA in a tris-buffer solution[6]. The difference is probably due to Mg^{+} ions in the SBF-K9 solution, which retard crystallisation of HCA[8]. A critical step in the nucleation of HCA crystallites is incorporation of carbonate anions[10,11]. The C-O vibrations at 1062 cm^{-1} are sharp in the carbonate mineral *and* the HCA reaction layer, where as the mode at 710 cm^{-1} is less intense.

DISCUSSION:

FT-Raman spectroscopy of natural bone samples (Fig. 1) reveals numerous peaks associated with an organic phase; i.e., the peaks at 1651-1656 cm^{-1} are likely to be due to the amide I band of collagen (a combination of C=O stretching, C-N stretching, and N-H bending modes)[10,12]. Removal of the bone protein results in an FT-Raman spectrum of the bone mineral phase (Fig. 3A) which is dominated by the P-O (stretch) vibration at 962 cm^{-1}. The mixed phosphate-carbonate peak at 1060-1080 cm^{-1} and phosphate(bend) peak at 577-584 cm^{-1} are also well resolved in the deproteinated bone. The HCA layer which grows on the surface of 45S5 Bioglass® in SBF-K9 has similar spectroscopic features to that of bone mineral. The 952 cm^{-1} phosphate(stretch) mode is a major peak in the spectrum of the HCA layer (Fig. 3B), but the peak is broadened and shifted slightly compared with the bone mineral spectrum. The carbonate modes present in the polycrystalline HCA layer (Fig. 3A & 3C) are more intense than those observed for the bone samples, (Figures 1, 2 and 3). The strong carbonate FT-Raman signals are not present in the synthetic HA material (Fig. 3C) and provide an important differences between HA ceramics and the bioactive glasses that form a HCA layer in SBF solutions and in vivo.

The results agree with previous studies on bioactive glasses[5,7,8]. The FT-Raman method shows that an amorphous CaP- rich phase forms on top of the silica-rich gel layer, but provides no information on the structural changes responsible for chemisorption of calcium phosphate on the reacting surface. However, the FT-Raman data show that carbonate modes are associated with formation of the crystalline HCA phase. This study shows that the HCA phase that forms on the bioactive glass is very similar to the mineral phase of all the bone samples studied. The study also indicated that all the bone sample from different speciess have similarspectra.

CONCLUSIONS:

Fluorescence has not been a problem, allowing considerable detail to be extracted from the spectra, in contrast to the situation encountered with conventional Raman spectroscopy. The quality of the structural and compositional detail which can be revealed by the technique, combined with the minimal sample preparation required, makes FT-Raman spectroscopy a very attractive method for the analysis of the organic phase of natural bones, as well as to compare reaction layers formed on a bioactive glass with the mineral phase of bone samples. The bioactive glass develops an amorphous calcium phosphate-rich surface layer which contains few carbonate ions, within two hours in simulated body fluid . Within six to 24 hours the α-CaP phase crystallises to an HCA layer with strong carbonate Raman modes. The spectrum of the HCA layer grown on the bioactive glass closely matches that of bones mineral.

ACKNOWLEDGMENTS:

The authors greatfully acknowledge the support of SERC fot the IRC programme on Biomedical Materials

REFERENCES:

1 I. Rehman, N. Chowdary and R. Smith, Signal to News (Spectral lines), (1993) 6(1), 5-7.

2. M. A. Walters, Y. C. Leung, N. C. Blumenthal, R. Z. LeGeros, K. A. Konsker, J. Inorgan. Biochem., 1990, 39 193-200.

3. B. Z. Gevorkyan, N. Y. Arnotskaya, Y. Fedorova, Biofizika, 1984, 29(6), 1140-1148.

4. R. M. Gendreau, Trends in Anal. Chem., 1986, 5(3), 68-71.

5. D. G. A. Nelson, and J. D. B. Featherstone, Calcif. Tissue Int., 1982, 34 S69-S81.

6. L. L. Hench, in Bioceramics; P. Ducheyne and J. Lemons (eds.) Annals N. Y. Acad. Sci., 1988, 54-71.

7. M. R. Filgueiras, G. LaTorre, L. L. Hench, J. Biomed. Mater. Res. 1993, 27, 445-453.

8. T. Kokubo, H. Kushitani, S. Sakka, T. Kitsugi and T. Yamamuro, J. Biomed. Mater. Res. 1990, 24, 21-734.

9. B. B. Doyle, E. G. Bendit and E. R. Blout, Biopolymers, 1975, 14, 935-957.

10. L. L. Hench and G. P. LaTorre, Bioceramics, Vol. 5, T. Yamamuro et. al., (eds.) 1992,75-86

11. R. LeGeros, N. Balmain and G. Bonel, Calcif. Tissue. Int., 1987, 41 137-144 .

Table 1: Raman peak positions (cm^{-1}) for Bioglass (45-S-5), normal bone, deproteinated bone and synthetic hydroxyapatite(HA) with assignments[1, 5].

Peak Assignment	HCA Layer on 45S5 Bioglass®	Human Bone	Bovine Bone	Baboon Bone	Deprotein -ated Human Bone	Synthetic HA
CH stretch		3000-2900	3000-2900	3000-2900	2915	
C = O		1636	1636	1636	1636	
U						1578
U						1507
C - C		1443	1442	1443		
C - 0		1262	1261	1262		
C - 0		1242	1241	1242		
CO_3^{2-} (stretch)	1064	1062	1062	1062	1062	
CO_3^{2-} (bend)	711	1036	1036	1036	1036	
PO_4^{3-} (stretch)	954	952	952	952	952	952
PO_4^{3-} (bend)	595	584	583	584	577	
PO_4^{3-} (bend)						436
PO_4^{3-} (bend)		422	421	422	422	

U = unassigned peaks

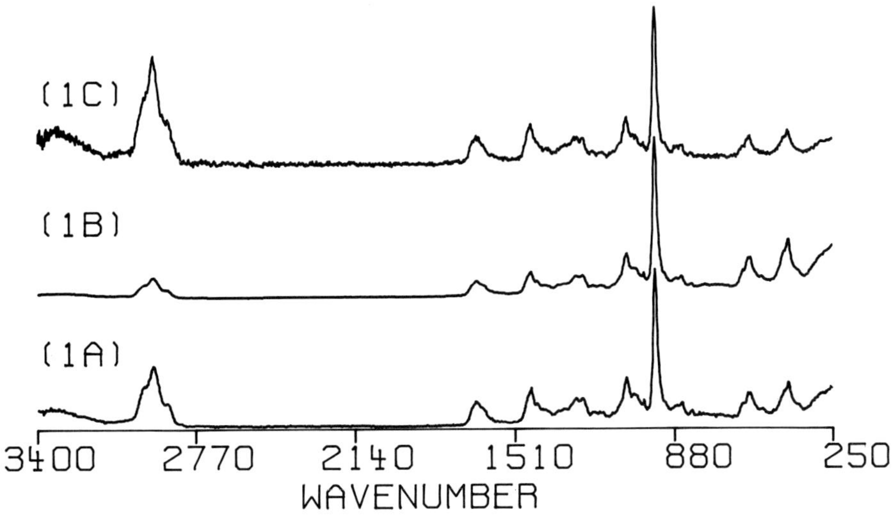

Figure 1; FT- Raman specta of 1A = human bone, 1B = baboon bone and 1C = bovine bone

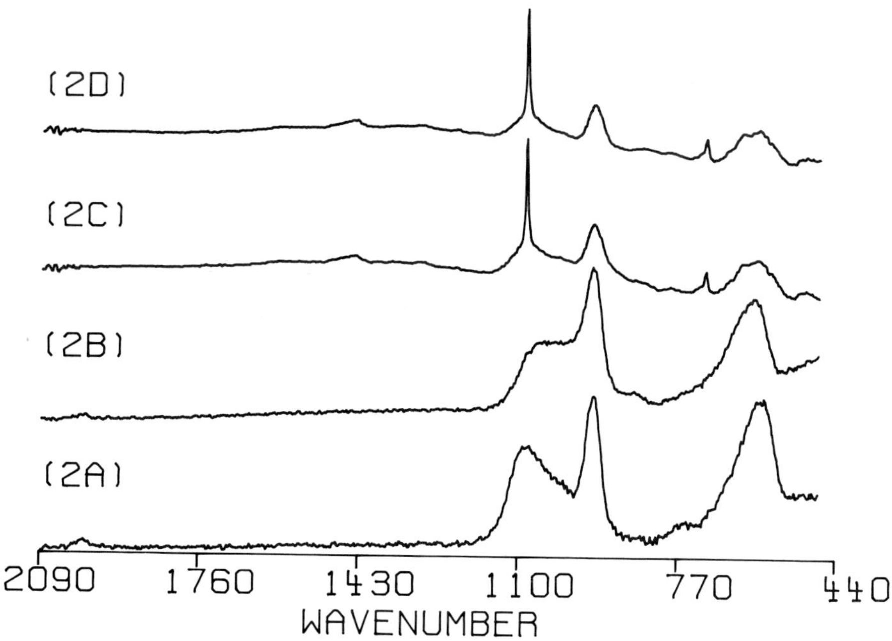

Figure 2; FT- Raman specta of Bioglass® 45S5, 2A = initial point, 2B = after 2 hours, 2C = after 6 hours and 2D = after 24 hours in SBF

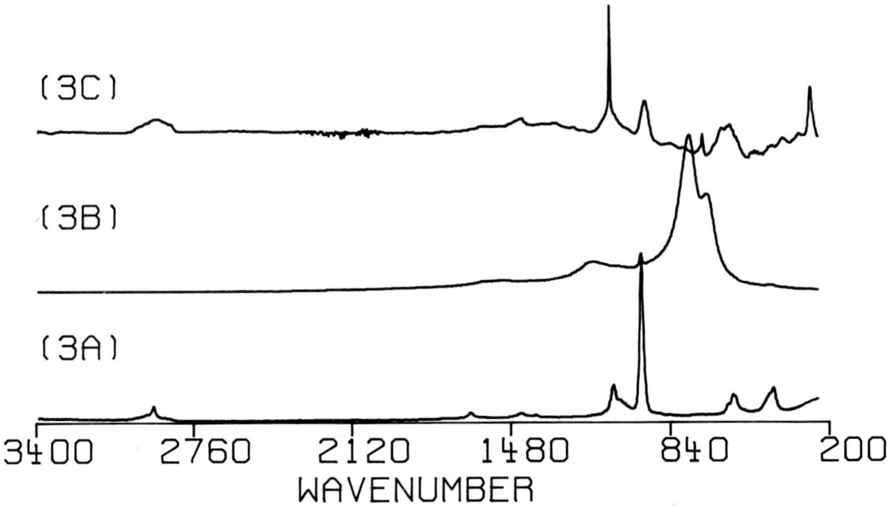

Figure 3; FT- Raman specta of 3A = deproteinated human bone, 3B = synthetic hydroxyapatite and 3C = Bioglass® 45S5 after 24 hours in SBF

Bioceramics, Volume 7 Edited by Ö.H. Andersson and A. Yli-Urpo
(*Proceedings of the 7th International Symposium on Ceramics in Medicine, Turku, Finland, July 1994*)

Spectroscopic and Crystallographic Analysis of the Solution Kinetics of a Range of Soluble Phosphate Based Bioactive Glasses

J.C. Knowles, I. Rehman and W. Bonfield
IRC Biomedical Materials, Queen Mary and Westfield College, Mile End Road, London E1 4NS, UK

ABSTRACT

Glass discs using P_2O_5 to act as a network former were produced with differing solubility rates, varying from 0.015 mg/cc/hr to 0.25 mg/cc/hr. The glass discs were placed in simulated body fluid (SBF) and at various time intervals, were removed carefully washed and dried and then measured using FT-Raman spectroscopy, X-Ray diffraction with a thin film analyser and also weight loss was monitored. At two hours, all specimens, irrespective of composition, showed the formation of an hydroxy-carbonate layer. At time periods of four hours and above, the layer had considerably reduced and at forty hours was completely absent. The dissolution of this layer was ascribed to the solubility of the glass.

INTRODUCTION

Phosphate based glasses have considerable potential for use as biomaterials (1,2), due to the unique property of being completely soluble. This solubility is controlled by the composition of the glass. The solubility may be accurately controlled, such that a mass loss per unit area per unit time may be measured. Because the composition of the glasses are based mainly on P_2O_5 , with additions of other oxides, when implanted into hard tissue, they may release beneficial ions into the surrounding hard tissue. However, previous work, examining the solubility characteristics of the glasses (2) only examined the solubility in distilled water. The solubility characteristics may be considerably altered if placed in a solution containing physiological ions, for example with a simulated body fluid (SBF) solution. Furthermore, the interaction of the ions from the dissolving glass and the ions in solution, may result in the formation of crystalline layers on the glass surface. The aim of this study was to examine some of the subtle short term surface effects, by FT-Raman spectroscopy and longer term by monitoring weight loss and with X-Ray diffraction thin film analysis. This has important consequences, because from in vivo work performed with this material, good bonding with hard tissue was seen (2), whereas, it would be expected that a soluble material, releasing ions into the surrounding tissues would promote the formation of fibrous tissue formation, yet this was not seen. It may be surmised that this result is due to a dissolution/reprecipitation mechanism that is occurring.

MATERIALS AND METHODS

Glasses were produced by conventional glass forming techniques. Glasses were melted in a platinum crucible at temperatures from 1000 to 1200°C, depending on composition. The glasses were cast into rods using a graphite coated steel mould, which was preheated to the appropriate temperature, the temperature for annealing being determined by the Tg of the particular glass being melted. The glass compositions are detailed in table 1.

Oxide (mol.%)	0.25 mg/cc/hr	0.18 mg/cc/hr	0.05 mg/cc/hr	0.015 mg/cc/hr
P_2O_5	45	45	45	45
Na_2O	31	27	29	26
CaO	24	28	24	24
Al_2O_3	0	0	2	5

Table 1. Composition of glasses used and corresponding solubility rates

Glass rods were prepared and discs were cut from these rods using a diamond circular saw. Once cut, the disc surfaces were polished to 5μm using diamond paste.

The simulated body fluid (SBF-K9) was prepared by mixing sodium chloride, sodium bicarbonate, potassium chloride, calcium chloride, dibasic potassium phosphate, and magnesium chloride in deionised water, according to Kokubo et al. (3), yielding the SBF solution. All the solutions were buffered at pH 7.25 with hydrochloric acid and tris buffer, as per Kokubo et al. The discs were placed in the SBF in a Nalgene container under 37°C static test conditions. At the appropriate time, they were removed and washed in distilled water and then allowed to dry for a minimum of 24 hours prior to analysis. Once analysed, they were returned to the same solution from which they were removed.

Fourier transform Raman spectroscopy (FT-Raman) was performed on the glass discs before and after exposure to the SBF solution using a Nicolet FT-Raman spectrometer, Nicolet 910, equipped with a Nd:YVO4 laser, which eliminates the problems of sample fluorescence and photodecomposition characteristic of conventional Raman instruments with high power lasers.
Thin film analysis was carried out on a Siemens D5000 diffractometer in detector scan mode, with a thin film analyser attached on the diffracted beam side. Samples were aligned at 0.5° to the incident beam. A step size of 0.02° 2θ was used and a scan time of 12 seconds used. Data was collected from 25 to 35° 2θ.

RESULTS

Figure 1 shows the *in vitro* change in mass with time. The mass loss approximately correlates with the predicted values for solubility.
Figure 2 shows the overlaid specta for glass of solubility 0.25 mg/cc/hr at different time periods. At time zero, (A) only the characterisitc phosphate peaks may be seen. However at two hours, a strong peak is detected at $1084cm^{-1}$. A further related peak may also be detected at $710cm^{-1}$. The C-O vibration at $1084cm^{-1}$ is sharp and is due to the HCA layer, whereas the mode at $710cm^{-1}$ is less intense.

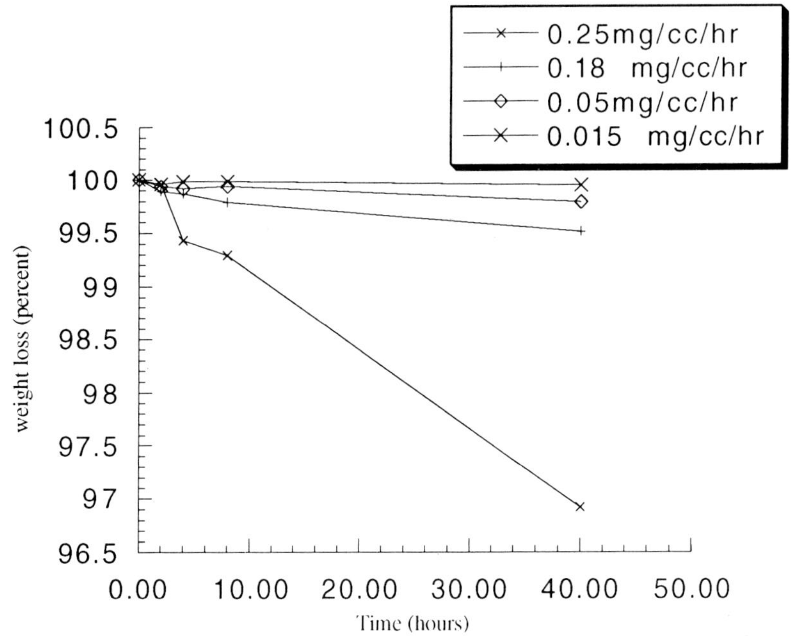

Figure 1. Percentage change in mass with time for glass specimens in SBF.

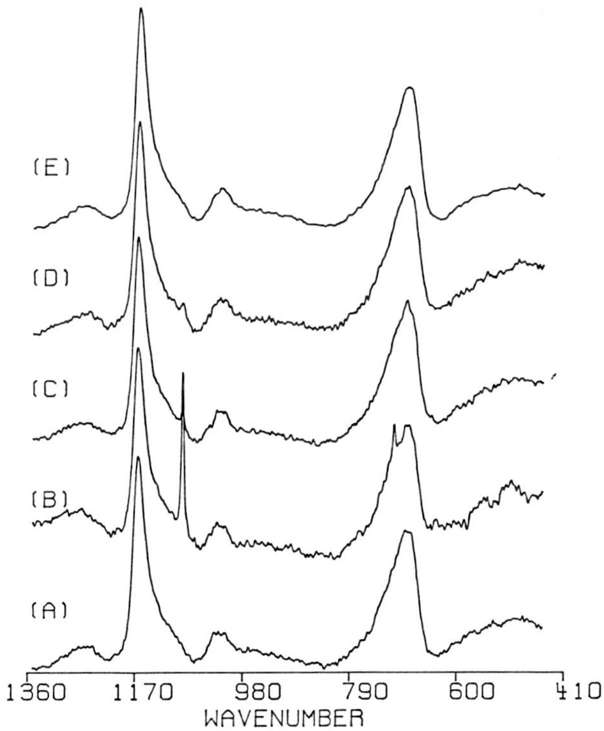

Figure 2. FT-Raman traces of glass 0.25mg/cc/hr at different time periods in SBF. (A=0 hours, B=2 hours, C=4 hours, D=8hours and E=16 hours).

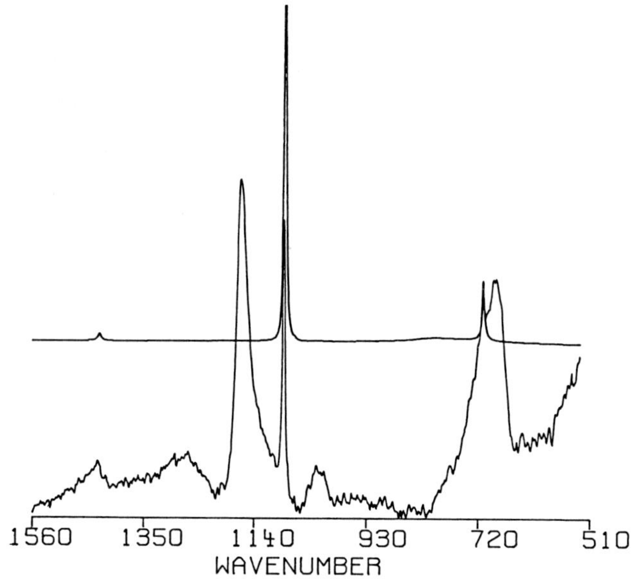

Figure 3. FT-Raman trace for glass of solubility 0.25mg/cc/hr after two hours in SBF. Overlaid (upper trace) is a standard trace for calcium carbonate, showing three characteristic carbonate peaks

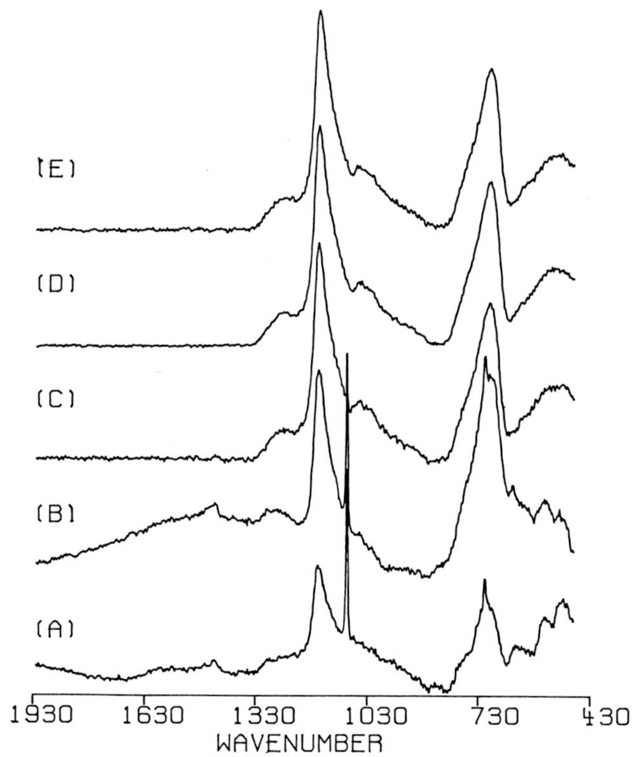

Figure 4. FT-Raman traces of glass 0.015mg/cc/hr at different time periods in SBF. (A=0 hours, B=2 hours, C=4 hours, D=8hours and E=16 hours).

Comparison with a standard trace for calcium carbonate (Figure 3) indicated that these peaks were due to carbonate bonds, indicating that at two hours an hydroxycarbonate layer is being formed on the glass surface. Analysis of the four glass types, indicates that this effect is produced for all glass types. After 2 hours, the hydroxycarbonate layer may be seen to be disappearing with increasing soak time. For the less soluble glasses (0.18, 0.05 and 0.015mg/cc/hr), it should be noted that prior to placing in solution, traces of an hydroxycarbonate layer are present on the surface of the sample, particularly for the least soluble glass. As shown in Figure 4. With time however, the hydroxycarbonate layer disappears as seen with the other glass types.

DISCUSSION AND CONCLUSIONS

From the results for mass loss, all specimens show a decrease in weight with time, indicating solubilisation of the specimens. It was perhaps expected with the low solubility specimens, to see evidence of deposition of phases from solution as seen by other workers (4), but the dominant mechanism is dissolution of the specimens. Whilst all specimens do show a decrease in mass with time, the values do not correlate exactly with calculated figures for mass loss with time, indicating that some precipitation may be occurring.

FT Raman however, reveals subtle changes occurring, at very early time periods, not previously seen (4) with the formation of a carbonate compound on the glass surface. The presence of this carbonate layer has been shown to be important (4,5) as it is the precursor layer formed onto which bone may attach. For the soluble glasses, all show the development of this layer at 2 hours. However, due to the soluble nature of the glass, this layer is short lived and subsequently disappears.

Samples were analysed by thin film X-ray diffraction and even at a very low incident angle, no surface layers were detected at 40 hours in solution. The physical appearance of the samples indicated that surface changes were occuring, ie a surface layer was being formed, but was not detected by the thin film analyser. Samples are currently in solution, to be analysed at longer time periods, to establish whether a thicker detectable layer may be formed.

In conclusion, small but significant surface changes occur with these phosphate based glasses at very early time periods. However, due to the solubility of the glasses, the surface layers formed underwent re-dissolution and were thus very short lived. Longer time period studies are underway to deduce whether any precipitation does occur at longer time periods.

ACKNOWLEDGEMENT

The continued support of the SERC for the IRC programme is gratefully acknowledged

REFERENCES

(1) Knowles, J.C. and Hastings, G.W., J. Mater. Sci., 1993 4, 102-106.

(2) Burnie, J., PhD Thesis, University of Strathclyde, 1982

(3) Kokubo, T., J. Non. Cryst. Solids, 1990 120, 138-151.

(4) Rehman, I., Hench, L.L. and Bonfield, W., in Bioceramics Vol. 6, Butterworth-Heinemann, Oxford, UK., 1993, 123-128.

(5) Filgueiras, M.R., LaTorre, G. and Hench, L.L. J. Biomed. Mat. Res., 1993, 27, 445-453.

Bioceramics, Volume 7 Edited by Ö.H. Andersson and A. Yli-Urpo
(*Proceedings of the 7th International Symposium on Ceramics in Medicine, Turku, Finland, July 1994*)

The Early Dissolution-Deposition of Calcium Phosphate on the Biphasic Calcium Phosphate Ceramics Implanted in Dogs and Rabbits

S. Qu, W. Chen, J. Weng and X. Zhang
Institute of Materials Science and Technology, Sichuan University, 610064 Chengdu, P.R. China

ABSTRACT

The porous biphasic calcium phosphate ceramics composed of hydroxyapatite (HA) and β-tricalcium phosphate (β-TCP) had been implanted the muscles of dogs and rabbits, respectively. We have investigated the early dissolution and deposition of calcium phosphate on the biphasic ceramics in varied animal species. The micrographs of biphasic ceramics after implanted were different between in dogs and rabbits. The X-ray diffraction patterns of biphasic ceramics showed the greater degradation in dogs than in rabbits the early stage in vivo. These results argue that the equilibrium between dissolution and deposition of biphasic calcium phosphate are different among various amimal species, which may result in the differential tissue response to the biphasic ceramics.

INTRODUCTION

Calcium phosphate ceramics are clinically for bone repair in the past two decades. Most investigations have already demonstrated that the calcium phosphate ceramics show good biocompatiblity and biodegradallity.[1-4] It is known that mineral constituent of human hard tissue showed a porous biphasic (HA/TCP) structure after sintering at high temperature. For this reason, the bionic porous biphasic or multiphase (HA+α-or/and β-TCP)ceramics are found to have better bone-bonding ability than pure HA and can accelerated new bone formation in bony sites.[5-7] Recently, some reports even showed that bone formation associated with the crystals of biphasic ceramics in a nonsseous tissue sites.[8-10] Our groups have investigated and reported that the porous biphasic ceramics (HA/β-TCP) had been implanted in the muscles of dogs for 2 months and there were bone formation in the pores of the ceramics.[11] The bioperformonce of biphasic ceramics is related to the dissolufion-deposition of apatite calcium phosphate and may be influenced by the physical and chemical factors. However, the real reaction process is still unclear. Especially so far the bone formation induced in nonosseous tissues is only observed in muscles or subcutaneous sites of dogs, not in mice, rabbits and goats.[10] What is the inducing factor, certain proteins or tissues? It is necessary to understand the early response of ceramics in nonosseous sites, which directly influence on the bone formation. In this study, porous biphasic ceramics (HA/β-TCP) were implanted the muscles of dogs and rabbits. The samples after implanted several weeks were evaluated using Scanning Electron Microscopy (SEM) and X-ray diffraction (XRD).

MATERIAL AND METHODS

Starting powders were prepared by precipitation of $Ca(NO_3)_2$ and $(NH_4)H_2(PO_4)$.[9] The biphasic ceramics had macroporous structure by the addition of hydrogen peroxide. Then the green bodies were sintered at 1250°C for 4 hours. The X-ray diffraction pattern of ceramics (figure 1) revealed that they had a HA/β-TCP ratio of 65/35. The total porosity of materials was 47% and the average size of macroporous was 300μm or so. These pores appeared to be interconnected. There was an amount of microporous, size 5μm or so, in the ceramics (fiqure 2a).

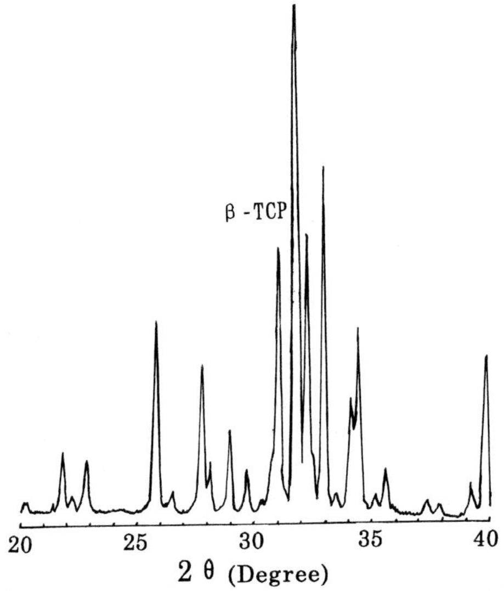

Figure 1. X-ray diffraction pattern of the biphasic calcium phosphate ceramics before implantation

Porous cylinders were fabricated (4mm diameter, 7mm length in dogs, 3mm diameter, 5mm length in rabbits) and implanted in muscels of dogs and rabbits, respectivily. The animals were scarificed postoperatively and the implants were recovered. The soft tissue which surround the implants was removed and the implants which were separated from any solution to avoid the dissolution of deposition on the ceramics during this process were dried. The sections were obstained from these implants and were sputter coated with gold for SEM in order to examine the micrograph of these implants. XRD were performed on the dried implants without soft tissue to determined the multiple phase transformation.

RESULT AND DISSCUSSION

SEM analysis of implants 7 days in dog revealed that the degradation/dissolution had mainly occured at the inter-grain boundaries accompanying with some grains of biphasic ceramics been partially or completely dissolued in different areas. In some typical region, there were crystal granules deposited and needle-like crystals less than 5μm in length were clearly visible. Moreover, the plate-like crystals occured in marcopores nearly the surface of the implants (figure 2b). The X-Ray Energy Dispersion Analysis (EDAX) showed that the needle-like crystals were composed of calcium phosphate. These crystals had grown into large plate-like. In dog after implanted 30 days, abundance of plate-like crystals associated with the crystals of the biphasic ceramics. The relative abundance of plate-like crystals appeared to be related to the distance from the surface of implant: the outer region, the greater the relative abundance of the plate-like crystals. There were, however, no be nove crystals formation besides the dissolution of implanted ceramics grains in rabbits as observed by SEM from 7 to 23 days after implanted (figure 2c). While, there were plate-like crystals being connecting with collagen fibers were closely to the ceramic surface after 30 days in rabbit by SEM (figure 2d).

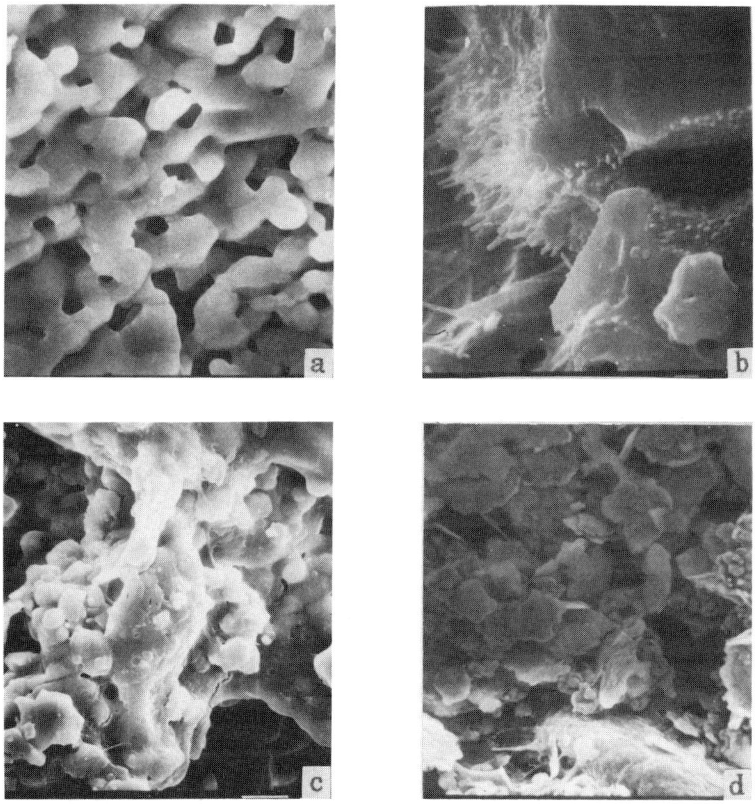

Figure 2. SEM photographs of the biphasic calcium phosphate ceramics. (a) before implantation, there was an amount of micropore, size 5μm or so. (b) after implanted 7 days in dog, there were apatite-like crystal deposition on ceramics. (c) after implanted 7 days in rabbit, no be nove crystal deposition besides the dissolution of biphasic ceramics. (d) after implanted 30 days in rabbit the plate-like crystals being connecting with collagen fibers were closely to the ceramic surface (Bar=5μm).

XRD peaks of HA and β-TCP in expended the implanted biphasic ceramics demonstrating the decreasing of grain size. Further more the XRD peaks of β-TCP were reduced, indicating the degradation of β-TCP in vivo, but this degradation were somewhat greater in dogs than in rabbits the early time after implanted (table 1).

Table 1. HA/β-TCP ratio of the biphasic ceramics before and after implantation by XRD analysis.

	HA/β-TCP	HA/β-TCP	HA/β-TCP
	Before Implantation	After 7 Days	After 30 Days
in dogs	65.0/35.0	69.8/30.2	73.4/26.6
in rabbits	65.0/35.0	67.7/32.3	72.3/27.7

Although it is known that bioactive materials form a bond with bone tissue, the new bone formation have remained obscure. Epitaxial growth[13] or seeded growth[6] has also been postulated as a possible mechanism in the formation of the bond. The partial dissolution of calcium phosphate ceramics caused increasing the local concentration of calcium and phosphate ions, thereby their microenvironment may attaint the degree of saturation resulting in the formation of the microapatite crystals. M. Neo et al. had reported that deposition of acicular crystals similar to those in bone on the surface apatite in vivo.[14] De Bruijn et al. demonstrated needle-like crystals were in direct contact with TCP in vitro.[15] We had studied the apatite crystals prepared by aqueous solution synthesis using Transmission Electron Microscopy (TEM) and observed the apatite are needle-like crystals. In our present studies, we observed that there were acicular or needle-like substance deposition on ceramic surface(Figure 2b). Base on above and EDAX analysis, we speculated that needle-like crystals are calcium phosphate crystal. The deposition occoured directly the surface with few degraded and phagocytized crystals, indicating the stable surface as one factor possibly contributing to the bone formation. The whole processes of epitaxial growth including the nucleation and growth have been revealed in our experiment(Figure 2b).

Generally, the formation of bone-like apatite in vivo is speculated to be both biologically and material-dependent to some extent. However, in the present study, the results obstained demonstrated that the deposition of apatite-like on the surface of cereamics in dogs and rabbit was not the same: the formation needle-like crystals on the ceramic surface is faster in dogs than in rabbits. This may be due to the complexity of the in vivo environment, such as the difference of body fluid, the PH value, the temperature of body, systematic circulation and so on. All of these may be resulting in varied degraded of biphasic ceramics which have be demonstated by XRD. These would be expected to exert the surface structure and the degree of supersaturation of the surrounding fluid with respect to ceramics, and finally affect the precipitation of new apatite-like crystals on the surface of implants. The varied early dissolution-deposition may result in a differential tissue response to the biphasic ceramics.

CONCLUSION

The formation of the bond seems to result from the physicochemical interaction between the ceramics and the interfacial fluid, in the sense of dissolution/deposition, and epitaxial growth. The extent of degradation in vivo of biphasic ceramics and the formation of the apatite-like crystals are different among varions animal species, which may result in the differential tissue response to the biphasic ceramics.

REFERENCE

1. E.B. NERY, K.L. LYNCH, W.M. HIRTHE and K.H. MUELLER, J. Periodontol 46 (1975)328
2. M. JARCHO, C.H. BOLEN, M.B. THOMAS, J. BOBICK, J.E. KAY, R.H. DOREMUS, J. Mater. Sci. 11(1976) 2027-2035
3. L.L. Hench and J. Wilson, Science 226(1984)630
4. K. De Groot, in "Bioceramics of Calcium Phosphate" (CRC Press, Bocca Raton, 1983)
5. David B. Burr, Satoshi Mori., Robert D. Boyd, T.C. Sun, J. David Blaha, Leon Lane, and Jack Parr, J. Biomed. Mater. Res. Vol.27, 645-655, (1993)
6. G. Daculsi, R.Z. Le Geros, E. Nery and K. Lynch, B. Kerebel. J. Biomed. Mater. Res. Vol.23, 883-894,(1989)
7. P. Frayssinet, J.L. Trouillet, N. Rouquet, E. Azimus and A. Autefage, Biomater. 14(1993)423
8. Yamasaki H, Sakai H. Biomater. 1992; 13; 308-312
9. Heughebeart M, Legeros RZ, Ginest M, Guilhem A, Bonel G., J. Biomed. Mater. Res.; Appl., Biomater. 1988; 22(A_3); 257-268
10. C. Klein, K. de Groot, Chen Weiqun, Li Yubao and Zhong Xingdong. Biomater. 1994, Vol.15, No.1, 31-34
11. Zhang Xingdong, Zhou Pin, Zhang Jiangguo, Chen Weiqun and Wu Chuong, BIOCERAMICS AND THE HUMAN BODY, Ed. by, A. RAVAGLIOLI and A. KRAJEWSKI, 408
12. Young R.A., Holcomb D.W., Calcif, Tissue Int, 1982, 34, 17-32
13. J.E. Kay, R.H, Doremus, and M. Jarcho. Trans. 4th Ann. Meet. Soc. Biomater. 10th Int. Biomater. Symp., 154-155(1978)
14. M. Neo, T. Nakamura, C. Ohtsuki, T. Kokubo and T. Yamamuro, J. Biomed. Mater. Res. Vol.27, 999-1006(1993)
15. J.D. de Bruijn, J.E. Davies, J.S. Flach, K. de Groot and C.A. Van Blitterswijk, in Tissue-Inducing Biomaterials, L.G. Cima and E.S. Ron(Eds), Mater. Res. Soc., Pittsburgh, 1992, 63-70

Bioceramics, Volume 7 Edited by Ö.H. Andersson and A. Yli-Urpo
(*Proceedings of the 7th International Symposium on Ceramics in Medicine, Turku, Finland, July 1994*)

In Vivo Reactions of the Porous Hydroxyapatite, β-TCP and β-TCP Coated Hydroxyapatite

K. Ioku, H. Kurosawa,* K. Shibuya,** H. Yokozeki** and T. Hayashi**

Dept. Advanced Materials Science and Engineering, Fac., Eng., Yamaguchi Univ., 2557 Tokiwadai, Ube-shi, Yamaguchi 755, Japan

* Dept. Orthop. Surg., Tokyo Teishin Hospital, 2-14-23 Fujimi, Chiyoda-ku, Tokyo 102, Japan

** Dept. Orthop. Surg., Fac. Med., Univ. of Tokyo, 7-3-1 Hongo, Bunkyo-ku, Tokyo 113, Japan

ABSTRACT

Porous hydroxyapatite(HA), β-tricalcium phosphate(β-TCP) and β-TCP coated HA, with 60% porosity, were implanted into the quadriceps muscles of adult rabbits. These rabbits were sacrificed 4, 12, 24 and 48 weeks after the implantation. The specimens were analyzed by X-ray diffractometry(XRD), scanning electron microscopy(SEM) and infrared spectroscopy(IR). The measurement of weight of the specimens was carried out after heating at 900 °C for 3 hrs. The weight of porous HA was increased obviously at 4, 12, 24 and 48 weeks. The biological hydroxyapatite containing carbonate was formed on the HA surface according to the IR. The weight of porous β-TCP was slightly decreased at 24 weeks. The weight of porous HA coated with β-TCP (33wt% β-TCP, 67% HA; 33TCP) was scarcely changed, but the weight tended to increase at 24 weeks. Thus 33TCP behaved like β-TCP initially after implantation, and then behaved like HA in vivo.

INTRODUCTION

Hydroxyapatite ($Ca_{10}(PO_4)_6(OH)_2$; HA) is one of the most bio-compatible materials with human bones. The new bone is formed in direct contact with the HA surface, therefore porous HA is required for bone substitute materials (1). In order to introduce the newly formed bone fully into the material and increase the bone mass, the HA must be designed as high porous one with inter-connecting pore structure with large surface area as much as possible. The porous HA, however, is fragile because of its high porosity. Common measures for the improvement of the mechanical strength of ceramics are to decrease their porosity and pore diameter, but such materials prevent osteoblasts which make new bones from entering the pores. One possible solution is to prepare the material with relatively lower porosity with a certain intention before implantation, then with relatively higher porosity after implantation (2). If the higher porosity is accomplishde by the resorption of the surface of the ceramics after implantation, the surface area of the ceramics as the scaffold for new bone formation must increase gradually. There are reports that β-tricalcium phosphate ($Ca_3(PO_4)_2$; TCP) has higher resorbability than HA when the material is implanted in a bone defect (3-6). To make a stronger bone substitute, porous HA coated with β-TCP was prepared by our unique method (7, 8). In this study, in vivo behavior of porous HA, β- TCP and β-TCP coated HA was analyzed quantitatively.

METHOD

The starting materials used in this experiment were porous HA, β- TCP and β-TCP coated HA (33wt% β-TCP, 67% HA; 33TCP) with inter-connecting pore structure of cylindrical shape of 5mm $\phi \times$5mmL which had parameters as follows (Table 1). The sample preparation method was descrived previously in detail (7, 8). The specific surface area of samples was measured by BET method. The compressive strength of samples was measured using a material testing machine (Shimadzu, REH-100) with a cross-head speed of 0.02 cm/min.

In order to examine the in vivo behavior of materials, cylindrical samples were implanted into the quadriceps muscles of twenty-three adult Japanese White Rabbits weighed at 3.5-4.0 kg. These rabbits were sacrificed 4, 12, 24 and 48 weeks after the operation. Retrieved and analyzed materials were 102 in number. The specimens from the muscle were analyzed by X-ray diffractometry (XRD; Rigaku, RAD-RC) and infrared spectroscopy (IR; Perkin Elmer, 1600 FT-IR). The weight measurement of the specimens was carried out after removal of organic matter by soaking them in NaOH solution and heating at 900 °C for 3 h. The morphology of these specimens without organic matter was observed by scanning electron microscopy (SEM; Hitachi, S530).

Table 1 Properties of implanted materials

	Porosity (%)	Pore size (μm) macro	Pore size (μm) micro	Specific surface area (m^2/g)	Compressive strength (MPa)
HA	60±5	150～400	0.2～0.5	10.0	12.5±4.0
β-TCP	57±5	150～400	—	2.0	21.4±2.6
33TCP	58±5	150～400	—	3.8	17.5±2.7

RESULTS AND DISCUSSION

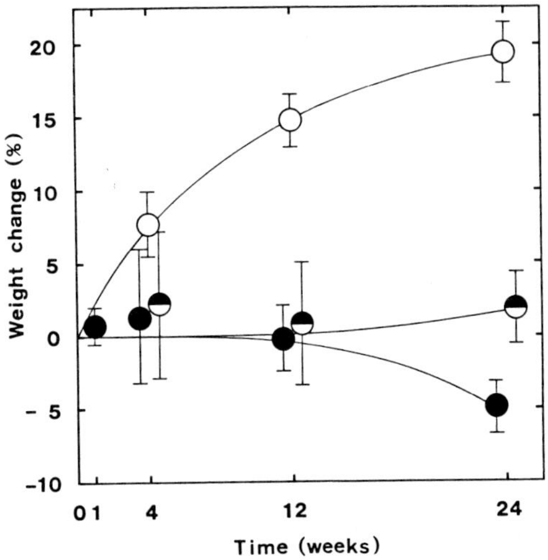

Figure 1 Weight change of the porous HA (○), β-TCP (●) and 33TCP (◒) implanted in muscle.

The weight of porous pure HA was increased obviously at 4, 12, 24 and 48 weeks (Figure 1). No other phase than apatite were revealed by XRD in any products studied. The biological hydroxyapatite containing carbonate was formed on the HA surface after implantation according to the IR measurement. The peaks of CO_3^{2-} of hydroxyapatite are seen at 12 weeks and they are more clearly observed at 24 and 48 weeks. Thus the increment of HA in weight after implantation is due to the carbonate-containing hydroxyapatite. The SEM observation (Figure 2) indicated that the apatite were deposited on the surface of the porous HA. At 12 weeks, some part of the surfaces were covered by hill-like precipitates without micro-pores. At 24 and 48 weeks, most of the surfaces was covered by mountain-like precipitates. The ions of Ca^{2+}, PO_4^{3-} and OH^- of the body fluid are supersaturated with respect to the apatite, therefore growth of the apatite was achieved on the porous HA. The gradual increase of the weight after implantation was due to the apatite precipitation. The apatite formation mechanism was descrived previously in detail (9). The weight of β-TCP was decreased at 12, 24 and 48 weeks (Figure 1). The SEM observation of β-TCP (Figure 3) at preimplantation shows smooth surface, which means the accomplishment of TCP coating on HA. At 12, 24 and 48 weeks, many round concavities was observed. The observation indicated that the surface of it was resorbed after implantation. The biological hydroxyapatite containing carbonate, however was revealed from the samples at 12, 24 and 48 weeks by XRD and IR. The weight of 33TCP was scarcely changed up to 12 weeks, but the weight slightly increased by 2% at 24 weeks (Figure 1). The SEM observation of 33TCP (Figure 4) at preimplantation shows smooth surface, which means the accomplishment of TCP coating on HA. At 12 weeks, many round concavities was observed like β-TCP. At 24 weeks, a few round concavities and some hill-like precipitates without micro-pores could be observed. The SEM observation of 33TCP indicated that the surface of it was resorbed just after the implantation, and then the apatite formed on the bare more-porous HA surface according to the IR measurement. Therefore 33TCP behaved as β-TCP with high resorbability in the early stage, and then behaved as HA in the latter stage in vivo.

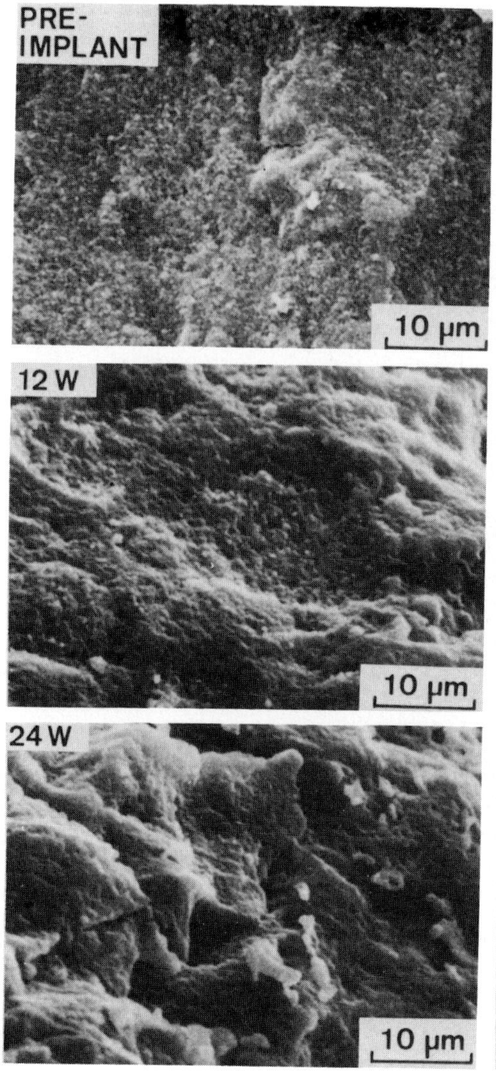

Figure 2 SEM observation of pre-implant HA and retrieved HA from muscle at 12 W and 24 W.

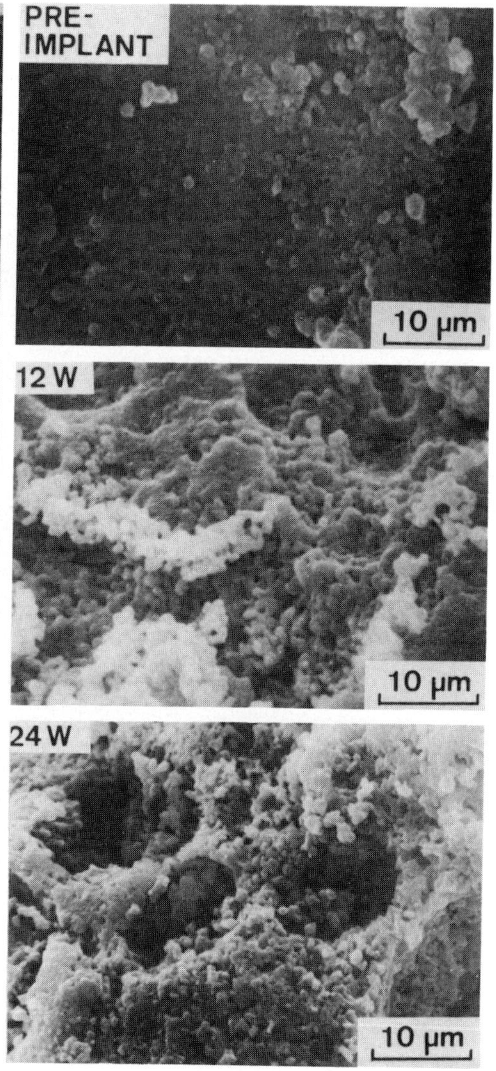

Figure 3 SEM observation of pre-implant β-TCP and retrieved β-TCP from muscle at 12 W and 24 W.

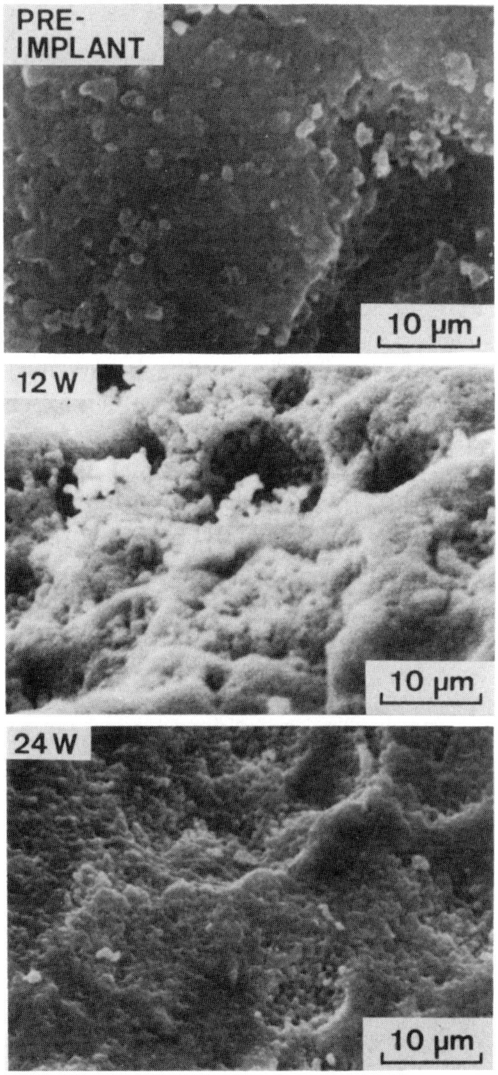

Figure 4 SEM observation of pre-implant 33TCP and retrieved 33TCP from muscle at 12 W and 24 W.

SUMMARY

Porous HA, β-TCP and HA coated with β-TCP were implanted into muscles of adult rabbits, then in vivo behavior of the materials was analyzed quantitatively. The following results were obtained;

1) In vivo behavior of the material could be analyzed quantitatively and easely by implantation into muscles.

2) The weight of HA increased steadily after implantation. The increment was found to be precipitates on the surface by SEM observation. It was proved to be carbonate hydroxyapatite by XRD and IR. Resorption was hardly noticed from weight change and on SEM observation.
3) The weight of β-TCP was decreased at 24 and 48 weeks. Resorption was found at 12, 24 and 48 weeks by SEM observation. Carbonate hydroxyapatite was revealed by XRD.
4) The weight of 33TCP was initially unchanged, but after 12 weeks slightly increased. Resorption was found at 12 weeks by SEM observation. Carbonate hydroxyapatite was clearly observed at 24 weeks. Therefore 33TCP behaved as β-TCP in the early stage, and then behaved as HA in the latter stage in vivo.

REFERENCES

1. Iwano, T., Kurosawa, H., Murase, K., Takeuchi, H. and Ohkubo, Y. In Clinical Orthopaedics and Related Research 1991, 268, 243-252

2. Ioku, K., Murakami, T., Ikuma, Y. and Yoshimura, M. In J. Ceram. Soc. Jpn., Int. Ed. 1992, 100, 1001-1005

3. Renooij, W., Hoogendoorn, H. A., Visser, W. J., Lentferink, R. H. F., Schmitz, M. G. J., Ieperen, H. V., Oldenburg, S. J., Janssen, W. M., Akkermans, L. M. A. and Wittebol, P. In Clinical Orthopaedics and Related Research 1985, 197, 272-285

4. Eggli, P. S., Muller, W. and Schenk, R. K. In Clinical Orthopaedics and Related Research 1988, 232, 127-138

5. Delecrin, J., Daculsi, G., Dard, M., Basle, M. and Legeros, R. In Apatite Vol.1 (Proc. First Int. Sympo. Apatite, Tokyo, Jpn.), Assoc. Apatite Sci., 1992, 107-110

6. Akao, M. In Ceramics Japan 1985, 20, 1096-1103

7. Ioku, K., Yanagisawa, K., Yamasaki, N., Kurosawa, H., Shibuya, K., Yokozeki, H. and Hayashi, T. In Phosphorus Research Bulletin 1993, 3, 73-78

8. Ioku, K., Yanagisawa, K., Yamasaki, N., Kurosawa, H., Shibuya, K. and Yokozeki, H. In Bio-Medical Materials and Engineering 1993, 3,

9. Ioku, K., Yamasaki, N., Yokozeki, H., Iwano, T., Shibuya, K., Kurosawa, H., Kurokawa, T. and Kawahara, H. In Rep. Res. Lab. Hydrothermal Chem. 1991, 7, 42-45

Induction of Bioactivity

Bioceramics, Volume 7 Edited by Ö.H. Andersson and A. Yli-Urpo
(*Proceedings of the 7th International Symposium on Ceramics in Medicine, Turku, Finland, July 1994*)

The Effect of U.V. Irradiation on the Adhesive Strength of a Hydroxyapatite Layer on PMMA Bone Cement

A. Murakami,[1] A. Yamamoto,[1] I. Nishiyama,[2] T. Kokubo,[3] M. Tanahasi,[3] T. Yamamuro[4] and W. Bonfield[5]

[1] Himeji Institute of Technology, Shosha Himeji 671-22 Japan
[2] Mitubishi Electric Co. Ltd., Tsukaguti, Amagasaki 661 Japan
[3] Faculty of Engineering, Kyoto University, Sakyo-ku, Kyoto 606 Japan
[4] Faculty of Medicine, Kyoto University, Sakyo-ku, Kyoto 606 Japan
[5] IRC in Biomedical Materials, QMW College, Mile End Road, London E1 4NS, UK

ABSTRACT

The aim of the present study is to develop a new system for total hip replacement (THR). An apatite layer on cured PMMA bone cement was formed in an aqueous solution of 1.5 times the ion concentration of simulated body fluid (SBF), after keeping on G glass in SBF. The PMMA cement substrate was treated by U.V. irradiation before apatite coating. The adhesive strength of the apatite layer was measured by newly developed equipment (Daipla-Saicas). The effect of U.V. irradiation on the adhesive strength of the apatite layer is discussed.

INTRODUCTION

In the 1960's Charnley developed a method of total hip replacement (THR) using acrylic bone cement as an important component(1). However, it has been shown that failure due to loosening can be as great as 20% over 10 years(2). It is desirable that a modified system is found which improves this situation. More recently cementless fixation has been used, but clinical results suggest that the failure rate is greater, due the mechanical mismatch across the bone/metal interface(3). The aim of this study is to develop a new system for THR which overcomes the mechanical mismatch and also promotes bioactivity at the bone implant interface.

METHOD AND RESULTS

A coating of apatite was produced on cured PMMA by resting the polymer substrate on a layer of Glass G in a solution of simulated body fluid (SBF). The SBF is contained in an inert polystyrene bottle. After 24 hours apatite nucleation had taken place on the substrate surface. The substrate was then placed in a solution of 1.5 times SBF concentration (1.5 SBF) and an apatite coat developed over a period of 7 days with a supernatant change every 24 hours(4). The composition of Glass G, SBF and 1.5 SBF are given in table 1 and table 2, respectively. Before apatite coating the cured PMMA was irradiated using U.V. radiation of dual wavelength (184.9 and 253.7 nm).

Table 1 Composition of Glass G (wt%)

MgO	CaO	SiO_2	P_2O_5	CaF_2
4.6	44.7	34.0	16.2	0.5

Table 2 Concentration of simulated body fluid(SBF) and human blood plasma

	Na^+	K^+	Ca^{2+}	Mg^{2+}	Cl^-	HCO_3^-	HPO_4^{2-}	SO_4^{2-}
Human blood plasma	142	5.0	2.5	1.5	103	27.0	1.0	0.5
SBF	142	5.0	2.5	1.5	148	4.2	1.0	0.5
1.5SBF	213	7.5	3.8	2.3	222	6.3	1.5	0.8

Buffered with $(CH_2OH)_3CNH_2$ and HCl at pH 7.25

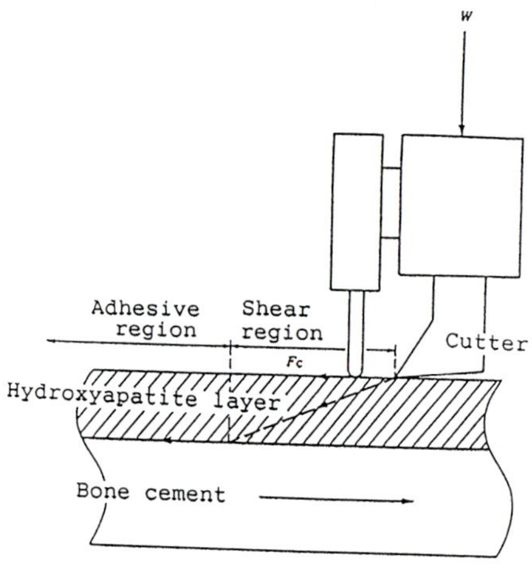

Fig.1
Surface and interface cutting method.

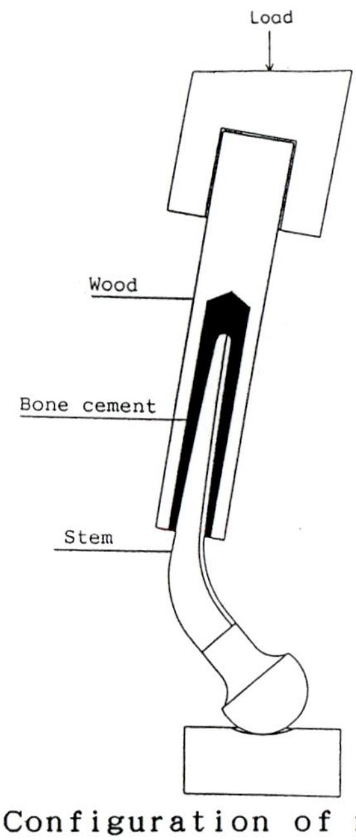

Fig.2 Configuration of model Specimen during fatigue testing.

The adhesion strength and shear strength of the apatite layer was measured using a new system (Daipla-Saicas)

developed by Mitsubishi Electric Co. Ltd. Figure 1 shows the surface and interface cutting method of the Daipla-Saicas equipment. Adhesion and shear properties were also measured on apatite coated PMMA which had been previously treated by low temperature plasma as opposed to U. V. radiation.

In addition to these studies the fatigue property of two model THR systems were evaluated. Using wood as a simulated bone shaft, a stainless steel prosthesis was insert in two ways. In the first model uncured PMMA was placed in the simulated bone cavity and the prosthesis introduced to this (here called stem to shaft+cement fixation). The PMMA was then allowed to cure. In the second model the prosthesis shaft was coated with PMMA which was cured prior to introduction into the cavity (stem + cement to shaft fixation). The two models were cycled as shown in figure 2. Maximum load range was 5 kN at a frequency of 1 Hz. The load angle along the stem axis was 10 degree.

RESULTS

The effect of the surface treatment on the adhesive strength and shear strength of the apatite layer on the acrylic bone cement is shown in Fig.3 and Fig.4, respectively.

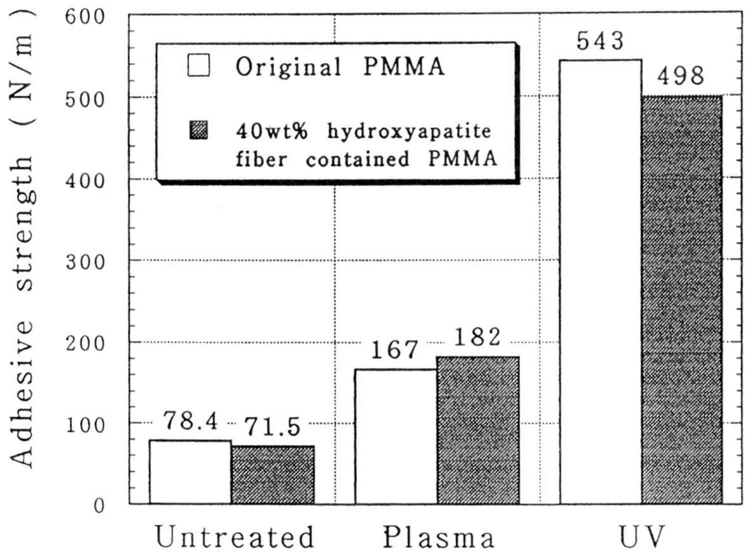

Fig.3 Effect of the surface treatment upon the adhesive strength of the apatite layer on acrylic bone cement.

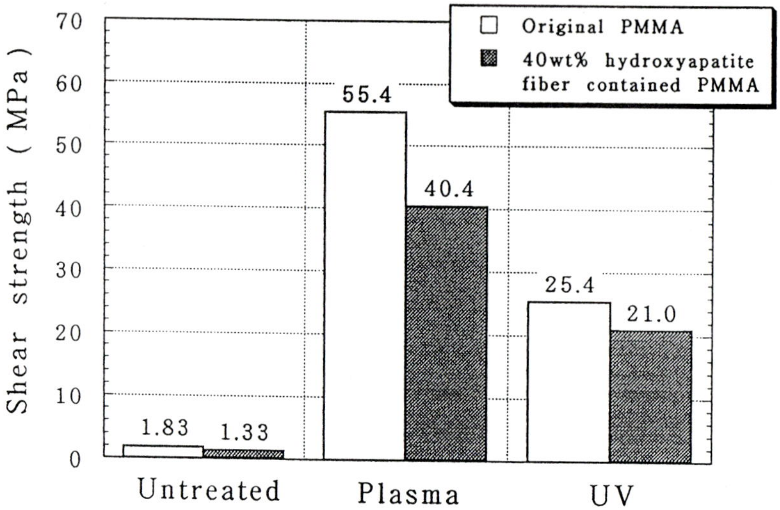

Fig.4 Effect of the surface treatment upon the shear strength of the apatite layer on acrylic bone cement.

The adhesive strength of apatite on U. V. treated PMMA can be seen to be approximately three times greater than in the case of the plasma treated PMMA. In the case of untreated PMMA the adhesive strength is lower; approximately half the value for plasma treatment. The addition of hydroxyapatite fibre to the polymer has no effect on the adhesive properties.

The situation for shear strength is different. Plasma treatment causes shear properties of the apatite to increase by approximately 30 times when compared to untreated PMMA, however U. V. irradiation causes an increase of only half this. As with adhesive strength the addition of apatite has no beneficial effect.

Figure 5 shows the apatite layer on each of the polymer treatment types; untreated, plasma treated and U.V. treated respectively. The scanning electron micrograph on the left of each set is taken perpendicular to the apatite/polymer interface and shows that for each polymer type the apatite layer thickness is similar at approximately 10 μm. This shows that the apatite thickness is not dependent on the polymer treatment. The right hand micrograph of each set is taken perpendicular to the apatite surface. The cracks have been caused by bending the apatite/polymer composite by the same amount in each case. In the untreated case the cracks are pronounced whereas with the U.V. treated polymer the

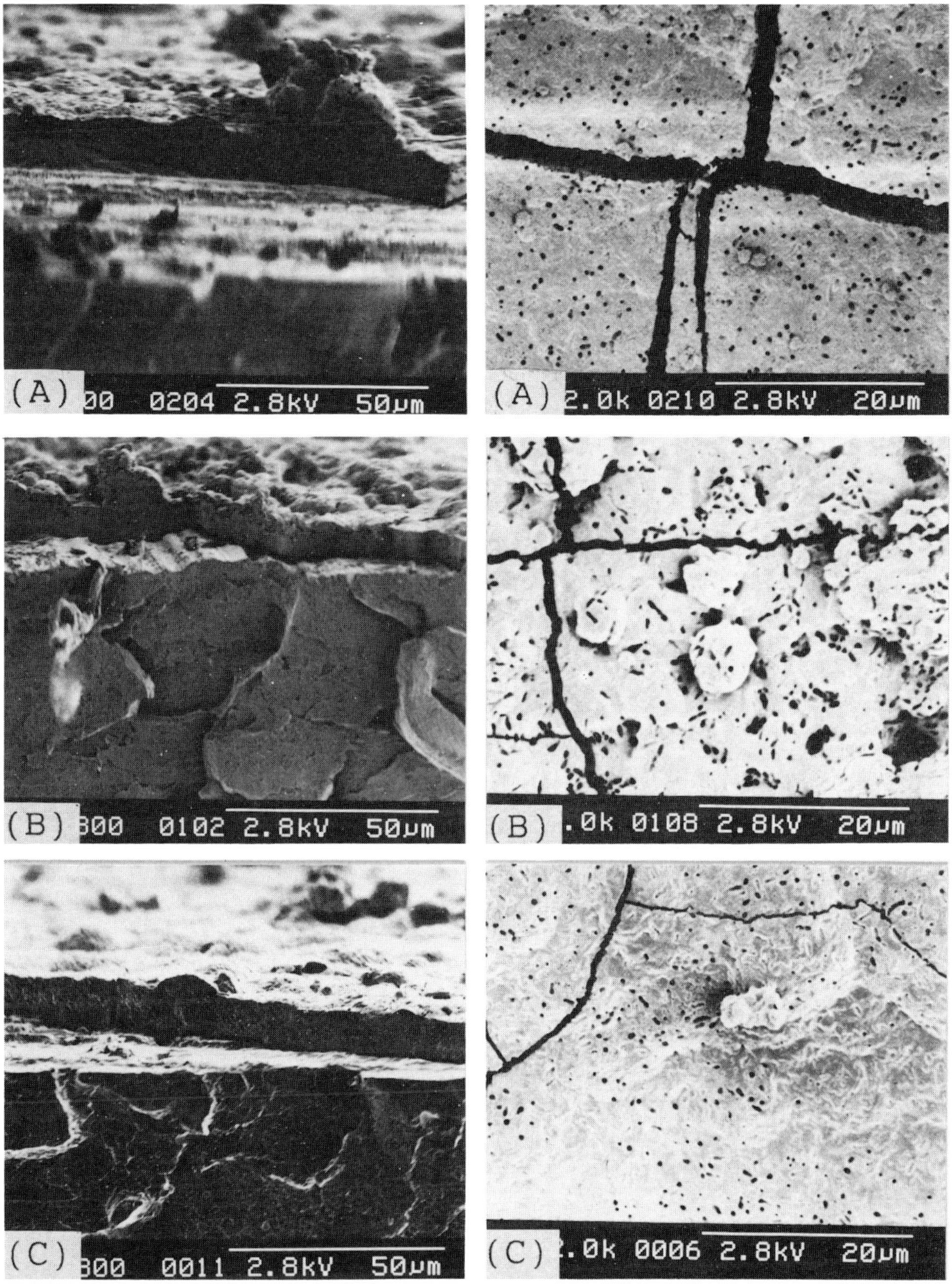

Fig.5 SEM of hydroxyapatite layer on the surface treated acrylic bone cement.
(A) Untreated (B) Plasma treated
(C) UV radiation treated

cracking is substantially reduced. The plasma treated polymer is intermediate. This shows clearly that apatite on the U.V. treated polymer has superior adhesive qualities.

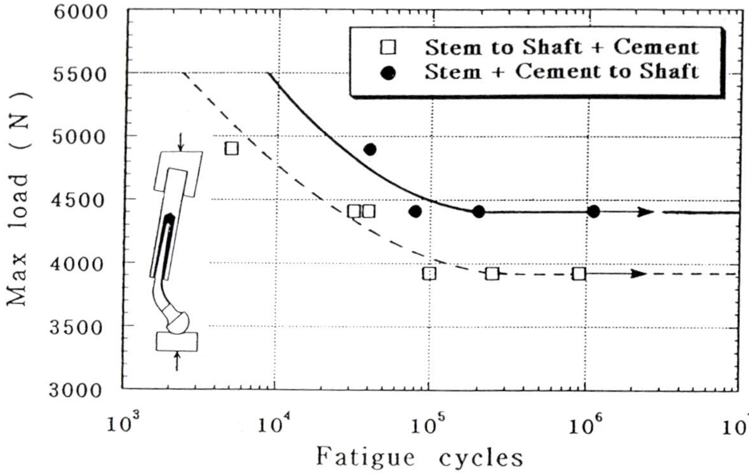

Fig.6 Fatigue behaviour of the two hip joint models.

Comparison of the fatigue behaviour of the two fixation models is shown in figure 6. It is clear that the stem + cement to shaft model has superior fatigue properties to the stem to shaft + cement model.

DISCUSSION AND CONCLUSION

Although we have measure both adhesive and shear strength in this study, the adhesive properties are more important when considering the suitability of implants for total hip replacement. From this work it is clear that U. V. irradiation of the substrate PMMA causes improved adhesion of the apatite layer. The reason for this is the introduction of carbonyl and carboxylic groups to the PMMA surface during irradiation. These groups are hydrophylic and so allow greater binding with the water rich apatite crystals (5).

We added 40% by weight of hydroxapatite (HAP) fibre to the PMMA as this is known to increase both the toughness of the polymer and also improve the fatigue crack propagation resistance (6). We expected that the presence of HAP fibres at the surface of the polymer would promote adhesion of the apatite coating. This does not appear to have occurred. The reason is most likely that the apatite fibres at the surface are coated with PMMA and so inert in terms of binding the apatite coating.

The results from fatigue testing the implant two models, where the metal prosthesis is introduced to the uncured PMMA in the shaft, or the stem of the prosthesis is coated with cured PMMA before inserting to the shaft, show that the new method of fixation produces better fatigue properties. The improvement is however slight. More importantly it was evident that in the case of stem to shaft + cement model loosening as determined by change in the stroke length during fatigue testing was more pronounced. The loosening of the prostheses fitted by this method was shown to be the result of cracking of the PMMA coat. These cracks were larger and more frequent in stem to shaft + cement model systems.

In conclusion we can say that the use of PMMA in THR is important in decreasing the loosening of prostheses by reducing the mechanical mismatch across the bone implant interface. We have shown here that precoating the prosthesis stem with PMMA prior to fixation improves adhesion between the bone and implant. We have also shown that U. V. irradiation of PMMA allows a more stable coating of apatite to be placed on the PMMA surface. As it is well known that apatite promotes bioactivity at the bone/prosthesis interface, this finding is important.
The use of prostheses with a precoat of cured PMMA with an apatite surface layer for total hip replacement is a possibility as a new "cementless" system.

REFERENCES

1. Charnley,J., Acrylic Cement in Orthopaedic Surgery Livingstone, Edinburgh, 1970
2. Saha,S. and Pal,S., J. Biomed. Mater. Res. 1984, 8, 435-462
3. Bonfield,W., Personal communication 1993
4. Kokubo,T., J. of The Ceramic Society of Japan 1991, 10, 965-973
5. Teramoto,K., Nishiyama,I., Okajima,T. and Moriwaki,T., Preprint of The Adhesion Soc. of Japan 1990, 28, 115-116
6. Murakami,A., Behiri,J.C. and Bonfield,W., Phase Interaction in Composite Materials Omega, 1992,460-466

Bioceramics, Volume 7 Edited by Ö.H. Andersson and A. Yli-Urpo
(Proceedings of the 7th International Symposium on Ceramics in Medicine, Turku, Finland, July 1994)

Structural Change of Hydroxyapatite Coated on Polymers Implanted in Soft Tissue

K. Suzuki, Y. Kageyama, Y. Yokoyama, Y. Harada and T. Kokubo*
First Department of Surgery, Hamamatsu University School of Medicine, 3600 Handa-cho, Hamamatsu, 431-31, Japan
* Division of Material Chemistry, Faculty of Engineering, Kyoto University, Sakyo-ku, Kyoto, 606-01, Japan

ABSTRACT

Structural changes of two types of hydroxyapatite (HA) which were coated on polymers and implanted in soft tissue were studied. In amorphous type of HA-coating, crystallization occurred rapidly and Ca/P ratio increased within 2~3 weeks. The surface of HA-coated layer became to fine granular crystallization. In highly crystalline type of HA-coating, only a little change was observed. These results suggest that it is possible to chose an ideal type of HA-coating for a specific clinical purpose.

INTRODUCTION

Hydroxyapatite (HA) shows high histocompatibility, and has been used in bone-repairing materials(1,2). We have been investigating the usefulness of HA-coated polymers as a material for prosthesis in soft tissues, and already reported successful results in the experimental study of HA-coated artificial trachea(3). Currently, HA-coating is carried out according to the biomimetic process reported by Kokubo et al.(4,5,6). Various types of coated HA are obtained by different conditions of biomimetic solutions in this process. However, little is known about the structural changes of coated-HA in each type. It is necessary to examine these changes to investigate the in vivo reactions of soft tissues against HA-coated materials and the usefulness of HA-coating in the preparation of artificial materials. In the present study, two types of coated HA were investigated to detect the structural changes after implantation in soft tissues.

MATERIALS AND METHODS

Polyethersulfone (PES) was coated with two types of HA: amorphous type and highly crystalline type. Coating was performed according to the biomimetic process reported by Kokubo et al.(4,5,6). For coating with amorphous type, a biomimetic solution with highly supersaturating concentrations of calcium and phosphate ions were employed; whereas for highly crystalline type, the solution contained relative low supersaturating concentrations of calcium and phosphate ions. (Fig. 1,2)

The plate (10 mmX10 mmX2 mm) made of PES were coated by two types of HA, and then implanted subcutaneously in the abdomen of rats. These plates were removed to investigate structural changes of HA with the passage of time (3 days ~ 40 weeks). The structure of coated-HA was analyzed by thin-film X-ray diffraction (XRD) and a scanning electron microscope (SEM). The ratio of calcium and phosphorus in the HA-coating layer was also determined by electron probe X-ray microanalyzer (EPMA).

Figure 1 SEM photograph of amorphous type (before implantation)

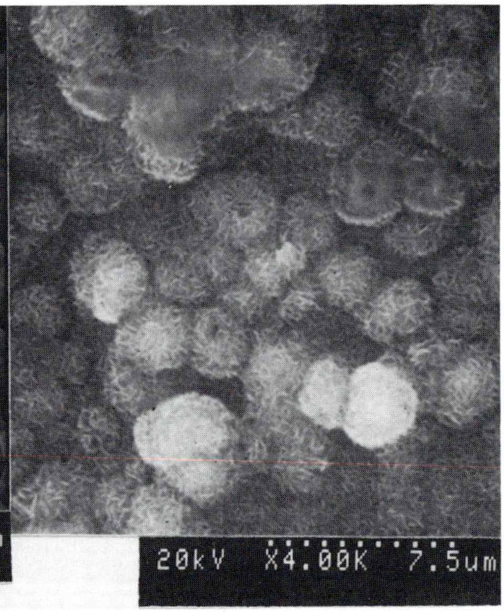

Figure 2 SEM photograph of highly crystalline type (before implantation)

RESULTS

SEM and XRD showed that both types of HA lasted for at least 40 weeks after implantation. In the amorphous type, the rapid crystallization occurred and became to highly crystalline type 2 or 3 weeks after implantation. In the highly crystalline type, coated HA showed only a little changes (Fig. 3,4).

The ratio of calcium to phosphorus in the amorphous type increased with the progress of crystallization within two weeks. The changes of this ratio were mild in the highly crystalline type (Fig. 5).

The surface structure of the amorphous type changed tremendously to the fine granular crystalline type (Fig. 6). In the highly crystalline type, the surface structure of coated HA did not change judging from SEM (Fig. 7).

With respect to the reaction in surrounding tissues, no strong inflammatory response were observed, and no histopathological differences were found between two types of HA. However, adhesion of surroundings tissues to the material appeared to be stronger in the amorphous type.

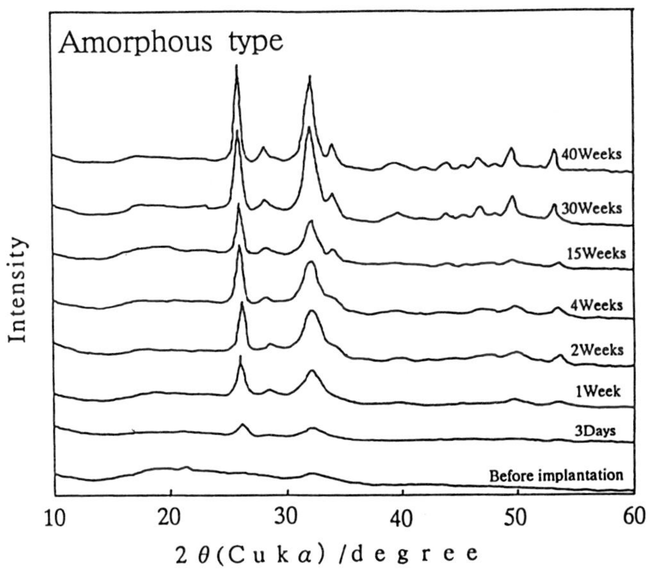

Figure 3
Thin film X-ray diffraction patterns before and after implantation.

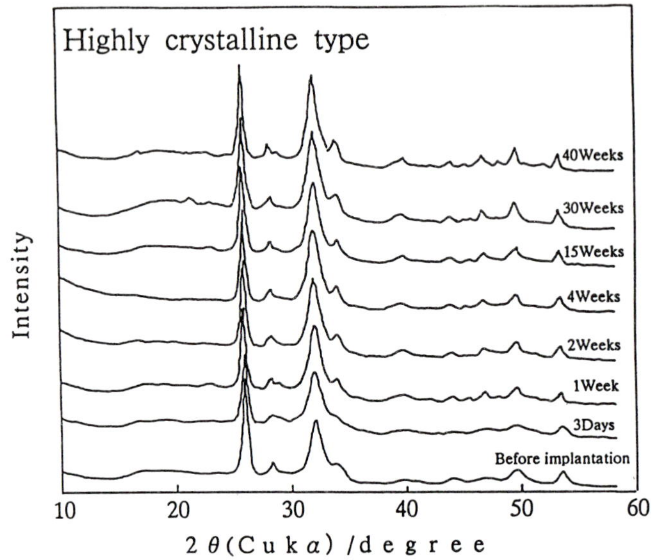

Figure 4
Thin film X-ray diffraction patterns before and after implantation.

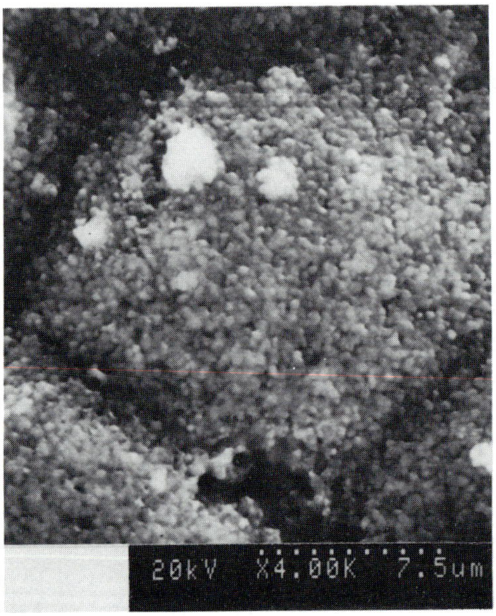

Figure 6 SEM photograph of amorphous type (20 weeks after implantation)

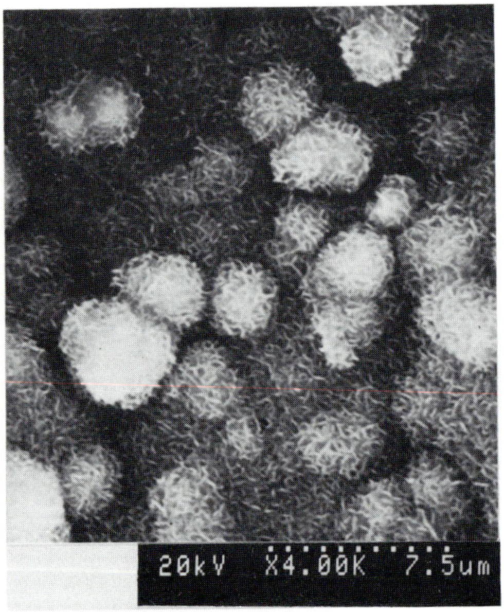

Figure 7 SEM photograph of highly crystalline type (20 weeks after implantation)

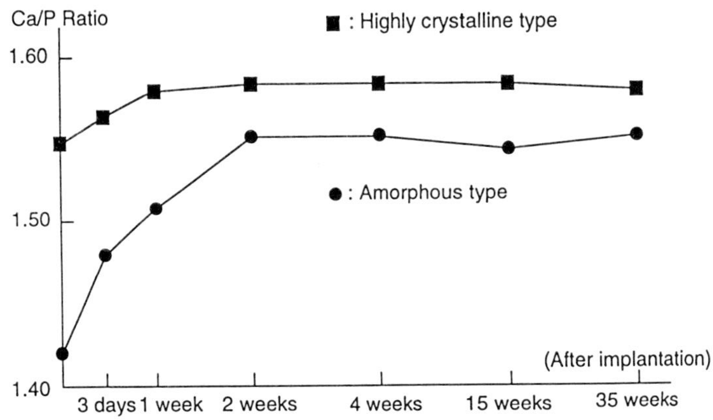

Figure 5 Change of ratio of calcium to phosphorus measured by EPMA.

DISCUSSION AND CONCLUSIONS

It is now possible to obtain various types of coated HA according to the different conditions of the solution used in the biomimetic process (4,5,6). Present study showed that the various types of HA-coatings underwent the different changes of surface structure in soft tissues. These structural changes of HA may allow us to select the ideal type of HA-coating for a specific prosthesis. That is, the highly crystalline type might be better in some cases, while gradual progress of crystallization in vivo might be preferable in others. Further investigations about the change of HA structure and about the tissue response against various type of HA should be carried out (3).

REFERENCES

1) Itiravivong, P., Lorprayoon, C., Sukpech, A. and Runruxsirivorn, S., In Bioceramics, Vol. 5, (1992) 157-164.

2) Misiek, D.J., Kent, J.N. and Carr, R.F., J. Oral Maxillofac. Surg., (1989) 42, 150-160.

3) Suzuki, K., Kobayashi, R., Yokoyama, Y., Harada, Y. and Kokubo, T., In Bioceramics, Vol. 6, (1993) 245-250.

4) Kokubo, T., Hata, K., Nakamura, T. and Yamamoto, T., In Bioceramics, Vol.4, (1991) 113-120.

5) Tanahashi, M., Hata, K., Kokubo, T., Nakamura, T. and Yamamoto, T., In Bioceramics, Vol. 5, (1992) 57-64.

6) Kokubo T., Biomaterials, (1992) 12, 155

Bioceramics, Volume 7 Edited by Ö.H. Andersson and A. Yli-Urpo
(*Proceedings of the 7th International Symposium on Ceramics in Medicine, Turku, Finland, July 1994*)

Chemical Treatment of Ti Metal to Induce its Bioactivity

F. Miyaji, Xu Zhang, T. Yao, T. Kokubo, C. Ohtsuki,* T. Kitsugi,** T. Yamamuro** and T. Nakamura†

Division of Material Chemistry, Faculty of Engineering, Kyoto University, Sakyo-ku, Kyoto 606-01, Japan
* Department of Bioengineering Science, Faculty of Engineering, Okayama University, Tsushima, Okayama 700, Japan
** Department of Orthopaedic Surgery, Faculty of Medicine, Kyoto University
† Research Center for Biomedical Engineering, Kyoto University, Sakyo-ku, Kyoto 606-01, Japan

ABSTRACT

A chemical treatment was applied to Ti metal in order to induce its bioactivity. The pure Ti metal was treated with 10M NaOH or KOH aqueous solution at 60°C for 1 to 24 h to form a hydrated TiO_2 gel layer with alkali ions on its surface, then subjected to a heat treatment at 400 to 600°C for 1 h to form amorphous alkali titanate layer. Thus treated Ti metal formed a dense and uniform bone-like apatite layer on its surface via forming a TiO_2 hydrogel layer when it was soaked in the simulated body fluid with ion concentrations nearly equal to those of human blood plasma. This indicates that thus treated Ti metal could form the bone-like apatite layer even in the body. It is expected that the apatite layer thus formed *in vitro* as well as *in vivo* could exhibit high bioactivity as well as strong bonding to the substrate.

INTRODUCTION

Ti metal and its alloys coated with hydroxyapatite by plasma spray method are widely used as hip joints, dental implants and so on (1). Thus formed apatite is, however, not easy to be controlled in its composition and structure, and not easy to be strongly bonded to the substrates. It is believed that the effective phase for artificial material to be bonded to living bone is the bone-like apatite, i.e., a kind of carbonate-containing hydroxyapatite formed on its surface *in vivo* or *in vitro* (2). The present authors previously showed that even pure TiO_2 hydrogel can form the bone-like apatite on its surface in the simulated body fluid (SBF) with ion concentrations nearly equal to those of human blood plasma (3). In the present study, it was attempted to induce bioactivity of Ti metal by forming thin TiO_2 hydrogel layer on its surface with a chemical treatment.

EXPERIMENTAL PROCEDURE

Commercially pure (99.5%) Ti metal (Nilaco Co.) plates with 15 X 10 X 1 mm^3 in size were polished with #400 diamond paste, washed with acetone and distilled water. They were soaked in 10M-NaOH or -KOH aqueous solution at 60°C for 1 to 24 h, washed with distilled water and dried at 40°C for 24 h. Thus treated Ti metal specimens were heated to 400, 500, 600 or 800°C with heating rate of $5°Cmin^{-1}$ and kept at the given temperatures for 1 h.

The surface structural changes of the specimens were characterized by thin-film X-ray diffraction [TF-XRD] (Rigaku Co., thin-film attachment CN2651A1) and scanning electron microscope [SEM] (Hitachi Co., S-2500CX) attached with an energy-dispersive electron probe X-ray microanalyzer [EPMA] (Horiba Co., EMAX-3700) after the above treatments.

Subsequently, the specimens were soaked in SBF (4) for various periods. The apatite formation on their surfaces after soaking in SBF was also examined by the above methods.

RESULTS AND DISCUSSION

Figure 1 shows the TF-XRD patterns of the surfaces of Ti metal treated with 10M-NaOH solution at 60°C for 1 to 24 h. It is found that the peak intensity ascribed to Ti metal decreases, whereas the broad halo due to amorphous phase appears at about 23 to 30° in 2θ as the soaking time increases. The similar results were observed for 10M-KOH-treated Ti metal. The SEM observations of the surface of the specimens suggested that such an amorphous phase may be formed even after 1 h-soaking in the alkaline solution.

Figure 2 shows the TF-XRD patterns of the surfaces of Ti metal treated with 10M-NaOH solution at 60°C for 24 h and heated at various temperatures for 1 h, in comparison with those without alkali-treatment. The Ti metal subjected to no alkali-treatment (a) precipitates rutile phase at 600°C and its layer wholly covers the metal surface at 800°C. On the other hand, the NaOH-treated Ti metal (b) does not precipitate the rutile phase at 600°C, and increases the amorphous phase represented by the halo at about 23 to 30° around this

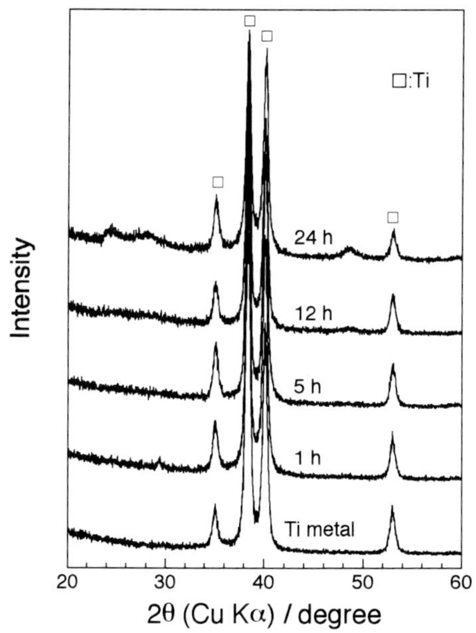

Figure 1. TF-XRD patterns of the surfaces of Ti metal soaked in 10M-NaOH solution at 60°C for various periods.

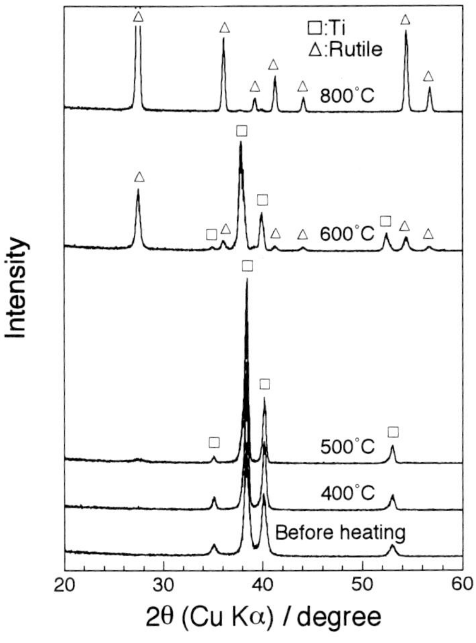

Figure 2(a). TF-XRD patterns of the surfaces of Ti metal heated at various temperatures for 1 h.

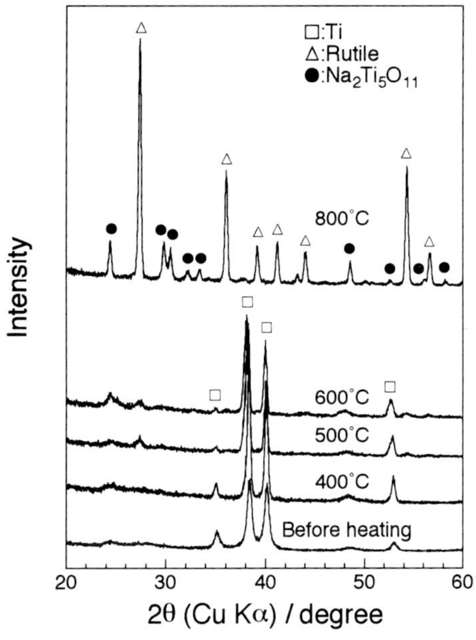

Figure 2(b). TF-XRD patterns of the surfaces of Ti metal treated with 10M-NaOH and heated at various temperatures for 1 h.

temperature. The crystalline sodium titanate as well as rutile phase, however, precipitates and the amorphous phase disappears at 800°C. The KOH-treated samples showed the similar tendency. Taking into consideration the precipitation of crystalline alkali titanate at higher temperatures than 800°C, the amorphous phase with alkali titanate composition might be formed on the surface of Ti metal by the alkali- and subsequent heat-treatments.

Figure 3 shows the SEM-EPMA photographs of cross-section of the Ti metal treated with 10M-NaOH (a) or -KOH (b) and heated at 600°C. It is found that the thickness of amorphous alkali titanate layer is about 1 μm and the Ti and K concentrations gradually increases and decreases from the metal surface to the inside, respectively. The distribution of Na concentration could not be measured owing to the limitation of the present EPMA instrument, but is expected to show the similar tendency to that of K.

It was confirmed by TF-XRD that no apatite is formed on the surfaces of pure Ti metal even after soaking in SBF for 3 weeks. Figure 4 shows the TF-XRD patterns of the surfaces of Ti metal treated with 10M-NaOH, heated at various temperatures for 1 h and soaked in SBF for 3 weeks. Only the 600°C-treated sample is found to form apatite on its surface. It can be therefore said that both the alkali-treatment and subsequent heat-treatment at 600°C are required to form apatite layer on the surface of Ti metal.

Figure 5 shows the TF-XRD patterns of the surfaces of Ti metal treated with 10M-NaOH, heated at 600°C and soaked in SBF for various periods. It is found that the apatite is formed within 2 weeks and its layer almost wholly covers the surfaces of substrates by 3 weeks, suggested by the disappearance of the peaks ascribed to Ti metal. The similar apatite formation and growth were

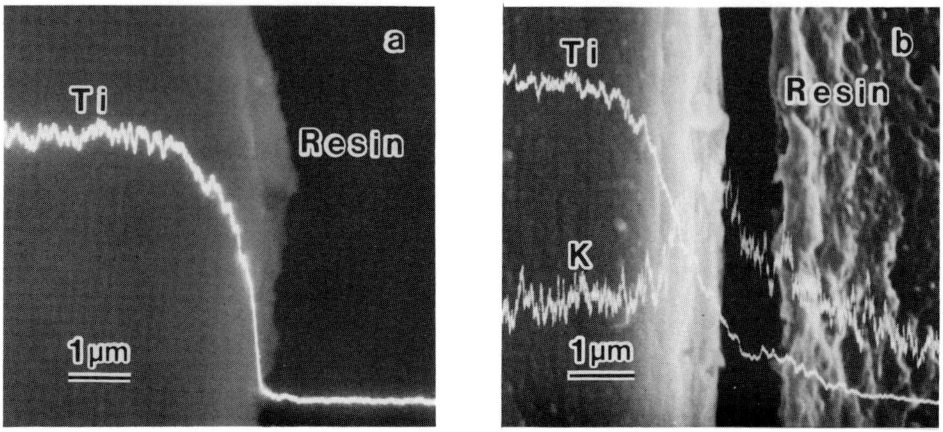

Figure 3. SEM-EPMA photographs of cross-sections of the Ti metal treated with 10M-(a)NaOH or -(b)KOH and heated at 600°C.

observed for 10M-KOH- and 600°C-treated Ti metal.

Figure 6 shows the SEM-EPMA photographs of the surface (a) and cross-section (b) of Ti metal treated with 10M-NaOH, heated at 600°C and soaked in SBF for 3 weeks. It is found that the apatite layer, corresponding to the Ca- and P-rich layer, with 5-10 μm thick was homogeneously formed and tightly bonded to the substrate. In fact, when a tensile stress was applied normally to the

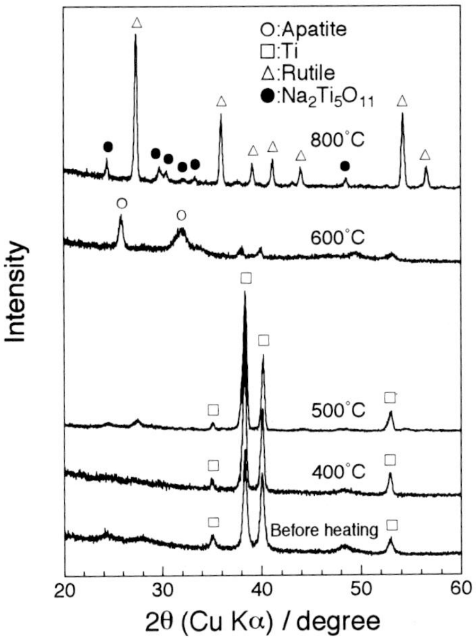

Figure 4. TF-XRD patterns of the surfaces of Ti metal treated with 10M-NaOH, heated at various temperatures for 1 h and soaked in SBF for 3 w.

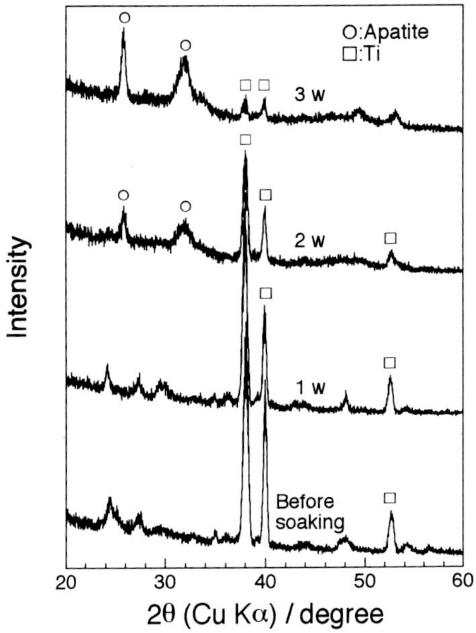

Figure 5. TF-XRD patterns of the surfaces of Ti metal treated with 10M-NaOH, heated at 600°C and soaked in SBF for various periods.

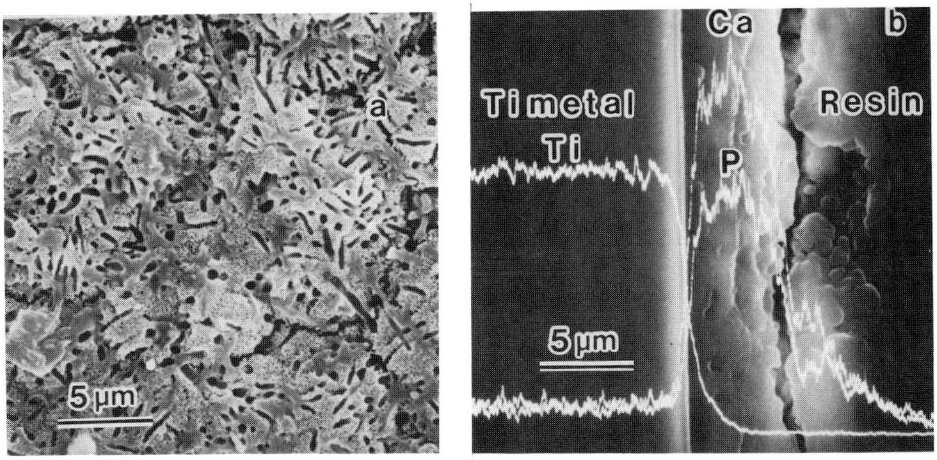

Figure 6. SEM-EPMA analyses of the (a)surface and (b)cross-section of Ti metal treated with 10M-NaOH, heated at 600°C and soaked in SBF for 3 w.

substrate, the fracture occurred not at the apatite-substrate interface but in the apatite layer.

The processes of surface structural changes of Ti metal due to alkali- and heat-treatments might be interpreted as schematically shown in Figure 7. Thin TiO_2 passive layer is usually present at the surface of Ti metal. When Ti metal is

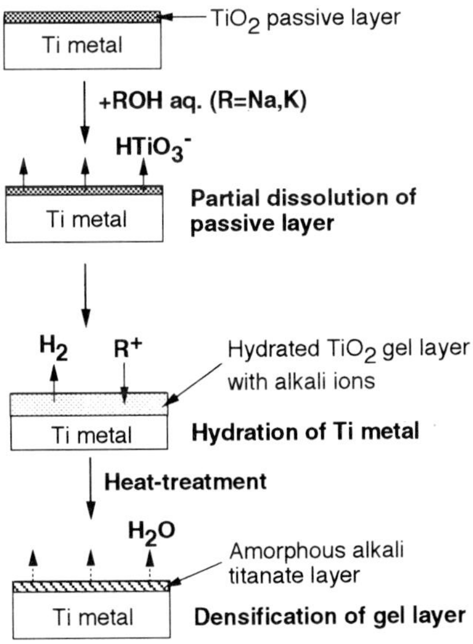

Figure 7. Structural changes of Ti metal due to alkali- and heat-treatments.

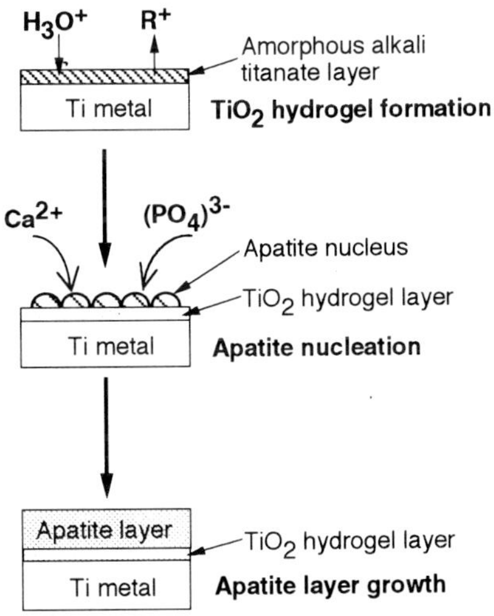

Figure 8. Aapatite nucleation and growth on Ti metal in SBF.

soaked in alkaline solution, the surface passive layer is partially dissolved into the solution (5).

$$TiO_2 + OH^- \rightarrow HTiO_3^- \qquad (1)$$

Such dissolution of TiO_2 passive layer and the hydration of Ti metal are assumed to be proceed simultaneously (5-7).

$$Ti + 3OH^- \rightarrow Ti(OH)_3^+ + 4e^- \qquad (2)$$

$$Ti(OH)_3^+ + e^- \leftrightarrow TiO_2 \cdot H_2O + \frac{1}{2}H_2 \uparrow \qquad (3)$$

$$Ti(OH)_3^+ + OH^- \leftrightarrow Ti(OH)_4 \qquad (4)$$

The hydrated titania ($TiO_2 \cdot nH_2O$) gel layer formed on the substrate will be in part attacked by OH^- ions (6).

$$TiO_2 \cdot nH_2O + OH^- \leftrightarrow HTiO_3^- \cdot nH_2O \qquad (5)$$

As a result, the negatively charged $HTiO_3^- \cdot nH_2O$ species may adsorb the alkali ions and the hydrated TiO_2 gel layer containing alkali ions is formed on the substrate. Such gel layer is densified to form an amorphous alkali titanate layer and tightly bonded to the substrate by heat-treatment. The subsequent apatite formation on Ti metal in SBF can be explained as follows (Figure 8). When the amorphous alkali titanate layer is exposed to SBF, the alkali ion is dissolved from the titanate layer and titania hydrogel layer is formed. The alkali ion increases the degree of supersaturation for apatite around the substrate, and the TiO_2 hydrogel induces the apatite nucleation on its surface (3). Consequently, the apatite nucleation on the surface of substrate is induced. If the apatite nuclei are once formed, they spontaneously grow by incorporating the Ca^{2+} and PO_4^{3-} ions from the surrounding fluid which is already supersaturated with respect to apatite (2).

A series of the present results suggests that thus chemically treated Ti metal can form a bone-like apatite layer even in the body. It is expected that the apatite layer thus formed *in vitro* as well as *in vivo* can exhibit high bioactivity as well as strong bonding to the substrates.

REFERENCES

1. Pilliar, R.M. and Filaggi, M.J. Bioceramics, Vol. 6 Ed by Ducheyne, P. and Christiansen, D. Butterworth-Heinemann, Oxford, (1993), 165
2. Kokubo, T. J. Ceram. Soc. Japan, (1991), 99, 965
3. Li, P. Ohtsuki, C. Kokubo, T. Nakanishi, K. Soga, N. and de Groot, K. J. Biomed. Mater. Res., (1994), 28, 7
4. Kokubo, T. Kushitani, H. Sakka, S. Kitsugi, T. and Yamamuro, T. J. Biomed. Mater. Res., (1990), 24, 721
5. Prusi, A.R. and Arsov, L.D. Corrosion Science, (1992), 33, 153
6. Tengvall, P. and Lundström, I. Clinical Materials, (1992), 9, 115
7. Healy, K.E. and Ducheyne, P.h J. Biomed. Mater. Res., (1992), 26, 319

Bioceramics, Volume 7 Edited by Ö.H. Andersson and A. Yli-Urpo
(Proceedings of the 7th International Symposium on Ceramics in Medicine, Turku, Finland, July 1994)

Potentiostatic Controlled Growth of a Very Stable Titanium Oxide Anodic Film on Ti-6Al-4V Alloy

R.L. Reis,[#] O.C. Paiva*[#] and J.D. Santos[#]
[#] INEB-National Institute for Biomedical Engineering, Department of Metallurgical Engineering, University of Porto, Rua dos Bragas, 4099 Porto Codex, Portugal
* ISEP-Polytechnic Engineering Institute of Porto, Rua de S. Tomé, 4200 Porto, Portugal

ABSTRACT

Ti-6Al-4V samples were anodised in a 5% H_2SO_4 solution, in a saturated calcium phosphate solution (SCP), in Hank's solution and in a simulated human blood plasma, under potentiostatic control at respectively 15-65V (H_2SO_4) and 2.5-7.5 V (other solutions), for 10-30 minutes. Corrosion behaviour was studied in Hank's solution. A decrease in the corrosion rate was achieved with increasing anodising potentials, as determined by Tafel analysis. Oxide layers were characterised using SEM/EDS and thin film XRD. Layers were porous and seemed to be composed of a TiO_2-Ti_2O_3 mixture. It was possible to correlate anodising parameters with film thickness and corrosion behaviour. The incorporation of Ca and P in the film was accomplished in SCP although a much less stable film than that produced in %5 H_2SO_4 was obtained.

INTRODUCTION

Several surface treatments have been attempted to enhance corrosion resistance and raise biocompatibility or stability of Ti and its alloys (1-3). It is well known that the titanium oxide film spontaneously formed in the atmosphere contributes to its good biological behaviour (4-5). The use of an anodising process allows the formation of a much more stable and thicker passive layer (3,6-7). In addition, this technique may be easily adapted to homogeneously cover any shaped form.

The use of hydroxylapatite plasma-sprayed coated titanium has extensively been studied over the past few decades, but many problems may still be pointed out to this solution in both orthopaedic and dental applications. Pilliar (8) has reviewed a number of interesting new calcium phosphate coating methods. Slip casting, electrophoretic deposition, sputtering or ion beam deposition, formation of sol-gel films and electrochemical deposition are some of the promising alternatives. Some authors (3,6) have tried to develop more stable and thicker titanium oxide anodic films. Some concern has been centered on controlling morphology of the films (3,6). The improvement in surface bioactivity of titanium alloys has also been claimed by several authors (1,9-11). Li (9) coated titanium with a gel titanium by a sol-gel process. Royer (10) studied the deposition of calcium phosphate salts onto several biomaterials. The formation of a titanium anodic film containing calcium and phosphorous (1) and the electrochemical synthesis of

calcium phosphates in simulated body fluids (11) were reported.
Nevertheless, studies dealing with the electrochemical growing of titanium oxide films containing Ca and P are very rare in literature (1,11). The aim of this work was to generate a very stable passive layer on Ti-6Al-4V surface which, at the same time, was microporous to allow bone ingrowth. The developed anodic oxidation route should enable the incorporation of Ca and P ions into the film or the electrochemical synthesis of apatite compounds within it. Under potentiostatic conditions several combinations of corrosion behaviour, and surface morphologies may be attained, by adjusting the applied potential. Roughness profile and porosity of the oxide layers, which are critical factors for bone ingrowth, also depend greatly on the anodising parameters.

MATERIALS & METHODS

Anodising Procedure

Commercially Ti-6Al-4V alloy matching the ASTM F136-84, was anodised in a 5% H_2SO_4 solution, under potentiostatic control at 15 to 65V, for 10-30 minutes. Aiming at nucleating Ca-P compounds within the film, three different solutions were used: a simulated body fluid (SBF), an Hank's balanced salt solution (HBSS) and a saturated calcium phosphate solution (SCP) prepared from commercial pure grade reagents. Compositions of the alternative solutions studied in this work are presented in Table I. The anodising potentials were in the 2.5-7.5V range for the three solutions.
Plates samples with 2.0 x 1.0 x 0.1 cm were anodised using a Ti-6Al-4V cathode placed 6 cm distant from the anode. Cathode dimensions were 10.0 x 7.0 x 0.1 cm. Surface preparation for anodising was as follows: grinding, polishing and etching in a HF-HNO_3 according to ASTM specification B-481-68, followed by rinsing in distilled water. A volume of 500 cm^3 of fresh electrolyte at room temperature was used in each anodisation.

Oxide Films Characterisation

Anodic oxide films were characterised by Scanning Electron Microscopy (SEM), in a Jeol JSM-35C, and thin film X-Ray Diffraction (XRD) analysis in a D 5000 Siemens Diffractometer Kristalloflex. Secondary electron images were obtained at 25 KeV. Composition of the films was determined by Energy Dispersive Spectroscopy (EDS), in a Noran Instruments device. Spectra were acquired for beam energies in the 15 to 25 KeV range. In XRD CuK_{α} radiation, at 40KV and 40 mA, with a graphite monochromator, was used.

Corrosion Testing

Potentiodynamic anodic and cathodic curves were determined for each condition. Experiments were run in a EG&G Potentiostat/Galvanostat model 273. Softcorr® software was used to control testing parameters, acquisition steps and calculate results. All tests were carried out in a cell designed according to ASTM G5-82 standard. Potential was scanned from -500 to 2500 mV vs. reference electrode (standard calomel electrode, SCE) at a scanning speed of 5 mV/s. Based on the obtained results, Tafel anodic and cathodic coefficients (β_a and β_c)

and polarization resistance (R_p) were computed. R_p was calculated in a E (I=0) ±10 mV range. Corrosion rate (i_{corr}) was then calculated using the well known Stern-Geary equation (12). All tests were carried out in HBSS, at 37±1 °C. Solution was not deareated prior to tests, aiming at improving the simulation of body conditions. Solutions were buffered at pH of 7.4 using trishydroxyaminomethane ($(CH_2OH)_3NH_2$) and hydrochloridric acid (HCl).

Table I - Composition of the tested anodised solutions. Hank's balanced salt solution (HBSS) was also used as corrosion testing environment.

Compound	H_2SO_4 (%v/v)	SBF (g/l)	SCP (g/l)	HBSS (g/l)
H_2SO_4	5.0	-	-	-
$Ca_3(PO_4)_2$	-	-	saturated	-
NaCl	-	8.1	-	8.0
$CaCl_2$	-	0.28	-	0.14
KCl	-	0.22	-	0.40
$NaHCO_3$	-	0.36	-	0.35
$MgCl_2 \cdot 6H_2O$	-	0.22	-	0.10
$Na_2HPO_4 \cdot 2H_2O$	-	-	-	0.06
KH_2PO_4	-	0.17	-	0.06
$MgSO_4 \cdot 7H_2O$	-	-	-	0.06
Glucose	-	-	-	1.0

RESULTS & DISCUSSION

Fig.1 presents a SEM photomicrograph of Ti-6Al-4V sample anodised at 65V for 10 min in the 5% H_2SO_4 solution. It may be seen that an homogenous film was grown during the anodising process. A rough morphology was attained. By EDS analysis only Ti and O were detected. XRD spectra (Fig.2) indicated that films were composed of a TiO_2-Ti_2O_3 mixture. Some TiS_2 was also detected revealing that some sulfur from the SO_4^{2-} was incorporated in the film. This seems to be independent of the applied potential within the 15-65V range. On the contrary, corrosion resistance of the anodised samples was deeply dependent on the anodising potential as plotted in Fig.3. All treatments were efficient and a decrease in the corrosion rate (i_{corr}) was achieved for anodised samples comparing to non-anodised ones. For the same anodising time (t=10 min), a decrease in i_{corr} values of 50.3% (15V), 53.8% (30V), and 58.7% (65V) was determined. This may be related with the thickness of the oxide layers which were estimated to be 120 nm (65V), 50 nm (30 V), 20 nm (15 V). This is in good agreement with the results obtained by Dunn (4), using Rutherford backscattered spectroscopy (RBS).

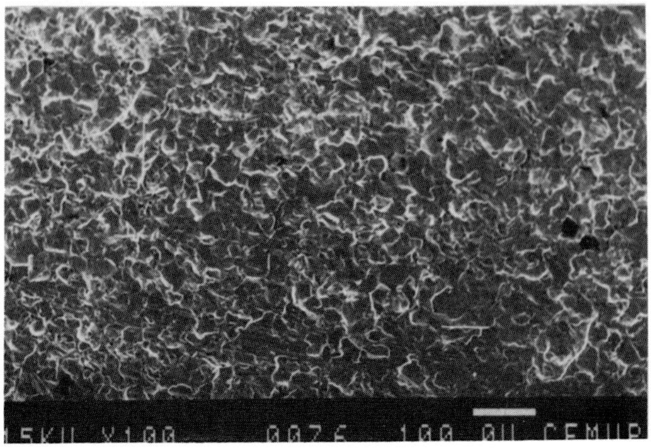

Fig.1 - An aspect of the surface of a sample anodised in H_2SO_4 at 65V for 10 min.

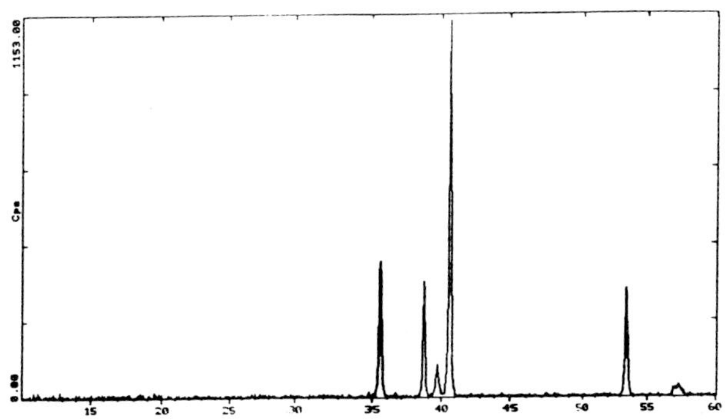

Fig.2 - Low angle thin film XRD spectrum of a sample anodised in H_2SO_4 at 65V for 10 min. A mixture of TiO_2 and Ti_2O_3 was detected for all potentials (15-65V).

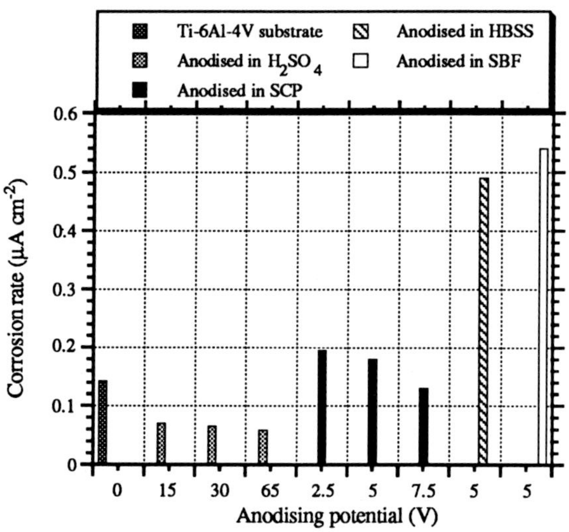

Fig.3 - Corrosion rates calculated for samples anodised at several potential values in the four different anodising solutions. Anodising time was kept constant (t =10 min).

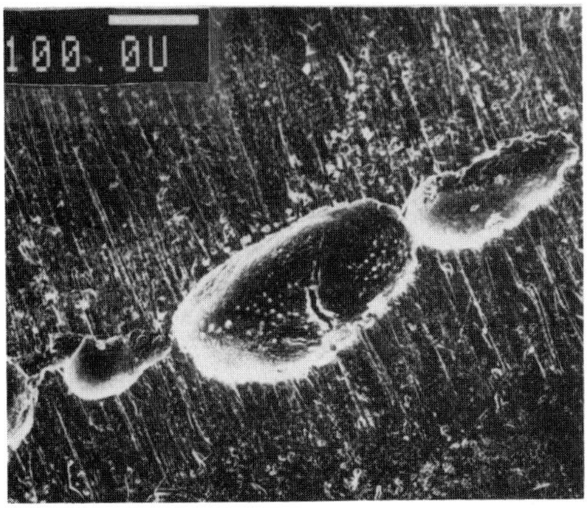

Fig.4 - Pitting observed after anodising in HBSS at an anodising potential of 7.5V (10 min).

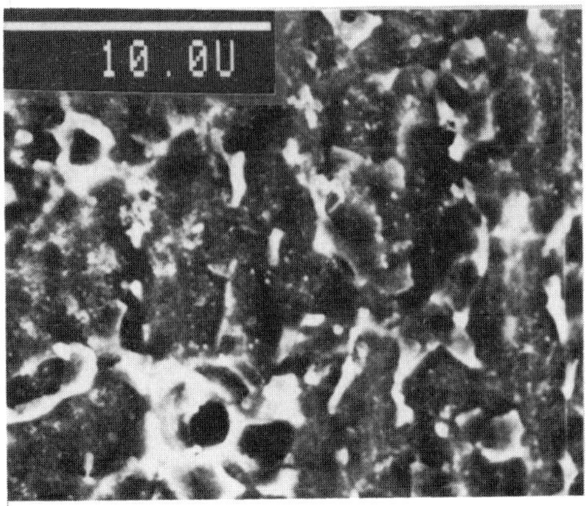

Fig.5 - Ti-6Al-4V anodised in SCP at 7.5V for 10 min. Several Ca-P particles were observed in several spots of the surface.

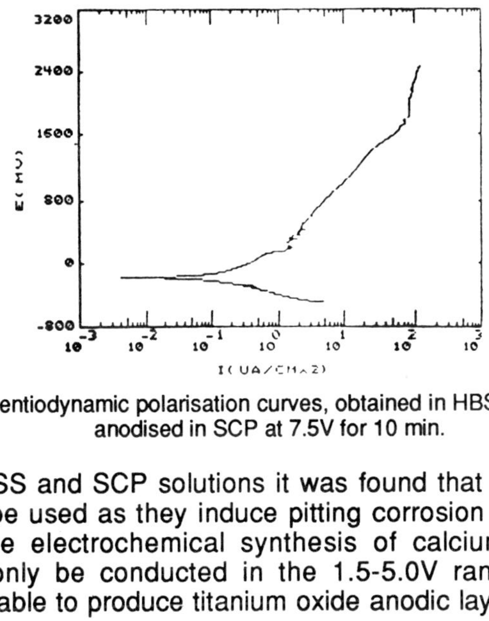

Fig.6 - Potentiodynamic polarisation curves, obtained in HBSS, for a sample anodised in SCP at 7.5V for 10 min.

For the SBF, HBSS and SCP solutions it was found that potentials higher than 7.5V should not be used as they induce pitting corrosion (Fig.4). Ban (10) also observed that the electrochemical synthesis of calcium phosphate in SBF solutions may only be conducted in the 1.5-5.0V range. On the contrary, Ishizawa (1) was able to produce titanium oxide anodic layers containing Ca and P at much higher potential levels, 250-350V, in a β-glycerophosphate solution. Unfortunatley our power source was not capable of working at such high potentials but it may be speculated that with these solutions, specially when Cl^- is present pitting might be observed. Furthermore, the presence of Ca and P in the solutions has been reported (13) to decrease the corrosion resistance of titanium both with and without film breakdown.

The incorporation of Ca and P in the oxide layer was only achieved in the SCP solution (Fig.5). The absence of Cl^- ions and the much higher concentration of Ca-P compounds have an important role in this result. However, corrosion rates were increased, as expected (see Fig.3), when compared to both samples anodised in 5% H_2SO_4 and non-anodised samples. Curves show a typical passive behaviour as plotted in Fig.6. i_{corr} is always lower than 0.20 μA cm^{-2}.

Samples anodised in SCP at the higher potentials (7.5V) corroded at rates very similar to those of the spontaneously (in air) passivated substrate. Plates anodised in HBSS and SBF showed a much faster corrosion kinetics (see Fig.3) and pitting was observed as reported in Fig.4.

It was possible to find a relationship between anodising parameters, film thickness, and corrosion behaviour. The incorporation of Ca and P in the film was accomplished although a much less stable film was obtained. Further work is being carried out to improve this technique, as it seems to be possible to nucleate Ca-P compounds in several spots of the oxide layer. Histological studies in a rabbit model are currently under way to assess biological behaviour of anodised materials.

CONCLUSIONS

Very stable and high corrosion resistant, passive films were produced on Ti-6Al-4V by an anodic oxidation procedure. A decrease in the corrosion rate was achieved with increasing anodising potentials. Passive layers seemed to be composed of a mixture of TiO_2 and Ti_2O_3. Incorporation of Ca and P in the film was accomplished in a saturated calcium phosphate solution, at potentials between 2.5 and 7.5V, although a much less stable film was obtained. Nevertheless, corrosion current density was always lower than 0.20 $\mu A\ cm^{-2}$.

ACKNOWLEDGMENTS

The authors wish to gratefully acknowledge the financial support of Fundação Calouste Gulbenkian. We also thank Dr. Carlos Sá for helping with the SEM/EDS studies and I.R.C in Biomedical Materials, University of London, for allowing the use of their XRD facilities.

REFERENCES

1. Ishizawa, H., Fujino, M. Ogino, M., in Bioceramics 5, Yamamuro, T. Kokubo, T. Nakamura, T., Editors, Kobunshi Kankokai, Kyoto, Japan (1992), 43

2. Zitter, H. Plenk jr., H., J. Biomed. Mater. Res., (1987) **21**, 881

3. Paiva, O.C., Santos, J.D., Reis, R.L., in 185th Electrochemical Society Meeting-Symposium on Compatibility of Biomedical Implants, Electrochemical Society Volume Series, in press

4. Williams, D.F., in Biocompatibility of Clinical Implants, Vol.1, p.13, CRC Press, Boca Raton, (1981)

5. Barbosa, M.A., in Biomaterials: Hard Tissue Repair and Replacement, Muster, M., Editor, , E-MRS Monographs, Vol.2, North-Holand, Amesterdam, The Netherlands (1992), 257

6. Dunn, D., Raghanavan, S., Surface & Coatings Technology, (1992) **50**, 223

7. Fraker, A.C., Ruff, A.W., Sung, P., van Orden, A.C., Speck, K.M., in Titanium Alloys in Surgical Implants, ASTM STP 796, Luckey, H.A., Kubli jr., F., Editors, ASTM, (1983) 206-219

8. Pilliar, R.M., Filiaggi, in Bioceramics 6, Ducheyne, P., Christiansen, D., Editors, Butterworth-Heinemann, Philadelphia, U.S.A. (1993), 165

9. Li, P., Kangasniemi, I., de Groot, K., in Bioceramics 6, Ducheyne, P., Christiansen, D., Editors, Butterworth-Heinemann, Philadelphia, U.S.A. (1993), 41

10. Royer, P., Amrah-Bouali, S., Freche, M., Rey, C., Rouquet, N., Bonel, G., in Bioceramics 5, Yamamuro, T., Kokubo, T., Nakamura, T., Editors, Kobunshi Kankokai, Kyoto, Japan (1992), 95

11. Ban, S., Maruno, S., in Bioceramics 5, Yamamuro, T., Kokubo, T., Nakamura, T., Editors, Kobunshi Kankokai, Kyoto, Japan (1992), 49

12. Stern, M., Geary, A.L., Corrosion, (1957) **13**, 139

13. Sousa, S., Barbosa, M.A., Clinical Materials, (1993) **14**, 287

Bioactive Ceramics and Tissue Reactions

Bioceramics, Volume 7 Edited by Ö.H. Andersson and A. Yli-Urpo
(*Proceedings of the 7th International Symposium on Ceramics in Medicine, Turku, Finland, July 1994*)

Healing of Periodontal Ligament after Implantation of Bioactive Glass in Surgically Created Periodontal Defects: A Pilot Study

A.M. Gatti,* Ö.H. Andersson,*** A. Piattelli,* P. Trisi,* E. Monari** and L. Chiarini**

* Dental School, University of Chieti, Italy
** Lab. dei Biomateriali, Dental School, University of Moderna, Italy
*** Department of Chemical Engineering, Åbo Akademi University, Turku, Finland

ABSTRACT

Granules of a bioactive glass were implanted in a surgically created hole in a sheep's mandibular bone. After one year the bone with the glass was explanted, processed and examined under optical and scanning electron microscope. All particles were surrounded by a layer of mineralized tissue with histological features very similar to root cementum; serial sections revealed that a small cementum bridge connected this mineralized tissue directly to root cementum. Under polarized light it was possible to observe oriented collagen fibers similar to the periodontal ligament inserting on the mineralized tissue.

INTRODUCTION

In the past decade there has been an extensive research concerning the use of alloplastic materials in reconstructive periodontal surgery. The use of bone grafts has been limited due to problems related to supply and to unpredictable behaviour in the postoperative period. Alloplastic materials used in the restorations of osseous defects due to periodontal disease are synthetic materials like ceramics and polymers, or natural ones like collagen. The calcium phosphate ceramics (1) are the most widely used. Hydroxyapatite (HA) is employed in angular periodontal defects in order to reduce the depth of the pocket, fill the bony lesion and gain in attachment (2,3). It has, however, been shown that with nonporous HA used in human periodontal defects it is not possible to obtain regeneration of cementum nor the presence of functioning periodontal fibers adjacent to the HA (2). The aim of the present study was to evaluate the histological responses to a bioactive glass in the periodontal ligament of sheep and to check if such a bone bonding material facilitates the regeneration of cementum and the periodontal ligament collagen fibers.

MATERIALS AND METHODS

A bioactive glass called S53P4 (Åbo Akademi University, Turku, Finland) was used in the present study. The composition of glass S53P4 is: SiO_2 53, Na_2O 23, CaO 20, P_2O_5 4 weight %. The glass was melted in a platinum crucible at 1350 °C for two hours annealed and subsequently crushed to a grain size of 300-500 µm. The particles were stuffed in a syringe and pushed into a surgically created hole in sheep's mandibular bone. After one year the bone segment with the particles was retrieved and fixed in 4% paraformaldeyde and dehydrated in ascending alcohols. After embedding in methylmethacrylate, the bone was sectioned with a diamond saw (Accutom by Struers, Denmark), and microradiographed with a Siemens radiograph (Germany) at 22 KV. The sections were polished with silicon carbide papers and diamond paste. Some sections were coated with carbon and observed with a Scanning Electron Microscope (Philips 500 and XL 40, the Netherlands) in normal and back-scatter mode (BSE). The SEM was equipped with an Energy Dispersive System (EDAX, PV9900) for the evaluation of the inorganic elements present in the material. Other specimens were fixed in 10% buffered formalin, dehydrated in an ascending series of alcohols and embedded in Technovit 7200 methacrylate resin (Kulzer, Germany)(4). The sections were stained with basic fuchsin-methylene blue and observed under a Leitz Laborlux S microscope (Leitz, Wetzlar, Germany) in normal transmitted and polarized light.

RESULTS

The SEM examination showed that the hole, drilled in the alveolar bone, was completely repaired and that it was difficult to recognize the initial position of it. White arrows approximately indicate the zone of the hole (Figure 1).

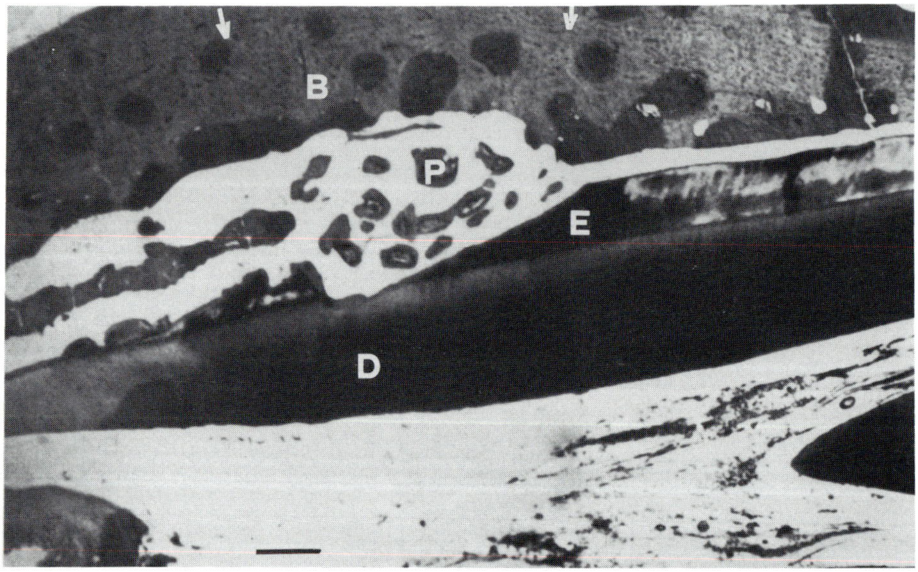

Figure 1 Histologic image of the glass granules immersed in the periodontal space close to the tooth root. B = Bone; D = Dentin; E = Enamel; P = Glass Particles. (Basic fuchsin-methylene blue). Bar = 600 µm.

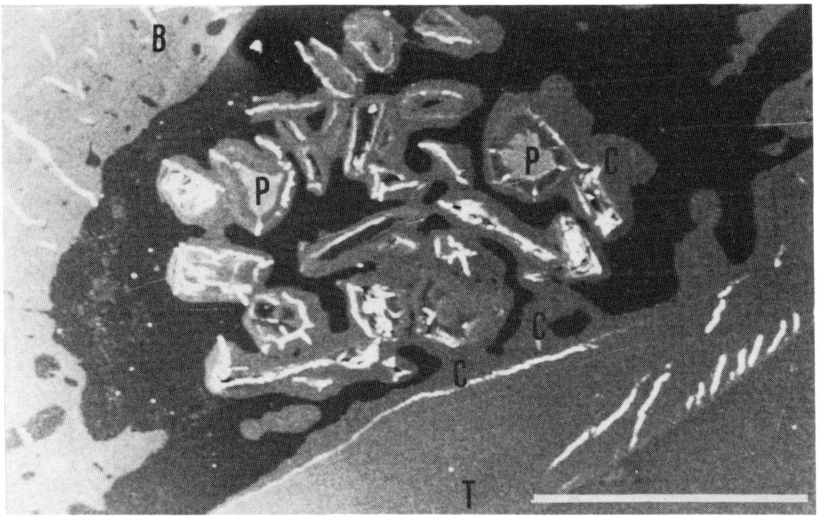

Figure 2 SEM image of the glass granules connected among them by small bridges of mineralized tissue. B = Bone; T = tooth; P = Glass Particles; C = Cementum. Bar = 2 mm.

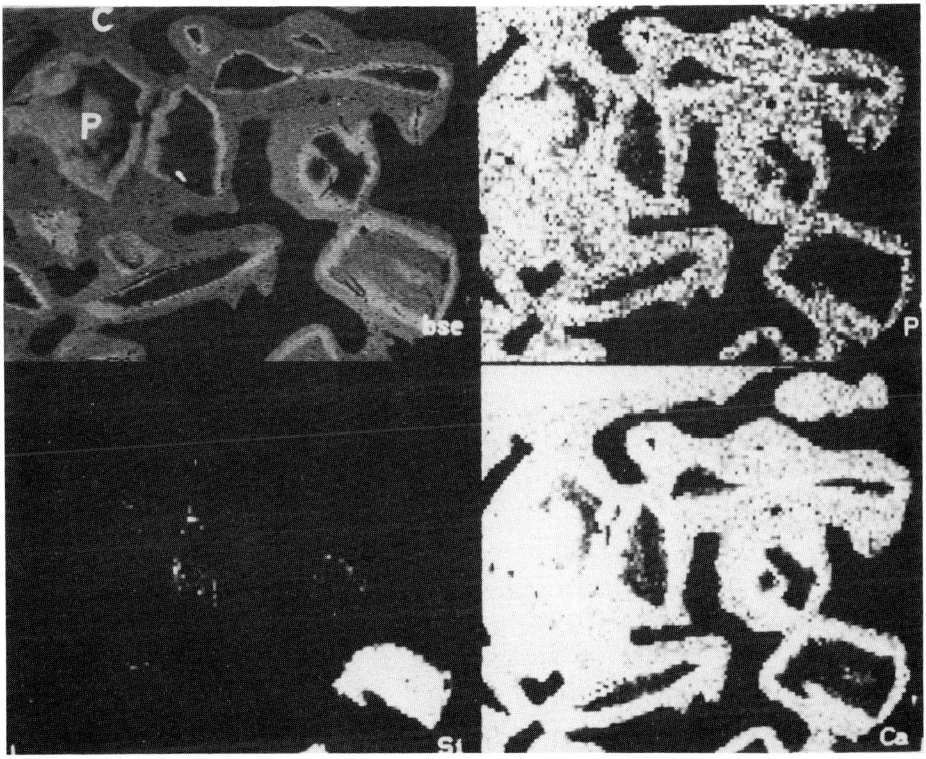

Figure 3 BSE image of the granules with the x-ray dot maps for Silicon, Phosphorus and Calcium.

Figure 4 BSE image of a granule with an excavated core replaced by new bone (Arrows). C = Cementum; P = Glass Particles; M = Core of the Glass. Bar = 100 μm.

Granules were found only in the periodontal ligament space and not embedded in bone. These particles were completely surrounded by a 100-200 μm thick layer of newly formed mineralized tissue of a plate-like nature (Figure 2). This tissue contained some cells similar to osteocytes or cementocytes and was seen in different section planes, to be directly connected to the root by small bridges of cementum or collagen fibers. Small bridges of mineralized tissue also connected the glass granules to each other. The mineralized tissue surrounding the particles had histologic features very similar to root cementum. The EDS analyses showed that some granules consisted of an unreacted core, surrounded by a leached Si-rich layer and a surface layer consisting of calcium and phosphorus. This layer is chemically bonded to the cementum bridges (Figure 3).

Many particles showed an excavation of the central core with cracks connected with the external environment (Figure 4). The excavated core of the particles showed the presence of mineralized tissue.

Externally fibers running from the alveolar bone inserted on one side of the particles and, on the other side, collagen fibers connected the particles to the tooth cementum or to other particles (Figure 5). Few cells were present in this connective tissue. Macrophages were never seen in contact with the active glass.

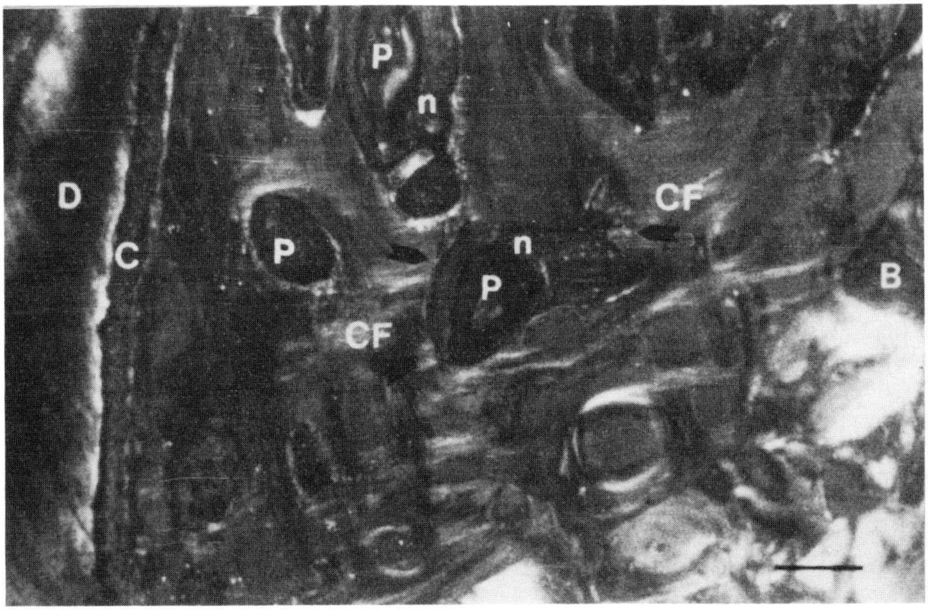

Figure 5 Histological image under polarized light. Fibers running from the alveolar bone insert on one side of the particles, while on the other side collagen fibers connect the particles to the tooth cementum. D = Dentin; C = Cementum; P = Glass Particle; n = newly formed cementum; CF = Connective fibers. (Basic fuchsin-methylene blue) Bar = 250 μm

DISCUSSION

During the drilling for the implantation of the granules, the bone wall was damaged and probably some granules were pushed into the proximity of a tooth. However it is also possible that a subsequent resorption or remodelling of the bone wall, or the new bone growth pushed some granules toward the inner part of the hole, close to the tooth area, as reported for small zirconia spheres (5). Regardless of the reason, this situation allowed us to check the behaviour of the glass in this particular tissue. No particles were found in the cortical bone. The reason for these findings could be that the particles lacked a primary stability in the surgical bed and so were not able to produce a direct bone contact. According to Mellonig one of the prerequisites for achieving direct bone contact is the primary stability of the biomaterials in bone (3,6). Moreover the particles were completely surrounded by a layer of mineralized tissue, the staining and morphological features of which resembled those of cementum. Furthermore this tissue was connected, through oriented collagen fibers, to the surrounding alveolar bone and through small cementum bridges to root cementum. The presence of these granules did not disturb the regeneration of the periodontal ligament. A striking finding was the complete coverage of glass particles with a layer of cement-like mineralized tissue. What we expected to see in a surgical site of a periodontal defect was the embedding of granules in newly formed bone, like an ankylosis, or a complete encapsulation by collagen fibers as has been observed in periodontal defects treated with HA or TCP particles (7-12). The basic differences between the studies published in the literature about periodontal defects repair and this study are in the material used, i.e. glass S53P4, and the site of the defect. The defect did not originate in the gingiva but was produced in the central part of the root without involvement of epithelium and oral cavity. The absence of the oral cavity environment involvement could explain this healing pattern of the periodontal ligament, while the bioactive properties of the glass explain the regeneration of the mineralized tissue around the particles with oriented collagen fibers and inside the excavated core; perhaps through the cracks cells penetrated into the excavated core finding a suitable environment for the mineralization.

This observation can lead to hypothesize that cementoblasts are able to produce cement-like mineralized tissue on biomaterials which do not present a primary stability, while bone cells need an almost complete absence of mobility for deposition of mineralized tissue (6). Further studies are needed to provide a better understanding of our preliminary findings, and to evaluate the possibility of a clinical application of the present results.

ACKNOWLEDGEMENTS

This work was supported by the National Research Council of Italy "Progetto finalizzato Biomateriali". The Authors thank Dr.P.Pinchiorri, Dr.D.Tanza and Miss S.Piaggi for their technical assistance.

REFERENCES

1. Shetty V., Han T.J., Dent Clin North Amer. 1991, 35, 521-529
2. Carranza F.A., Kenney E.B., Lekovic V., Talamante E., Valencia J., Dimitrijevic B., J Periodontol. 1987, 58, 682-688
3. Mellonig J.T., Int J Periodont Rest Dent 1991, 11, 217-223
4. Donath K., Breuner G., J Oral Pathol 1982;11:318-326
5. Zaffe D., Gatti A.M. in Williams K.W., Toni A., Middleton J., Pallotti G. (Eds.) Interfaces in Medicine and Mechanics, Elsevier, Amsterdam, NL, 1991,194-203
6. Hjorting-Hansen E., Worsaae N., Lemons J. Int J Oral Maxillofac Implants. 1990, 5, 255-263
7. Frame J.W., Rout P.G.J., Browne R.M. J Oral Maxillofac Surg, 1989, 18, 142-144
8. Sugaya A., Minabe M., Hori T. , J Periodont Res ,1990, 25, 60-63
9. Minegishi D., Lin C., Noguchi T., Ishikawa I., Int J Periodont Rest Dent 1988, 8, 51-63
10. Kramer G.M., Mattout P., Mattout C., Mellonig J.T. Int J Periodont Rest Dent 1989, 9, 9-21
11. Beck-Coon R.J., Newton C.W., Kafrawy A.H., Oral Surg Oral Med Oral Pathol 1991, 71, 483-488
12. Minabe M., Sugaya A., Satou H. et Al. J Periodontol 1988, 59, 671-677

Bioceramics, Volume 7 Edited by Ö.H. Andersson and A. Yli-Urpo
(*Proceedings of the 7th International Symposium on Ceramics in Medicine, Turku, Finland, July 1994*)

Bone Growth into Spaces Between 45S5 Bioglass Granules

H. Oonishi,[1] S. Kushitani,[1] E. Yasukawa,[1] H. Kawakami,[1] A. Nakata,[1] S. Koh,[1] L.L. Hench,[2] J. Wilson,[2] E. Tsuji[3] and T. Sugihara[4]

[1] Dept. of Orthopaedic Surgery, Artificial Joint Section and Biomaterial Research Laboratory, Osaka-Minami National Hospital, Kido-Cho, Kawachinagano-Shi, Osaka, 586 Japan
[2] Bioglass Research Center, University of Florida, Gainesville, FL 32611, USA
[3] Osaka Prefectural Industrial Engineering Research Institute, Enokoshima, Nishi-ku, Osaka, Japan
[4] Nitta Gelatin Inc., Research and Laboratory Center 2-22, Futamata, Yao, Osaka 581, Japan

ABSTRACT

Granules of Bioglass® 100-300 μm in diameter were implanted into 6mm diameter holes made in the femoral condyles of mature rabbits. As a control, HA granules of similar size were used. After 12 days new bone had grown into the defect to a depth of 1500 μm, equivalent to 10 layers of granules. Of these layers three or four were completely enclosed in new bone. By three weeks the whole defect was filled (to 3,000 μm) and approximately 10 layers of granules were completely enclosed. At five weeks this bone had continued to densify. In contrast only the outer layer of HA granules was surrounded by bone at two weeks although bone had infiltrated to the third or fourth layer. By three weeks bone had reached 1000 μm but even after six weeks had not reached the center, being only 2500 μm in thickness. In conclusion, the speed of bone growth around Bioglass® particles was faster and the new bone was denser than that associated with hydroxyapatite granules.

INTRODUCTION

Each type of bioactive ceramic has a different reaction in bone. Differences in bone growth behavior and reaction to the host bone vary not only with the difference in materials, but also with the different sizes of the granules.

Several types of bioactive ceramics - HAp, α-TCP, β-TCP, TeCP, OCP, etc. - have been studied in our laboratory, and results have been reported elsewhere.[1,2]

In this paper, the bone growth behavior and reaction of bone to 45S5 Bioglass® particulate are observed and compared with the reaction to HAp granules in the same rabbit tibial model.

MATERIALS

Granules of Bioglass® 100~300 μm in diameter, and HAp granules of a similar size, sintered at 1200°C were used. Mature rabbits were used for this experiment.

®Registered Trademark of University of Florida.

METHODS

Drilled holes 6 mm in diameter were made bilaterally in the femoral condyles of mature rabbits, and sufficient hemostasis was achieved by filling and compressing the holes with gauze for five minutes. Immediately following, granules of Bioglass® and HAp were placed in sufficient amounts to fill the holes in the femoral condyles, one material on each side, providing each with its own control. Three rabbits were used for each time period.

The animals were killed at intervals, and non-decalcified specimens were observed by light microscopy and backscattered electron imaging, and decalcified specimens by light microscopy.

RESULTS AND DISCUSSION

Hydroxyapatite

One week after implantation of the HAp granules, new bone entered into the spaces of the first and second layers of the HAp granules, and surrounded the periphery of the HAp granules (Fig. 1).

At three weeks, new bone growth reached the fourth and fifth outer layers of the HAp granules. New bone surrounded large parts of the surfaces of HAp granules of layers one and two (Fig. 2).

At six weeks, (Fig. 3) new bone entered into the spaces of the ten outer layers; that is to the center of the hole filled with HAp granules. New bone surrounded the surfaces of all granules of the outer three layers. However, in the deeper layers, the new trabecular bone only partially surrounded the HAp granules (Fig. 3A). In the deep layers, only a few of the spaces between the HAp granules were filled with new bone (Fig. 3B).

45S5 Bioglass®

Twelve days after implantation of the 45S5 Bioglass® granules, new bone entered into the spaces of the outer eight to ten layers of Bioglass® granules, and surrounded the most of the surfaces of the Bioglass® granules (Fig. 4).

In comparison, with HAp granules at three weeks, new bone had reached only the fourth and fifth layers, and surrounded only the larger parts of surfaces of the first two outer layers.

At three weeks with Bioglass® (Fig. 5), new bone reached the spaces of the tenth layer of Bioglass® granules, that is, to the center of the hole and surrounded most of the surfaces of the Bioglass® granules. At the center of the 6 mm defect, Bioglass® granules were surrounded by thick, new, trabecular bone.

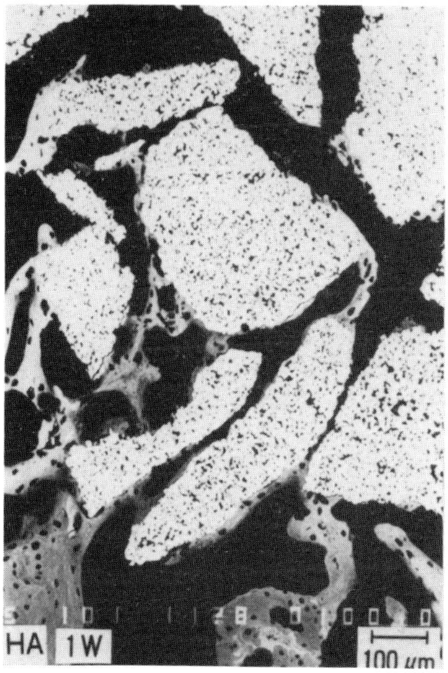

Fig. 1. HA, 1 week at the tibia periphery. Back Scattered electron image.

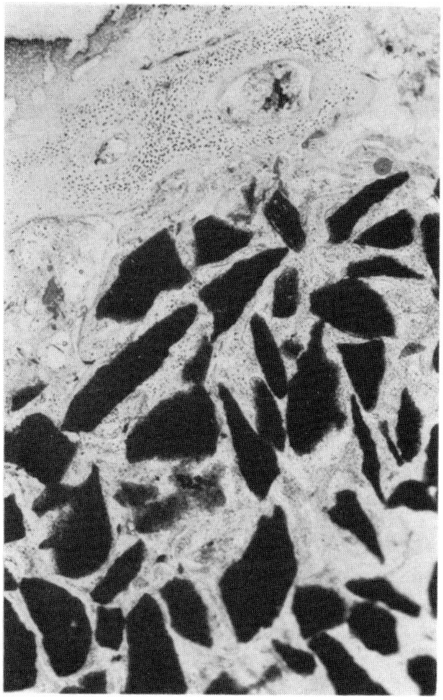

Fig. 2. HA, 3 weeks at the tibia periphery light microscopy of non-decalcified specimen, toluidine blue stain.

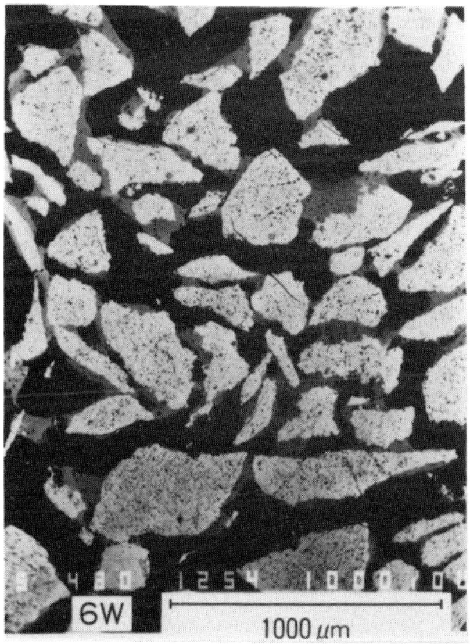

Fig. 3(A). HA, 6 weeks at center of the tibial defect. Back scattered electron image.

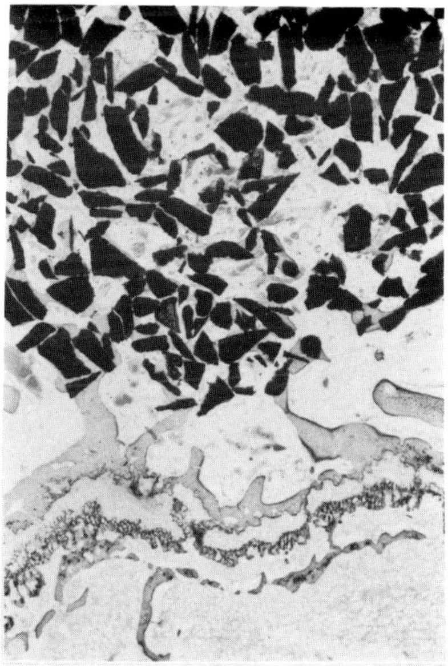

Fig. 3(B) HA, 6 weeks at center of tibial defect. Light microscopy of non-decalcified sample, toluidine blue stain.

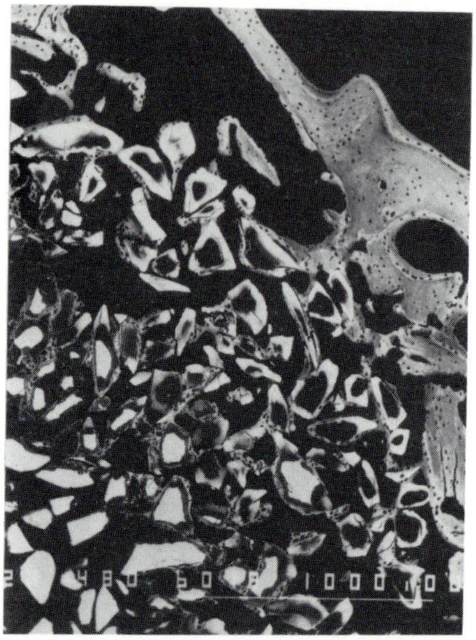

Fig. 4(A). Bioglass®, 12 days near center of tibia. Back scattered electron image.

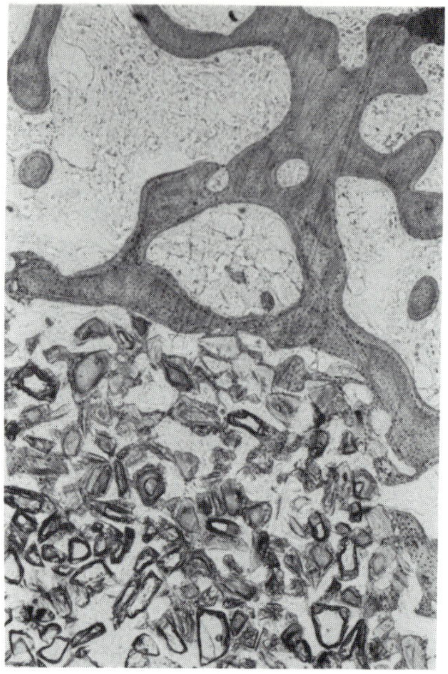

Fig. 4(B). Bioglass®, 12 days at tibia periphery. Light microscopy of non-decalcified samples, toluidine blue stain.

Fig. 5(A) Bioglass®, 3 weeks at center of tibia. Back scattered electron image.

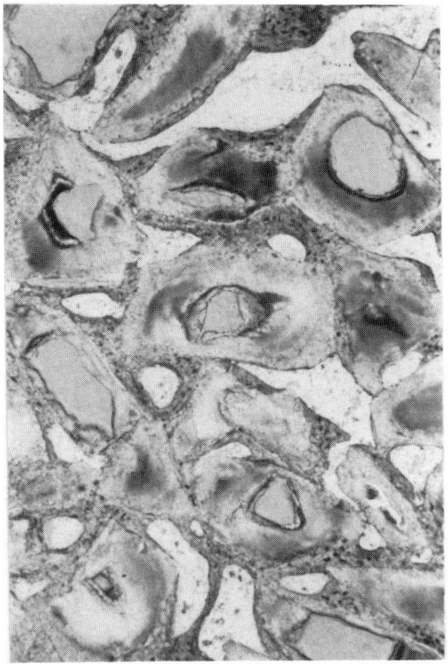

Fig. 5(B). Bioglass®, 3 weeks at center of tibia. Light microscopy of non-decalcified samples, toluidine blue stain.

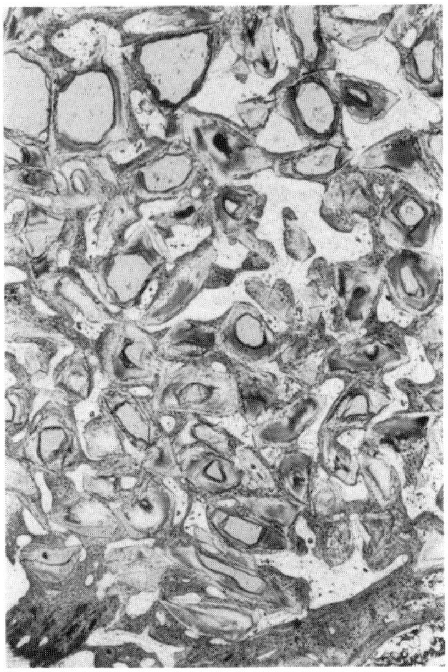

Fig. 5(c). Higher magnification of 5(B) at center of tibia showing extensive new bone formed around Bioglass® particles.

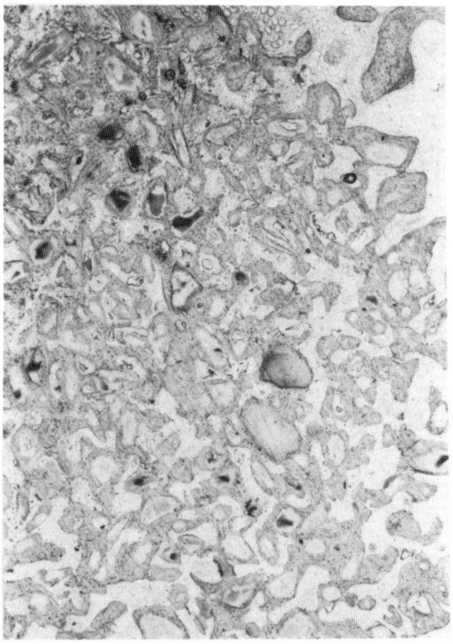

Fig. 6. Light micrograph of Bioglass® at 5 weeks showing extensive bone formation at center of the tibial defect. Non-decalcified sample, toluidine blue stain.

With HAp granules, new bone reached the center of the defect 6 weeks after implantation (Fig. 3), and new bone surrounded only small parts of the surfaces of the HAp granules. The new bone that filled the spaces between Bioglass® granules was more extensive and denser than the bone around HAp particles.

At five weeks, (Fig. 6) the bone growth between Bioglass® particles continued to increase in density and entire Bioglass® granules were surrounded with thick new trabecular bone and most of the spaces were filled with new bone.

CONCLUSION

The speed of bone growth around Bioglass® particles was much faster and the new bone was much denser than bone associated with HAp granules. These findings of rapid bone proliferation around Bioglass® granules are consistent with previous studies conducted in monkey alveolar bone[3] and canine rib bone.[4]

ACKNOWLEDGEMENTS

The authors gratefully acknowledge receiving the Bioglass® particulate from Dr. David Greenspan, U.S. Biomaterials Corp., Alachua, Florida.

REFERENCES

1. Oonishi, H., Tsuji, E., Mizukoshi, T., Kushitani, S., Aono, M. and Minami, K. In *Bioceramics 4*, W. Bonfield, G. W. Hastings and K. E. Tanner, eds., Butterworth-Heinemann Ltd., Guildford, England, 1991, 191-197.

2. Oonishi, H., Tsuji, E., Kushitani, S., Aono, M. and Minami, K. In *Bioceramics 5*, T. Yamamuro, T. Kokubo and T. Nakamura, eds., Kobunshi Kankokai, Inc., Japan, 1992, 139-146.

3. Wilson, J. and Low, S. *J. Appl. Biom.* 1992, 3, 123-129.

4. Wilson, J., Yu, L.T., Beatle, B.S. In *Bioceramics 5*, T. Yamamuro, T. Kokubo and T. Nakamura, eds., Kobunshi Kankokai, Inc., Japan, 1992, 139-146.

Bioceramics, Volume 7 Edited by Ö.H. Andersson and A. Yli-Urpo
(Proceedings of the 7th International Symposium on Ceramics in Medicine, Turku, Finland, July 1994)

Augmentation of Bonding of Porous-Coated Materials to Bone with Bioactive Ceramic Granules

T. Nakamura,[1,2] Y. Senaha,[2] T. Kitsugi,[2] T. Kokubo,[3] T. Yamamuro[2] and M. Oka[1]
[1] Research Center for Biomedical Engineering, [2] Department of Orthopedic Surgery, Faculty of Medicine, [3] Faculty of Engineering
Kyoto University, 53 Kawahara-cho, Syougoin, Sakyouku, Kyoto, Japan

ABSTRACT

Porous-coated titanium plates were implanted into the tibiae of rabbits with and without A-W glass-ceramic granules (A-W GC) inserted into the bone marrow to investigate whether the use of bioactive ceramic granules augment the bonding of bone to the porus-coated plate. The bonding strength of the plate to bone was evaluated by measuring the failure loads at 4, 8 and 24 weeks after surgery. The failure loads of the specimens with A-W GC granules and without A-W GC granules were: 9.2±1.7 kg (mean±S.D.) and 6.9±0.8 kg at 4 weeks, 9.7±2.7 kg and 6.7 ±0.6 kg at 8 weeks and 11.8±2.4 kg and 7.9±1.9 kg at 24 weeks after implantation, respectively. Histologically, bone formation was observed in the bone marrow around the A-W glass-ceramic granules and the cortical bone adhering to the plate was thicker in the specimens with A-W glass-ceramic granules than in the controls.

INTRODUCTION

Bone ingrowth into a porous surface is a potent method of fixation of metal to bone. Clinically, it has been used in the fixation of joint replacement prostheses as an alternative to PMMA cement(1,2). For fixation of porous-coated material to bone, initial immobilization and new bone formation around the material are very important and key factors for clinical success. It has been shown that new bone is formed in the bone marrow space when the granulated bioactive ceramics such as hydroxyapatite, bioglass or A-W glass-ceramic, are implanted(3).Therefore enhanced bone formation using bioactive ceramic granules implanted in the bone marrow may augment fixation of a porous-coated prosthesis. In this study, a porous-surfaced titanium plate was implanted into the proximal tibiae of rabbits with and without A-W GC granules in the bone marrow to determine the effect of the bioactive granules on the fixation strength of prosthesis.

MATERIALS AND METHOD

Porous-coated titanium plates (Kobe Steel Ltd., Kobe, Japan) with a porous layer having an average pore size of 350 μm and a porosity of 60% were manufactured using plasma-sprayed pure titanium. The plate dimensions were:

2.5 mm thick ,15 mm long and 10 mm wide. A-W GC granules (Nippon Electric Glass Ltd., Ohtsu, Japan) with an average particle size of 200 μm were made by pulverizing A-W GC produced by the method of Kokubo et al (4). The experimental procedure was performed as previous reported (5). In brief, a plate was transcortically implanted into the metaphysis of each tibia of mature white rabbits weighing between 2.5 kg and 3.0 kg. Using a dental burr a 16 mm x 2.6 mm slot was made from the medial cortex to the lateral cortex parallel to the longitudinal axis of the tibial metaphysis. The plate was implanted perpendiculary to the AP plane, perforating the tibia and protruding from both the medial and lateral cortices. Before inserting the plate, A-W GC granules were packed into the bone marrow space in the experimental group and nothing in the control group. The rabbits were sacrificed at 4, 8 and 24 weeks after surgery and the bone section containing the implant was dissected 2 mm above and below the implant by making parallel cuts with a disc cutter. The bone attached to the lateral sides and both ends of the plate was removed with a dental burr leaving bone only on the anterior and posterior surface of the titanium plate. The specimens were positioned in a special holder for testing in an Instron-type machine (Autograph 500, Shimadzu, Japan). The tensile load required to break the bone or the interface between the bone and plate was determined as the failure load. After mechanical testing the samples were fixed in a 10% phosphate-buffered formalin solution, dehydrated in serial concentrations of alcohol and then embedded in polyester resin for histological examination. Undecalcified 50μm thin sections were made using a band saw and grinding machine (BS-3000 and MG-4000, Exakt, Hamburg, Germany). Then, contact microradiographs were taken and the samples stained with Giemsa surface staining after which the bone-implant interface was observed using light microscopy.

RESULTS

The failure load results are presented in Figure. 1 and show that the failure load increased with length of implantation. At all intervals after surgery the mean failure loads of the samples with A-W GC granules were higher than those without A-W GC granules. The mean failure loads of samples with A-W GC granules and without A-W GC granules were: 9.2$\pm$1.7 kg (mean$\pm$S.D.) and 6.89$\pm$0.8 kg 4 weeks after implantation; and 11.8$\pm$2.4 kg and 7.9$\pm$1.9 kg 24 weeks after implantation, respectively. At each time interval the differences were statistically significant.

The contact microradiograms and Giemsa surface staining showed active bone formation around the edges of the dissected cortical bone in both groups. In the bone marrow, active bone formation was observed around the A-W GC granules, whereas no bone formation was observed in this region in specimens without A-W GC granules (Figure.2,3) Bone formation was most active at 4 weeks after surgery. At 8 and 24 weeks after surgery remodeling of the bone tissue was observed. In the specimens with A-W GC granules the width of the cortical bone attached to the plate was larger than in those without the granules (Figure 4, 5). Compared to the samples at 4 weeks those at 8 and 24 weeks showed a reduced number of A-W GC granules.

DISCUSSION

Bone ingrowth into the pores of a porous-coated material is an important factor in its fixation to bone. However, in recent research on retrieved prostheses bone ingrowth was observed only on a small percentage of the porous surface(6). Thus, increased bone ingrowth may be essential for permanent fixation of the prosthesis. In this region bone ingrowth is very difficult due to the lack of bone tissue. In this region there are preosteoblast cells which are capable of differentiating into osteoblasts. In the presence of bioactive ceramics these preosteoblast cells are known to differentiate into osteoblasts to produce bone tissue on (7). Therefore implantation of a bioactive ceramic into the bone marrow may enhance bone formation. As A-W GC has a greater ability to conduct bone in marrow than dense hydroxyapatite(1), it was the most suitable material to use in testing this hypothesis.

In this study, bone formation in the bone marrow space was observed only in specimen with the A-W granules. In addition, in the region of the porous surface within the marrow space, bone ingrowth was observed only in the specimens with A-W GC granules. This bone may have a role in enhancing the bonding strength between the plate and bone as shown by the mechanical test. At 8 and 24 weeks after surgery the newly-formed bone was remodeled and the new bone at the interface between the cortices and the plate changed into cortical

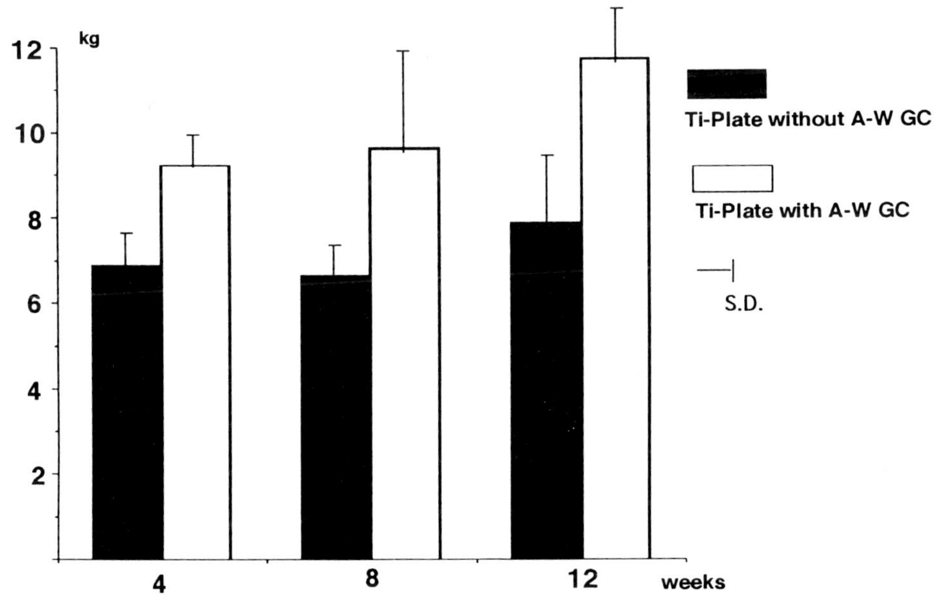

Figure 1 Failure loads at 4, 8, and 24 weeks after surgery.

bone in both experimental and control groups. However, in the group with A-W GC granules, the additional bone around the A-W GC granules fused with the cortical bone and widening of the cortical bone attached to the porous plate was observed. These histological findings are compatible with the results of the mechanical test.

In this study, the plate was in a non-loaded configuration and the results cannot be directly applied to the loaded situation. However, the results suggest the possibility that increased bone formation due to the presence of bioactive ceramic granules clinically may enhance prothesis fixation.

In conclusion, these results show that implantation of bioactive A-W glass-ceramic granules with a porous-coated plate can augment fixation. The increased failure load at 4 weeks may be partially due to immobilization by the granules, however, at 8 and 24 weeks it appears chiefly due to ne bone formation.

Acknowledgements

The contribution of materials by Kobe Steel Ltd. and Nippon Electric Glass Ltd. is greatly appreciated.

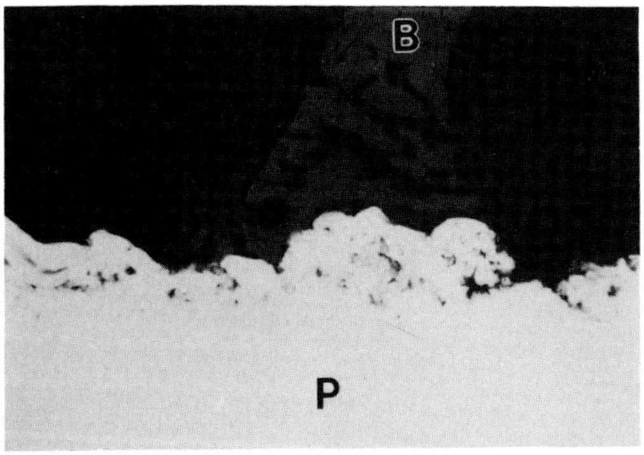

Figure. 2
Contact microradiogram of the plate without A-W GC granule at 4 weeks. original magnification x40;
P: plasm-sprayed titanium plate, B:original cortical bone

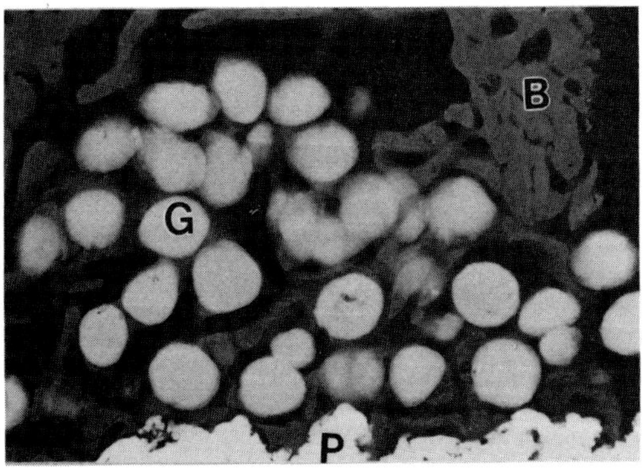

Figure. 3
Contact microradiogram of the plate with A-W GC granule at 4 weeks. original magnification x40;
P: Plasm-sprayed titanium plate, B:original cortical bone, G: A-W GC granule

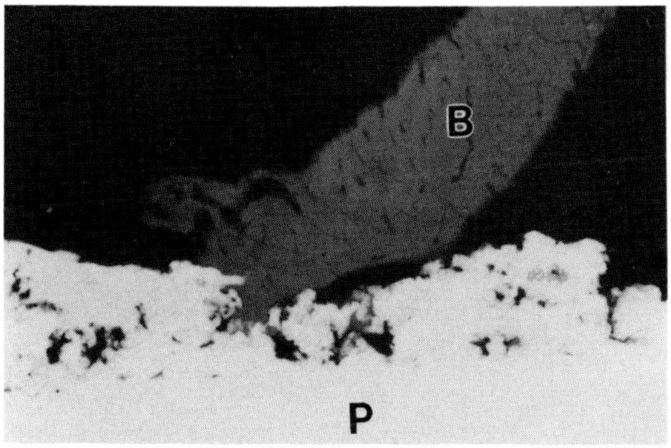

Figure. 4
Contact microradiogram of the plate without A-W GC granule at 8 weeks. original magnification x 40,
P: plasm-sprayed titanium plate, B:original cortical bone

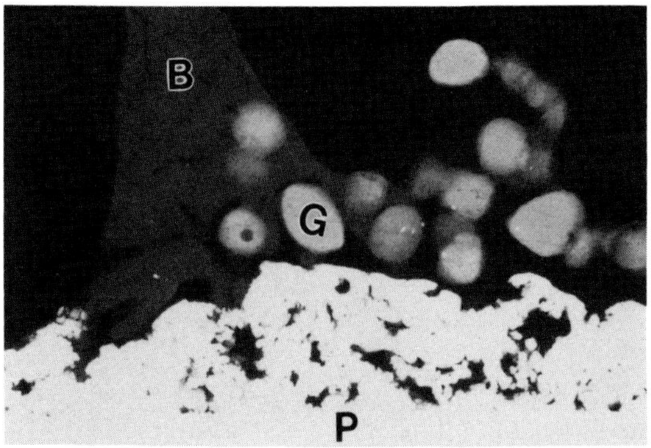

Figure. 5
Contact microradiogram of the plate with A-W GC granule at 8 weeks. original magnification x 40
P: plasm-sprayed titanium plate, B:original cortical bone, G: A-W GC granule

REFERENCE

1. Lord, G.A., Hardy, J.R. and Kummer, F.J. Clin. Orthop., 1979, 141, 2- 16
2. Engh, C.A., Bobyn, J.D. and Glassman, A.H. Porous coated hip replacement. J.Bone Joint Surg.(Br), 1987, 69, 45-55,
3. Ono, K.,Yamamuro,T., Nakamura, T. and Kokubo, T. Biomaterials, 1990, 11, 265-271
4. Kokubo, T.,Shigematu, M., Nagashima, M., Tashiro, H., Nakamura, T., Yamamuro, T. and Higashi, S. Bull. Inst. Chem. Res. Kyoto Univ.,1982, 60, 260-268
5. Nakamura, T., Yamamuro, T., Higashi, S., Ito, S. and Kokubo, T. J. of Biomed. Mater. Res., 1986,19, 685-698
6. Cook, S.D., Thomas, K.A. and Haddad, R.J.Jr. Clin. Orthop., 1988, 232, 90-101
7. Ohgushi, H., Okumura, M., Tamai, S., Shors, E.C. and Caplan A.I. J. of Biomed. Mater. Res., 1990, 24,1563-1570

Bioceramics, Volume 7 Edited by Ö.H. Andersson and A. Yli-Urpo
(*Proceedings of the 7th International Symposium on Ceramics in Medicine, Turku, Finland, July 1994*)

Bioceramics in the System CaO-Nb_2O_5-P_2O_5

L. Berzina, R. Cimdinš, J. Vētra,[1] M. Dobelis and O. Diktenko
Department of Chemical Technology, Riga Technical University, 14 Azenes St., LV-1048, Riga, Latvia
[1] Latvian Medical Academy, 16 Dzirciema St., LV-1007, Riga, Latvia

ABSTRACT

Investigation of the structure and properties of glasses in the CaO-Nb_2O_5-P_2O_5 system revealed that niobium ions substantially influence these characteristics. With increasing the content of Nb_2O_5 increases the density and chemical durability of glasses. *In vivo* studies of glass-ceramics (GC) of this system confirmed their biocompatibility with soft tissues in white rats. Morphological studies of the implants surface showed that the interaction between the implant and living body takes place. To promote ions exchange between the implant and natural bone GC were modified with 2 to 10 mol% Na_2O. The optimal composition with addition of 8 mol% Na_2O was implanted into the tibia of 12 rabbits. After 7 month of implantation the implants were removed and subjected to micro X-ray analysis to detect the distribution of chemical elements in the interfacial zones.

INTRODUCTION

Metallic niobium is known to be used as an endoprosthetic material. The biocompatibility of niobium implant with the bone is due to the formation of thin niobium oxide layer as a result of interaction between the implant and living body [1]. During the last years extensive investigations have been carried out to establish the possibility of using glasses as biomaterials. In 1969 it was discovered by Hench [2] that glasses of the Na_2-CaO-SiO_2-P_2O_5 system can bond chemically to natural bone. Promising bioceramics are glasses of the CaO-SiO_2-Al_2O_3, CaO-SiO_2-TiO_2, CaO-SiO_2-P_2O_5 systems investigated by Kokubo et al [3]. Gross et al [4] showed that numerous bioactive glasses and GC compositions (Ceravital) with low alkali content form a good bond with bone. They revealed that small additions of Al_2O_3, Ta_2O_5, TiO_2, ZrO_2 in glasses act as inhibitors in bone bonding. Vogel et al developed machineable mica containing phosphate GC [5]. Numerous compositions of bioactive glasses designed as coatings for dental implants, have been developed at Turku Abo Academi [6]. In this work an original glass forming system CaO-Nb_2O_5-P_2O_5 was chosen to develop biomaterials. The bioactivity of calcium and phosphorous ions is proved [2,3]. It is known that niobium ions strengthen the structure of

phosphate glasses due to the high charge and small radius increasing the mechanical strength and chemical durability.
In the CaO-Nb_2O_5-P_2O_5 system formation of glasses and crystalline phases and their structure and properties are investigated. By suitable heat-treatment from initially prepared glasses GC are developed. The biological grow together of GC with bone provides the strong ionic bond between these materials. To promote the ion exchange between the living body and bioceramic implants Na_2O is added to glasses of the CaO-Nb_2O_5-P_2O_5 system. Although the investigations in the area of bioactive glasses are widely performed, the method of detection bioactivity and the influence of various factors on it are far from being settled. *In vitro* and *in-vivo* investigations of materials obtained were also performed.

METHOD AND RESULTS

For glasses melting as raw materials were used $NH_4H_2PO_4$, $CaCO_3$, Na_2CO_3 and Nb_2O_5. Glasses were molten in alumina crucibles for 1 h at 1400°C by using an electric furnace. By adding 2-10 mol% Na_2O to promote ions exchange the temperature of melting decreased to 1200°C. To eliminate the loss of P_2O_5, the batches were preliminary heated for 4 h at 600-700°C.

Selected glass compositions were subjected to chemical analysis. It was estimated that alumina slightly dissolved and made a contribution of 2.4-2.9 mass% to the total weight. Two glass forming regions were found in the CaO-Nb_2O_5-P_2O_5 system (Figure 1). The first one with 15 mol% Nb_2O_5 content leans to the binary CaO-P_2O_5 system. The second one is with a lower P_2O_5 content (30-40 mol%) and with a higher concentration of Nb_2O_5 (25 mol%).

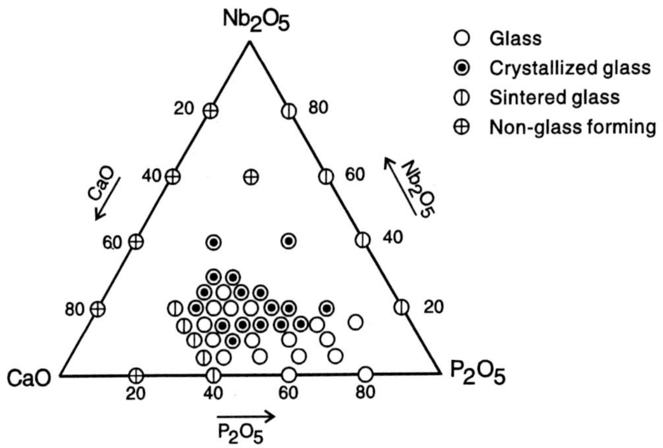

Figure 1 Glass forming in the CaO-Nb_2O_5-P_2O_5 system.

The difference in the structure of glasses of these two regions has been characterised in IR-spectra (Figure 2). Decreasing of the absorption band at

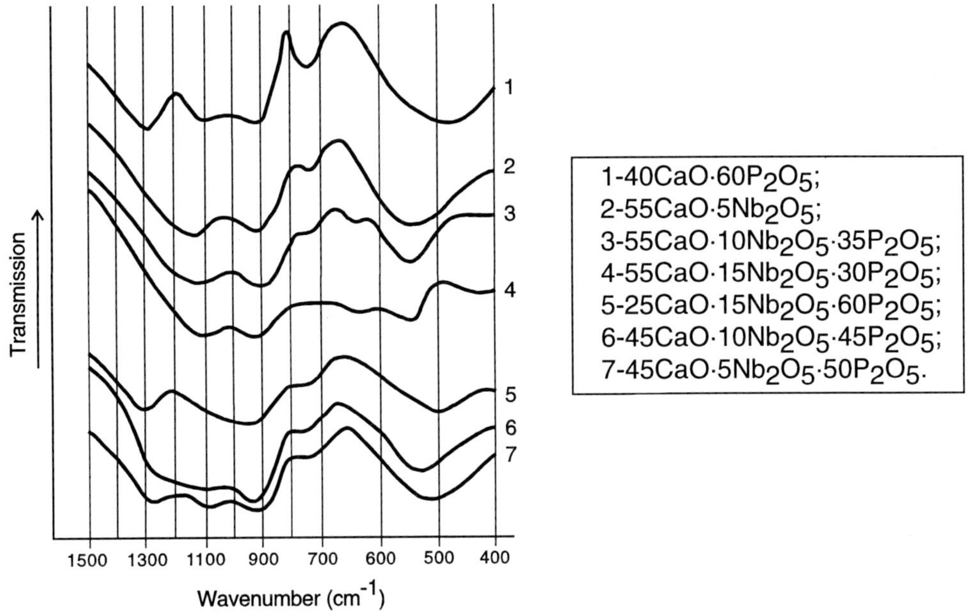

Figure 2 Infrared transmittance spectra for glasses.

750cm^{-1} and disappearance it at 1300 cm^{-1} is due to the change of the bonds in the PO_4 tetrahedron and partial substitution of phosphorous to niobium. The absorption at 640 cm^{-1} corresponds to the NbO_6 groups, which connect the neighbour phosphate groups via oxygen ions. With increasing the content of Nb_2O_5 from 5 to 25 mol% the density of glasses increases from 2.7 to 3.4·10^3 kg/m^3, but microhardness from 3,6 to 5,4 GPa. The chemical durability of the above shown glasses is influenced by the Nb_2O_5 content and CaO/P_2O_5 ratio. Nevertheless, chemically more resistant glasses have higher Nb_2O_5 content. All glasses investigated tend to crystallise. For GC processing according to crystallisation studies two glass compositions were chosen: 45CaO 5Nb_2O_5 50P_2O_5 (2Nb) and 45CaO 15Nb_2O_5 30P_2O_5 (11Nb). GC were produced by powder technology. The main crystalline phase of GC 2Nb is $Ca(PO_3)_2$, but of GC 11Nb - $Ca_2P_2O_7$. GC 11Nb has an advantage to GC 2Nb in producing and machining. GC 11Nb characterises with fine crystalline structure and relatively wide interval of the thermal treatment. GC 2Nb has higher shrinkage and short interval of crystallisation.

In-vitro. Tests of GC were performed for one month in distilled water at 70°C. The weight ratio of sample/water was 1:10. Water extract was removed by static and dynamic methods and analysed every week. After the month pH of water extract of GC 2Nb decreased from pH7 to pH3, but pH of GC 11Nb remained unchanged. In the both water extracts of GC increased concentration of Ca and

P ions was revealed while the concentration of Nb ions remain constant. The ion concentrations were detected by atom absorption spectroscopy.

In vivo. Investigations of GC 2Nb and 11Nb were carried out by implanting 4 parallel samples (1 mm thick and 8 mm in diameter) in thigh muscles of white rats. The samples were retrieved at intervals of 1, 2, 4, 6 and 7 months. During autopsy the tissue reaction around implants and changes in heart, lungs, kidneys, liver and spleen were evaluated visually and histologically. Histological samples were stained by hematoxyline, eosin picrofuxin after Veigert, and after Perls. The thickness of fibrous capsule, the presence of foreign body cells and micro implant parts, and toxic changes in leukocytes were evaluated using optical microscope. After 1 and 2 months of the implantation time of GC a fibrous capsule was observed: GC 2Nb 80μm in size and GC 11Nb 150μm. The size of the capsule decreased with the time. After 6-7 months it decreased to size 50 μm for GC 2Nb and practically disappeared for GC 11Nb. No aseptic irritation and macroscopic changes of the internal organ's tissue were visible during the whole period of implantation. The same samples preheated at 500°C were used for investigation of GC surface morphological changes using SEM JSM 35. GC 11Nb surface before implantation had irregular structure - on the smooth GC surface separate globules covered with smooth grain were detected (Figure 3).

a) b)

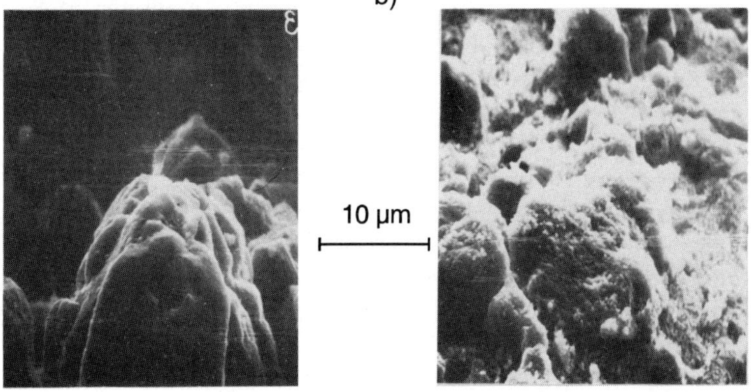

Figure 3 GC surface separate globules before (a) and after (b) implantation.

After implantation on the smooth grain surface traces of biodegradation were seen and the surface of globules was a little swelled. According to X-ray micro analysis data Ca and P ions concentration before and after implantation on 11Nb GC smooth surface was without essential changes. On the globule surface and in the places of biodegradation after implantation Ca and P concentration have decreased approximately 2 times. X-ray diffraction patterns and infrared spectra of these surfaces before and after implantation were identical. It means that different ions concentration could be related to GC 11Nb glass phase partial solution in a living body.

Modification. The optimal content of modifier was defined. It was shown that bending strength in 4-point test can be increased by adding up to 8 mol% Na_2O. At 10 mol% Na_2O the bending strength of GC rapidly decreased. In further investigations GC modified with 8 mol% Na_2O (mark 4N) was used to establish the relationships between the bending strength, temperature of the synthesis and cooling rate. Prismatic samples were pressed under pressure of 400 MPa. Twenty samples were sintered at 8 of each temperature values from 770°C to 840°C with the step 10°C. Half of the parallel samples were cooled with rate 2° C/min while the rest of them - with rate 20°C/min. Finally samples were grounded to the size 3x5x30 mm. The total number of prepared samples was 160.

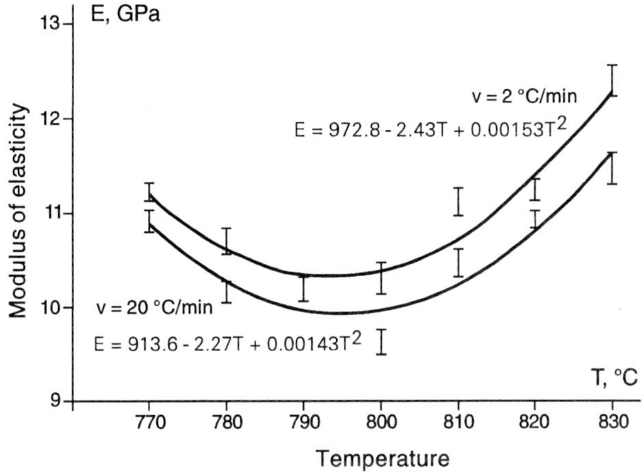

Figure 4 Relationships between modulus of elasticity of GC 4N composition and the syntering temperatyure and cooling rate. Bars represent mean quadratic error of average values.

Interaction. For the study of GC 4N and osseous tissue interaction a truncated cylindrical specimens (diam. 5mm, length 4,5mm) were produced. After 7 months of implantation in rabbit tibia the concentrations of chemical elements in the interfacial zones GC-bone callus and bone-bone callus were detected (Figure 5).

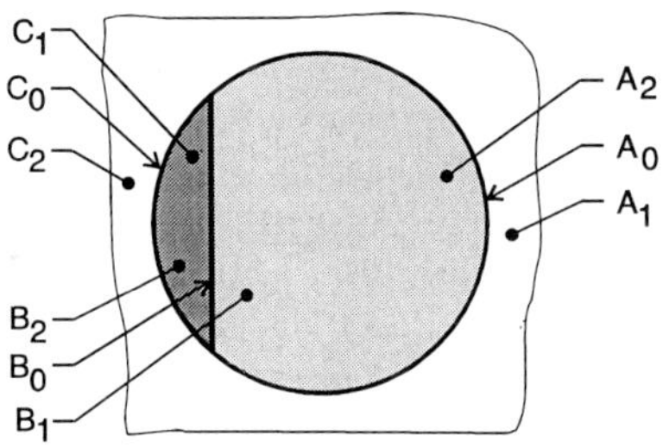

Figure 5 Microregions of the investigated interfacial zones:
bone-GC (A_1, A_0, A_2),
GC-bone callus (B_1, B_0, B_2),
bone callus-bone (C_1, C_0, C_2).

Cracks in the GC implant and in the interfacial zone bone callus-bone developed during the preparation process. The most clearly visable is the interfacial zone GC-bone (5-8 μm) of the region A. Crack is observed in the all the region. There were no possibilities to find such areas of interfacial zone with statical distribution of chemical elements.

In the interfacial zone natural bone-bone callus (region C), as proved in the morphological analysis, exists a longditudal crack. But the interfacial zone was denser than in the region B. This made reliable testing of the chemical element distribution in this area possible. Four parallel measurements were carried out in each point, totally 36 measurements. The results are shown in the Table.

Table Concentration of chemical elements (%) in the investigated interface zones

Zones	P	Na	Ca	Nb	Ca/P
4N	13.86	3.68	10.84	6.92	0.77
A_1	13.94	4.04	9.96	7.30	0.71
A_0	8.36	0.72	11.93	2.23	0.78
A_2	13.86	3.34	10.60	6.95	0.76
B_1	14.09	2.54	11.23	6.64	0.79
B_0	n	n	n	n	n
B_2	16.34	0.31	22.40	0.79	1.37
C_1	12.37	?	20.92	0.88	1.69
C_0	14.57	?	22.64	0.79	1.59
C_2	15.86	?	24.18	?	1.52

? - concentration is less than sensitivity of the determination method
n - obtained results are not statistically significant

Analysis of the results shows a correspondence between those results and the known values in the C_2 bone point as well as the A_2 and B_1 points at the GC implant; the values being natural bone (Ca/P=1.67) and GC 4N (Ca/P=0.78). On the other hand, the results of the measurements taken at point A1 do not correspond to the known values. The measurements taken at point A1 were reliable, but needed to correspond to those of the interzones with GC, which they did not. This demonstrates that the point was incorrectly chosen.

The Ca and P correspondence values found in region B0 (Ca/P=1.37) compared with these of C_0 (Ca/P=1.69) are lower. This prevents a more extensive exchange of ions between the natural bone and the bone callus than between glass-ceramic and the bone callus.

From the comparison of Na ion concentration in the implant (3.34), in the interzone (0.72) and the bone callus (0.31), it follows that Na ions are taking part in the process of exchange between the implant and the bone callus. Future experiments have to determine weather enough Na ions are taking part in order to supply the strong bond needed between the bone and the implant. Furthermore, the dynamics of the migration processes of Ca, P and Na ions must be studied in much longer glass-ceramic implantation periods of one or more years.

CONCLUSIONS

Biocompatible glass-ceramics (GC) are developed on the base of glasses of the $CaO-Nb_2O_5-P_2O_5$ system. The migration of Ca and P ions is an evidence of interaction between GC and soft tissue. GC are modified by Na_2O. The optimal composition GC 4N was estimated by the testing of mechanical properties. The results of 7 month implantation studies in rabbits tibia shows the change of concentration of Na and Ca ions, connected with interaction between the natural bone and GC.

1. Plenk,H., Pfluger, G., Schneider, S., et.all. In Transactions of the Seventh Annual Meeting of the Society for Biomaterials New-Yourk, USA, 1981, 40.

2. Hench, L. J. Amer. Ceram. Soc. 1991, 14(7), 1487-1510.

3. Kokubo, T., Ito, S., Sakka, S. and Yamamuro, T. J. Mater. Sci. 1986, 21, 536-540.

4. Gross, U.M. and Strunz, V. J. Biomed. Mater. Res. 1985, 19, 251-271.

5. Holand, W., Vogel, W., Neumann, K. and Gummel, J. J. Biomed. Mater. Res. 1985, 19, 303-312.

6. Andersson, O.H., Karlsson, K.H. and Kangasniemi, J. J. Non-Cryst. Solids 1990, 119, 290-296.

7. Bērziņa, L. In Glass-Ceramics in the Calcium and Phosphorus containing Systems, PhD. Theses, Riga Technical University, Latvia, 1989.

Bioceramics, Volume 7 Edited by Ö.H. Andersson and A. Yli-Urpo
(*Proceedings of the 7th International Symposium on Ceramics in Medicine, Turku, Finland, July 1994*)

Sol-Gel-Produced Ti-Coatings on Titanium Implants in Bone

C. Müller-Mai, I. Kangasniemi,[2] P. Li,[2] A. Yli-Urpo,[2] C. Voigt, K. Kandilakis[1] and U. Gross[1]

[1] Department of Traumatology and Reconstructive Surgery and Institute of Pathology, Klinikum Steglitz, Freie Universität Berlin, Hindenburgdamm 30, 12200 Berlin, Germany
[2] Institute of Dentistry, University of Turku, Lemminkäisenkatu 2, 20520 Turku, Finland

ABSTRACT

Sol-gel-produced Ti-Coatings on titanium Implants have been investigated for their bioactive behaviour in a bony implantation bed at 84 and 168 days after implantation by applying light microscopy, histomorphometry, and tensile tests of the interface. Implants were smooth to avoid mechanical interlocking. At both time intervals implants were surrounded by a bone frame separating the implants from the epiphyseal bone trabeculae. Focal bone contacts were recorded between areas of osteoid contact. There was no tensile strength of the interface. The amount of bone in contact to the implants was higher as compared to commercial pure Ti implants with similar surface morphology. Thus, the implants seem to be biocompatible as other Ti implants, but they do not express bioactivity which allows the establishment of interfaces being resistant against tensile forces.

INTRODUCTION

Load-bearing Ti implants are widely used in bone surgery. The material can be used under load-bearing conditions and is almost inert, i.e. it is one of the materials being most highly resistant to corrosion and allows direct bone contact or even bone connection due to chemical bonds (1). Ca/P-deposits have been observed on Ti implants without intervening soft tissue (2,3). An unsolved problem is the aseptic loosening of loaded Ti implants, e.g. hip prostheses. This problem could probably be solved if the implants would be bone bonding. Recently it was shown that titanium implants coated with sol-gel-produced Ti-layers formed an apatite like layer on their surfaces in simulated body fluid in vitro (4). This points to the possibility in achieving a strong chemical bond between such implants and living bone as known from bioactive glasses, glass-ceramics, and other bioactive materials (5-8). The bonding relies on the active formation of a carbonated hydroxyapatite layer on the implant surface (9). It is still unclear, whether such sol-gel Ti-coated Ti implants will bond to bone in vivo. Therefore, the aim of the present investigation was to test sol-gel-produced Ti-coatings on Ti implants in a standardized animal model (10) by applying light microscopy (LM), histomorphometry, and tensile tests of the interface. All implants had smooth surfaces (roughness $< 1\ \mu m$) to avoid mechanical interlocking.

METHOD AND RESULTS

Method

Sol-gel Ti-coatings on cylindrical commercial pure Ti implants (8 mm in length, 4 mm in diameter, surface roughness < 1 μm) were produced by a dipping technique as described previously (4). In brief, the following substances were mixed to produce a Ti-sol-solution: tetraisopropylorthotitanate ($Ti(OC_3H_7i)_4$), anhydrous ethanol (C_2H_5OH), ethylenmonoethylether ($C_4H_{10}O_2$), hydrochloric acid (HCl) and distilled water. The plugs were dipped two times into the solution and were withdrawn at a speed of 10-15 mm/min. The coated plugs were then fired at 600°C for 10 min.. This procedure provided a coating of about 2 μm thickness containing many OH groups as described earlier (4). The implants were sterilized for 1 hour at 180°C in dry heat and were inserted press-fit below the patella sliding plane into the trabecular bone of female Chinchilla rabbits. A total of 21 implants was used. Ten implants were explanted at 84 days and at 168 days. At each time period it was intended to use 6 implants for tensile tests and 4 for LM and histomorphometry. The surface morphology of one implant was analysed prior to implantation in the scanning electron microscope.

The rabbits were kept in pairs in wire-cages under standardized conditions as described earlier (10). The implantation was under general anesthesia as described in detail previously (7). The animals received 20 mg Gentamycine and 0.5 ml Aubikal (Atarost, 2832 Twistringen 1, Germany) before operation. There was no infection. The animals were sacrificed in general anesthesia and the abdominal aorta was cannulated and perfused with 10 mg Regitin for 30 seconds (Ciba, Wehr, Germany) and 5% buffered formaldehyde solution according to Lillie using pH 7.3 and 100 ml perfusion solution per animal.

The preparation for LM specimens has been previously described in detail (10). In brief, the fixed specimens were rinsed in tap water, dehydrated in ethanols, chloroforme/methanol 50/50, infiltrated with PMMA monomer, and polymerized at 40°C in an oven. The blocks were then microtomed and the sections stained either by the Giemsa mixture or the von Kossa-Fuchsin staining procedure, shortly dehydrated in ethanols and mounted in corbit balsam under a glass cover slip. The length of bone, osteoid, chondroid, and soft tissue in contact with the implant circumference was measured in four sections by means of histomorphometry and the values reported as percent of the total implant circumference in the section evaluated.

The tensile test procedure was intended to be carried out as described previously (10). After rupture of the implant tissue interface the tissue sides were embedded for LM as described above or in Spurr's resin as described earlier (7).

Results

The implant surface prior to implantation was smooth as intended and appeared amorphous in higher magnifications in the scanning electron microscope. There were some tiny clefts in the coating of about 1 μm in width and up to 10 μm in length (Fig.1). The coating was homogeneous all over the cylinder's surface.

In LM, the most striking phenomenon was the development of a bone frame surrounding the implants at 84 and 168 days of implantation (Fig.2). The frame contacted the implant surface and separated the implant from the epiphyseal bone trabeculae. There was a complete bony encapsulation of the implants. The frame seemed to be thicker at 168 days. In higher magnification of the interface, the bone frame and implant surface were partially separated by a few μm wide seam of osteoid and focally chondroid (Figs.3,4). These tissues covered significant amounts of the implant surface (9 and 11.5 % at 84 and 168 days, respectively)(Table 1). Mineralized bone contacted the implants focally (20,7 and 31 % at 84 and 168 days, respectively) and other parts of the implant circumferences were in contact to soft tissues (Table 1)(Fig.4). There was more bone in the interface at 168 days. The arrangement of different tissues in the interface was identical with regard to the two time intervals.

At both time intervals there was no tensile strength in the interface. Five of the 168 days implants and two at 84 days were separated from the tissue in contact to the distal implant circumference without any force, so that the tensile testing procedure could not be carried out. Therefore, the remaining implants were used for LM and histomorphometry. Thus, a total of eight implants at 84 days and of five at 168 days was used for LM.

DISCUSSION AND CONCLUSIONS

The mineralization proceeded to the sol-gel Ti-coated Ti implant surface in the same manner but to a higher extent as known from commercial pure Ti implants. A bony encapsulation comparable to the frame observed here was already observed around other metal implants (3,11). Such a reaction was considered to reflect the biomechanical interaction between the implant and the surrounding tissue and was interpreted as the result of forces and strains around the implant, i.e. that load is not transmitted by the implant (11).

The relatively high amount of osteoid which did not decraese with time indicates a delayed mineralization, since high amounts of osteoid do not occur in contact to bioactive materials at 84 days, e.g. bioglass. On the other hand, less bone in contact to commercial pure Ti implants with smooth surfaces was observed (10 and 15 % at 84 days, respectively)(10,11). This points to a higher bone formation rate at sol-gel Ti-coated Ti implants as compared to non coated specimens. The amount of osteoid was comparable to other Ti implants (11).

In conclusion, because of the delayed mineralization and the failure in the tensile testing procedure the sol-gel-produced Ti-coatings of the present investigation must be considered to be not bioactive. The amount of bone in contact to the implant or the type of implant-bone contact did not allow for the establishment of a resistance against tensile forces. This might be related to the very smooth surface morphology or to the coating thickness of 2 μm which might have been not strong enough to enhance the bone formation as required. In a former study a coating thickness of 5 μm was used (4). Smooth surfaces are known to delay the development of bone tissue as compared to rough surfaces (3,11). Therefore, the in vitro formation of an apatite like layer in simulated body fluid does not guarantee the formation of bone bonding in vivo. Further developments of sol-gel coatings are necessary to improve the surface properties of implants and to provide bone bonding properties and to explain the mechanism of delayed mineralization.

Table 1

Percentage of bone, osteoid, chondroid and soft tissue ± standard error of the mean (SEM) in contact to sol-gel Ti-coated Ti implants. d = days, n = number of implants. * Measurements not completed.

d	n	bone	osteoid	chondroid	soft tissue
84	6*	20.7 ± 3	8.8 ± 1	0.2 ± <1	70.3 ± 4
168	2*	31.0 ± 2	11.5 ± 1	0 ± 0	57.5 ± <1

REFERENCES

1 S.R. Morrison, in: Electrochemistry at Semiconductor and Oxidized Metal Electrodes, chap. 2, The Solid/Liquid Interface, Plenum Press, New York, 1980.
2 Albrektsson, T. in: CRC Critical Reviews in Biocompatibility, Vol. 1, CRC Press, Boca Raton, Fl, USA, 1984, pp 53-84.
3 Müller-Mai, C., Voigt, C., Ashrafi-Habibabadi, B., and Gross, U.M. to be submitted.
4 Li, P. PhD Thesis, Leiden, The Netherlands, 1993.
5 Hench, L.L., Splinter, R.J., Allen, W.C., and Greenlee, T.K. J. Biomed. Mater. Res. Symp. 1971, 2, 117-141
6 Gross, U.M. and Strunz, V. J. Biomed. Mater. Res. 1980, 14, 607-618
7 Müller-Mai, C.M., Voigt, C., and Gross, U.M. Scanning Microsc. 1990, 4, 613-624
8 Neo, M., Nakamura, T., Ohtsuki, C., Kokubo, T., and Yamamuro, T. J. Biomed. Mater. Res. 1993, 27, 999-1006
9 Daculsi, G., LeGeros, R.Z., Heughebaert, M., and Barbieux, I. Calcif. Tissue Int. 1990, 46, 20-27
10 Müller-Mai, C., Schmitz, H.J., Strunz, V., Fuhrmann, G., Fritz, T., and Gross, U.M. J. Biomed. Mater. Res. 1989, 23, 1149-1168
11 Gross, U., Müller-Mai, C., Fritz, T., Voigt, C., Knarse, W., and Schmitz, H.J. in: Clinical Implant Materials, G. Heimke, U. Soltész, A.J.C. Lee (eds.), Advances in Biomaterials, Vol. 9, Elsevier, Amsterdam, The Netherlands, 1990, pp 303-308

Figure 1 Implant surface prior to implantation with smooth, amorphous appearence and tiny clefts of less than a μm in width and several μm in length. SEM, bar 10 μm.

Figure 2 Implant at 168 days after insertion (black, bottom) with bone frame separating the implant and the trabeculae of the epiphyseal bone. LM, von Kossa-Fuchsin staining, bar 200 μm.

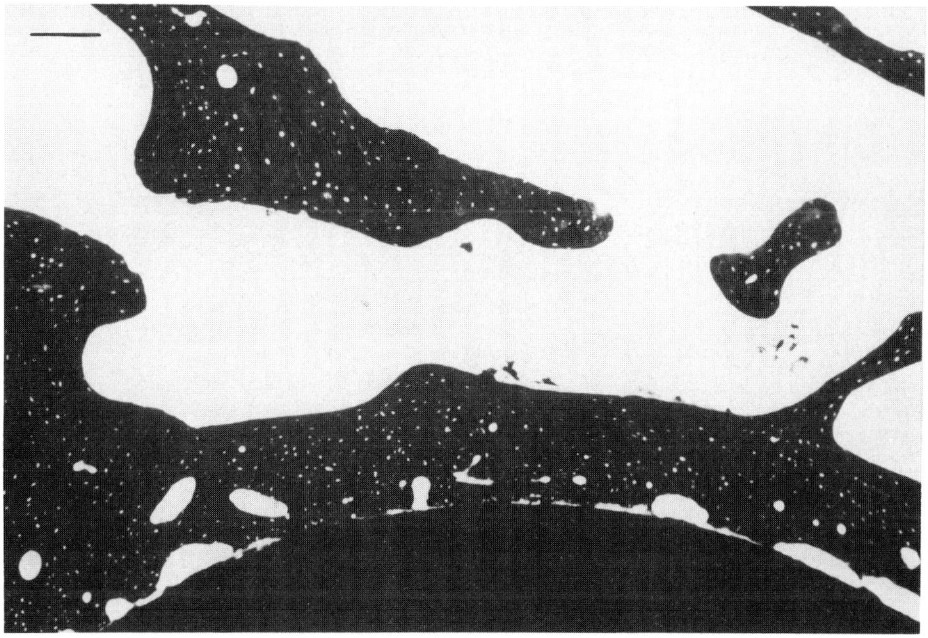

Figure 3 Higher magnification of an implant tissue interface at 84 days with osteoid (O) and chondrocytic cells (arrows) in contact to the surface. LM, Giemsa staining, bar 20 μm.

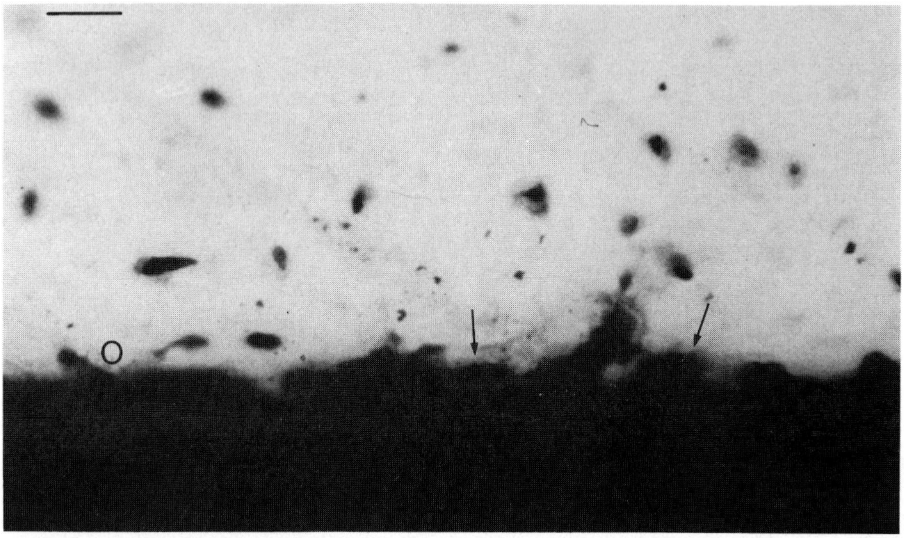

Figure 4 Semithin section of an implant at 168 days after implantation. Focal bone contacts (B), soft tissue (S) and osteoid (O, black) in the interface. Implant removed prior to sectioning. Bar 100 μm.

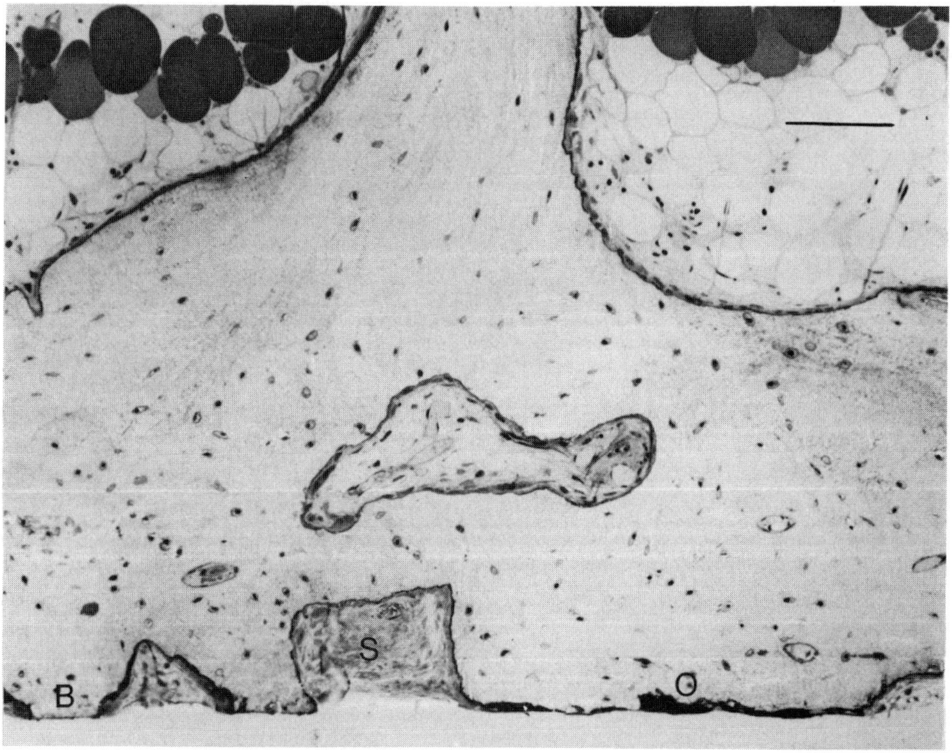

Bioceramics, Volume 7 Edited by Ö.H. Andersson and A. Yli-Urpo
(*Proceedings of the 7th International Symposium on Ceramics in Medicine, Turku, Finland, July 1994*)

Response of Soft Tissue Against Hydroxyapatite Coated on Polymers

Y. Kageyama, Y. Yokoyama, K. Suzuki, Y. Harada and T. Kokubo*
First Department of Surgery, Hamamatsu University School of Medicine, 3600 Handa-cho, Hamamatsu, 431-31, Japan
* Division of Material Chemistry, Faculty of Engineering, Kyoto University, Sakyo-ku, Kyoto, 606-01, Japan

ABSTRACT

The reaction of soft tissue against experimentally implanted polyethersulfone (PES) coated with hydroxyapatite (HA) was examined. PES, silicon-coated PES, and HA-coated PES were implanted subcutaneously in the abdomen of rats. Tissue reaction was examined weekly after implantation. The surrounding tissue firmly adhered to the HA-coated PES, and showed no sign of strong inflammatory reaction. Proliferations of fibroblasts were observed in contact with the surface of the HA-coated PES. Because of its good histocompatibility and the adhesiveness to soft tissues, this HA-coating technique may be very useful in the production of biomaterials.

INTRODUCTION

Hydroxyapatite (HA) is an important component of osseous tissue. Its clinical application as a bone-repairing material has been reported extensively and fundamental studies have demonstrated its good histocompatibility (1,2). Our study of artificial trachea coated with HA showed excellent results (3). But the interaction which occur between HA and soft tissues have not been fully clarified. It is important to examine the reaction of soft tissues against coated HA in order to apply the HA-coating technique to a prosthesis for soft tissues (4).

MATERIALS AND METHODS

PES was coated with uniform apatite layer according to biomimetic process reported by T. Kokubo et al. (5,6). PES was placed on CaO, SiO2 based glass particles in a stimulated body fluid (SBF) for nucleating apatite at 36.5 °C for 2 days. Subsequently PES was soaked in a solution with ion concentrations 1.5 times those of SBF at 36.5 °C for 8 days for growing the apatite. The three types of plates (10 mm X 10 mm X 1 mm; PES, silicone-coated PES, and HA-coated PES) were implanted subcutaneously in the abdomen of rats. The surrounding soft tissues were then examined histopathologically at weekly intervals. Immunostaining using anti-leukocyte common antigen (LCA) antibody was also performed to assess whether the infiltrating cells are inflammatory cells or not.
Four types of sticks (2 mm in diameter, 20 mm long; smooth surface PES, rough surface PES, silicone-coated PES, and HA-coated PES) were implanted subcutaneously in the abdomen of rats. Rough surface PES proved to have much rougher surface than HA-coated PES. One end of the stick was pierced to pass a thread through for pulling. Four and eight weeks after implantation, the stick was pulled out by a thread passed through the hole, and the degree of adhesion between soft tissues and materials was measured by a pull tester.

RESULTS AND DISCUSSION

Inflammatory cell infiltration into capsule surrounding HA-coated plate was as mild as PES and silicone-coated PES. Proliferations of immature cells were observed around the surface of the HA-coated plate even at long intervals after implantation, whereas these phenomena were not observed in PES and silicon-coated PES (Fig. 1,2,3). Immunostaining with LCA antibody revealed that these cells were not inflammatory cells but seemed to be fibroblasts (Fig. 4,5,6). The materials, which do not induce inflammatory reaction and facilitate proliferation of fibroblasts for a long period, have not been reported. Furthermore, in the extraction test using a pull tester, coated HA showed the strongest adhesiveness to soft tissues (Fig. 7). These results suggest that proliferating fibroblasts may have close relation to the adhesiveness. In general, if tissue reaction is not strong, materials may not adhere to the soft tissues. Coated HA, however, seems to have the strong adhesiveness in spite of the less inflammatory reaction (7). Therefore coated HA can be useful for a prosthesis which need tight attachment to soft tissues; ex. artificial trachea, artificial esophagus and so on.

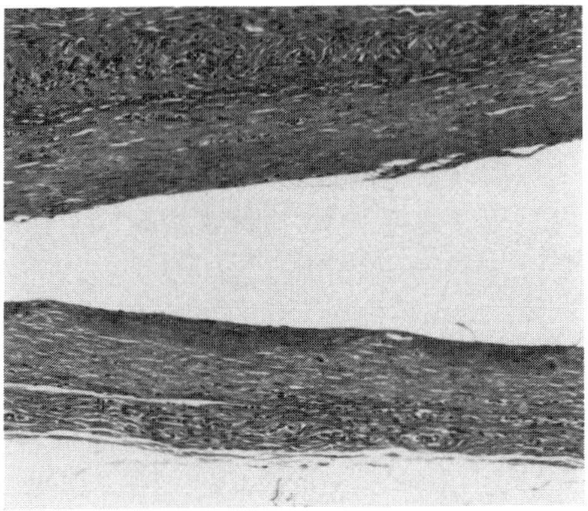

Fig. 1 (a) Pathological findings of the capsule around PES 5 weeks after implantation. (H. E., X50)

Fig. 1 (b) PES (30 weeks) (H. E., X50)

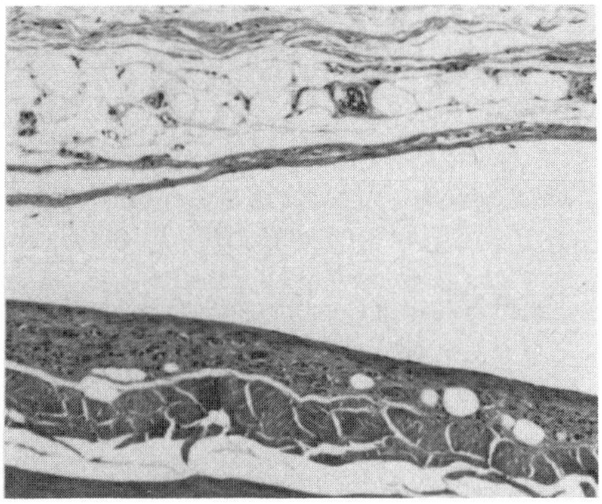

Fig. 2 (a) Pathological findings of the capsule around silicone-coated PES 5 weeks after implantation. (H. E., X50)

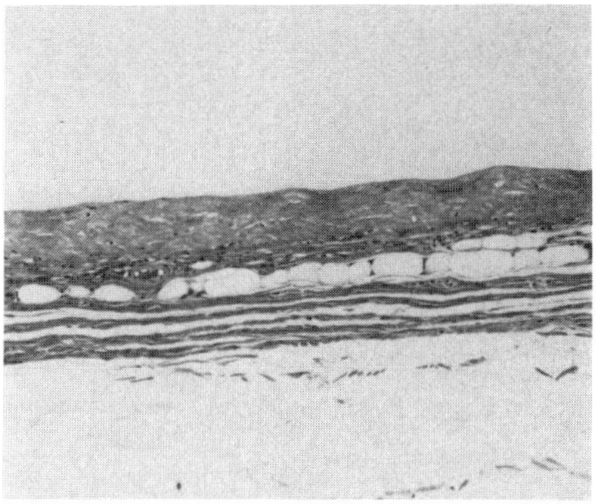

Fig. 2 (b) Silicone-coated PES (30 weeks) (H. E.,X50)

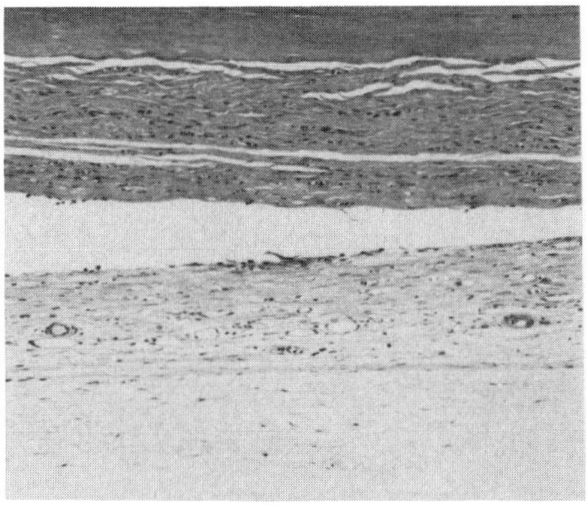

Fig. 3 (a) Pathological findings of the capsule around HA-coated PES 5 weeks after implantation. (H. E., X50)

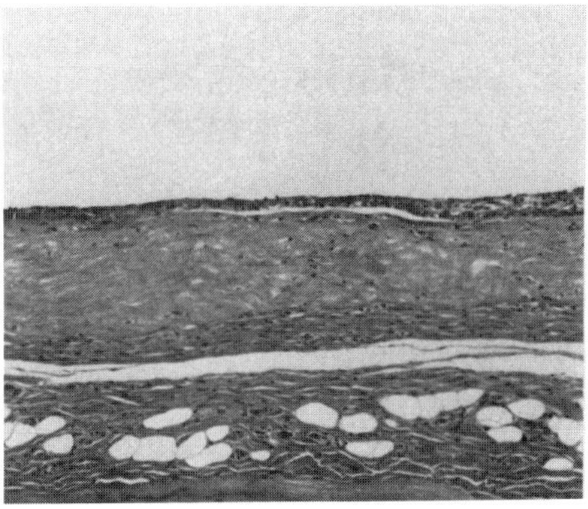

Fig. 3 (b) HA-coated PES (30 weeks) (H. E., X50)

Fig. 1 (c) PES (45 weeks)
(H. E., X50)

Fig. 2 (c) Silicone-coated PES
(45 weeks) (H. E., X50)

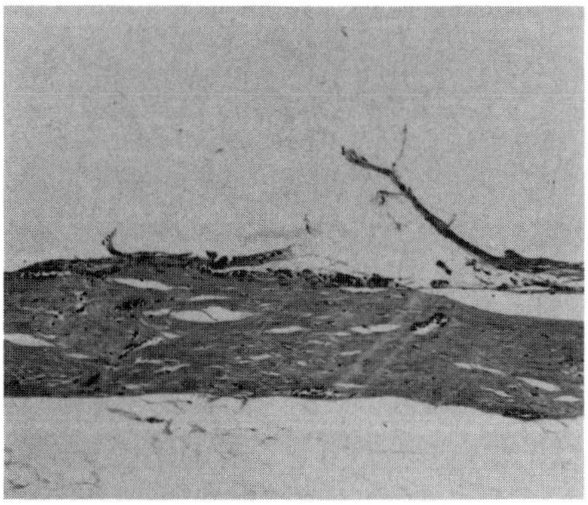

Fig. 3 (c) HA-coated PES
(45 weeks) (H. E., X50)

Fig. 3 (d) HA-coated PES
(45 weeks) (H. E., X200)

Fig. 4 The same specimen as Fig. 3 (d). Immunostaining with LCA antibody shows that proliferating cells around the surface of HA-coated PES are not inflammatory infiltrating cells. (X 200)

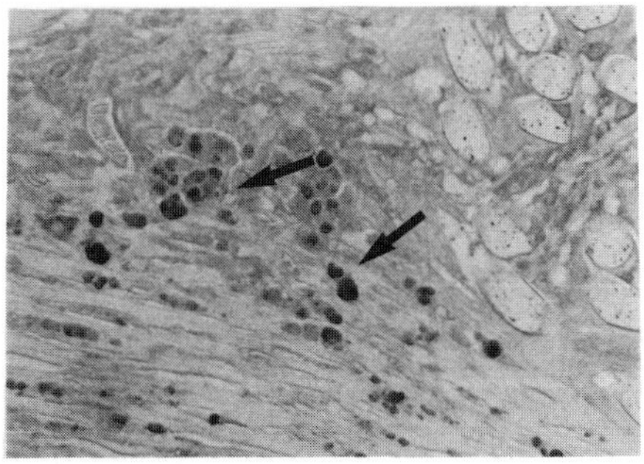

Fig. 5 Positive control (immunostaining with LCA antibody). Arrows show inflammatory cells. (X200)

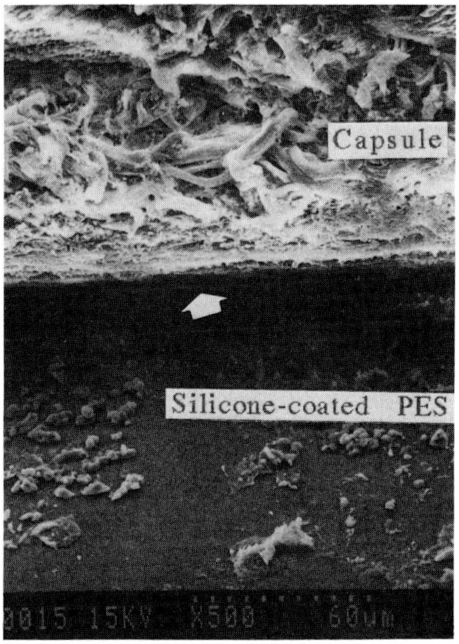

Fig.6 (a) SEM photograph of the interface of silicone-coated PES and capsule.

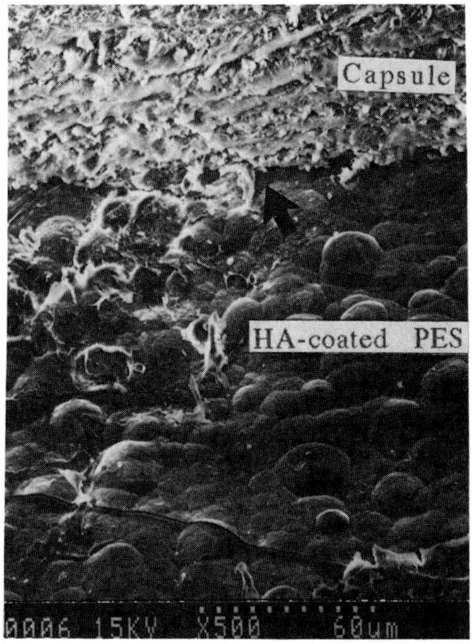

Fig.6 (b) SEM photograph demonstrated continuity between HA-coated PES and surrounding tissues.

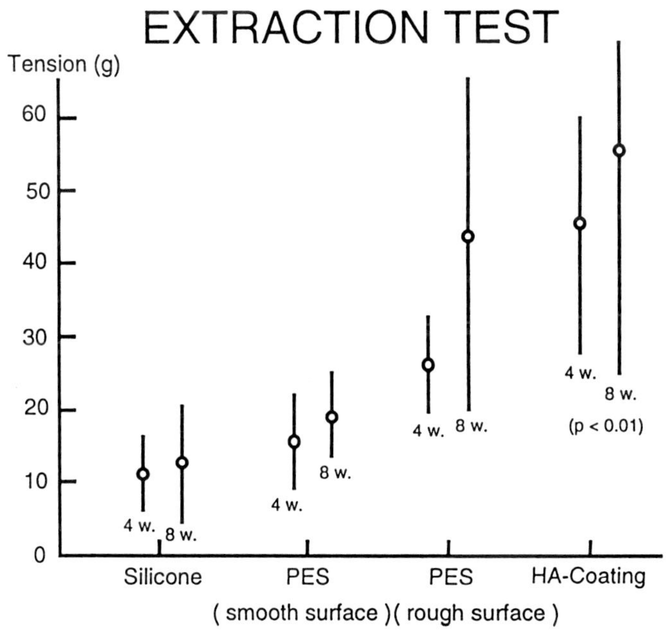

Fig. 7 The extraction test using a pull tester showed that HA-coated sticks needed the strongest tension to remove out.

REFERENCES

1) Itiravivong, P., Lorprayoon, C., Sukepetch, A., and Runruxsirivorn, S., In Bioceramics Vol. 5, 1992, 157-164.

2) Ogilvie, A., Frank, R. M., Benque, E. P., Gineste, M., Heughebaert, M., and Hemmerle, J., J. Periodont. Res., 1987, 22, 270-283

3) Suzuki, K., Kobayashi, R., Yokoyama, Y., Harada,Y., and Kokubo, T., In Bioceramics Vol. 6, 1993, 245-250

4) Misiek, D. J., Kent, J. N., and Carr, R. F., J. Oral Maxillofac. Surg., 1984, 42, 150-160

5) Kokubo, T., Hata, K., Nakamura, T., and Yamamoto,T., In Bioceramics Vol. 4, 1991, 113-120.

6) Tanahashi, M., Hata, K., Kokubo, T., Nakamura, T., and Yamamoto, T., In Bioceramics Vol. 5, 1992, 57-64.

7) Praeger, F. C., and Gilchrest, B. A., Proc. Soc. Exp. Biol. Med., 1989, 190, 28-34

Bioceramics, Volume 7 Edited by Ö.H. Andersson and A. Yli-Urpo
(*Proceedings of the 7th International Symposium on Ceramics in Medicine, Turku, Finland, July 1994*)

Reaction of Porous Hydroxyapatite, β-TCP and β-TCP Coated Hydroxyapatite in the Bone

H. Yokozeki, H. Kurosawa,* K. Shibuya,** K. Ioku,*** T. Hayashi**** and H. Kawahara****

Dept. Orthopaedic Surgery, Asama General Hospital, 1859-1, Iwamurada, Saku-shi, Nagano 385, Japan
* Dept. Orthopedic Surgery, Tokyo Teishin Hospital, 2-14-23, Fujimi, Chiyoda-ku, Tokyo 102, Japan
** Dept. Orthopedic Surgery, Kosei Hospital, 5-25-15, Yayoi-cho, Nakano-ku, Tokyo 164, Japan
*** Res. Lab. Hydrothermal Chem., Fac. Science, Kochi University, 2-5-1, Akebono-cho, Kochi-shi, Kochi 780, Japan
**** Dept. Orthopedic Surgery, Fac. Medicine, University of Tokyo, 7-3-1, Hongo, Bunkyo-ku, Tokyo 113, Japan

ABSTRACT

This study was conducted to analyze and compare the bone ingrowth (amount of regenerated bone) and biodegradation of low-porosity (60%) hydroxyapatite (HA), 100 wt% β -tricalucim phosphate (TCP) and 33 wt% β -tricalucim phosphate-coated HA (33TCP). The results suggest that in the early in vivo stages 33TCP behaved similarly to β -TCP by showing high resorption and rapid bone ingrowth, but in the later stages behaved like HA by maintainig a high rate of the bone ingrowth, making it a more effective bone substitute material.

INTRODUCTION

Previous research has established that porous hydroxyapatite (HA) and β — tricalcium phosphate (β — TCP) show good tissue tolerance, producing no immunological or toxic reactions (1). These materials have therefore been used as bone graft substitutes not only in experimental models but also in clinical applications. Several previous studies have also confirmed that β — TCP is a more biodegradable implant material than HA (2). However, it has yet to be established which of the two materials has the greater potential for bone formation. Only a few studies have addressed this question and their findings were contradictory (3) (4). Rather than different bone substitute materials, the present study emphasized the different properties of materials (e. g. porosity and pore size). Bone tissue may be deposited directly onto the surface of

porous (5) (6). The higher the material's porosity, however, the greater its fragility. An ideal material, therefore, would be one whose porosity was low at implantation but thereafter gradually increasing through resorption of the surface area of the ceramics. Prepared using our unique method (7), the 33 wt% β — tricalcium phosphate-coated low-porosity HA tested in this study may be just such a superior bone substitute material.

MATERIALS AND METHOD

Table 1

Properties of implanted materials (5mm ϕ × 5mmL)

	Porosity (%)	Pore size (μm) macro	Pore size (μm) micro	Supecific surface area (m^2/g)	Compressive strength (MPa)
HA	60±5	150～400	0.2～0.5	10.0	12.5±4.0
33TCP	58±5	150～400	——	3.8	17.5±2.7
TCP	57±5	150～400	——	2.0	21.4±2.6

The HA used in this study was prepared by the Mitubishi Material Co., Ltd., sintered at 900℃ and with a porosity of 60±5%. The β－TCP coated HA was prepared as follows: the pure HA was soaked in a solution of diammonium hydrogen phosphate and then sintered at 900℃ for three hours. Controlling the percentage of β－TCP formed by varying the concentration of the solution, we obtained 33 wt% β－TCP coated HA (33TCP) and 100 wt% β－TCP (TCP). Table 1 gives the properties of HA, 33TCP and TCP, showing that the β－TCP coated porous HA prepared in this way was considerably less brittle than the original HA.

We implanted a column of each of the three materials, 5 mm in diameter and 5 mm in length, into the distal femurs of each of 20 male rabbits of five to six months age and of a mean weight of 3.0 kg. This was done by drilling a 5 mm hole in the bilateral condyles of the bilateral distal femurs, through the cortex and into the medullary canal, using a specially prepared trephine (5.2 mm in outer diameter) under continuous saline irrigation. The implants were inserted tightly into the holes.

Table 2

Number of Columns Implanted in 20 Rabbits

Material	4 weeks	12 weeks	24 weeks
HA	7	7	6
33TCP	7	10	6
TCP	6	9	6

The rabbits were sacrificed by anesthetic overdose after periods of 4, 12 and 24 weeks (Table 2). The implant specimens were removed, dehydrated in a 70% solution of ethanol and embedded in methyl methacrylate. Each was then sliced into sections approximately 150-200 μm thick by means of a diamond blade saw. The thickness of the sections was then further reduced to 95-105 μm by hand polishing, finally producing 6-8 slides of each specimen. To measure the area of the ceramic material and amount of regenerated bone in each specimen, contact micro-radiography was performed and morphometric analysis carried out using the hit-point method on a microscope at 20-times magnification. The average of these amounts for each time group was then calculated from the average area of each specimen, thus allowing comparison of the average areas of the respective materials.

RESULTS AND DISCUSSION

The ceramic area of the HA was significantly greater both at 12 weeks ($p<0.01$) and at 24 weeks ($p<0.05$) than the control material, while that of the TCP showed a significant decrease at 24 weeks compared to the same at 4 weeks ($p<0.05$). The area of the 33TCP showed no significant change over the three time periods. There was considerable difference in area between HA and 33TCP and between HA and TCP both at 12 weeks ($p<0.05$) and at 24 weeks ($p<0.05$; Figure 1).

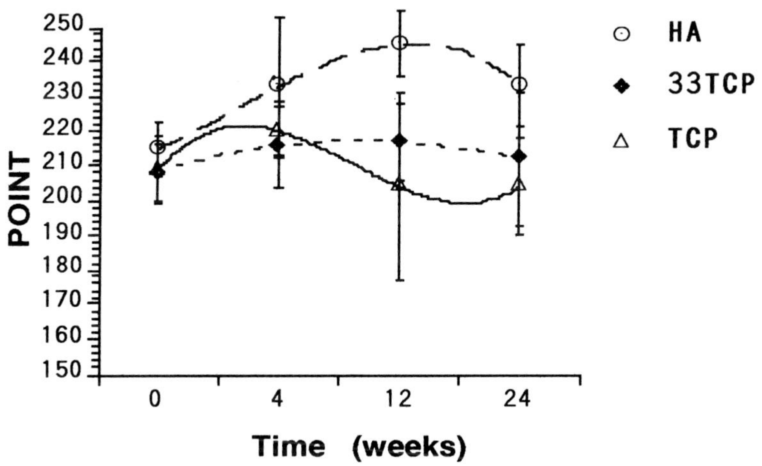

Figure 1 Ceramicus area

In the HA specimens, bone volume had increased significantly at 24 weeks compared with that at 4 weeks ($p<0.01$) and at 12 weeks ($p<0.05$). At 4 weeks, the bone volumes in both the 33TCP and the TCP specimens were significantly greater ($p<0.05$ and $p<0.01$, respectively) than that of the HA specimens. At 24 weeks, the bone volume in the 33TCP specimens was significantly greater than that in the TCP ($p<0.05$) but not significantly different to that in the HA (Figure 2).

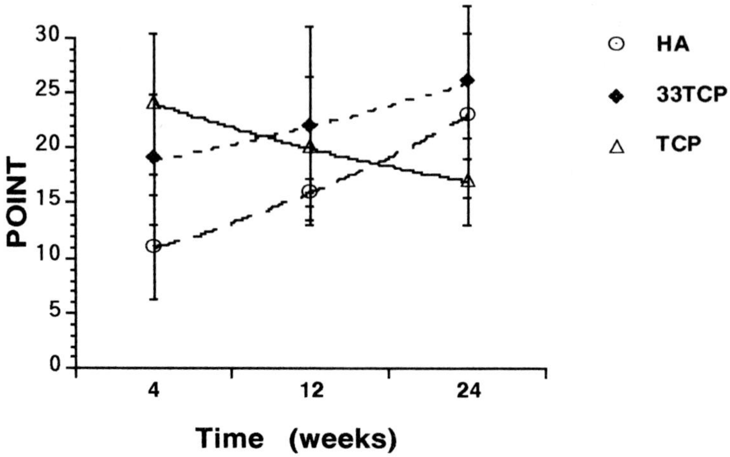

Figure 2 Regenerated bone area

In 1988, Eggli (4) reported TCP to be more biodegradable than HA. In the present study, too, the ceramic area of the TCP was significantly lower at 24 weeks than at 4 weeks. Meanwhile, there was a significant difference in ceramic area between HA and TCP at both 12 and 24 weeks. We may say, therefore, that 33TCP behaved like β—TCP (showing high resorption) in the early stages of the in vivo test period, but then behaved like HA in the later stages.

The effects of these materials on bone formation are still unclear. Shimazaki and Mooney (3) reported bone ingrowth from 3 to 24 weeks after implantation to be greater in HA than in TCP. Eggli (4), meanwhile, reported that the amount of bone was similar in HA and TCP for pore diameters of 50-100 μm, but significantly higher in TCP for pore diameters of 200-400 μm. In our study, the bone ingrowth in TCP was significantly greater than that in HA at 4 weeks but from then on decreased until 24 weeks. The bone ingrowth in the HA, on the other hand, significantly increased up to 24 weeks. Meanwhile, the bone ingrowth in the 33TCP was less than the TCP but greater than the HA at 4 weeks, but greatest of all three materials at 24 weeks, behaving, as was mentioned, like the TCP at first but like the HA thereafter.

These results suggest that, because of its initial similarity to β—TCP (good resorption and rapid bone ingrowth) and later similarity to HA (continued high rate of bone ingrowth), 33TCP may be a superior bone substitute material.

REFERENCES

1. K. Kato, H. Aoki, T. Tabata, and M. Ogiso : Biomat. Med. Dev. Art. Org., 7 : 291-297, 1979.

2. H. U. Cameron, I. Macnab, and R. M. Pilliar, : J. Biomed. Mater. Res., 11 : 179-186, 1977.

3. K. Shimazaki, and V. Mooney : J. Orthop. Res., 3 : 301-310, 1985.

4. P.S. Eggli, W. Muller, and R. K. Schenk : Clin. Orthop., 232 : 127-138, 1988.

5. K. Ioku, M. Yamasaki, H. Yokozeki, T. Iwano, K. Shibuya, H. Kurosawa, T. Kurokawa, and H. Kawahara : Rep. Res. Lab. Hydrothermal Chem., 7 : 42-45, 1991.

6. T. Iwano, H. kurosawa, K. Murase, H. Takeuchi, and Y. Ohkubo : Clinical Orthopaedics and Related Research, 268 : 243-252, 1991.

7. K. Ioku, T. Murakami, Y. Ikuma, and M. Yoshimura : J. Ceram. Soc. Jpn., Int. Ed., 100 : 1001-1005, 1992.

Bioceramics, Volume 7 Edited by Ö.H. Andersson and A. Yli-Urpo
(*Proceedings of the 7th International Symposium on Ceramics in Medicine, Turku, Finland, July 1994*)

Correlation between Clinico/Histological Results and the Hydroxyapatite/Phosphate Ratio of Implanted Ceramic Granules

A. Piattelli,* C. Mangano,** A. Krajewski,*** A. Ravaglioli,***
R. Martinetti**** and M. Fabbri****

* Clinica Odontostomatologica, Dip.di Istologia, Università di Chieti, Chieti, Italia
** Clinica Odontoiatrica, Università La Sapienza, Roma, Italia
*** IRTEC – CNR Faenza, Faenza, Italia
**** FIN-CERAMICA FAENZA, Faenza, Italia

ABSTRACT

This study was carried out on porous ceramic granules incorporating different hydroxyapatite (HA) / β-tricalcium phosphate (β-TCP) ratios. The size (450–600 μm), the porosity (90–100 μm) and the sphericity of the granules were controlled. The aim of the study was to assess the bio-physico-chemical behaviour of the granules grafted in bone and of the tissue towards them. A quantification of directly bonded areas between bone and surface of the granules was performed and related to the content of β-TCP.
Implantation in humans was carried out to effect light mandibular corrections.

INTRODUCTION

This study deals with ceramic granules made with hydroxyapatite (HA) as the main material and β-tricalciumphosphate (β-TCP) as additional substance, to evaluate the influence of β-TCP and if threshold limit exists to allow an infiltration of living tissue.
Microscopy examinations on bioptic thin sections from rabbits displayed how the newly formed bone is allowed to adhere on the surface of these granules and that microvolumes of bone tissue are able to growth inside suitably interconnected porosity.

MATERIALS AND METHODS

The granulates were prepared through a new forming process, dried and then fired at 1200°C for 2 hours. The starting green powders were either pure HA or HA with addition of pure β-TCP. Samples containing 0% (pure HA), 25%, 50%, 75% of β-TCP were prepared for this study, and coded as HA*n*, with *n* = 1, 2, 3 and 4 respectively. A whole surface porosity of 25% was determined with a 2% of pores ranging around 2 μm^2 and a 23% around 60 μm^2. Twelve rabbits were used for this study, that were anesthetized with intramuscular injections of fluanizone (0,7 mg per Kg b.wt.) and diazepam (1.5 mg per Kg b.wt.), and local anaesthesia was given using 1 mL of 2% lidocaine/adrenaline solution. Bone sites were prepared with burs under saline irrigation: 4 mm diameter holes were made in the cortical bone. The granules were then implanted and the periosteum and fascia were sutured with catgut and the skin with silk. Postoperatively the animals were administered intramuscular injections of penicillin (2 Millions IE/5 mL; 0.1 mL per Kg b.wt.). The animals were killed half after two months and the other half after six months with an overdose of intravenous

pentobarbital. The blocks of the bone with the surrounding tissues were removed and washed in saline solution and fixed in 10% neutral buffered formalin.
The specimens were dehydrated in an ascending series of alcohol rinses and embedded in a glycolmethacrylate monomeric resin (Technovit 7200 VLC, Kulzer AG., Germany). After a suitable lapse of time useful to reach a homogeneous imbibition, a photopolymerization process was activated. The obtained hard specimens were sectioned with a diamond saw at a thickness of about 200 μm and electronically controlled microground down (EXAKT Apparetebau, Norderstedt, Germany) to obtain thin sections of about 30 μm, according to Donath and Breuner's technique. After polishing the slides were stained with basic fuchsin-methylene blue, toluidine blue and von Kossa. A double staining layer was performed on one slide per implant, first von Kossa and then with basic fuchsin. After polishing, the slides were dipped in $AgNO_3$ for thirty minutes and then exposed to sunlight. The slides were then washed under tap water, dried and dipped in basic fuchsin for five minutes, and finally washed and mounted. The resulting slides so prepared were watched by microscope (Leitz Laborlux S, Leitz AG., Wetzlar, Germany) with both normal and polarized transmitted light. A morphometric analysis was conduced under a Leitz Microvid system connected to a computer (IBM 70) so that to quantify the percentage of contact between the growing bone tissue and the whole surface of the granules. The granules were also utilized for correction of mandibular defects in human beings to evaluate their fitness in terms of dimensions, round shape and porosity.

RESULTS

Histological analysis of samples coming from rabbits femurs showed that most of the granules were embedded in bone tissue coming from the repair of the cortical, while some of them were surrounded by fibrous compact subperiosteal tissue without any linking with the mineralized bone. It was also possible to evaluate the influence played by the different amounts of β-TCP that were present. On the different granules, according to their composition and arrangement, areas of localized resorption were noted. In particular these ceramic materials display a direct adhesion to the mineralized matrix, without any interposition of connective tissues. The external surface of each granule was contemporary covered by proteic layers, by medullary lacunae and by the mineralized matrix directly adhering. The percentage of the granules surface coated by mineralized tissue changed with the kind of implanted ceramic and obviously with the implantation time. Values are reported in Table 1.

Table 1 - Contact percentage between mineralized bone of rabbit femur and surface of the granules. The values are connected with the implantation time and with the amount of β-TCP.

	Time of contact	
Sample	2 months	6 months
HA1	59.36 (± 11.36)	57.45 (± 7.25)
HA2	48.96 (± 9.40)	49.37 (± 8.08)
HA3	38.56 (± 5.25)	44.12 (± 5.14)
HA4	30.-- (♣)	35.-- (♣)

(♣) value deduced on a statistically insufficient sampling of data, however agreeing with extrapolation.

The granules not incorporated in the newly formed bone tissue, but instead dislocated in the fibrous subperiosteal tissue, were not taken into consideration in this calculation. In this case granules were surrounded by a compact connective fibrous tissue without inflammation cells with the exception of some giant cells due to a foreign body response. Such cells showed a phagocytic activity since in the areas where they were localized, the granule surface presented resorption zones similar to the osteoclastic Howship's lacunae. The cellular analysis showed the presence of some macrophagic cells with small particles of calcium phosphate at the interior of the kitoplasma. This kind of cells were often localized at a distance of approximately 10 μm from the granules, testifying the resorption phenomena. When the ceramic surface is crumbled from macrophages, they gulp down into the detached grains. Macrophagic cells were clearly present in numerous medullary lacunae closed to the surface of the grains that were visibly engaged in phagocytic activity similar to that observed in the periosteal soft tissues. An observed localized erosion occurred in the zone shown by photo of Figure 1 in which ceramic grains are detached from the surface of a granule as a consequence of the attack by aggressive acid substances secreted from phagic cells. The observed number of the medullary lacunae that are in contact with the ceramic surface increases with the content of β-TCP. An erosion-reconstruction process seems to occur in the border line between newly formed bone and zone of the medullary lacunae with an association of macrophages and osteoblasts. Some lines of advancement of newly formed bone were visible in the same border and around grains detached by phagic activity. However such lacunae not always contained phagic cells. A view of newly formed bone grown around a granule is visible in Figure 2. The external morphology of the surface of the biphasic ceramic granules proved to be remarkably much more altered at 6 months after implantation than that of the granules of pure HA, showing also a very different shape compared to the initial one and widest areas with more depth erosion. The zones having direct contact between mineralized bone and ceramic material showed a variable morphology. Use of a non-specific colouring agent (*e.g.* the methylene basic bleu fuxina), made one believe that a direct bone-ceramic contact existed, whereas the use of a histochemical colouring method specific for calcium salts (*e.g.* the Von Kossa's one), made one realize that wide non-mineralized layers (thickness 2-5 μm) interposed between mineralized bone and ceramic surface. These layers were much wider in the biphasic ceramics, in this case represented and uniformely distributed at the interface, than in the granules of only HA. The morphological aspect of such interfacial structure displays to be similar to the ostecytic *lamina limitans* and to the bone reversal lines. An infiltration of chromophylic organic material, probably of the same nature, is observed at the interior of the porosity even over 50 μm from the surface of all the grains (Figure 3). However, there is also the formation of some mineralized spots which are directly joined with the surface of the ceramic. An interesting result is represented by the presence of cells and mineralized tissues at the interior of the granules. Some cavities that are present at the interior of the granules showed to contain mineralized bone colonized by osteocytes (Figure 4). In Figure 5 is showed the surface of HA3 sample in contact with medullary space: a layer of newly formed bone is visible in the centre, with some cells (osteoblastic-osteocytic) during their active deposition of new bone. Below the newly formed bone zone, some macrophagic cells are visible. The implantation of the granules in human beings was satisfactory in reaching the results expected both under a functional and esthetical profile without any problem.

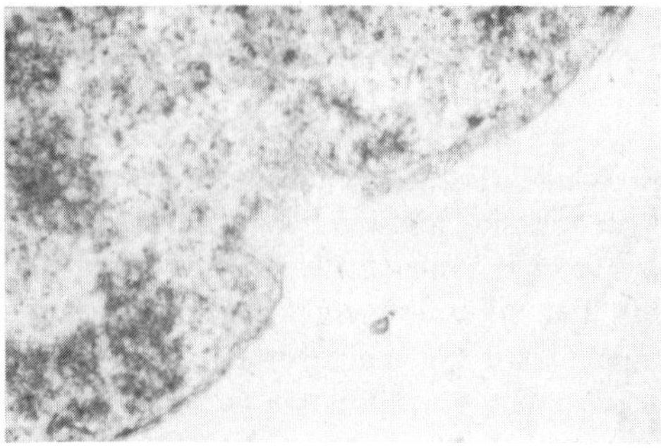

Figure 1

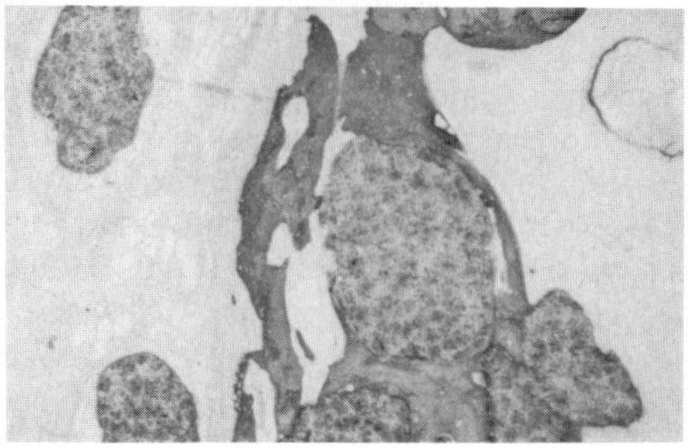

Figure 2

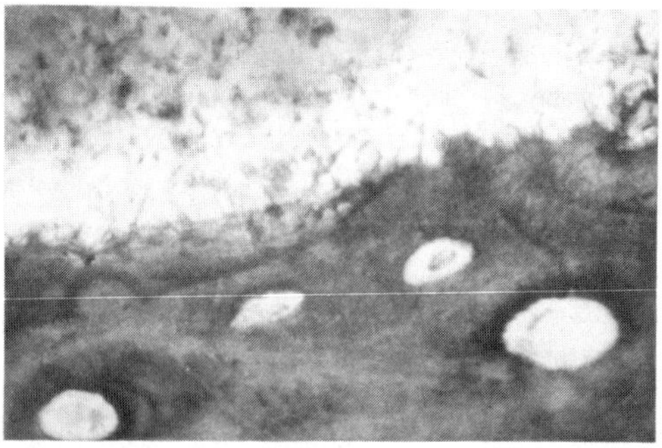

Figure 3

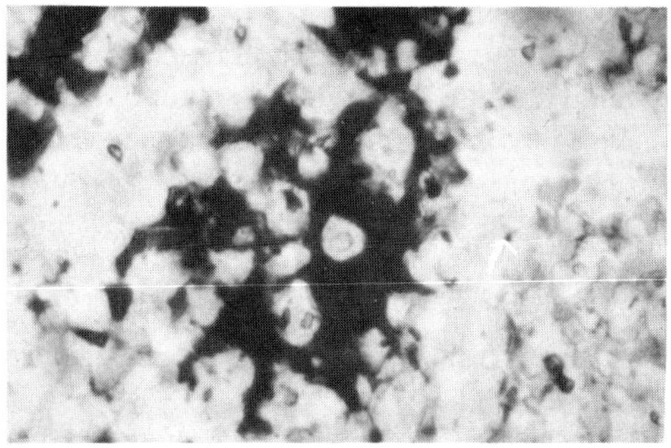

Figure 4

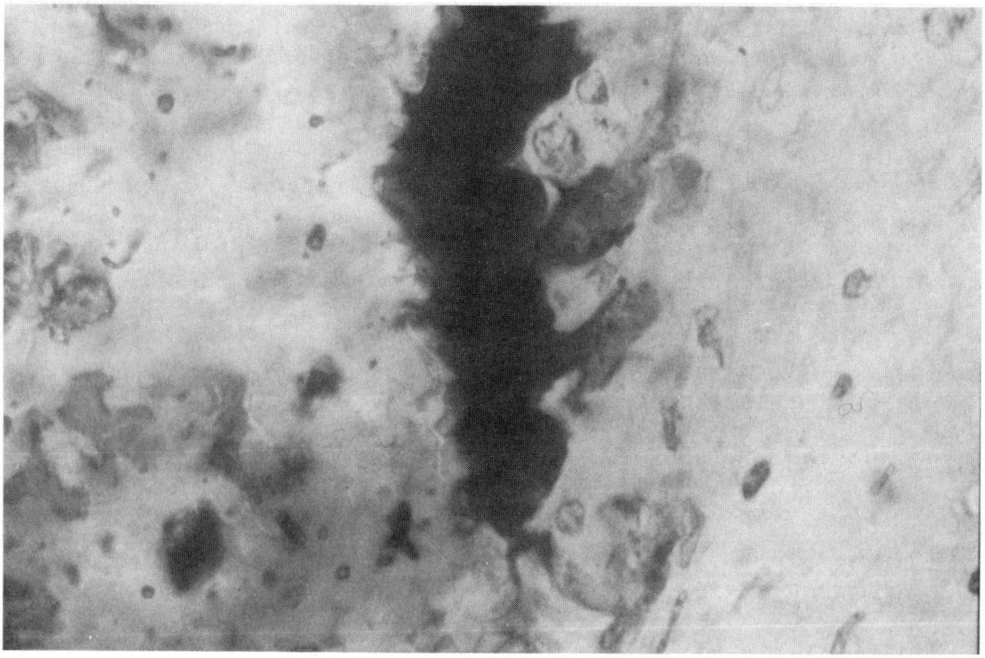

Figure 5

DISCUSSION

Bone growth all around the surfaces of the granules shows to be influenced by the presence of β-TCP in the ceramic composition. The direct contact degree of the newly formed bone decreases as β-TCP percentage increases. The remainder of the surface is covered by proteic layer and by medullary lacunae (many with macrophages).The number of the medullary lacunae and the zones where they can develop, the zones of the interposition of the chromogenic proteic layers and those of direct mineralization are not predictable and on many granules are absent. A more accurate and reproducible evaluation of these variables can be made only on a statistical base, *e.g.* on a wide interested area that is to say on a wide number of granules; this kind of evaluation has yielded to the values reported in the Table 1.
The internal colonization of the granules from osteoblasts, with formation of spots of tissue inside the existing macroporosity, appeared to be uncorrelated with the composition of the ceramic. Most likely some more decisive probabilistic factors overpower it, since *e.g.* this colonization is connected with the real possibility that an osteoblast could penetrate effectively within the ceramic bulk. This event may be linked to different concomitant factors: the presence of a pore sufficiently large (>10 μm) let the cells penetrate inside, preventive leaching action of macrophages or of their aggressive secretion, a medullar lack containing osteoblasts close to the zone of surface in which a pore wide enough is present. In trying to find best comfortable sites, an osteoblast may grow thinner and stretch out elastically along the pore, exactly as a flexuous cat does in passing through a too narrow passage for him. However, not to remain an isolated cell scarcely nourished, there need presence of a diffused microporosity that allows a permeation of physiological liquids that bring the nutritional factors and a right degree of oxygenation and also that allows the constitution of an intercellular network of filaments of connection. A cavity wide enough (>30 μm) is on the other hand necessary to give rise to reproduction, to associate with other cells for the formation of an isle of tissue. As concerns the formation of an organic layer apparently similar to that of the ostecytary *lamina limitans*, we consider its formation a natural event like that one of the natural resolution of a fracture. The thin proteic layers formed on the ceramic are the residual of the lamina of advancement of the bone growth. Where newly formed bone is grown up on a zone of the ceramic surface, that zone is not attached by macrophagic action anymore.
The difficulty of digestion of HA ceramic grains from the macrophages, unlike it occurs for bone HA crystallites (*e.g.* during the phagic stage of a fracture repair), can be explained by the surface/volume ratio of the phagocitated particles. Such ratio is far smaller in the case of ceramic grains.This is a parameter that makes ceramic HA different from bone crystallites HA. The macrophages that have engulped an excess of ceramic grains are not eliminated by systemic methabolism and this can even further negatively reflect on remote distances. They often are not able to dissolve them since they are too big and roundish in comparison with HA bone crystallites, that are in fact smaller and shaped as thin needles. This can explain why macrophages containing still not completely digested ceramic grains were sometimes found, in previous clinical experience of mandibular corrections with the use of trade calciumphosphate granules, even accumulated inside glands of the neck (lymphonodes and salivary glands) of the patients, with the effect of deposition of HA, giving so rise to a sort of glandular stones. On the other hand, other negative factors can arise in their turn in case this ratio is too high (*e.g.* micronic powders), like those of micrograins that bring about to adhere on the cell membrane, affecting its normal movements and functions. Another

negative effect is the toxic one due to the excessive amount of Ca^{2+} and PO_4^{3-} ions leached in the physiological environment from too fine powders. A too high concentration of Ca^{2+} ions in the environmental physiological liquids can undermine the Ca/Mg pump of the cells by osmothic decompensation. Consequently, the cells are going to swell of water till to break of their membrane [1], that leads to their necrosis. As regards the erosion-reconstruction process observed in some medullary lacunae, a partial side-effect interpretation can be done which is based on the chemical environment establishing inside everyone of these lacunae. The substances secreted from both macrophages and osteoblasts are trapped inside by the membrane of the lacuna. This does not allow free diffusion of molecules and ions in both the directions. Consequently, an accumulation of lactic acid, phosphatases, and H^+ (decreasing of pH) together with other hypothetical substances (*e.g.* enzymatic factors, informative molecular messangers) could be expected, whereas a corresponding decrease of ATP and O_2 should take place. The combined action of both types of cells could allow an energetical saving and an integration of the reciprocal needs. Without possibility to receive Ca^{2+} and phosphate ions from outside, the only way is the phagic erosion of the ceramic surface to allow osteoblasts to calcify in another side of the lacuna that borders with bone, so completing the mentioned integration.
In the mandibular implantation on human beings, these granules play a double role in the post operative course as expected as an osteoconductive and a sustaining material of the gingival membrane avoiding its collapse. The clinical results were always good and no particular histological differences were observed with the use of the different compositions. The resorption rate showed to be only weakly proportional to the content of β-TCP.

CONCLUSIONS

The process of adhesion to bone tissue of these granules shows first of all the possibility to succeed in graduating the bioreadsorption through calculated additions of β-TCP to the HA main component. The neoformation of bone takes place in a good extent, even if grains of the ceramic implanted can be lost by the remodelling action during phagic stage. The acquired information show a general tendency to the integration of all these samples, also with formation of isles of bone tissue at the interior of the porosity wide enough. The speed and amount of direct joining of bone tissue to the surface of the samples appears inversely connected to the amount of β-TCP added to the HA.
The use in odontology was successful; the differences in the dissolution rates are attributed to the too high density of the granules. This study was in fact addressed mainly to the quality of the material from the chemical point of view. Further optimization of dimension, density and porous structure is going to be taken into consideration.

BIBLIOGRAPHY

1. K. Donath; "*Reaction of tissue to calcium phosphate ceramics*" in "*Ossointegrated implants*", book edited by G. Heimke, CRC Press (Boca Raton, Fl, 1990), Vol. 1, pp. 99-125.

Bioceramics, Volume 7 Edited by Ö.H. Andersson and A. Yli-Urpo
(*Proceedings of the 7th International Symposium on Ceramics in Medicine, Turku, Finland, July 1994*)

Hydroxyapatite Prototypes Under Physiological Loads

A. Ravaglioli, A. Krajewski, A. Piancastelli, M. Torijama, R. Giardino, M. Fini and G. Biagini

Institute for Technological Research on Ceramics of the Italian National Research Council – Faenza, Italy
Univ. of Bologna, Orthopaedic Institute Rizzoli, Italy
Univ. of Ancona, Faculty of Medicine, Italy
Government Industrial Research Institute, Nagoya, Japan

ABSTRACT

Prototypes of plaques for osteosynthesis made with hydroxyapatite and the relative nails were manufactured. These devices for the strengthening of the fracture were subsequently tested. The utilized powders based on hydroxyapatite were prepared according to a new method of mechano-chemical synthesis. This method showed to be decisive in obtaining ceramic products more and more mechanically resistant in respect to the ones prepared by the traditional methods.

Previous works shew that powders of hydroxyapatite obtained by the mechano-chemical method gave rise to ceramic samples able to reach easily high values of flexural strength, even 150 MPa, higher than the ones produced by traditional methods (< 80 MPa).

The production of these plaques at so high mechanical strength allows to place devices with very high biocompatibility towards the bone at the disposal of the surgeons for implants which before were possible only utilizing other materials. Differently from the similar prostheses based on metal or inert ceramic used up to now, these do not require their subsequent explantation.

The material exhibited successful behaviour in prototypal implants transverse to the rabbit femur compared to that of the ones sintetized by present usual critera, with an excellent bone reconstruction in the neighbourhood of the implanted samples.

INTRODUCTION

Recently a new technique was proposed to synthesize hydroxyapatite that gives rise to sinterized samples at very high mechanical strength [1,2]. This new technique was called mechanochemical, a term that recalls to mind the real processes with which it is carried on. The target of the present work was to verify if this mechanochemically prepared hydroxyapatite still retains the high capability of biological interactivity with tissues that theoretically would be expected, together with the exhibited higher mechanical performances. In order to obtain this information, it was important to carry out a histological evaluation: of the post-surgery course, of the fixation to tissues and of the tissural interaction, compared with that taken from similar samples produced with powders obtained by present usual procedures.

MATERIALS AND METHODS

Powders of the new mentioned production type were utilized for the preparation of the samples, the object of interest in this work. These powders were carefully produced by ourselves through the mechano-chemical method. They were prepared with a reactive grinding within containers made by teflon using ZrO_2 balls by wet milling a mixed slurry of $CaHPO_4 \cdot 2H_2O$ and $CaCO_3$, in proportions that give the correct calcium and phosphorous ratio of hydroxyapatite. Commercial powders synthesized according to traditional criteria were used (Riedel de Haïen) for the preparation of the comparative samples. Prototypes of small nails of hydroxyapatite (3 mm diameter and length ranging from 5 to 8 mm) were prepared through slip casting and by pressing using powders coming from the two sources. It is underlined that, at a mechanical resistance level, the differences existing among the samples prepared through slip casting or pressing differ from each other since they statistically give a spreading of values closer in the case of slip casting while it was much wider in the case of those made by pressing. This is to be ascribed to the lesser number of defects induced by the first method in respect to the second.

The samples prepared with the traditional powders were fired at 1280°C, the temperature that appeared to be the best for these powders. The samples prepared with powders obtained by the mechano-chemical reaction were fired at a temperature of 1250°C that appeared to be the best. At their optimal temperatures of firing, the ceramic samples gave values of flexural strength on four points of 50 MPa for the samples made with traditional powders and 150 MPa for those made with hydroxyapatite mechano-chemical prepared. All the samples were subsequently implanted in the femurs of rabbits after a previous dry sterilization at high temperature (300°C).

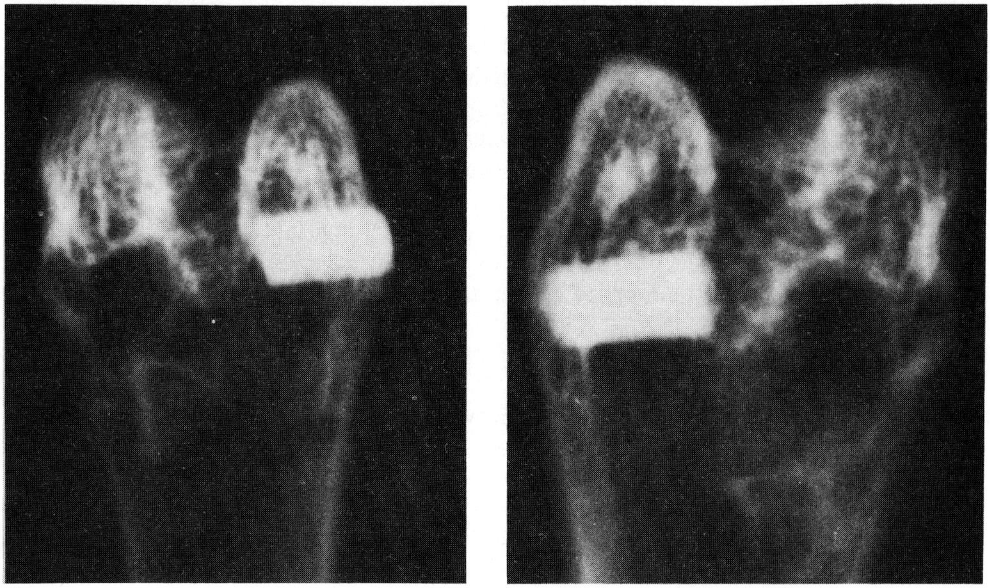

Figure 1 - Microradiographies of the extracted rabbit femura (right and left) containing the HA nails according to what described in the text. Sample from mechanochemical powderss is on right.

As far as the implantation surgery is concerned, an experimental *in vivo* model was preliminarily set up. Six adult male New Zeland rabbits, 3.00 (±0.25) Kg in weight, were utilized. Rabbits were obtained 1 week before the beginning of the experiment, in order to let them rest for a period of acclimatization to laboratory conditions. They were housed in single cages (3150 $cm^2 \times 40$ cm) and received food and water *ad libitum*. After this period, which

allowed also control of the animal health by performing random faeces exams and pharyngeal tampons to avoid the use of animals with associated diseases, the rabbits underwent surgery with a starve of 12 hours. Under general anaesthesia, with Ketamine (rate 25 mg per Kg b.w.) and assisted ventilation with nitrogen peroxide (3 dm^3 n.c.), Halothane (2%) and oxygen (2 dm^3 n.c.), the femoral condyles were bilaterally exposed by means of a lateral longitudinal incision.
In the lateral codylar face a substance defect was gently created by means of 3 mm drill and the nails were positioned. The wounds were sutured atraumatically in 2 layers with interrupted stitches. The samples made with traditional powders were introduced into right femura, whereas those made with powders mechano-chemically produced were introduced in the left femura. An antibiotic therapy (cefamezin i.m.) was applied in the first 3 days after surgical intervention. The postoperative course was followed by radiological analyses.
The animals were sacrificed under general anaesthesia after 2 months and the femura were explanted to carry on all the scheduled examinations. Bioptical samples taken away from these femura were analyzed with a light microscope operating at phase contrast and with a scanning electron microscope (SEM). Radiological analyses of the extracted femura were carried out and the result of a couple (right and left femura of the same rabbit) is reported in the radiographies of Figure 1.
Samples either decalcified or not were produced for the histological tests. The ones not decalcified were put in methacrylate resin for the production of thin sections.

RESULTS

Table 1 - Salient physical properties of the two types of HA powders.

Powder preparation	Firing temperature (°C)	4-points flexural strength (MPa)	porosity (%)	density (g cm^{-3})
Traditional	1280	<80	0	3.03
Mechanochemical	1250	≈150	0	3.04

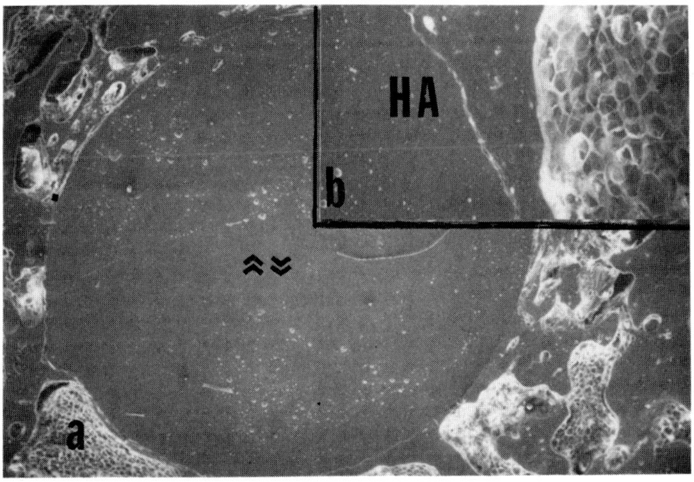

Figure 2 - HA nail implant observed under a SEM: (a) low magnification (×240); (b) higher magnification (×780) of the interface between bone and ceramic.

To have suitable elements to formulate a whole judgement on the need to choose ceramic HA coming from powders produced mechanochemically instead of those coming from traditionally prepared powders, it is highly useful to look at the data reported on the Table 1.
The implanted samples showed to give rise to a good adhesion with the

bone of neoformation grown normally around all the periphery in both types of the samples. No intraoperative and postoperative complications were observed. No fibrotic or flogistic inflammatory reactions of the tissues surrounding the site of the implant were observed at the moment of the sacrifice.
The surface base of the implant was still macroscopically visible only for some implanted samples coming from traditional powders (Figure 2). The implants appeared still well evident from radiological inspections with presence of normal bone and with a good ceramic-bone joining. The anchoring to the tissue already showed to be extremely efficacious only by examining the radiographic analyses following the evolution of the osteointegration during the time thanks to the clearness of the image at the interface. The ceramic material obtained by the mechanochemical method shows to induce a higher forming of bone lamellae.

DISCUSSION AND CONCLUSIONS

The test carried out exhibit a perfect biophysico-chemical behaviour of the implanted hydroxyapatitic prototypes. The histological results show perfect osteointegration, apparently a bit better for the material prepared with the mechanochemically produced powders. The physical parameters are substantially the same among the samples produced with the two different powders, with the exception of a higher mechanical strength. The reasons for such a higher strength were discussed previously [1,2]. Therefore, on the base of the histhologic results and of the best mechanical performances , applications previously not hypothisized in zones subjected to an average load may be expected. Since the mechano-chemical method is still developing as a technology and the involved parameters are now considered to generalize the possibility of production with reactors of different dimension, at present it is only at the stage of technological transfer to the industry.

BIBLIOGRAPHY

1. M. Torijama, A. Ravaglioli, A. Krajewski, G. Celotti, A. Piancastelli; "*Synthesis of hydroxyapatite by mechano-chemical method and sintering of obtained powders*", <u>in press</u> on J. Mat. Sc.
2. M. Torijama, A. Ravaglioli, A. Krajewski, G. Galassi, E. Roncari and A. Piancastelli; "*Slip casting of Mechano-chemically synthesized hydroxyapatite*", <u>in press</u> on J. Mat. Sc.

Bioceramics, Volume 7 Edited by Ö.H. Andersson and A. Yli-Urpo
(*Proceedings of the 7th International Symposium on Ceramics in Medicine, Turku, Finland, July 1994*)

Osteocompatibility and Biocompatibility of Tetracalcium Phosphate Cement Containing Cellulose

M. Yoshikawa,[1] T. Toda,[1] H. Oonishi,[2] S. Kushitani,[2] E. Yasukawa,[2] S. Hayami,[1] Y. Mandai[3] and F. Sugihara[3]

[1] Department of Endodontics, Osaka Dental University, 5-31, Otemae 1-chome, Chuo-ku, Osaka 540, Japan

[2] Department of Orthopaedic Surgery, Artificial Joint Section and Biomaterial Research Laboratory, Osaka-Minami National Hospital 677-2, Kido-cho, Kawachinagano-shi, Osaka 586, Japan

[3] Nitta Gelatin Inc. Research Laboratory Center, 2-22, Futamata, Yao, Osaka, 581 Japan

ABSTRACT

We performed *in vitro* and *in vivo* experiments on the cytotoxicity and histotoxicity of the cement prepared by kneading an equimolar mixture of tetracalcium phosphate and dicalcium phosphate dihydrate with modified McIlvain's buffer solution containing cellulose. The colony formation of cells (Balb/3T3) was not inhibited. The cement showed desirable biocompatibility in the dorsal subcutaneous tissue of rats and in the femoral condyles of rabbits. The cement placed in rat molar canals caused apical closure with a cementum-like calcified product.

These observations suggest that the cement is useful for application to bone defects and as a dental cement.

INTRODUCTION

There has recently been extensive research and development into inorganic biomaterials for body hard tissues. In particular, calcium phosphate-type ceramics such as hydroxyapatite, tricalcium phosphate (TCP) and tetracalcium phosphate (TeCP) have attracted considerable interest. It is well known that when α-TCP or TeCP powder is kneaded with water or suitable acidic solutions, it sets like a cement, and that an equimolar mixture (TeDCPD) of TeCP and dicalcium phosphate dihydrate (DCPD) kneaded with water or diluted phosphoric acid converts into hydroxyapatite upon setting (1,2). We have reported that application to the rat dental pulp of a kneaded product of TeDCPD powder with physiological saline promotes formation of hard tissue on the canal wall, and that a TeDCPD powder kneaded with McIlvain's buffer solution shows weakly basic pH values throughout the setting process from the start of kneading to completion of setting (3,4). We were able to change the setting time of these cements by modifying the concentration of the liquid (3,4). It was also found that a blend of sodium chondroitin sulfate and the buffer solution showed good biocompatibility and, when used as a root canal sealer, induced osteoblasts on the surface of alveolar bone (5).

In this study, we investigated the cytotoxicity and histotoxicity of a new cement based on the previous materials, and studied its potential clinical use as a bone filling material and root canal sealer.

MATERIALS AND METHODS

TeCP was prepared by dry synthetic method of sintering at 1400°C for 8 hours. It had particle sizes from 0.6 to 44.0μm, with an average of 12-13μm. DCPD was passed through a 32μm-sieve. The cement examined in this study was obtained by kneading a TeDCPD powder with McIlvain's buffer solution containing sodium carboxymethyl cellulose (powder to liquid ratio = 1.4 g/g) . Major components of the cement (wt %) were ; as powder, TeCP (68 %) and DCPD (32 %); and as liquids, Citric acid (2.1 %), Disodium hydrogen phosphate (7.2 %), Sodium carboxymethyl cellulose (1.5 %), Distilled water (88.0 %), and Others (1.2 %).

Cytotoxicity test

The test was based as closely as possible on the Japanese Guidelines for Basic Preclinical Toxicity Testings for Medical Materials and Devices published in 1992 (6). After setting, the cement was clashed and passed through a 75μm-sieve to obtain the granules. They were immersed in Earle's MEM at 37°C for 12 hours and the supernatant fluid was used as the cement extract. For negative control extract, a 2 X 15 mm piece of plastic sheet for tissue culture was immersed in the medium at 37°C for 24 hours, and the supernatant fluid were used. The standard material A [zinc diethyldithio carbamate (ZDEC)] and B [zinc dibutyldithio carbamate (ZDBC)] were respectively dissolved in the medium and used as positive control solution. The Balb/3T3 Clone A31 cell line was isolated by conventional methods and cultured in each extracted solution, and also in Earl's MEM as non-treated control, in respectively several concentration for 11 days. The cultured cells were Giemsa-stained, and the colony numbers were counted (only colonies of greater than 50 cells were counted).

Histocompatibility test

The test cement and zinc oxide eugenol cement (ZOE) were subcutaneously implanted in the dorsum of 7-week old male Sprague Dawley rats. The rats were sacrificed at one and four weeks after implantation, and the implants, together with surrounding tissue, were removed, fixed, decalcified, embedded in paraffin, serially sectioned at 6μm, stained by hematoxylin and eosin (H.E.stain), and histopathologically examined. The cement was also implanted into the femoral condyle of mature rabbits to evaluate tissue reaction. The rabbits were sacrificed at 3 and 6 weeks after implantation, and the implants together with surrounding tissue were removed and processed as described above. Then, the cement and ZOE were respectively placed in the canals of the mandibular first molars of Sprague Dawley rats. The rats were sacrificed at 1, 2, 3, 4, and 5 weeks after application, the mandibles removed, and the above procedures carried out.

RESULTS AND DISCUSSION

Cytotoxicity test

Table 1

Effect of negative control extract on Balb/3T3 colony formation

	colony numbers/dish	plating efficiency (%)	colony forming efficiency (%)
non-treated control	80 95 109 102	64.3±12.8	100
negative control	111 95 94 104		104.7±8.3

(Plating efficiency = colony cell numbers/cell numbers in a plate)

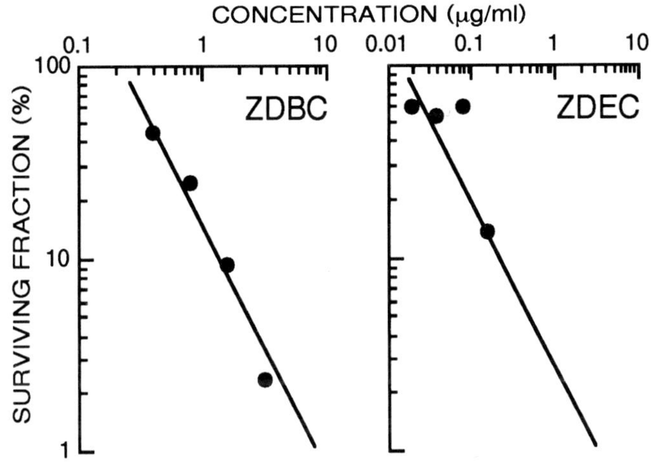

Figure 1 Cytotoxicities of reference standard material A (ZDEC ; right) and B (ZDBC ; left) on Balb/3T3 cells

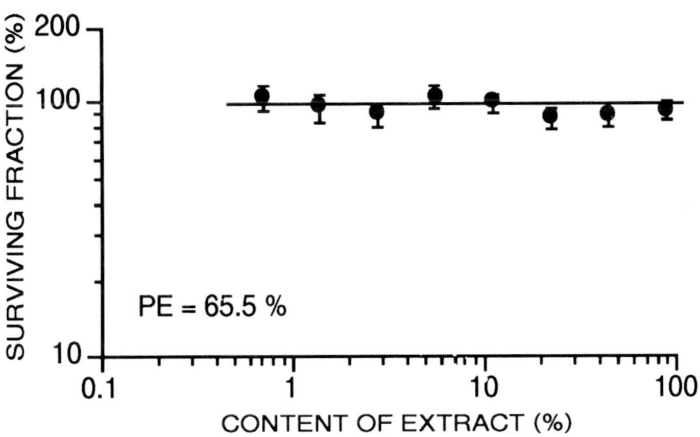

Figure 2 Effect of cement extract on Balb/3T3 colony formation
PE:Plating efficiency(%)

Colony inhibition was the same in the negative controls as in the non-treated controls (Table 1). This suggests that the procedures and conditions used in the test are appropriate. Colony inhibition in the positive controls was concentration-dependent (Figure 1). The 50% inhibitory concentrations were 0.031μg/ml for ZDEC and 0.42μg/ml for ZDBC. The results show that the cells used have satisfactory anti-reagent characteristics. The cement extract did not decrease the colony inhibition in every concentration range (Figure 2). The safety of the cement extract, therefore, was confirmed.

Histocompatibility test

Accumulation of inflammatory cells was observed in the dorsal subcutaneous tissue adjacent to the implanted cement one week after implantation. The cell infiltration almost completely disappeared at four weeks. Although the safety of an extract of the test cement had been confirmed by cytotoxicity testing, a relatively large piece of the hardened cement was surrounded by a thin layer of fibrous connective tissue, as shown in Figure 3. This suggestes that although the cement implanted in the rat dorsal subcutaneous tissue did not irritate the tissue, it was too large to be tolerated if left intact in the body. Also the relatively small granules scattered around the implanted cement were surrounded by granulation tissue and would sooner or later be absorbed or organized. On the other hand, severe inflammatory reactions observed at one week in the tissue surrounding the ZOE did not disappear even at four weeks (Figure 4), and no encapsulation was seen during the experimental periods. ZOE was highly amyctic and toxic. Therefore, we

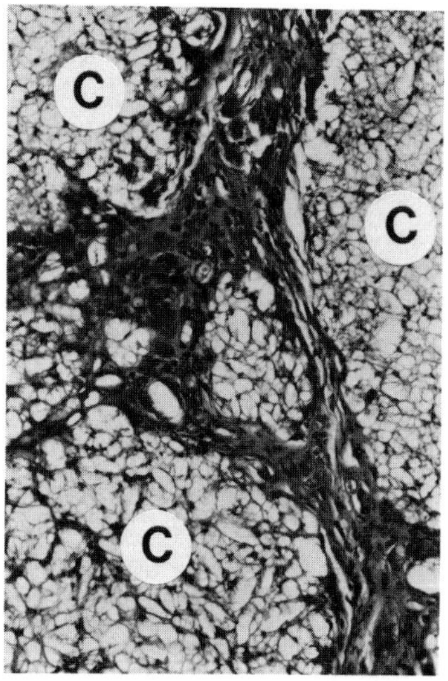

Figure3 Four weeks after subcutaneous implantaiton of cement C : Cement H.E. stain (2.5 X 20)

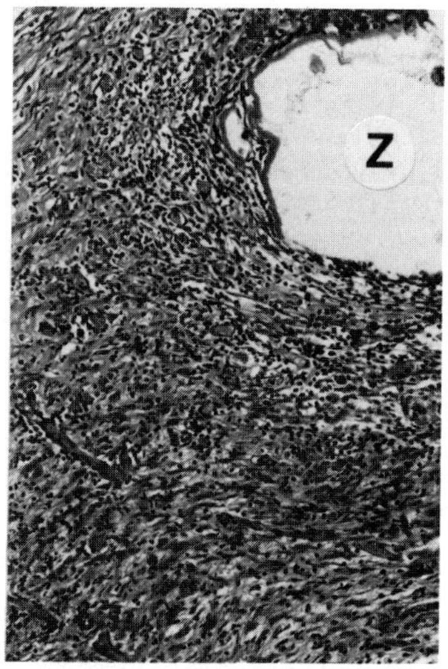

Figure4 Four weeks after subcutaneous implantation of ZOE Z : ZOE H.E. stain (2.5 X 20)

concluded that the cement was highly biocompatible and may be quickly replaced by granulation tissue when implanted in connective tissue. In bone, inflammatory reactions did not occurred three weeks after implantation. The presence of multinucleated giant cells was observed in places at the interface of the cement

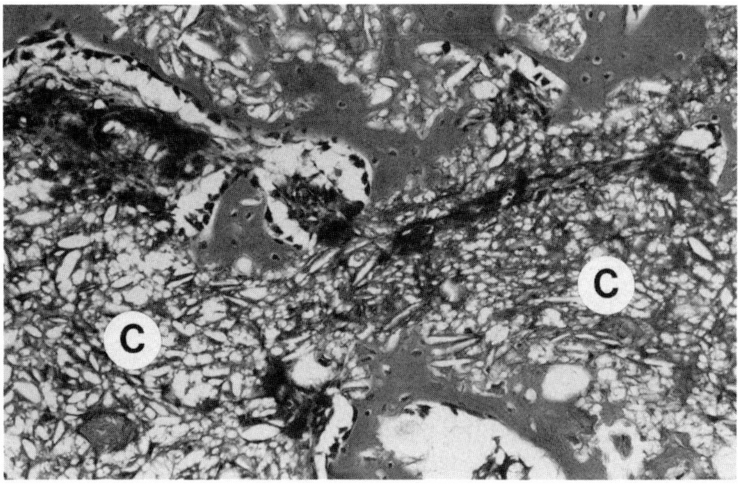

Figure 5 Six weeks after placing of cement in the bone
C : cement H.E. stain (2.5 X 20)

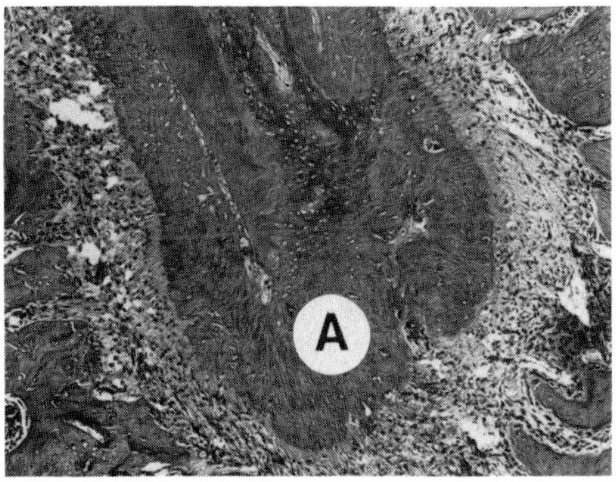

Figure 6 Five weeks after placing of cement in root canal A: Apical foramen H.E. stain (2.5 X 10)

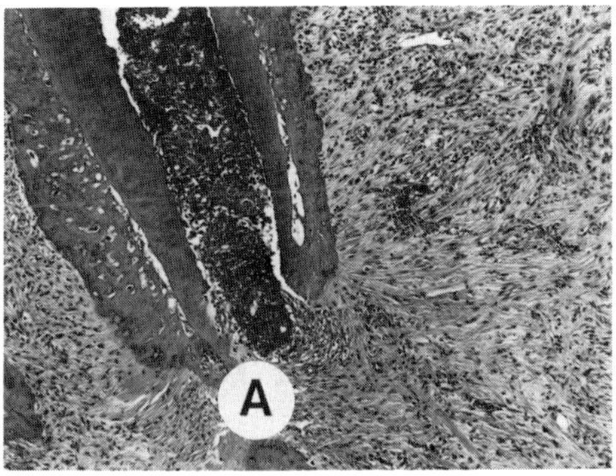

Figure 7 Five weeks after placing of ZOE in root canal A: Apical foramen H.E. stain (2.5 X 10)

and cancellous bone after three weeks. They were still seen after six weeks and the amount of bone penetrating the cement after six weeks was larger than that after three weeks (Figure 5). This seems to indicate that the cement has high osteoconductivity and is gradually metabolized by normal remodeling when implanted in bone. Accumulation of infiltrated inflammatory cells was observed in the pulp remnants and the periapical tissue near the apex at one, two and three weeks findings after placing the cement in the root canals. The apical foramen of the root canal had been completely closed by cementum-like calcified product (Figure 6). On rare occasions, there was still a weak inflammatory reaction in the pulp remnants and periapical tissue near the apex at five weeks, and no cementum-like calcified product deposition in the apical portion of root. In contrast, ZOE caused severe inflammatory reactions in the pulp remnants and periapical tissue for one to five weeks (Figure 7). In addition, alveolar bone resorption were seen. These reactions gradually increased in severity during later observation periods. Free eugenol from the set ZOE might have caused these severe inflammatory reactions (7).

CONCLUSIONS

These results indicate that the cement we tested is valuable as a root canal sealer and bone filling material because of its excellent biocompatibility and osteocompativility. However, some problems must be solved before it can be applied clinically. The fate of the cement extruded into the periapical tissue through the apical foramen is a problem. However, its clinical use in bone defects might be possible without further modification.

REFERENCES

1 Monma, H., Goto, M. and Kohmura, T. Gypsum & Lime, 1984, 188, 11-16

2 Brown, W.E and Chow, L.C. In United States Patent, 1985, 4518430

3 Yoshikawa, M., Oka, Y., Toda, T. and Sugihara, F. Japan. J. Conserv. Dent, 1991, 34 (spring issue), 119

4 Mandai, Y., Sugihara, F., Nagatomi, K, Minamigawa, K., Hayami, S, Yoshikawa,M. Oka, Y. and Toda, T. Japan. J. Conserv. Dent, 1993, 36(2), 627-637

5 Tsuji, I. and Yoshikawa, M. Japan. J. Conserv. Dent, 1993, 36(3), 845-867

6 Pharmaceutical Affairs Bureau Ministry of Health and Welfare In Approvals and Licenses of Medical Devices in Japan Yakuji Nippo, Japan, 1992, 245-251

7 Webb, J.G.Jr. & Bussell, N.E. J. Dent. Res., 1981, 60, 1724-1728

Bioceramics, Volume 7 Edited by Ö.H. Andersson and A. Yli-Urpo
(Proceedings of the 7th International Symposium on Ceramics in Medicine, Turku, Finland, July 1994)

Characteristics and Bone Tissue Reaction of Tetracalcium Phosphate-Collagen Conjugated Sponge

F. Sugihara,[1] K. Minamikawa,[1] H. Oonishi,[2] Y. Mandai,[1] K. Nagatomi,[1] E. Yasukawa,[2] S. Kushitani[2] and E. Tshuji[3]

[1] Nitta Gelatin Inc. Research Laboratory 2-22, Futamata, Yao-shi, Osaka, 581 Japan
[2] Department of Orthopaedic Surgery, Artificial Joint Section and Biomaterial Research Laboratory, Osaka-Minami National Hospital, Osaka, Japan
[3] Osaka Prefectural Industrial Engineering Laboratory, Osaka, Japan

ABSTRACT

A tetracalcium phosphate-collagen conjugated sponge was prepared by the reaction of tetracalcium phosphate having hydration activity with type I collagen giving a gel in a short time under the physiological condition. Complete hemostasis was obtained immediately after pressing with the sponge on bleeding bone on surgery. Histologically the sponge conducted massive bony tissue early and was then replaced by new bone. This bioactive sponge may be used as both hemostatic and filling materials, since it is adaptable even to defected bleeding bone with any type of shape.

INTRODUCTION

Dried fibrous collagen has been used clinically as a local hemostatic material because of the astriction effect as well as the hemostatic coagulation mechanism due to its platelet aggregation ability (1). Collagen itself, however, has no osteoconductivity.

On the other hand, granular or blocky hydroxyapatite (HA) has been employed clinically as a filling material for bone defects because of its good osteoconductivity. However, it is difficult to fit HA to the shape of defected bone cavity to be filled, and hence there are some problems for clinical use. HA is not replaced by bone, but it has been reported that granular tetracalcium phosphate was replaced by bone when filled in bone defects (2).

Therefore, the composite of collagen and tetracalcium phosphate might be quite useful as a hemostatic filling material which is fittable to any type of bone defects.

In addition, if the composite has high biocompatibility, it would be expected that the filling is replaced partially or completely by regenerating bone or new bone.

From the point of view, we prepared a sponge-like product by complexing type I collagen, which sets to gel in a short time under the physiological condition, and tetracalcium phosphate [$Ca_4(PO_4)_2O$; TeCP], which has hydration activity. In the present study the product is filled into bone defect to examine the hemostatic effect while surgery, as well as the bone tissue reaction.

MATERIALS AND METHODS

The method of preparing tetracalcium phosphate (TeCP)-collagen conjugated sponge is shown in Figure 1. A pepsin-solubilized type I collagen solution [Cellmatrix I-P®, Nitta Gelatin Inc., Osaka Japan], BGJb medium [SIGMA Co., St. Louis U.S.A], and a reconstitution buffer solution of HEPES [Dojin Chemicals Co., Kumamoto Japan] were used. TeCP was prepared by sintering the stoichiometric mixture of $Ca_2P_2O_7$ and $CaCO_3$ (1:2 mol) in air at 1400 °C for 8 hours and mixed with BGJb medium to prepare the sponge (TeCP-sponge). A control (HA-sponge) was prepared using hydroxyapatite (HA) sintered at 1200 °C in place of TeCP. The particle sizes of TeCP and HA used were 10 μm. The obtained sponge-like composites were analyzed by means of XRD and SEM.

These were filled into a 5 mm hole drilled on femoral condyle of mature rabbits, and the hemostatic effect was evaluated during surgery. At 6 weeks after operation, the condyle was trimmed out *en bloc* including the implant, and decalcified and undecalcified specimens were made therefrom and observed by SEM, backscattered electron image and light microscopy.

(A) collagen solution [0.3% (w/v)]
(B) 5x BGJb medium containing TeCP particles [5% (w/v)]
(C) reconstitution buffer (HEPES) solution
mixed ice cold A, B and C at a ratio of 7:2:1
↓
Incubated in a 37°C incubator for 3 days
↓
lyophilized

Figure 1. Method of preparing TeCP-collagen conjugated sponge

RESULTS AND DISCUSSION

Characteristics of sponges

TeCP-sponge was found to be a low crystalline apatite-collagen conjugate with an unreacted trace of TeCP (Figure 2.). The crystal structure of HA was unchanged in HA-sponge. The both sponges had a honeycomb structure with pores of several microns in diameter. In higher magnification of the both sponge matrices, low crystalline apatite microcrystals were observed on TeCP particles conjugated with collagen fibrous bundles (Figure 3A), whereas aggregated HA particles were covered by mesh-like collagen bundles (Figure 3B).

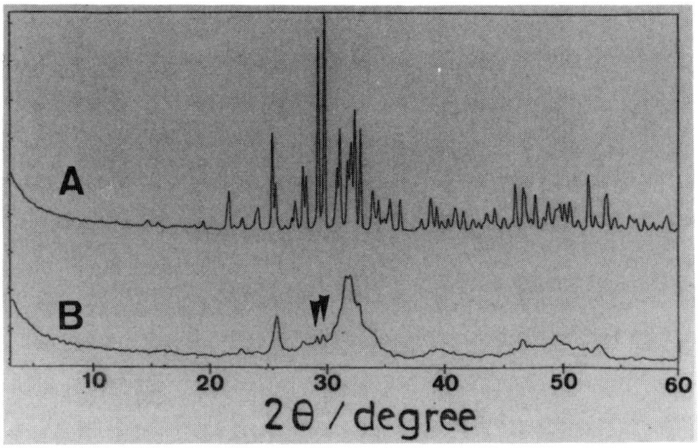

Figure 2. X-ray diffraction patterns of TeCP (A) and TeCP-sponge (B). Note trace of TeCP (arrowheads) in TeCP-sponge.

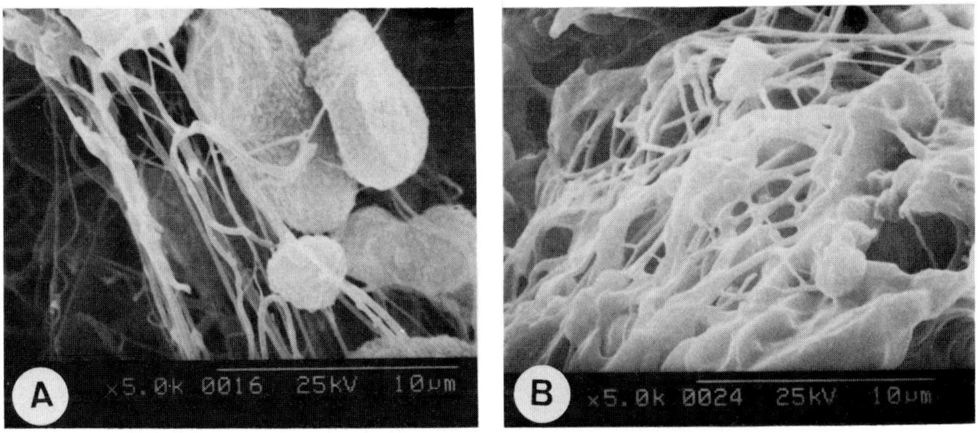

Figure 3. Scanning electron micrographs of matrices of TeCP-sponge (A) and HA-sponge (B).

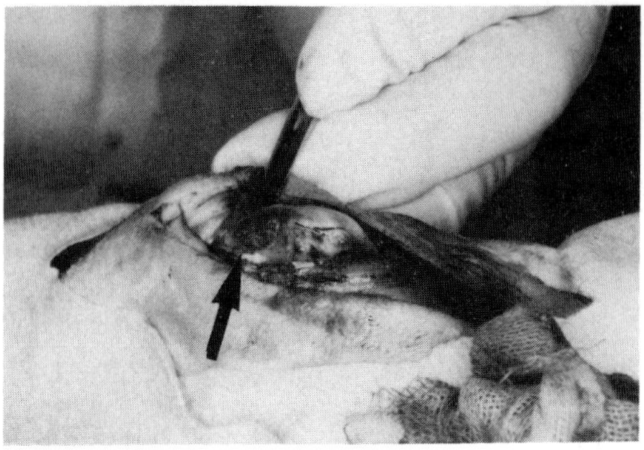

Figure 4. Hemostatic effect of TeCP-sponge (arrow) on bleeding bone.

Hemostatic effect and histological observation

Complete hemostasis was obtained immediately after pressing with TeCP-sponge on bleeding bone (Figure 4), and the similar effect was obtained by HA-sponge.

Microscopically, ingrowth of massive bony tissue was observed along the matrix of TeCP-sponge (Figure 5A), and ossification was beginning from the edge of the matrix (Figure 5C). Backscattered electron image disclosed that TeCP-sponge was replaced by new trabecular bone from the periphery, and most of low crystalline apatite particles were taken into the bone. In the region where TeCP-sponge was replaced by new bone, only a small amount of remaining low crystalline apatite particles was observed (Figure 6A). It is considered that most of the particles were absorbed and replaced by new bone.

In contrast, HA-sponge was replaced by new bone only at the periphery and a large amount of HA still remained at the central area (Figure 5B). Similar to the result of implantation of HA particles alone (3), the presence of a large amount of HA in marrow cavity suggested that HA particles taken into trabecular bone began to be eliminated from the bone with remodeling (Figure 6B).

CONCLUSION

TeCP-collagen conjugated sponge has both desired osteoconductive characteristics and hemostatic effectiveness, and hence this may be useful as a filling material in bony defected area.

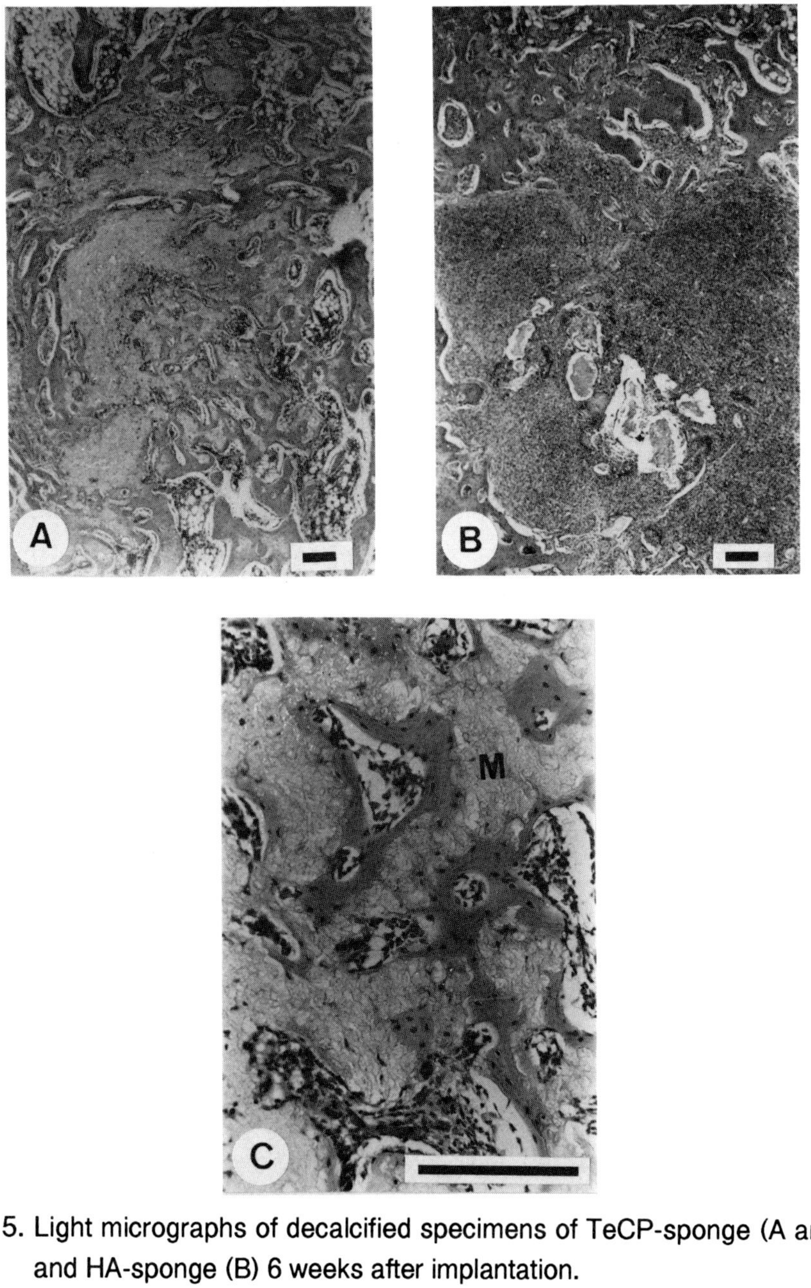

Figure 5. Light micrographs of decalcified specimens of TeCP-sponge (A and C) and HA-sponge (B) 6 weeks after implantation.
Note that ossification was observed at edge of TeCP-sponge matrix (M).
H.E. stain, Bar = 200 μm.

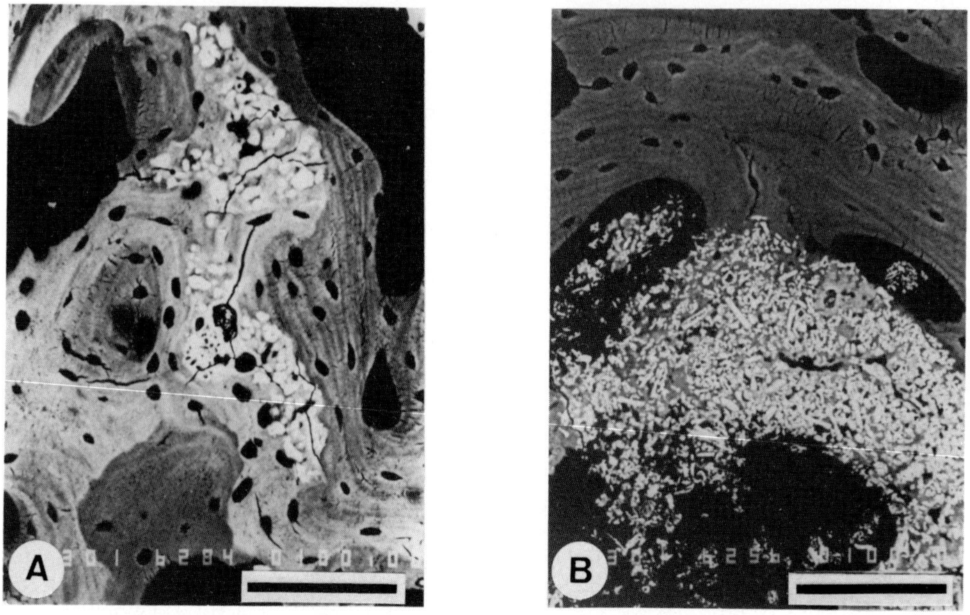

Figure 6. Backscattered electron micrographs of TeCP-sponge (A) and HA-sponge (B) 6 weeks after implantation. Bar = 100 μm.

REFERENCES

1. Blanchy, B. G., Cowlett, P. R. and Gantheron, D. C. J. Biomed. Mat. R. 1986, 20, 469-479.

2. Oonishi, H., Tsuji, E., Mizukoshi, T., Kushitani, S., Aono, M., Minami, K., Watanabe, A., Ogino, A. and Fujisawa. N. In Bioceramics, Butterworth-Heinemann, 1991, 4, 191-197.

3. Oonishi, H., Tsuji, E., Kushitani, S., Aono, M. and Minami, K. In Bioceramics, Kobunshi Kankokoi, 1992, 5, 147-153.

Hydroxylapatite Coatings

Bioceramics, Volume 7 Edited by Ö.H. Andersson and A. Yli-Urpo
(*Proceedings of the 7th International Symposium on Ceramics in Medicine, Turku, Finland, July 1994*)

Bone Apposition on Apatite Coatings Inhibited by Large Gaps

J.A.M. Clemens,[1,2] C.P.A.T. Klein,[1] W.J.A. Dhert,[3] P.M. Rozing[2] and K. de Groot[1]
[1] Dep. of Biomaterials, University Hospital, Bldg 55, Rijnsburgerweg 10, 2333 AA Leiden, The Netherlands
[2] Dep. of Orthopaedics, University Hospital, Leiden
[3] Dep. of Orthopaedics, University Hospital, Utrecht

ABSTRACT: The bone formation in 1 and 2 mm. gaps around implants was studied in goats. Implants were coated with fluorapatite, 30 and 62% crystalline hydroxylapatite respectively; gritblasted titanium was used as control. After 6 weeks bone tissue covered large parts of the coated implants with 1 mm gaps. The coated implants with a 2 mm. gap showed only minimal bone ingrowth. In these cases fibrous tissue was seen on the coating. The different apatite coatings did not show significant differences. Histometry confirmed these findings.
It can be concluded that gaphealing is significantly reduced by large gaps.

INTRODUCTION

Hydroxylapatite is known for its good biocompatibility and osseointegration. The material has been studied extensively in vitro and in vivo. Clinical applications can be found in dentistry and orthopaedics[1]. Originally, HA was studied and used as a bulkmaterial. However, its brittle character made it less suitable for load-bearing conditions.
One of the major contributions of hydroxylapatite coatings is its capacity to give bone fixation despite non-pressfit implantation, so-called gap-healing.
Nowadays research focusses on the application as a coating on a strong substrate material, for example titanium alloys. Plasma spraying appeared to be the most suitable coating proces.
The extreme temperatures during plasma spraying are known to change the crystalline structure of HA[2]. This can quantified by the degree of crystallinity. Materials with lower crystallinity are known to dissolve faster, which on the short term may show a faster osseointegration[3].
Fluorapatite is more thermostable and shows less dissolution, due to the crystal structure[4,5]. It shows good bone bonding, and might be an alternative for hydroxylapatite[6].
Gaps up to 1 mm. are known to fill with bone and direct bone apposition on the hydroxylapatite coated surface is seen within 6 weeks, in contrast to non-coated titanium implants which show only a minimal gaptolerance[7].

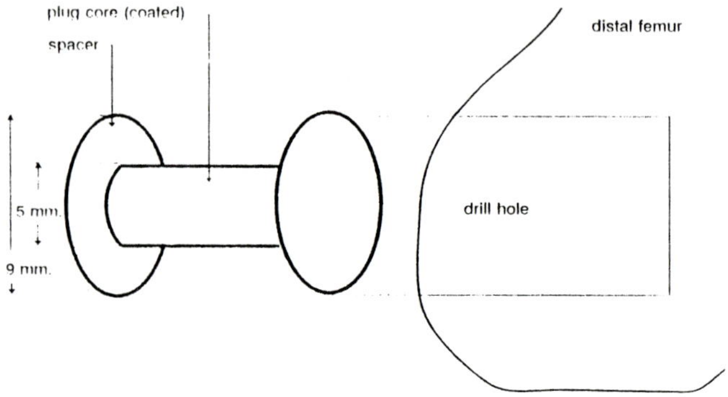

Fig. 1: gapmodel according to Soballe

We studied the gap healing character of three different calciumphosphate coatings, and studied whether an upper limit of gaphealing exists.

MATERIALS AND METHODS

Sample preparation

Ti-Al6-Va4 cylindric implants were machined (See Figure 1). Gaps of 1 or 2 mm. were created by spacers on the extremities of the implants.

After gritblasting, the implants were plasma sprayed.

High crystalline hydroxylapatite (HA) or fluorapatite (FA) powder was used for plasma spraying. 30% and 62% crystalline HA coatings and a 70% crystalline FA coating were obtained.

Except for the plasma spray procedure, the uncoated controls were given the same treatment.

Before implantation the implants were cleaned and sterilised by autoclavation.

Surgical and histological procedure

16 Healthy female Dutch goats were operated under general anaesthesia. Each goat received four implants, each with a different surface.

7 or 9 mm. holes were drilled in the medial and lateral precondylar trabecular bone of both femora and the implants were inserted with the spacers pressfit.

After six weeks, the goats were killed by barbiturate overdose.

The plugs with surrounding bone were removed, dehydrated in ethanol and embedded in MMA.

Histological sectioning was done by a special microtome, allowing approximately 10 μm thick sections of undecalcified bone and titanium implant. Methylene blue and basic fuchsine stainings were used.

Qualitative light microscopy was performed.

Histometric analysis was done with VIDAS, measuring the percentage of bone contact on the coated or gritblasted surface. Also, the relative amount of bone ingrowth between the spacers was determined.

RESULTS

Two goats needed open reduction of a dislocated patella. No further complications

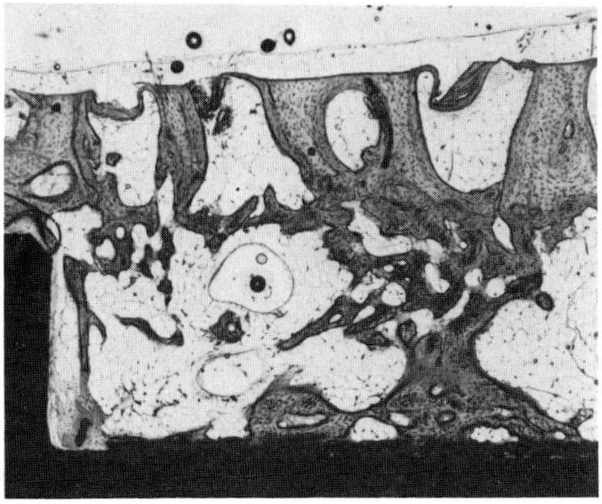

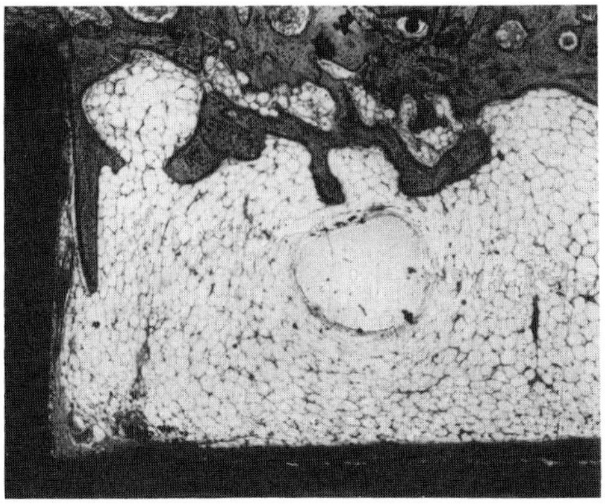

Figure 2 (left) one millimeter gap model, demonstrating bone contact with the apatite coating (bottom). 1 mm. spacer on the left.
Figure 3 (right) two millimeter gap model demonstrating lack of bone contact despite the apatite coated surface (bottom). 2 mm. spacer on the left.

were seen. The animals walked without a limb after 2-3 weeks.
At sacrifice, all the animals were healthy. Gross inspection of the implantation site and the implants showed a normal appearance.

Uncoated implants
Uncoated titanium implants with a 1 millimeter gap showed a fibrous tissue lining over the implant surface.
Bone tissue filling the gap itself grew mainly from the surrounding bone in the periphery of the gap. Bone tissue close to the titanium surface remained separated from this surface by fibrous tissue. Minimal bone contact at the titanium surface was seen only in exceptional cases.
Most of the tissue filling the gap consisted of normal marrow tissue with large quantities of fat tissue.
Signs of wound healing phenomena were seen: erythocytes, lymphocytes and macrophages. Bone tissue and marrow tissue had a normal aspect.
Titanium implants surrounded by a 2 millimeter gap showed an identical tissue response: peripheral bone ingrowth, fibrous tissue covering the implant surface, and hardly any bone contact.

Apatite coated implants
In contrast to the uncoated specimen, bone tissue was in direct contact with the coated surface of the implants. No fibrous tissue interposition was seen. The amount of bone layed down on the coating varied, demonstrating gaphealing initiating from the coating. However, quantitavily, the amount of bone ingrowth from the periphery seemed more impressive. See Figure 2.
In areas left uncovered with bone, a varying tissue response was seen: some case showed fibrous tissue, others showed fatty marrow tissue lining the coating. Also, giant cells were seen, espescially in recessions in the coating.

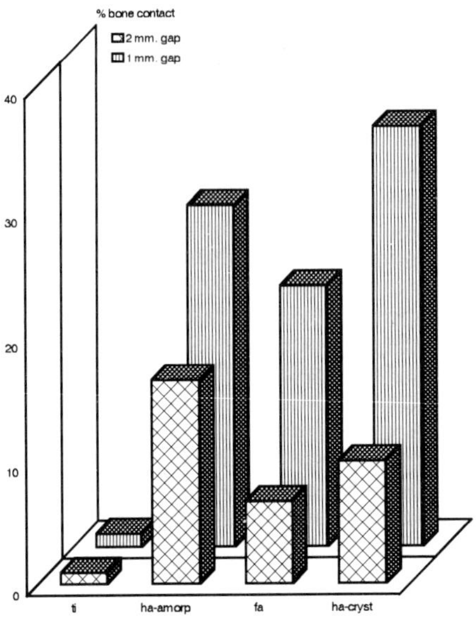

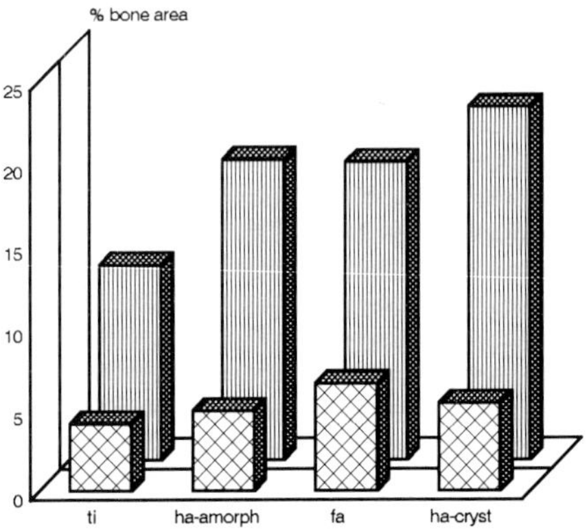

Figure 4 (left): Bone contact measurements.
Figure 5 (right): Bone area measurements.

Next to the bone tissue, the gap was filled with normal appearing marrow tissue, with large quantaties of fat tissue. Some wound healing phenomena were seen, identical to those around the uncoated implants.

Coated implants surrounded by a two milimeter gap showed much less bone contact and ingrowth into the gap. Large parts of the coating were covered with fibrous tissue or fatty marrow tissue. See Figure 3.
In some cases, bone tissue approached the coating closely, but remained separated by a fibrous tissue layer.
Gaphealing hardly arose from the coating, but initiated from peripheral bone tissue.

No large differences were seen between the three different coatings.

Histometry by VIDAS image analysis confirmed the differences in both bone contact and bone ingrowth (Figure 4 and.5). The differences between coated and uncoated implants were significant. The differences between the three different coatings were not significant.

DISCUSSION AND CONCLUSIONS

In this research project gaphealing phenomena around calciumphosphate coated titanium implants were studied. The two different gapsizes allowed us to study whether an upper limit in gaphealing exists.
This is clinically relevant for the orthopaedic surgeon in prosthetic surgery, in which case large gaps can arise, especially in revision surgery.

Søballe[7] investigated relatively small gaps of one millimeter and described good bone apposition if a hydroxylapatite coating was used.
In our study we noted a reduced bone apposition if the gaps were enlarged from one to two millimeter. Instead of bone growth on the different apatite coatings, fibrous tissue was often seen covering the coating in the two millimeter cases. Bone was often seen laying separated from the coatings by a fibrous tissue lining.
This means that in orthopaedic surgery, gaps should be avoided.
Whether a longer follow-up would demonstrate the amount of bone contact to increase is unknown. In that case a delayed bony fixation of the orthopaedic implant can be expected, causing a longer time needed for recovery of the patient.
Filling the gap with allograft bone tissue might enhance bone growth in direct contact with the coating, by osteoconduction. Søballe found no contribution of this in combination with hydroxylapatite coatings, but he used a 1 millimeter gap, demonstrating already good bone apposition without the allograft. In these larger gaps a positive effect might be possible[8].
Varying the coating solvability by crystallinity or composition did not cause significant differences in gaphealing in this goat study.
Although a strongly reduced bone contact was seen if large gaps surrounded coated implants, in all cases they showed more bone contact compared to the uncoated implants.
It can be concluded that large gaps (≥1mm.) strongly reduce bone apposition on apatite coated bone implants in the early postoperative period.
The differences between different coatings used in this study are unsignificant.
It was confirmed that apatite coatings on bone implants cause a higher tolerance for inaccuracy. This effect does not seem infinite.

REFERENCES

1 Geesink, R.G.T., Hydroxylapatite coatings in orthopaedic surgery Raven press, USA, 1993.
2 Wolke, J.G.C., de Blieck-Hogervorst, J.M.A., Dhert, W.J.A., Klein, C.P.A.T., de Groot, K. J.Thermal Spray Technology, 1992, 1(1), 75-82.
3 de Bruijn, J.D. Calcium Phosphate Biomaterials: bone bonding and biodegradation properties. PhD-Thesis, Leiden, 1993.
4 Lugschneider, E., Weber,T., Knepper, M. Production of biocompatible coatings of hydroxyapatite and fluorapatite In: Proceedings of National Thermal Spray Conference, Cincinatti, 1988.
5 Klein, C.P.A.T., Wolke, J.G.C., de Groot, K. In: An introduction to bioceramics. World Scientific Publishing Co., Singapore, 1993, 199-222.
6 Dhert, W.J.A. Plasma-sprayed coatings and hard tissue compatibilty, PhD-Thesis, Leiden, 1993.
7 Søballe K., Hansen, E., Brockstedt-Rasmussen, H., Pedersen, C. and Bünger C. Acta Orthop. Scand. 1990:61(4):299-306.
8 Søballe, K., Hansen, E., Brockstedt-Rasmussen, H., Pedersen, C. and Bünger C. Clin. Orthop. 1991:272:282-293.

Bioceramics, Volume 7 Edited by Ö.H. Andersson and A. Yli-Urpo
(*Proceedings of the 7th International Symposium on Ceramics in Medicine, Turku, Finland, July 1994*)

Hydroxyapatite Surface Coatings for Biomaterials: A Report of a Multi-Centre Investigation – Brite-Euram Project BREU/O172C

G.W. Hastings,[1] S. Morrey,[1] M. Davies,[2] F. Monteiro,[3] M. Barbosa,[3] S. Sousa,[3] C. Ribeiro,[3] R. Reis,[3] P. Tranquilli Leali,[4] A. Merolli,[4] D. Muster,[5] B. Demri,[5] M. de Sousa,[6] A.J. Coita,[6] M. Santos,[6] F. Arosa,[6] C. Faleiro,[6] D. Anderson,[7] P. Evans[8] and P. Harrison[8]

[1] London University, IRC in Biomedical Materials, Queen Mary & Westfield College, Mile End, London E1 4NS, UK. [2] Department of Pharmaceutical Sciences, University of Nottingham, University Park, Nottingham, UK. [3] Departamento de Engenharia Metalurgica, University of Porto, P-4099 Porto Codex, Portugal. [4] Universita Cattolica del Sacro Cuore, Istituto de Clinica Ortopedica, Agostino Gemelli 8, Roma, Italia. [5] L.E.E.D. Biomateriaux, Laboratoire de Recherches de la Chaire de Stomatologie, CHRU BP 426 67091 Strasbourg Cedex, France. [6] University de porto, Mestrado de Immunologia, Instituto de Ciencias Biomedicas de Abel Salazar, 4000 Porto, Portugal. [7] Plasma Coatings Limited, Derbyshire, UK. [8] H.E. Associates, Leeds, UK.

ABSTRACT

Results from a multi-centre investigation of hydroxyapatite coatings applied to titanium alloy, stainless steel and chromium-cobalt alloy are presented and include surface characterisation, corrosion, immunological and in vivo studies. The protective effect of coatings is shown and titanium can form compounds with HA replacing calcium. The importance of immunological screening and the specification of the crystallinity of the coatings is emphasised.

INTRODUCTION

The project addressed the aim of providing the orthopaedic surgeon with a well characterised bioactive coating on hip joint replacement prostheses, based on hydroxyapatite ceramic. The requirement for such a coating arose from the concern with the use of prostheses secured into the bone with a cold-curing acrylic cement, which were showing an incidence of loosening of around twenty percent of all prostheses inserted. The considered opinion now, however, is that the causative factor in the loosening is the polyethylene of the acetabular component , and that the wear particles developed lead to progressive osteolysis. In fact the acrylic cement may even have a protective effect by blocking the polymer particles from entry to the medullary space. However, bioactive ceramic coatings do provide a very effective means for securing a device into bone.

It was the purpose of the project to evaluate a hydroxyapatite - HA - coating material and the system used to apply it in a multi-centre investigation using physical and biological methods.

MATERIALS AND METHODS

Coatings

All the HA coatings were produced by a proprietary thermal spraying method developed by Plasma Coatings Ltd. to form highly crystalline, pure hydroxyapatite coatings from a hydroxyapatite powder manufactured by Plasma Biotal Ltd.. The substrates used were titanium 6-Al,4-V alloy, 316 & 350 stainless steel , and a cobalt chrome surgical alloy. Samples of all coatings were examined using powder X-ray diffraction methods on a Phillips diffractometer, using copper Ka radiation, at a power level of 50kV, and 35mA.

Surface analysis

SIMS and SIMS imaging methods were applied to the study of HA coatings and powders. For the first time, near-infrared Fourier transform Raman spectroscopy (FTRS) was applied to the study of the chemistry and structure of apatitic calcium phosphate coatings [1].

Corrosion

Degradation of HA coated metals in the presence of a simulated physiological solution, and of changes occurring in the HA itself, in contact with the solution, and with ions being released from the substrate metal was studied. Electrochemical tests included cyclic polarisation under D.C. conditions and potentiostatic measurements of current vs. time, electrochemical rest potential in open circuit conditions. Also electrochemical A.C. impedance spectrometry was used to characterise the behaviour of these complex interphases. Long term immersion tests were used, with periodic measurements of Ecorr, pH, and ion concentrations in solution.

Immunological studies

Iron (FeC_6H_5O7) and zinc ($ZnCl_2$) were tested in comparison with nickel ($NiCl_2$) and cobalt ($CoCl_2$) for their effect on six different surface molecules, known to be involved in recognition and activation processes, namely, CD4, CD2, CD3, CD8, HLA-ABC and HLA-DR. Peripheral Blood Mononuclear Cells (PBMC) obtained from venous peripheral blood of 14 female and 13 male healthy donors, after appropriate preparation, at a concentration of 1.2 $x10^6$/ml in a !ml volume were placed in wells where previously the metal alloy disk under testing had been introduced. Control cultures were set in wells without a disk, or without the disk and the antibody. A monoclonal anti CD3 antibody (DAKOPATTS; M 756) titrated previously, was added to the cell culture at a final concentration of 1/1000. [^{3}H]-Thymidine incorporation was measured after 72 hours culture after harvesting. Results are expressed in c.p.m.(counts per minute). The percent inhibition/ stimulation for each condition was calculated from the [^{3}H]-Thymidine incorporation in cpm..

In vivo work

The in vivo study was designed to evaluate the effect of the HA crystallinity and thickness of the coating and the effect of the metallic substrate. Crystallinity was greater than 90% in one group and between 70% and 60% in the other. Thickness was 50 and 100 μm. Forty-four cylinders of 25x3mm (length x diameter) Ti6Al4V and duplex 350 stainless steel were implanted in the distal femoral canal of young adult New Zealand White rabbits weighing approximately 2700 g without preference regarding their gender and retrieved at: 4 , 8 , 26 , and 34 weeks. Uncoated cylinders were controls for each

substrate material. Un-operated controls and "sham-operated" controls were included. Ethylene Oxide sterilisation was used.
After retrieval sections were analysed by polarised light microscopy and the surface of the blocks by back scattering electron microscopy (BSEM) after preliminary wet grinding and carbon coating of the block.

RESULTS

Coatings preparation
From the XRD examination the coatings were characterised as:

Coating	Relative crystallinity	Composition
CAPTAL 60, 150 μm thick	92%	HA + traces calcium oxide phosphate & a tri-calcium ortho phosphate
CAPTAL 30 50μm thick	91.5%	HA + traces calcium oxide phosphate
CAPTAL 30 150μm thick	90.7%	HA+ traces a-tri-calcium ortho phosphate

There is currently no agreed standard for the specification of HA coatings on implants.
The mean shear bond strength of HA coatings determined by loading at a constant cross head speed of 0.5 mm/min until failure was 29.3 MPa (SD 3.71) for freshly produced samples. After aging similar HA coatings in Ringer's solution for three months the average shear strength had fallen to 15 MPa (SD3.15). (ASTM C633 - 79 1985)

Surface characterisation

Changes in structure related to the corrosion resistance of BIOTAL powders on exposure to titanium (Ti) solutions were studied by monitoring the changes in the surface chemistry of the powders by SIMS and XPS. The relative amount of titanium present at the surface of these samples increased as the concentration of titanium in solution and also the time of exposure increases, in both the SIMS and XPS data.

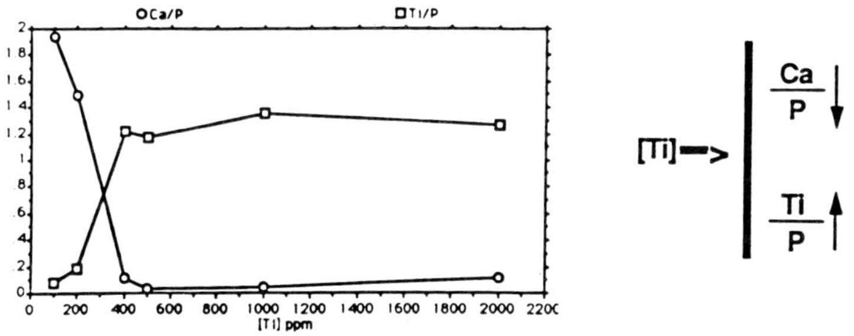

Figure 1: Ca/P and Ti/P versus (Ti) - EDS results of solid samples

This could be due to the preferential competition of titanium with calcium for sites within the hydroxyapatite crystal structure which is confirmed by the molecular nature of the SIMS data. For 500 ppm titanium solutions, most of the calcium is replaced by titanium at the surface of the hydroxyapatite after only 30 minutes. Titanium is rapidly incorporated into the surface of HA, and then diffuses into the bulk as time progresses. While the P/O ratio remained constant, the Ti/Ca ratio increased and the Ca/P ratio decreased as titanium exposure was increased.

The results from other ions may be summarised. Aluminium seems to be incorporated into HA without destruction of its crystal structure. Vanadium phosphate was found for high concentrations of V in solution. In the case of Fe and Cr, the substitution of Ca by the cations may occur but Ni does not seem to interact with HA. The titanium forms a very stable, amorphous compound identified as a titanium phosphate.

Corrosion studies[2]

Titanium alloy substrates for samples coated with Biotal 60 hydroxyapatite

Corrosion potentials E_{corr} for HA coated samples are more noble than for passivated ones.The 200 mm thick coating produces more noble corrosion potentials than the thinner ones. Grit blasting of metal surfaces promoted corrosion after cyclic polarisation. However HA coatings proved to be efficient barriers to solution attack on the metal, since no passive oxide film breakdown was observed in such cases. All coated and passivated samples have shown good resistance to crevice corrosion.

Six months immersion degrades HA, with the coating disappearing completely at some sites due to detachment.

Stainless Steel substrates for samples coated with Biotal 60 hydroxyapatite

200mm thick coatings were the most protective with respect to corrosion, and no passive film breakdown was found either for coated or passivated samples. After 6 months immersion, HA showed a lower degree of degradation and detachment than that seen for Titanium. The reason for that is the greater homogeneity of the HA coating systematically observed, in the case of stainless steel substrates.

Co-Cr-Mo alloy substrates coated with Biotal 30 and Captal 60 Hydroxyapatite

This alloy degraded under cyclic polarisation, with the formation of a Co,Ca and P rich precipitate, which presents small amounts of Cr. Co and Cr were able to migrate, crossing the Hap coatings towards the solution.

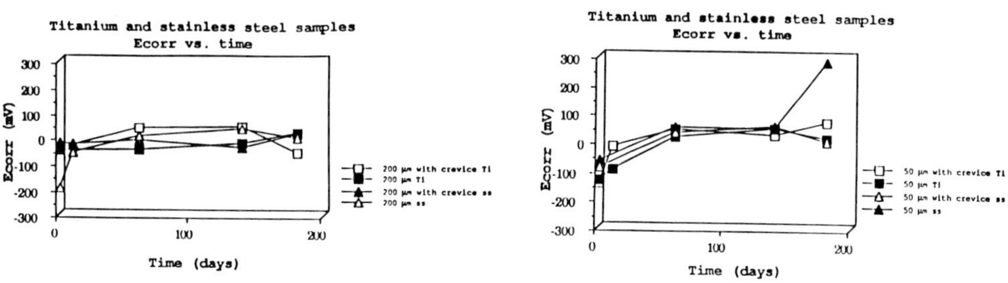

Figure 2 Charge versus time curves for titanium coated and uncoated samples

Immunology

T cell activation in vitro in the presence of Co-Cr-Mo and Ti6Al4V alloy disks showed that different metal alloys affect T cell activation differently.

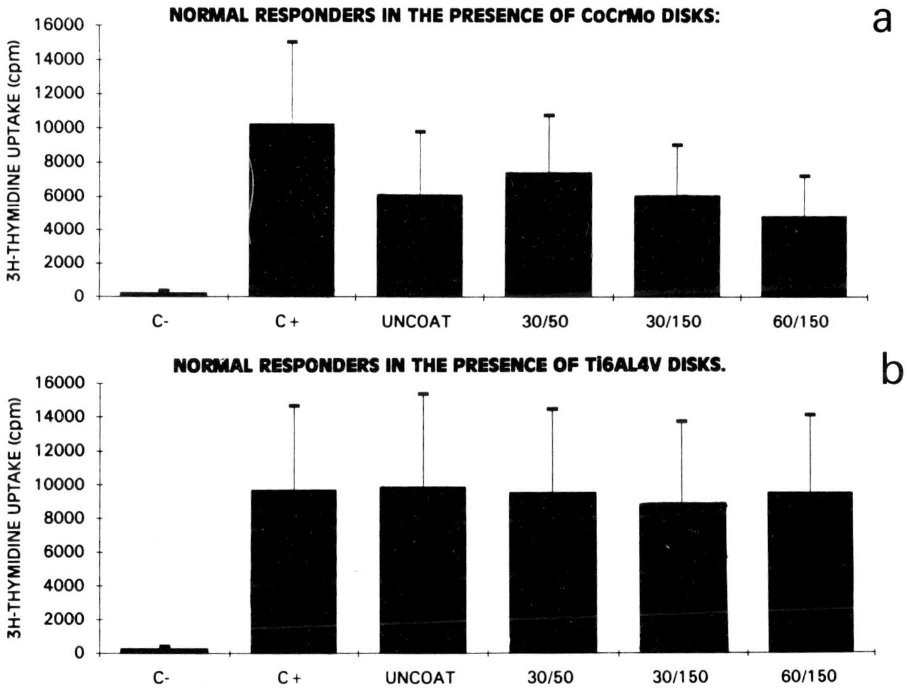

Figure 3 Immunology of HA coatings; cell response to metals showing low and normal responders.

As shown in Figure 3 the control responses could be divided into two groups. Twenty donors responded with high stimulation indices and cpm ranging between 3000 and 23000. Seven donors displayed low stimulation indices with cpm values ranging between 600 and 3000. The donors could thus be divided into "low" and "normal" responders according to their lymphocyte proliferation capacity measured by the uptake of [^{3}H]-Thymidine.

A distinctive trend repeated itself in all the donors, suggesting that the presence of such metal alloys could be "discriminated" by T-lymphocytes.

Activation in the presence of the uncoated Co-Cr-Mo alloy was diminished, relative to the control in most donors studied. It appeared that the HA coating did not significantly modify the response. The Co-Cr hydroxyapatite coating of Biotal 30/50mm seemed, however, to have a " protective effect ", since the inhibitory effect observed in the presence of the uncoated disk seemed "somewhat" recovered. In contrast to the results observed in the presence of Co-Cr-Mo, no inhibition of lymphocyte proliferation was observed in the presence of the uncoated or coated Ti6Al4V metal alloy discs

In vivo study

The full results have been reported elsewhere [3]

A highly crystalline coating seems to be more durable than a less crystalline one . Fragmentation and degradation of the highly crystalline coating is seen and detachments of the crystalline coating have been observed. Small localised areas of fragmented, detached and degraded coating can be observed in highly crystalline coatings up to 8 weeks though the majority of the coating has remained intact. Detachments appear between the coating and the metallic substrate, but seldom between coating and bone, and if so specimens were never older than 4 weeks. For a less crystalline coating localised areas of far advanced degradation can be observed in the early stages despite a nearly intact coating present all around. A less crystalline coating gives a different picture in the long-term often presenting a thinning of the coating and giving BSEM pictures of coating vanishing during integration with bone.

Correlation of thickness to time of degradation is not possible because areas of well preserved coating were seen together with areas where local removal/fragmentation had already completely exposed the metallic substrate, even in early stages. In general the majority of coatings continued to be adequately preserved in time as expected.

Figure 4. In vivo results showing fragmentation ofcrystalline coating at 8 months

Clinical X-rays were of no value in showing up different bone responses.

DISCUSSION

The results from corrosion studies and on HA powders/metal ion interactions suggest that hydroxyapatite may prevent the release of metal ions by acting as a barrier to their transfer from the underlying substrate. Since a chemical interaction between hydroxyapatite and titanium seems to have occurred, if titanium is released from a metallic substrate it could interact with the hydroxyapatite coating.

Setting up a reproducible in vitro method to test immunological functions in the presence of the metal alloy disks to be tested is possible. Implantation of devices should take into account the individual for whom it is being made. Only carefully designed long term serial clinical studies will decisively uncover the impact of a metal implant on a host's immunological system.

The implantation show that coating favours a more physiological integration of the implant compared with uncoated metal. Detachment at the metal-coating interface reflects a weaker bond strength in comparison with the bone-coating interface where detachments have been observed only during the first 4 weeks. Increase in coating thickness may probably lead to a reduction of detached areas but, at the same time, would increase problems related to the mechanical strength of the coating.

It is important to define the crystallinity : a highly crystalline coating (> 90%) is more stable in time but can give rise to fragmented bulky particles; a less crystalline coating (70%- 60%) is subject to slow degradation in the long-term but facilitates its substitution by newly formed bone.

ACKNOWLEDGEMENTS The assistance of the Commission of the European Communities, DGXII Brite-Euram Programme, is gratefully acknowledged.

REFERENCES

1. A. M. Tudor, C. D. Melia, M. C. Davies, D. Anderson, G. Hastings, J. D. Santos, F. Monteiro - "The analysis of biomedical hydroxyapatite powders and hydroxyapatite coatings on metallic implants by near-IR Fourier transform Raman Spectroscopy" - Spectrochimica Acta 49A, 4/6 675-680 (1993).

2. S. R. Sousa, M. A. Barbosa, "Corrosion Resistance of Titanium C.P. in Saline Physiological Solution with Calcium Phosphate and Proteins" - 6th Annual General Meeting of BIOSIS, 17-19 June 1993, Modena (Italy)

3. P. Tranquilli Leali, P. Gabbi, A. Merolli, O Palmacci, A. Cacchioli, G. Gonizzi - "Evaluation of different preparations of plasma-spray hydroxyapatite coating on titanium alloy and duplex stainless steel in the rabbit" - accepted for publication in Materials in Medicine.

Bioceramics, Volume 7 Edited by Ö.H. Andersson and A. Yli-Urpo
(*Proceedings of the 7th International Symposium on Ceramics in Medicine, Turku, Finland, July 1994*)

In Vivo Evaluation of Hydroxyapatite (HA) Sprayed by Different Coating Methods – One Year After Implantation

H. Oguchi[1] and G.W. Hastings[2]

[1] First Department of Oral and Maxillofacial Surgery, Tsurumi University School of Dental Medicine, 2-1-3 Tsurumi, Tsurumi-ku, Yokohama 230 Japan

[2] University of London, Queen Mary and Westfield College, Interdisciplinary Research Centre in Biomedical Materials, Mile End Road, London E1 4NS, UK

ABSTRACT

We have evaluated by different coating methods (plasma coating, high-velocity flame-coating) in vivo one year after implantation.

1. Thickness of Coated HA: The thickness decreased compared with the initial values. However, there were no significant intergroup difference in thickness. The minimum thickness was 0 μm in each group.
2. Biological Findings: Basically, there were no biological differences; sprayed HA come into direct contact with bone without fibrous tissues.

INTRODUCTION

Since the 1987 report by De Groot and Geesink of plasma-sprayed coating technique (PSCT) of hydroxyapatite (HA) on Ti-6A1-4V cylindrical rods (1), many researchers have evaluated its bone biocompatibility, shear strength, fracture surface after push-out testing in vivo and in vitro, and clinical performance. On the other hand, Jansen et al. have reported that progressive dissolution was observed for all coatings (various HA coating particle size distribution), and occasionally in some areas the HA coating totally disappeared for 12 weeks after implantation in the tibia of rabbits (2). Dhert (3), Klein (4), Bruijn (5) have also reported the resorption of PSCT-HA. In 1988, we developed a high-velocity flame spraying (HVFS) technique for applying HA onto titanium alloys (6). At the 10th European Conference on Biomaterials (1993) in Davos, we reported the characteristics of HA applied by different coating methods and biologically evaluated in vivo six months after implantation and in vitro (7). From these results, we have understood as follows: 1). From X-ray diffraction analysis, the HA layer of both Plasma-HA and HVFS-HA showed a well maintained structural pattern characteristic of HA. However, the plasma-HA on the plate was oriented to the C axis of HA.

2). In regard to surface microstructure, HVFS-HA was much more porous than Plasma-HA and Plasma-HA showed microbreakage due to the rapid reduction in temperature. 3). Solubility tests indicated the pH values of Plasma-HA and HVFS-HA to be comparable. The solubility of Ca^2 was higher for Plasma-HA than HVFS-HA. 4). In vivo, the thickness of the coated HA layer tended to decrease at 1 month, 3 months and 6 months for both Plasma-HA and HVFS-HA. However, Plasma-HA showed a greater decrease in thickness than did HVFS-HA, although the difference was not significant. 5). From biological findings, in the first month, Plasma-HA was superior to HVFS-HA and prominent bone tissue was noted on the surface of Plasma-HA. However, at 3 and 6 months, there were no differences biologically.

This study evaluated the thickness of sprayed HA and the response of bone to two different coatings (PSCT and HVFS) one year after implantation under the same conditions.

MATERIALS AND METHODS

For experimental studies, cylinders (2.0 mm diameter, 5.0 mm length) sprayed with HA by Plasma technique (Plasma Biotal Co., Ltd.; thickness: 48. 46 $\pm$ 6.71 μm) and HVFS technique (Asahi Optical Co., Ltd.; thickness: 47.16 $\pm$ 5.27 μm) were respectively prepared. White adult male rabbits of approximately 3.0 kg body weight were anaesthetized by intravenous injection of a pentobarbital sodium solution. The animals were sacrified 12 months after implantation.
The bone, including the samples, was fixed in a 10% formalin solution for light microscopy and measurement of the thickness of sprayed HA. Samples were embedded in polyester resin. Sections were cut perpendicularly in the middle with a diamond cutter. One half was polished and coated by Au by the authentic method. The thickness of sprayed HA at four sites on the cortical bone side and two sites on the marrow side, randomly selected, were observed by scanning electron microscopy and later photographed at a magnification of 350 times and measured. The remaining half was prepared at a thickness of 50 μm and stained with toluidine blue statin. Statistical analysis was done using Student's T-test. Significance was defined as $p<0.05$.

RESULTS

1. Thickness of Sprayed HA

The thickness of sprayed HA dated for one year is summarized in Table 1. Table 2 indicates the distribution of thickness over one year. One year after implantation, HVFS-HA was 41.45 ± 10.73 μm (N = 456) and Plasma-HA was 41.10 ± 10.27 μm (N = 457). The thickness decreased compared with the initial values. However, there were no significant intergroup differences in thickness. The minimum thickness was 0 μm in each group. There were no differences in the distribution of HA thickness between Plasma-HA and HVFS-HA. However, some values less than 10 μm (HVFS-HA: 3.5%, Plasma-HA: 2.6%) were noted for the first time at one year after implantation.

	1M-H	1M-P	3M-H	3M-P	6M-H	6M-P	1Y-H	1Y-P
AVERAGE	46.06 μm	44.66	45.92	41.61	49.16	45.17	41.45	41.1
S.D	6.73 μm	7.35	6.87	9.25	5.33	7.58	10.73	10.27
MIN	15.75 μm	14.7	15.75	21	28.86	13.65	0	0
N	446	455	362	168	243	261	459	459

(H:HVFS P: PLSAMA)

Table 1: Thickness of Coated HA

	1M-H	1M-P	3M-H	3M-P	6M-H	6M-P	1Y-H	1Y-P
<10	0 (0%)	0 (0%)	0 (0%)	0 (0%)	0 (0%)	0 (0%)	16 (3.5%)	12 (2.6%)
10=<20	3 (1.0)	2 (0.4)	4 (0.1)	0 (0)	0 (0)	1 (0.4)	2 (0.4)	6 (1.3)
20=<30	4 (1.0)	19 (4.1)	9 (2.5)	25 (14.9)	3 (1.2)	12 (4.6)	28 (6.1)	25 (5.4)
30=<40	74 (16.6)	104 (22.9)	56 (15.7)	48 (28.6)	9 (3.7)	43 (16.5)	143 (31.2)	115 (25.1)
40=<50	206 (46.2)	206 (45.3)	165 (45.6)	52 (30.9)	92 (37.9)	119 (45.6)	182 (39.7)	222 (48.4)
50=<55	159 (35.6)	124 (27.3)	128 (35.4)	43 (25.6)	139 (57.2)	86 (33.0)	88 (19.2)	79 (17.2)

Table 2: Distribution of Coated HA

2. Biological Findings

Basically, there were no biological differences; sprayed HA came into direct contact with the lamellar bone without intervening fibrous tissue (Figures 1, 2). We did not observe any multinucleated cells or osteoclasts at the HA-bone interface. Although sprayed HA maintained most of its thickness, we observed complete resorption of both types of sprayed HA at some areas on the bone marrow side (Figures 3, 4).

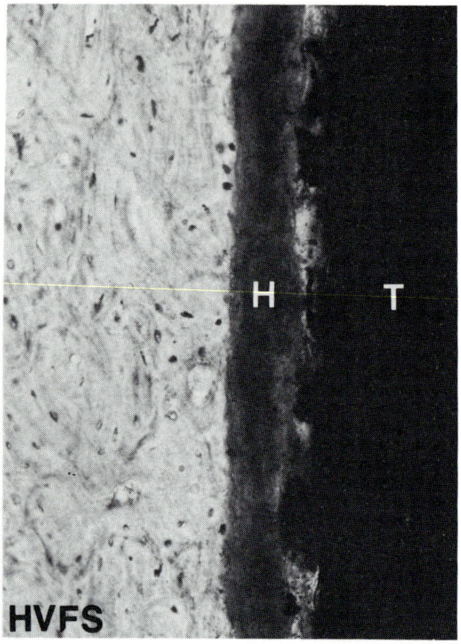

Figure 1

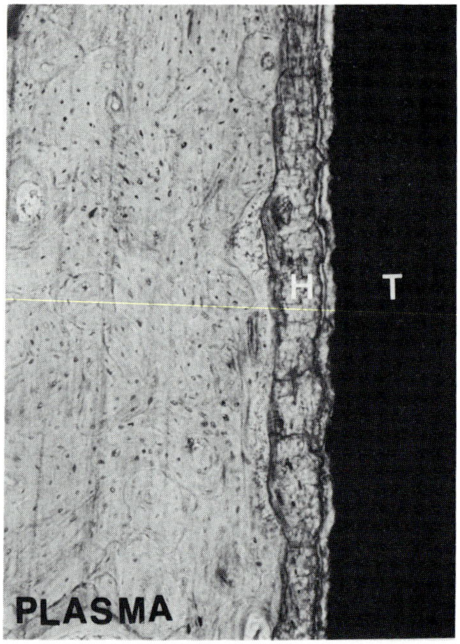

Figure 2

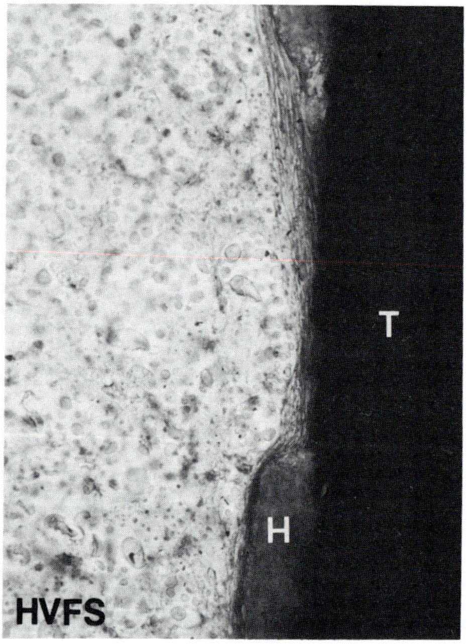

Figure 3

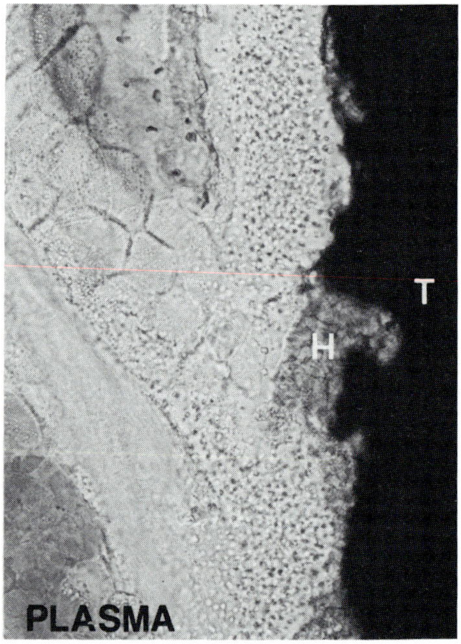

Figure 4

CONCLUSION AND DISCUSSION

The purpose of coating implants with HA is to promote the formation of chemical bonding to bone, thereby enhancing the growth of bony tissue at the implant surface and preventing metal-ion release from the implants. Recently, many investigators have described the properties of Plasma-HA applied under different conditions (thickness, Ca/P, size of starting powder, crystallinity, etc.). At present, the application and benefit of HA-coating is especially based on an initial increase in bone ingrowth. Regarding the resorption of Plasma-HA, Maixan et al. have reported that poorly crystallized HA coating underwent the most degradation, whereas an amorphous coating showed greater stability in a saline environment (8). Munting have also reported that over 85% of the coating was resorbed from long-term implants under functional loading in dogs (9). Klein et al. have reported that HA particle size influences the dissolution rate of HA coating in vitro (10). On the other hand, Bauer et al (11). and Bruijn et al (5). have reported the active cellular resorption of coated-HA.

This time, we have evaluated the thickness of coated-HA applied by different coating, methods and its biological behaviour. HA applied by these methods differed with regard to Ca dissolution rate, density and microsurface structure. Our results have indicated that there were no differences in the thickness of coated HA and no biological behaviour difference in vivo at one year.

The reasons for the observed decrease in the thickness of HA-coating are still not completely understood. Several factors have been implicated in this process. Wolke et al. have reported that the plasma temperature, nature of the plasma gas, particle size of the powder, and chemical nature of the ceramic powder are important factors (12). Maxian et al. have also reported that the amount of Ca dissolution from HA-coating implants was strongly dependent on the chemistry of the coating and less dependent on pH or time of incubation (8). Wang et al. have reported that the crystallinity of calcium phosphate (13), and Chae et al. have reported the crystallinity and porosity of coated-HA influence dissolution (14). We have generally observed a decrease in thickness and focal resorption of HA on the bone marrow side. We have concluded that the bioresorption of coated-HA might occur gradually in vivo, and partial and focal resorption might influence the chemical properties of coated HA.

In conclusion, our results indicate that if the crystallinity is the same, density, microsurface structure and Ca dissolution rate were not related to the resorption of coated-HA. However, our study was done under stable conditions. On the other hand, the resorption rate of our plasma-HA was less than that previously reported for Plasma-HA. We have concluded that the reason for this is the different plasma coating technique. Although we have not performed push-out tests, we have concluded that there is not much difference between Plasma-HA and HVFS-HA. Maxian et al. have suggested that the mechanical interlock around uncoated rough implants was as strong as the bond provided by Ca/P-coated rough implants (15). We have also concluded that the shear strength to bone is determined by the structure of the implant itself irrespective of whether or not it is coated. We conclude that it is necessary to reinvestigate the responsible factors for optimal HA-coating conditions in vivo and in vitro and to evaluate the significance of HA-coating.

ACKNOWLEDGEMENTS

This study was carried out through the NIBB Cooperative Research Program (93-120) and was supported by a Grant-in-Aid Scientific Research (General Research Category B, Project Number 04454515) and in a part by a Grant-in-Aid for Special Project Research fro the Ministry of Education, Culture and Science to Goro Eguchi.

REFERENCE

1. De Groot K., Geesink, R.G.T, Klein, C.P.A.T. and Serekian.P, In J.Biomed. Mater. Res. 1987, 21, 1375-1381

2. J.A.Jansen., J.P.C.M van der Waerden and J.G.C.Wolke. In J.Biomed.Mater.Res. 1993, 27, 603-610

3. W.J.A.Dhert., C.P.C.M.Klein., J.G.C.Wake., E.A.van der Velde., K.de Groot and P.M.Rozing. In J.Biomed.Mater.Res. 1991, 25, 1183-1200

4. C.P.A.T.Klein., P.Patka., H.B.M.van der Lubbe., J.G.C. Wolke., and K.de Groot. In J.Biomed Mater.Res. 1991, 25, 53-65

5. J.D.de Bruijn., Y.P.Bovell., J.E.Davis and C.A.van Blitterswijk. In J.Biomed.Mater.Res. 1994, 28, 105-112

6. H.Oguchi., K.Ishikawa., S.Ojima., Y.Hirayama., K.Seto and G.Eguchi. In Biomaterials. 1992, 13, 471-477

7. H.Oguchi and G.W.Hastings. In Proceeding of 10th European Conference on Biomaterials. 1993

8. S.H.Maxian., J.P.Zawadsky, and M.G.Dunn. In J.Biomed.Master.Res. 1993, 27, 111-117

9. Munting.E. In Proceeding of 10th European Conference on Biomaterials. 1993, 98

10. C.P.A.T.Klein. In proceeding of the IMechE Seminar on Future Trends in Design, Materials, and Technologies of Orthopaedic Implant. London, 1991

11. T.W.Bauser., R.C.T.Geesink., R.Zimmerman., and J.T.Mcmahon. In J.Bone and Joint Surg. 1991, 73-A, 1439-1452

12. J.G.C.Wolke., J.M.A.de Blieck-Hogervorst., W.J.A.Dhert., C.P.A.T.Klein, and K.de Groot. In J.Thermal Spray Technology. 1992, 1, 75-82

13. B.C.Wang., E.Chang., C.Y.Yang., D.Tu. In J.Materials science:Materials in Medicine. 1993, 4, 394-403

14. J.C.Chae., J.P.Collier., M.B.Mayor., and V.A.Surprenant. In Ann.NY Acad. Sci. 1988, 523, 81-90

15. S.H.Maxiam, J.P.Zawadsky, and M.G.Dunn. In J.Biomed. Mater.Res. 1993, 27, 717-728

Bioceramics, Volume 7 Edited by Ö.H. Andersson and A. Yli-Urpo
(*Proceedings of the 7th International Symposium on Ceramics in Medicine, Turku, Finland, July 1994*)

Crystallinity of the Coating Directs the *In Vivo* Response to Hydroxyapatite in the Rabbit

A. Merolli and P. Tranquilli Leali
Clinica Ortopedica dell'Universita' Cattolica, l.go Gemelli 8, I-00168 Roma, Italia

ABSTRACT

Authors investigated biological response to some of the most relevant variables in hydroxyapatite (HA) coating of metals; they were studied in two options: a) crystallinity of coating (>90% and 70-60%); b) thickness of coating (50 and 100 microns); c) metallic substrate (titanium alloy and duplex stainless steel). Analysis of results highlights the importance of defining the crystallinity of coating to make forecasts on its *in vivo* behaviour: a highly crystalline coating is more stable in time but can give origin to fragmented particles; a less crystalline coating is subject to slow degradation in the long term but facilitates its substitution by newly formed bone. Aspects of bone remodelling at bone-HA interface suggest a direct cellular role which is more relevant in highly crystalline coatings while, with lower crystallinity, hydrolytic dissolution plays a major role.

INTRODUCTION

Hydroxyapatite coating of metallic implants for orthopaedic and dental use by plasma-spray technique has become a common procedure for several manufacturers. It should be pointed out that many variables are involved in the accomplishment of the final result [1].

Fixation of an implant by enhanced bone growth favoured by hydroxyapatite coating requires persistence of the coating and adequate bond strength to bone and metal. Crystallinity of the coating is important in relation to its persistence. Highly crystalline and less soluble coatings are generally preferred for implant fixation [2,3]; experiments have been performed in relation to this aspect [4,5,6] and favourable initial clinical results have been reported [7,8,9,10,11]. However the long term survival of the coating seems to be not assured and, apart from hydrolytic dissolution, cellular remodelling of the coating has to be considered [12,13,14]. Also the mechanical competence of the metal-coating interface in the long term remains a concern [15].

Thickness is another parameter which is subject to variation between different manufacturers and the range of coating thickness has been reduced from 200 to 50 microns in recent times. A greater thickness of a highly crystalline coating will give a brittle material prone to cracking under bending or shearing forces. Also recently new metallic substrates like cobalt-chrome alloy have been tested in substitution of the classical titanium alloy.

The aim of the Authors was to investigate the biological response to hydroxyapatite-coated cylinders when the following variables were fixed in two options: a) crystallinity of the coating; b) thickness of the coating; c) metallic substrate.

MATERIALS AND METHODS

Two sets of values were chosen for each variable. Crystallinity was greater than 90% in one class and between 70% and 60% in the other. Thickness was 50 microns in one class and 100 microns in the other. Metallic substrate was Ti-6A1-4V in one class and duplex 350 stainless steel (a highly corrosion resistant steel) in the other. Forty-four cylinders of 25 x 3 mm (length x diameter) were implanted in the distal femoral canal of young adult New Zealand White rabbits weighing about 2,700g without preference regarding their sex. Four retrieval steps have been designed at: -4 weeks (1 month); -8 weeks (2 months); -26 weeks (6 months); -34 weeks (8 months). Duplex 350 stainless steel cylinders were provided by SEIPI SpA, Milano (I); titanium alloy cylinders, hydroxyapatite powder and all the coatings were provided by PLASMA BIOTAL LTD, Tideswell (UK).

Crystallinity was derived from X-ray diffraction examination of sprayed material in comparison with the starting powder. Uncoated cylinders are present as controls for each substrate material. Un-operated controls and "sham-operated" controls (where the surgical procedure was performed but no cylinder was actually implanted) are also present.

The implantation protocol consists of the following steps: a) cylinders are sterilized in ethylene oxide and single-packaged in sterile envelopes; b) intramuscular Valium (5mg/kg) and Ketalar (50mg/kg) is administered; c) a lateral parapatellar incision of about 15 mm is performed, the joint capsule is opened and the patella dislocated medially exposing the femoral inter-condylar groove; d) a cylinder is implanted in the distal femoral canal. This protocol proved to produce no artificial detachment of the coating during the insertion of the cylinder.

The retrieval protocol consists of the following steps: a) the rabbit is placed in a special sealed chamber where the atmosphere is quickly saturated with CO and left there for 3 minutes; b) the femur is carefully dissected and sealed in a plastic tube in 70% ethyl alcohol; c) x-ray films of the retrieved femur are taken in antero-posterior and latero-lateral views.

The histological protocol consists of several steps; the basic phases are: a) the femur is squared in the distal portion; dehydrated in serial passages in alcohol and embedded in methyl-methacrylate; b) the specimen is divided in three blocks: proximal, middle and distal; c) 2 sections of a 100 micron of thickness are taken with a rotating diamond-saw microtome: a diaphyseal section and a meta-epiphyseal section; d) sections are analysed by polarized light microscopy; e) surface of the blocks is examined by back scattering electron microscopy (BSEM).

Morphological analysis is used to investigate the biological response to implantation procedure; bone growth towards HA coating; possible resorption of the HA coating; possible interactions of bone with the metal substrate.

RESULTS

For the sake of simplification specimens with crystallinity higher than 90% will be referred as "highly crystalline" while specimens with crystallinity between 70% and 60% and presenting an amorphous phase will be referred as "less crystalline". For the same reason specimens with a coating thickness of 100 microns will be referred as "thick" while specimens with a coating thickness of 50 microns will be referred as "thin".

Differences between uncoated and coated substrates. The response to uncoated samples has a morphological character of properly formed bone but growth is mostly directed to encase the implant and, furthermore, it is more pronounced where a rough surface is present while the largest area of the metal has no direct contact with bone. In contrast, the response to coated samples has a morphological character of tight apposition and pictures are present of bone substituting areas of coating particularly in the less crystalline samples. Morphological pictures of physiological bone-turnover (deposition of new bone and resorption of the old one) are clearly distinguishable at the bone-coating interface.

Differences between highly crystalline coating and less crystalline coating; regardless of thickness. Highly crystalline coating seems to be more durable in time than less crystalline coating where a reduction in volume is detectable. Fragmentation and degradation of highly crystalline coating is not, anyway, an uncommon finding and true detachments of the crystalline coating have been also observed. What seems to be relevant is that little localized areas of fragmented, detached and degraded coating can be observed in highly crystalline coating even in the early stages (that is to say within 8 weeks) despite the fact that the vast majority of the coating has remained intact. Detachments appear between the coating and the metallic substrate; when they have been very seldom observed between coating and bone, specimens were never older than 4 weeks. When fragmentation occurs, the highly crystalline coating gives rise to bulky particles whose subsequent fate is unknown: they could be swept away since they do not seem to slowly dissolve on site (figure 1A).

In less crystalline coatings, areas are occupied and remodelled and then substituted by newly formed bone. Even in less crystalline coating, localized areas of far advanced degradation can be observed in the early stages, despite a nearly intact coating present all around. Anyway less crystalline coating gives a different picture in the long term often presenting a thinning of the coating and giving BSEM pictures of coating vanishing during integration with bone (figure 1B).

To summarize it can be stated that while a clear cut interface is detectable in BSEM between highly crystalline coating and bone, this is not often true for less crystalline coating.

Differences between thick and thin coating regardless of crystallinity. Thick coating is more uniformly present in the early stages (4-8 weeks) in respect to thin coating. Anyway an equation *thickness = time of degradation* cannot be applied because areas of well preserved coating may be present together with areas where local exfoliation/fragmentation has already completely exposed the metallic substrate even in early stages.

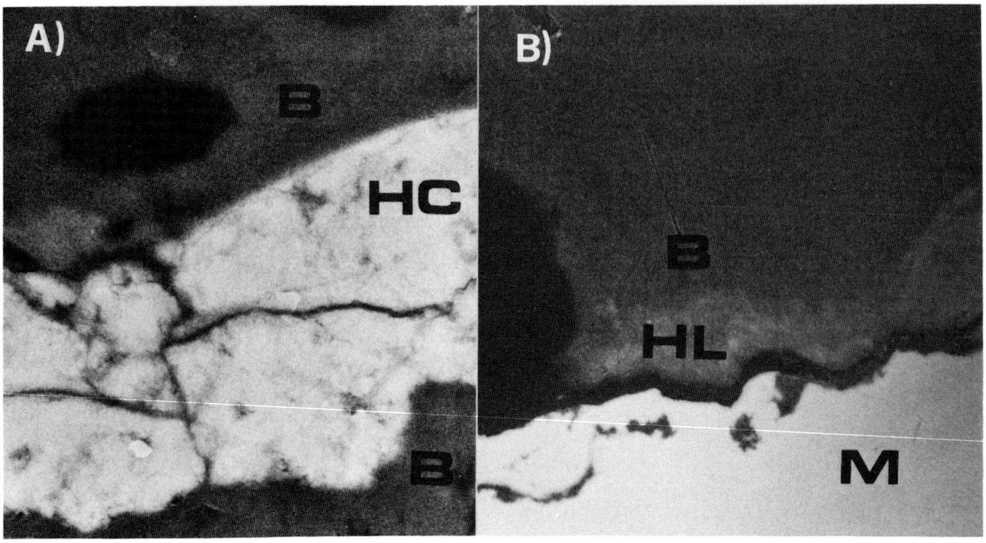

Figure 1. A) Fragmented bulky particle of highly crystalline coating (HC) embedded in bone (B) after 6 months. B) Less crystalline coating (LC) on metal (M) in the long term (6 months) presents a thinning and a BSEM picture of vanishing.

As a general rule the majority of coatings continue to be adequately preserved in time as should be expected. When degradation occurs it is more uniform along the circumference of the less crystalline coating and then a thicker coating is preserved longer; in highly crystalline coating the overall decrease in volume is associated with the fragmentation process already described and then thickness seems to play a minor role.

Differences between the same classes of coating on the two different substrates. As long as a good coating is present a difference in bone response related to the substrate does not seem to be present. When the coating has degraded or detached, bone response seems to be related primarily to the surface topography of the metal (more rough for steel than for titanium alloy). Pictures of adequate bone growth towards metal surface have been recorded both for stainless steel and for titanium alloy.

DISCUSSION AND CONCLUSIONS

The response to coated samples has a morphological character of tight apposition which can be followed by substitution of areas of coating by newly formed bone. Coating, then, favours a more physiological integration of the implant while uncoated-metal just produces an encasing response that confines the implant from the rest of remodelling bone. In this respect a coated implant should perform better from the biological point of view; anyway problems may arise with the mechanical competence of the coating because detachment at the metal-coating interface is often present. Detachments are produced in the large majority of cases as artifacts of embedding (due to retraction of the methacrylate). Anyway they are also observed *in vivo* and their presence, negligible in the first 4 weeks, increases in the long term (this fact discourage their explanation as artifacts of the implantation procedure).

Both as artifacts of embedding and as produced *in vivo*, these metal-coating detachments reflect a weaker bond strength in comparison with the bone-coating interface. Detachments at the bone-coating interface have been observed only, and very seldom, during the first 4 weeks when, even if a tight apposition between bone and coating is present, an ultramicroscopic interlocking has not always matured.

Thickness of the coating may play a role when a process of uniform degradation is expected; anyway the majority of the coatings continue to be adequately preserved in time. Even if a more uniform degradation can be ascribed to less crystalline coating while higher crystalline coating can give origin to fragmented bulky particles [16], a far more relevant finding is that localized areas of degraded or fragmented coating can be observed regardless of crystallinity even in the early stages (within 8 weeks). A further increase in coating thickness may probably lead to a reduction of these areas but, at the same time, would increase problems related to the mechanical competence of the coating.

When metal has been exposed, growth is more pronounced where a rough surface is present and this seems to highlight a primary bone response to topography. In our series, surface characteristics of coated duplex stainless steel cylinders differ from the uncoated controls because the latter are mirror-finished while the former have a more rough surface suited for HA coating; this accounts, in part, for the good response towards duplex stainless steel once the coating was lost.

The presence of sites of far advanced degradation close to a nearly intact coating was often observed and commonly in areas of active bone metabolism (both deposition and resorption). This stresses the importance of local cellular metabolism and seems to claim a role for a mechanism of active cellular remodelling of the coating probably more relevant than pure hydrolytic dissolution alone.

In conclusion, the present work highlights the importance of defining the crystallinity of coating in predicting its *in vivo* behaviour: a highly crystalline coating (crystallinity greater than 90%) is more stable in time but can give origin to fragmented bulky particles; a less crystalline coating (crystallinity between 70% and 60%) is subject to slow degradation in the long term but facilitates its substitution by newly formed bone. Aspects of bone remodelling at bone-HA interface suggest a direct cellular role which is more relevant in highly crystalline coatings while, with lower crystallinity, hydrolytic dissolution plays a major role. No relevant differences can be ascribed to a variation in coating thickness between 50 and 100 microns.

Furthermore it has been confirmed that bone response is related to the coating, then no differences are present regarding the *in vivo* behaviour when duplex stainless steel is used as metallic substrate instead of titanium alloy.

The authors would like to acknowledge project EU BE3048/89; Professors G Hastings, London (UK), M Barbosa, F Monteiro, C Sa', Porto (P), C Gabbi, Parma (I) and Dr. S Sousa, Porto (P).

REFERENCES

1. JE LEMONS, Clin Orthop, 235(1988) 220
2. RX LEGEROS, I ORLY, M GREGOIRE, et al, Trans Soc Biomaterials, 10(1987) 84
3. C GABBI, A MEROLLI, A CACCHIOLI Biomateriali, 1(1993) 17
4. SD COOK, K THOMAS, J KAY, Clin Orthop, 232(1988) 225
5. K DE GROOT, R GEESINK, CP KLEIN, et al, J Biomed Mater Res, 21(1987) 1375
6. B KOCH, JG WOLKE, K DE GROOT, J Biomed Mater Res, 24(1990) 655
7. TG ABRAHAMS, OD CROTHERS, Invest-Radiol, 27(1992) 779
8. JA D'ANTONIO, WN CAPELLO, OD CROTHERS, WL JAFFE, MT MANLEY, J Bone Joint Surg [Am], 74(1992) 995
9. RG GEESINK, K DE GROOT, CP KLEIN, Clin Orthop, 225(1987) 147
10. RG GEESINK, Orthopedics, 12(1989) 1239
11. RG GEESINK, Clin Orthop, 261(1990) 39
12. K GOMI, B LOWENBERG, G SHAPIRO, JE DAVIES, Biomaterials, 14(1993) 91
13. N TAKESHITA, T AKAGI, M YAMASAKI, T OZEKI, T NOJIMA, Y HIRAMATSU, N NAGAI, J Electron Microsc, 41(1992) 141
14. GJ WANG, TE BOMAN, KC CHUNG, et al, Trans Soc Biomaterials, 12(1989) 20
15. SD COOK, JF KAY, KA THOMAS, et al, Int J Oral Maxillofac Implants, 2(1987) 15
16. JD de BRUIJN et al, in "Bioceramics 5" (Butterworth-Heinemann, London, 1992) 291

Bioceramics, Volume 7 Edited by Ö.H. Andersson and A. Yli-Urpo
(*Proceedings of the 7th International Symposium on Ceramics in Medicine, Turku, Finland, July 1994*)

Enhancement of the Bone External Fixation Pin Interface: a Biomechanical and Morphological *In Vivo* Experimental Study

A. Moroni, V.L. Caja, C. Maltarello, N. Nicoli Aldini, S. Stea and M. Visentin
Third Department of Orthopaedic Surgery, Rizzoli Orthopaedic Institute, via G.C. Pupilli 1, 40136, Bologna, Italy

ABSTRACT

Eightyfour bicylindrical stainless steel external fixation pins were used. Half of the pins were coated with hydroxyapatite (HA). Six pins of the same type (HA coated or uncoated) were implanted monolaterally in the left tibiae of fourteen sheep, and a monolateral external fixator was assembled on the pins. Then the medial tibial mid-diaphysis was exposed and a bone cylinder five mm long was removed. Six weeks after surgery sheep were euthanized. Considering all the pins, independently from their position, mean extraction torque was 1485$\pm$1308 Nmm in the HA coated pins and 298$\pm$373 Nmm in the uncoated ($p=0.0001$). Histology showed extensive bony coverage of the HA coated pins, on the contrary severe fibrous tissue encapsulation was observed in the uncoated pins. The percentage of bone implant contact was 85.7$\pm$8.9 and 43.1$\pm$26.6 respectively in the HA coated and uncoated pins ($p<0.001$).

INTRODUCTION

The use of external fixation continues to spread. The bone pin interface however has remained an area of concern, because pin loosening may occur from mechanical or thermal damage to bone during pin insertion (1, 2, 3). Statistics on pin tract problems are difficult to interpret from literature, since both the threshold for diagnosing a pin tract problem and the quality of record keeping vary greatly among investigators. Edwards (4), estimated the incidence of pin problems reported from only those English language studies that discuss pin problems in some detail, use current generation fixators, and list more than thirty cases. For both the tibia and femur, infection or loosening developed in ten per cent of the pins, affecting thirty-four per cent of the patients. Pin problems are somewhat more common in the humerus and pelvis. In these locations, pin problems occur in sixteen per cent of the pins and affect almost forty per cent of the patients.

Aro et al. (1), and Pettine et al. (3), in recent experimental studies, demonstrated that a severe progressive reduction of pin extraction torque occurs with time independently of the pin position. On the other hand, synthetic hydroxyapatite (HA) has shown the ability to bind chemically to bone (5). The high biocompatibility and the low degradation of this bioceramic material were also demonstrated (5). For this reason, HA was proposed as a suitable coating material to provide stronger early fixation of uncemented prostheses (6).

The hypothesis of this study was that HA coating of external fixation pins could improve the bone pin interface strength (7). The purpose was to compare the radiographic, biomechanical, morphological and histomorphometric results of two different groups of external fixation pins, HA coated and uncoated, implanted into sheep tibiae.

MATERIALS AND METHODS

Eightyfour bicylindrical stainless steel external fixation pins were used. The outer thread pin diameters were 4 and 5 mm. Pins were divided in two groups of fortytwo pins each (group A and B). Group A pins, named Biopins were plasma sprayed with HA (Biocoatings, Flametal, Fornovo Taro, Pr, 43045, Italy), and group B pins remained uncoated. The cristallinity ratio of HA, calculated after X ray spectroscopy, was more than 70%, and purity by mass spectroscopy, more than 97%. The Ca/P ratio was 1.67$\pm$0.01. The thickness of the HA coating was 30-60 µm. The HA powder presented small traces of heavy elements below the limits set by the ASTM F1185-88 standard test (As < 2ppm; Cd < 1ppm; Hg < 1ppm; Pb < 1ppm; total heavy elements < 30 ppm; Cu < 1 ppm; Mn < 100 ppm; Fe < 100 ppm).

Fourteen adult female sheep of similar size and weight (45 kg average) were selected for this study. Maturity was determined by radiographic examination of the physes of the proximal tibiae and distal femora to confirm closure. General anesthesia was induced and maintened with intravenous Pentobarbital Sodium (55-65 mg/kg). Perioperative antibiotics (Cefazolin 22 mg/kg) were given starting the day before surgery, for three days. Using predrilling and tapping insertion technique, six pins of the same type (HA coated or uncoated) were implanted monolaterally in the left tibiae of fourteen mature sheep, and a monolateral fixator was assembled on the pins. During pin implantation, insertion torque was measured with a torque measuring machine (BLM, Cusano Milanino, Mi, 2095, Italy). Then the medial tibial mid-diaphysis was exposed and a bone cylinder five mm long was removed in order to highly stress the bone pin interface. Following surgery, sheep were placed in recovery rooms in heated cages and closely monitored over twentyfour hours. Butorphanol (0.4 to 0.8 mg/kg i.m.) was given for pain relief during this period. Twentyfour hours after recovery the sheep were returned to regular housing facilities where they were monitored twice a day and allowed normal activity. Six weeks after surgery sheep were euthanized.

X-rays were taken immediately after surgery and when the animals were euthanized. Each radiograph was examined for the presence of pin tract rarefaction which was defined as any radiolucency around a pin at the entry or exit cortex or both. Extraction torque was measured on four pins from each animal (pins with position number one, two, five, six in the external fixator). Two pins from each animal (pins with position number three and four in the external fixator) were processed for morphological analyses. The retrieved bone pin blocks were embedded in methylmetacrilate and processed for

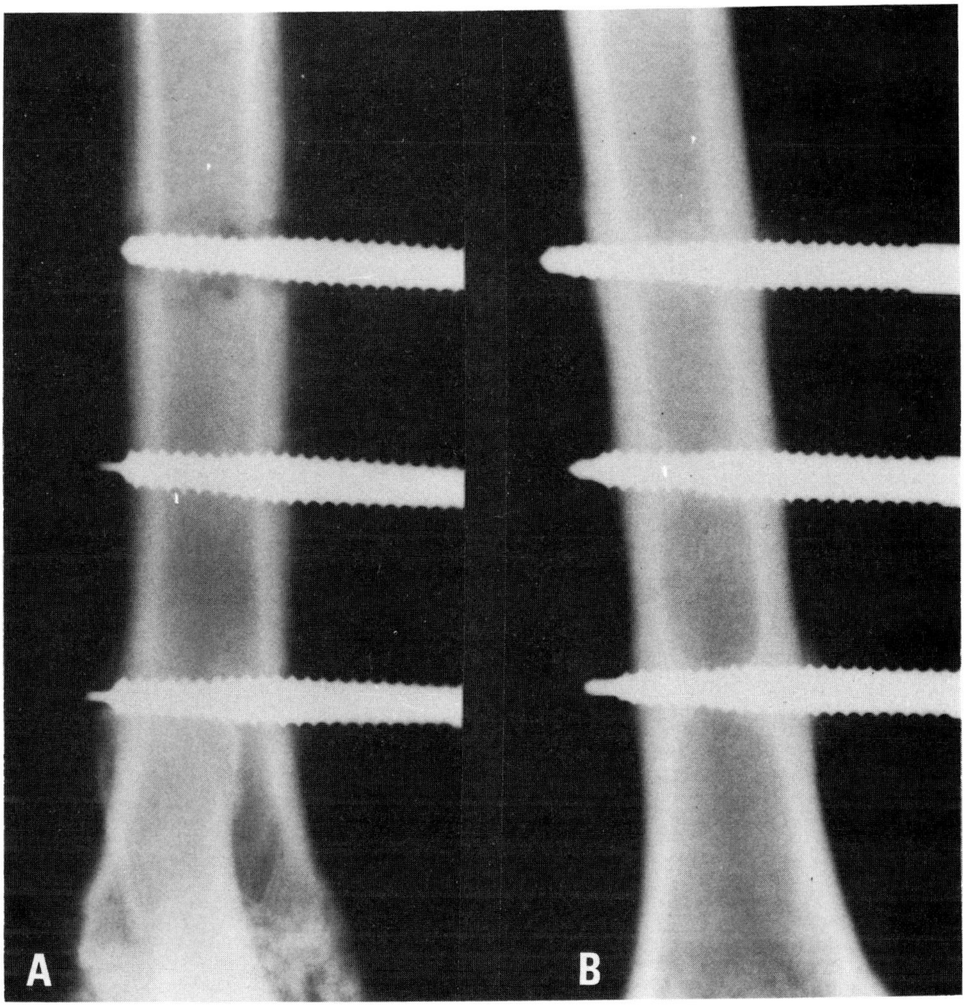

Figure 1 A: X-rays of uncoated pins. Severe bone pin tract rarefaction of the more proximal pin and less severe rarefaction of the remaining two pins are showed.

Figure 1 B: X-rays showing nicely osteointegrated HA coated pins. No sign of bone pin tract rarefaction can be seen.

undecalcified sectioning. Transverse sections were cut with a diamond saw (boron nitride blade, Exact system, Kulzer). After grounding to 30 μm thickness, sections were stained with fuchsin basic and light green. Photographs were taken at X16 magnification and the percentage of bone pin contact was measured on the entry cortex of the pins with position number four. BEI-SEM analyses were also performed to analyze the bone-pin interfaces in the pins with position number three and four.

Comparison of coated versus uncoated pins was performed by Student's t-test at a $p<0.05$ significance value. A Chi-squared test was used when the data being analyzed were either present or absent.

RESULTS

All the animals recovered from surgery and were walking normally. X-rays showed a very good appearance of the bone HA coated pin interfaces, on the contrary many areas of osteolysis were seen in the bone surrounding the uncoated pins. Radiographic rarefaction was observed in 16 (38%) HA coated pins and in 27 (64%) uncoated pins ($p<0.001$).

Considering all the pins, independently from their position in the external fixators, mean final insertion torque was 960 ± 959 Nmm in the HA coated pins and 709 ± 585 Nmm in the uncoated pins ($p=0.24$), mean extraction torque was 1485 ± 1308 Nmm in the HA coated pins and 298 ± 373 Nmm in the uncoated pins ($p=0.0001$).

Histology and BEI-SEM analyses showed extensive bony coverage of the HA coated pins, on the contrary severe fibrous tissue encapsulation and gap at the bone pin interface were observed in the uncoated pins. The percentage of bone implant contact was 85.7 ± 8.9 and 43.1 ± 26.6 respectively in the HA coated and uncoated pins ($p<0.001$). Histologically, no bone pin infection was observed in the HA coated pins. On the contrary, one (14%) infected pin was observed in the uncoated ones. The HA coating did not show either degradations or debondings from the metallic substrate.

DISCUSSION AND CONCLUSIONS

Our results demonstrate that HA coating significantly improved the bone pin interface. As expected, in the uncoated pins a severe decrease of mean pin extraction torque compared to the corresponding mean insertion torque occured. On the contrary, in the HA coated pins a remarkable increase of mean pin extraction torque compared to the corresponding mean insertion torque was observed. To our knowledge a similar biomechanical result had never been reported before.

The beneficial effects of the HA coating were confirmed by the histological, BEI-SEM and histomorphometric analyses which showed nicely osteointegrated HA coated pins with direct contact between bone and the HA coating. This nice osteointegration was responsible of the higher bone pin interface strength of the HA coated pins compared to the uncoated

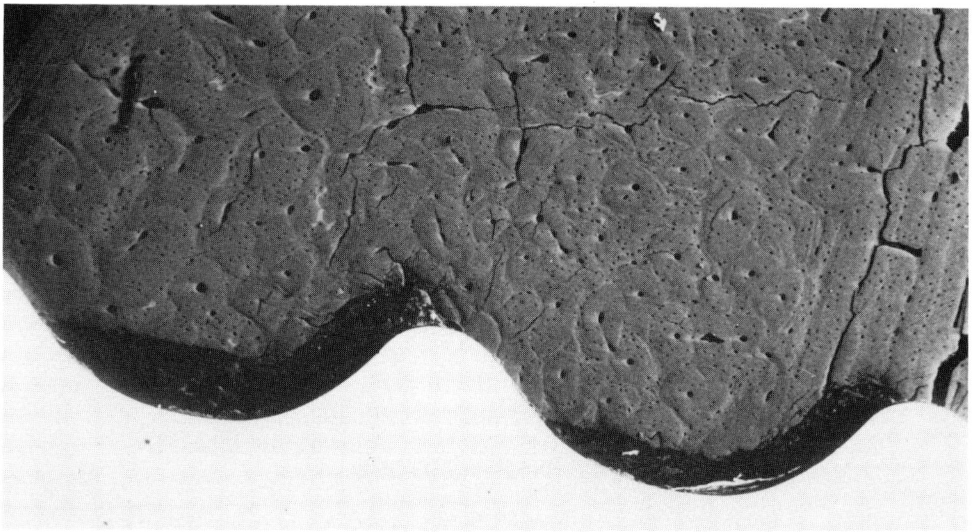

Figure 2 BEI-SEM of an uncoated pin. Extensive gap between the bone and the pin surface is showed.

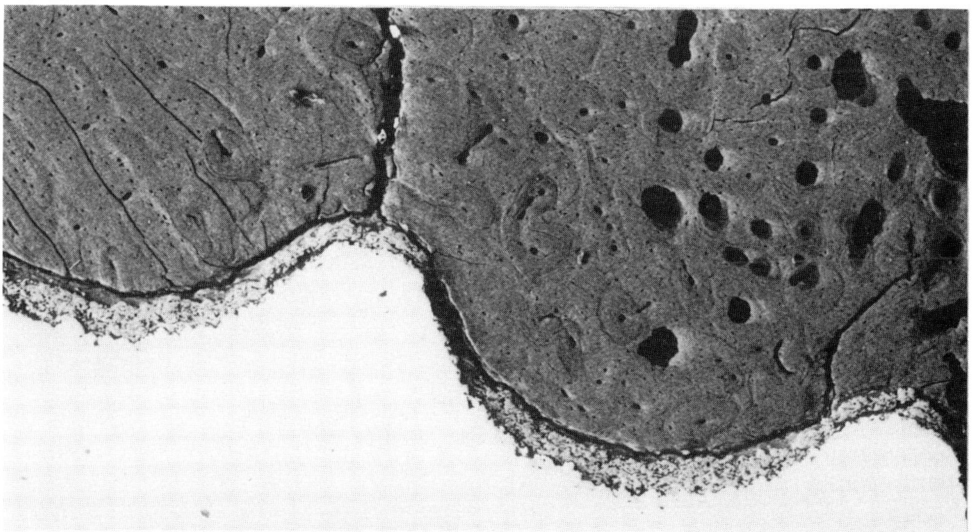

Figure 3 BEI-SEM of a HA coated pin. Direct contact between the bone and the HA coating without any interface gap is showed.

ones. This interface improvement could also be responsible for the reduction of the pin infection rate observed in the HA coated pins. On the contrary, in the uncoated pins histology showed an extensive fibrous tissue layer which was responsible of the severe deterioration of the interface strength.

On the basis of these results it can be concluded that clinically pin loosening should be significantly diminished by the use of the HA coated pins (Biopins), and consequently a remarkable improvement of the external fixation technique clinical results obtained.

REFERENCES

1 Aro, H.T., Markel, D.M., Chao, E.Y.S. In: Transactions of the 37th Annual Meeting. Orthopaedic Research Society, Anaheim, CA, US (1991) 447

2 Chao, E.Y.S., Wikenheiser, M.A., Pettine, K.A., Huiskes, R. In: Biomechanical and biological response of pin-bone interface. Proceedings of the 14th International Hoffman External Fixation Conference, Granada, Spain, (1992) 303-307

3 Pettine, K.A., Chao, E.Y.S., Kelly, P.J. Clin. Orthop. (1993) 293 18-27

4 Edwards, C.C. In: Complications in orthopaedic surgery 2nd Edition, Lippincott, Philadelphia, US (1986) 103-125

5 Cook, S.D., Thomas, K.A., Kay, J.F. Clin. Orthop. (1991) 265 280-290

6 Geesink, R.G.T. Clin. Orthop. (1990) 261 39-58

7 Moroni, A., Caja, V., Stea, S., Visentin M. In: , V., Chao, E.Y.S. In: Proceedings of the 61st Annual Meeting American Academy of Orthopaedic Surgeons, New Orleans, LO, US (1994) 104

Bioceramics, Volume 7 Edited by Ö.H. Andersson and A. Yli-Urpo
(*Proceedings of the 7th International Symposium on Cerámics in Medicine, Turku, Finland, July 1994*)

Recrystallization of Apatite in Plasma-Sprayed Coatings

J. Weng and X. Zhang
Institute of Materials Science and Technology, Sichuan University, Chengdu 610064, P.R. China

ABSTRACT

The starting hydroxyapatite (HA) powders with different initial characteristics, concerning absorbed water contents, OH^- groups in HA lattice and vacancies on the missing OH^- sites etc, were employed to determine the intrinsic factors influencing the recrystallization of apatite from the fused state. The influence of substrate properties, such as substrate type and temperature etc, on the recrystallization was also investigated. The results show that the retention of OH^- groups in HA lattice and the substrate temperature are two important factors in the nucleation of apatite from the fused state and dispersion of the resultant crystallites during the plasma-spraying coating.

INTRODUCTION

Hydroxyapatite (HA) used as implants can bond to bone without bone resorption around it.[1,2] The brittleness of bulk HA ceramics makes them incapable of being used in load bearing so that the HA coatings on metal implants have been developed.Among the methods of coating,the most widely employed is the plasma-spraying technique.[3]There were many investigations about in vitro and in vivo behaviours of the coatings in literatures,[4-7] showing the different dissolution of coatings from that of bulk HA ceramics. This is ascribed to the plasma-spraying-induced changes of HA.

One of factors which influence the stability of HA coatings is their crystallinity, which reflects not only the crystalline/amorphous (c/a) proportion but also the dispersion of the crystallites in the coatings. During plasma-spraying coating, there expected two results: first, the nucleation of a multitude of crystallites from the fused state and their rapid growth yielding a small grained crystalline phase; and second, the induced amorphous phase because fused droplets may be quenched so rapidly that some part will not have time to crystallize at the crystallization temperature. The well-crystallized HA coating showed no evidence of resorption in vivo[6] so that post-heat-treatment on as-received HA coatings was employed to promote the conversion of the amorphous to the crystalline phase.[5] Up to now, few researchers have systematically investigated the formation of super-fine crystallites or recrystallization of apatite from the plasma-sprayed droplets. The present study is focused on the intrinsic properties of starting HA powders and the substrate factors, such as substrate type and temperature etc, influencing the recrystalliztion of apatite in plasma-sprayed coatings.

MATERIALS AND METHODS

HA was wet-chemically synthesized in our laboratory and had no decomposition by sintering it at 1250 °C for 2 hours. Five starting powders for plasma spraying were obtained by grinding up the HA powders calcined for 2 hours at different temperatures: (1)100 °C; (2)700 °C; (3)900 °C; (4)1250 °C and (5)1250 °C then kept in the supersaturated atmosphere of water at 70 °C for 14 hours. Their particle sizes were all less than 70μm.

Both of titanium buttons(φ8×2mm) and glass plate(10×8×2mm)were used as substrate for the plasma-spraying coating. Titanium buttons were polished, sandblasted by SiC to obtain a rough surface. Both of them were degreased before the plasma-spraying coating.

The coating was obtained by injecting the starting powder into plasma through holes in the anode and spraying it onto substrates. The arc current and voltage were 70-120A and 60-80V respectively with nitrogen as plasma gas. The structure, OH^- content and morphology of the surface coatings were investigated by X-ray diffraction (XRD), infra-red analysis (IR) and scanning electron microscopy (SEM).

RESULTS

The used five starting HA powders were different in their crystallinity and content of OH^- groups in HA lattice. The crystallinity increased while the content of OH^- group decreased with the elevation of the calcining temperatures. According to the calculation by the well-known Scherrer formula,[8] the grain sizes (R_{211}) for starting powders calcined at different temperatures and corresponding plasma-sprayed coatings on titanium are listed in Table 1. After calcined at 900 °C for 2 hours, the starting HA powder possessed the perfect crystalline structure with little loss of lattice OH^- groups while grains grew bigger clearly and without absorbed water. On calcining at 1250 °C for 2 hours, there were considerable amounts of vacancies in the lattice because of the loss of OH^- groups while the crystallinity was quite high. This powder was a kind of oxyhydroxyapatite (OHA).

Table 1 shows that after the coating process, crystallites in all coatings were super-fine grains with sizes 10-100 times less than well-crystallized HA and different from those in the corresponding starting powders. The initially dispersed crystallites in the powders calcined at 100 °C and 700 °C grew bigger while the initially well-grown grains in the powders calcined at 900 °C and 1250 °C scattered after plasma-spraying. The bigger the initial crystallites and the less the contents of OH^- groups in the lattice, the bigger the resultant grains and the more the amorphous component after the plasma-spray coating.

By passing the glass substrates quickly through the plasma torch, single seperated particles of the sprayed HA were obtained to investigate the melt of sprayed droplets. Figure 1 gives the morphologies of the

Table 1 Crystallite sizes (R_{211}, Å) in five starting powders and corresponding coatings

Type	1	2	3	4	5
In powders	210	300	1280	30000	30000
In coatings	750	740	850	1810	1440

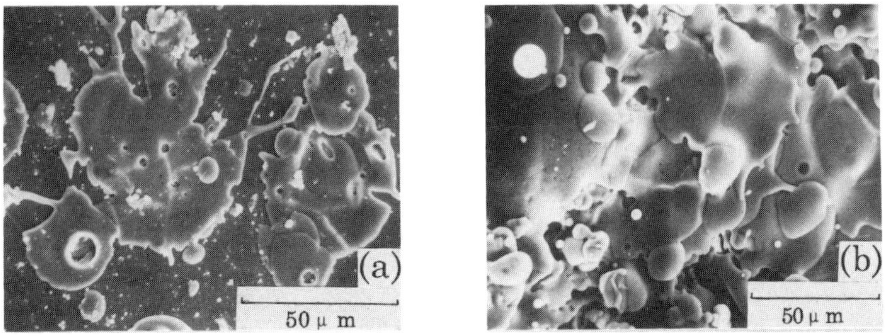

Figure 1 The SEM morphologies of as-sprayed particles on the glass plate (a) and the coating on the titanium button (b) respectively.

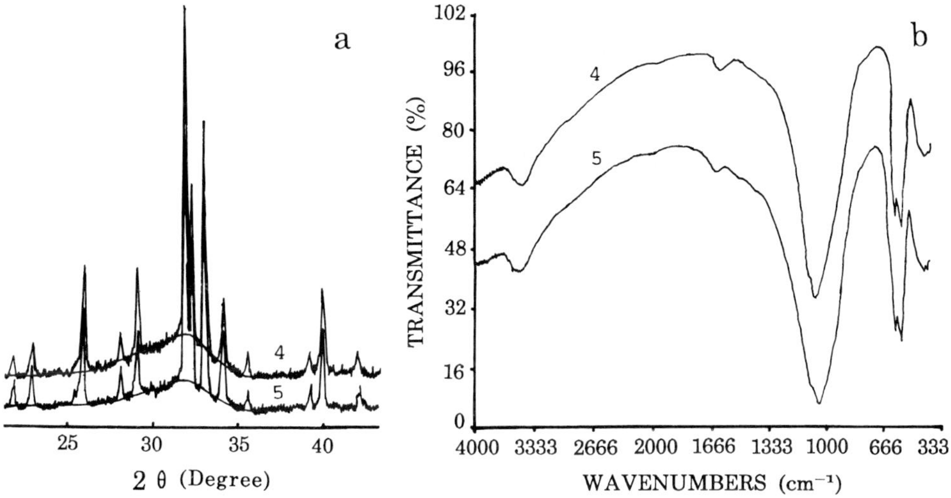

Figure 2 XRD patterns(a) and IR spectra(b) for two coatings with starting powders (4) and (5).

sprayed particles and coating, showing that all the sprayed particles were completely melted, which exhibited typical pancake-like morphology with no projecting cores owing to the unmelting parts.

The absorbed water influences the recrystallization of fused HA particles more or less. Figure 2 is the XRD patterns and IR spectra of two coatings prepared with the starting powders (4) calcined at 1250 °C for 2 hours alone and (5) then kept in the supersaturated atmosphere of water at 70 °C for 14 hours. Some OH^- groups existed in the coating (5) with the less amorphous component than in the coating (4).

Two coatings on the titanium button and glass plate prepared by the same plasma-spraying condition with approximate thickness of 50μm exhibited the same c/a proportion and crystallite sizes in their XRD patterns. Therefore, the recrystallization of apatite is little influenced by underlying substrates in the present coating process.

The influence of substrate temperature on the recrystallization was investigated by spraying the starting powder (4) onto titanium buttons with different initial substrate temperature. Figure 3 is the XRD

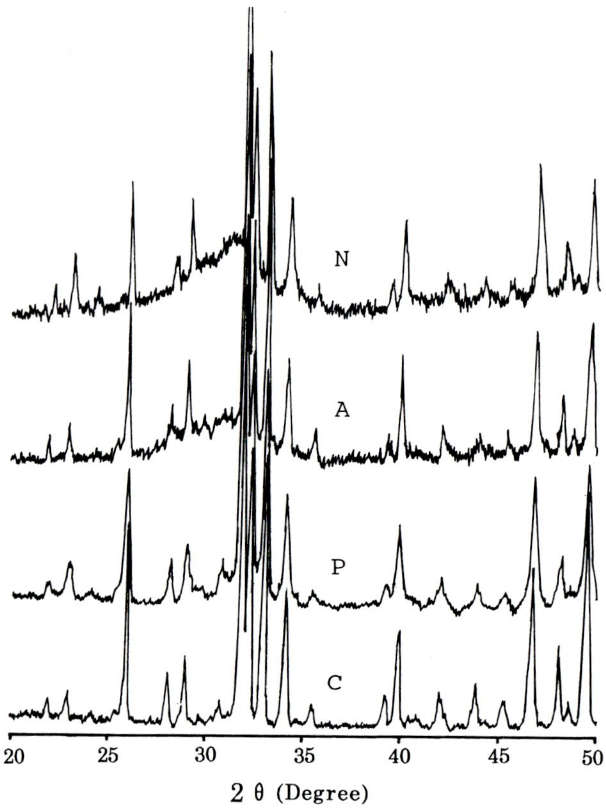

Figure 3 The XRD patterns of as-received coatings prepared with different initial substrate temperatures and by continuously plasma-spraying. (N: liquid-nitrogen-cooled; A: ambient; P: initially plasma-heated; C: continuously spraying)

Table 2 Crystallite sizes (R_{211}) in the four coatings

coatings*	N	A	P	C
R_{211}(Å)	2700	1800	600	1100

* Refer to the note in Figure 3

patterns of coatings with thickness about 50μm on the liquid-nitrogen-cooled, ambient and initially plasma-heated titanium substrates. There also gives the XRD pattern of the coating by plasma-saraying the powder (4) continuously on the titanium button initially with ambient temperature to thickness of 500μm. The c/a proportion increased with the elevation of substrate temperatures. The c/a ratio in the continuously sprayed coating with the thickness of 500μm was nearly the same with that in the coating formed on the preliminarily plasma-heated substrate. The crystallite sizes (R_{211}) in the four corresponding coatings are given in Table 2 by calculating from Scherrer formula. The crystallites were more scattering in less amorphous coatings when the starting powder was plasma-sprayed on titanium buttons with different initial temperatures to the thickness of 50μm.

DISCUSSION

During the plasma-spraying coating, starting particles experienced two processes: quickly melted as liquid droplets dispersed in the plasma exhaust gas, and rapidly solidified upon striking the substrate and being flattened. In the former process, all the particles, injected into the plasma gas, reached the same fused state before striking the substrate. During the second process, the liquified droplets were rapidly solidified with approximately cooling rate and finally formed the coating on the ambient titanium button or glass plate. This produced approximately the same properties of coatings with thickness of 50-100 μm on different substrates. Therefore, the properties of the resultant coatings were related to the nature of starting powders in the present condition with the ambient substrate temperature.

Rapid solidification when the starting powder was sprayed on ambient titanium buttons resulted in the partial recrystallization and amorphization of apatite from the fused state. The complete structure of apatite makes it tend to keep its amorphous or liquified state after the present plasma-spray coating process. The observations by kijima et al[9] indicated that the atoms in the HA structure are difficult to diffuse and move. Moreover, the recrystallization of apatite from the fused state related obviously to the intrinsic properties of starting powders. The starting powder with perfect apatite structure without vacancies on missing OH^- sites in the lattice was easy to nucleate from the fused state and resulted in the more scattering crystallites (Table 1). The absorbed water in the injected powder was ionized during plasma-spraying and provided OH^- groups to incorporate into the HA lattice and to form the coating with the existence of some OH^- groups (Figure 2). The more OH^- groups in the lattice, the easier the nucleation of apatite from the fused state so resulted in the more scattering crystallites and the less amorphous comperment (Table 1 and Figure 2).

The incorporation of OH^- groups to reoccupy the vacancies in the lattice is helpful to the recrystallization of sprayed apatite from the fused state. In our recent publication,[5] the OHA droplets were sprayed on the ice surface and well-crystllized HA was obtained because highly reactive OHA reacted with water to have OH^- incorporated into its structure to reoccupy the vacancies. Water was also reported to accelerate the conversion of the amorphous precusor phase to microcrystalline HA in the precipitation of HA.[10] Moreover, de Groot et al[11] found that water molecules reduced the plasma-sprayed amorphous phase so that to increase the crystallinity after as-received HA coatings were immersed in distilled water for weeks.

The different substrate temperatures produced the different cooling rates when the liquided droplets striked the substrates and solidified at about the same time interval. It is obvious that the higher cooling rate restrains more effectively the conversion of the amorphous to crystalline phase or the nucleation of apatite from the fused state. This was seen clearly from XRD patterns (Figure 3) of coatings with different substrate temperatures that the amorphous component increased with the decrease of substrate temperature. The lower cooling rate with the coating of the initially plasma-heated substrate was beneficial to the nucleation of apatite and resulted in the less amorphous phase and the more scattering crystallites (Figure 3 and Table 2).

The amorphous phase is metastable and converts into the crystalline phase when heat-treated in the temperature interval 600 to 630 °C. [11,12] During the continously spraying coating, the plasma gas heated the already-formed coating to a higher temperature while the following droplets striked up it. The plasma heating made the amorphous phase transform into the crystalline phase in one hand, and the hot already-formed coating promoted the recrystallization of apatite from the fused state. It was reported that the heat-treatment of as-received HA coatings leaded to the recrystallization of amorphous phase to form more dispersive crystallites.[12] Therefore, there was little amorphous phase with more scattering crystallites in the continously plasma-sprayed coating than in the as-received coating on the ambient titanium button (Figure 3 and Table 2).

CONCLUSION

The recrystallization of apatite from the fused state is influenced by intrinsic properties of starting powder such as its containing absorbed water, OH^- groups in HA lattice and vacancies on the missing OH^- sites etc. The retention of OH^- groups in HA lattice is an important factor to cause easy nucleation of fused apatite. More OH^- groups containing in HA lattice after plasma-spraying result in more crystalline phase with more dispersive crystallites in the as-received coating. The higher substrate temperature also results in the easy nucleation of apatite from the fused state with more super-fine crystallites. The continuously plasma-spraying heats the already-formed coating by the plasma gas, and causes the easy nucleation of apatite from the liquified droplets and the conversion of the amorphous to crystalline phase in the HA coating, which company the more scattering crystallites. In addition, the type of underlying substrates has little influence on the recrystallization of apatite by plasma-spraying HA powders onto them.

REFERENCES

1. Denessen, H.W., Walk, W., Veldhuis, A.A.H. and Van der Hooff, A., J. Prosthet. Dent. 1989, 61, 706-712
2. Jarcho, M., Clin. Orthop. Rel. Res. 1981, 157, 259-278
3. de Groot. K., Geesink, R.G.T., Klein, C.P.A.T. and Serekain, P., J. Biomed. Mater. Res. 1987, 21, 1375-1381
4. Maxian, S.H., Zawadsky, J.P. and Dunn, M.G., ibid. 1993, 27, 111-117
5. Weng, J., Liu, X., Zhang, X., Ma, Z., Ji, X. and Zyman, Z., Biomaterials 1993, 14, 578-582
6. Cook, S.D., Thomas, K.A., Kay, J.F. and Jarcho, M., Clin. Orthop. Rel. Res. 1988, 232, 225-243
7. Jansen, J.A., Van der Waerden, J.P.C.M. and Wolke, J.G.C., ibid. 1993, 27, 603-610
8. Klug, H.P. and Alexander, L.E., X-ray diffraction procedures for polycrystalline and amorphous material 2nd Edn., Wiley, NY,USA, 1971, 687-690
9. Kijima, T. and Tsutsumi, M., J. Amer. Cera. Soc. 1979, 62, 455-460
10. Boskey, A.L. and Posner, A.S., J. Phy. Chem. 1973, 77, 2313-2317
11. de Groot, K., Klein, C.P.A.T., Wolke, J.G.C. and de Blieck-Hogervorst, J.M.A., In CRC Handbook of Bioactive ceramics Vol.2, CRC Press, Boca Raton, FL, USA, 1990, 133-142
12. Zyman, Z., Weng, J., Liu, X., Zhang, X. and Ma, Z., Biomaterials 1993, 14, 225-228
13. Zyman, Z., Weng, J., Liu, X., Li,X. and Zhang, X., ibid.1993, accepted

Bioactive Composites and Cements

Bioceramics, Volume 7 Edited by Ö.H. Andersson and A. Yli-Urpo
(*Proceedings of the 7th International Symposium on Ceramics in Medicine, Turku, Finland, July 1994*)

Glass Reinforced Hydroxyapatite Composites: Physical Properties and Preliminary Histological Studies in Rabbits

J.D. Santos,* M. Vasconcelos,* R.L. Reis,* A. Afonso,* F.J. Monteiro* and G.W. Hastings‡

* INEB- Nacional Institute for Biomedical Engineering, Department of Metallurgical Engineering, University of Porto, Rua dos Bragas, 4099 Porto Codex, Portugal

‡ IRC in Biomedical Materials, Queen Mary and Westfield College, University of London, Mile End Road, London E1 4NS, UK

ABSTRACT

Glass reinforced hydroxyapatite composites were fabricated by a liquid phase sintering process. These composites showed to have much higher fracture toughness than sintered hydroxyapatite, 1.2 compared to 0.5 MPa $m^{1/2}$. *"In vitro"* tests in simulated physiological solution indicated potential bioactive behaviour, with formation of an apatite layer after 1 week immersion. Preliminary *"in vivo"* studies in rabbits showed a very close contact between bone and implanted materials.

INTRODUCTION

Several attempts have been made in order to reinforce hydroxyapatite by using glass additions (1-5). The mechanical properties of these composites have to match those of bone, namely bending strength and fracture toughness, and should allow bone growth. Glass-ceramics fabricated by controlled crystallisation proved to be a successful way of improving the mechanical properties of dense hydroxyapatite materials, although having an elemental composition different of that of bone (3,5). Furthermore, these hydroxyapatite composites exhibit a bioactive behaviour in body (6-8).

Recent works demonstrated that hydroxyapatite could also be reinforced by adding phosphate glasses to its sintering process (9-11). These glasses are composed of elements commonly found in the mineral part of bone, such as P, Ca, Na, etc. The amount of glass added to hydroxyapatite should be small in order to hamper phase changes into β and α tricalcium phosphate. The quantity of tricalcium phosphate phases present in the microstructure also depends on the chemical composition of the glass added to hydroxyapatite and on its physical properties, particularly viscosity and melting point.

The present work reports the most significant mechanical properties determined for phosphate glass reinforced hydroxyapatite composites, *"in vitro"* results in a simulated physiological solution and preliminary *"in vivo"* studies in rabbits.

MATERIALS & METHODS

Phosphate based glasses were produced using reagent grade chemicals heated at 1300°C in a platinum crucible. Table 1 shows the prepared chemical compositions.

Table 1 Chemical composition of the glasses, (mol%)

Ref.	P_2O_5	CaO	Na_2O	Al_2O_3
P1	45.5	54.5	-	-
P2	45.0	28.0	27.0	-
P3	62.9	10.1	10.1	16.9

After pouring, glasses were milled and wet mixed with hydroxyapatite powder in a proportion of 2.5 wt% of glass and 97.5 wt% of hydroxyapatite. Powders were uniaxially pressed at 288 MPa and sintered at 1300°C for 1h. Composites fabrication is fully reported elsewhere (9). Bending strength was calculated from 4-point biaxial bending test in a concentric ring test with 20mm supporting span and 10mm loading span at a cross head speed of 5mm/min. Fracture toughness measurements were performed using a strength-indentation technique proposed by Chantikul (12). In order to test *"in vitro"* bioactivity, samples were machined to the dimension of 10X10X1 mm, ground to 200 μm and immersed in a simulated physiological solution, buffered at pH of 7.4, with the following chemical composition (mol/dm^3): NaCl-8.0, $CaCl_2$-0.14, KCl-0.40, $NaHCO_3$-0.35, $MgCl_2.6H_2O$-0.10, $Na_2HPO_4.2H_2O$-0.06, KH_2PO_4-0.06, $MgSO_4.7H_2O$-0.06 and glucose-1.0. Samples were immersed for 1 and 2 weeks at 37+1°C without stirring.

In order to study *"in vivo"* behaviour of these composites, pins of 3 mm in diameter were implanted in the right posterior tibiae. of ten adult rabbits. Three holes were made per rabbit. Comparative studies were also conducted on sintered hydroxyapatite pins. Histological characterisation was performed under light microscopy after dehydration with alchool, impregnation in PMMA, cutting and colouration with haematoxylin and eosin.

RESULTS & DISCUSSION

The microstructure of phosphate glass reinforced hydroxyapatite was composed of a hydroxyapatite matrix and well dispersed β and α tricalcium phosphate (TCP) phases. The following phase contents were found in the microstructure of the composites: 50% hydroxyapatite, 20% β-TCP and 30% α-TCP for P1 sample, 45% hydroxyapatite, 15% β-TCP and 40% α-TCP for P2 sample and 40% hydroxyapatite, 15% β-TCP and 45% α-TCP for P3 sample. Calculation method is described in a previous work (10). The amount of TCP present in the microstructure of these composites may be controlled through sintering temperature, amount of glass added to hydroxyapatite and its chemical composition.

Fracture toughness measurements indicated 1.1-1.2 $\pm$ 0.05 $MPam^{1/2}$ range values for phosphate glass hydroxyapatite composites comparing to 0.5 $\pm$ 0.02 $MPam^{1/2}$ for sintered hydroxyapatite. This increase in toughness should be attributed to the higher densification and smaller grain obtained by phosphate glass additions, as reported previously (10,13). Although a significant improvement was achieved fracture toughness results values are still well below those usually reported for bone (14-15).

SEM analysis equipped with EDS, showed that after 1 week immersion in the simulated physiological solution an apatite layer covered the whole surface of all composites. An identical layer was also observed on sintered hydroxyapatite samples. Figure 1 a) and b) shows the surface of sintered hydroxyapatite and P1 respectively, after 1 week immersion.

As the formation of a Ca-P rich layer *"in vitro"* seems to be an essential requirement for bone bonding to occur (16-17) the bioactive behaviour of these composites suggest that they have great potential for bone ongrowth. The bioactive behaviour of glass reinforced hydroxyapatite composites might be adjusted to the implantation situ by varying the chemical composition of the glass added. As a consequence, these materials might offer several advantages over sintered hydroxyapatite in a wide range of biomedical applications.

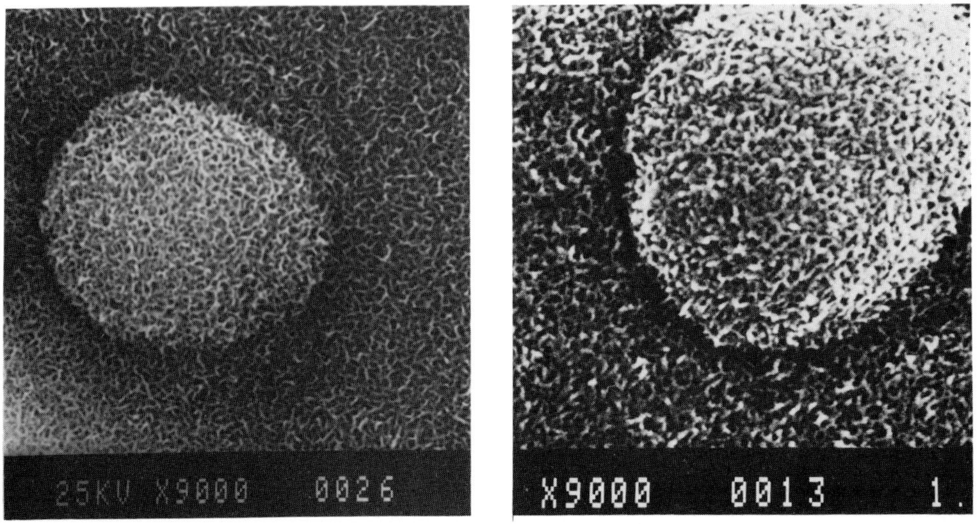

Figure 1 - Apatite layer on the surface of sintered hydroxyapatite after 1 week immersion (a). An identical layer was also formed on glass reinforced hydroxyapatite P1 composite (b).

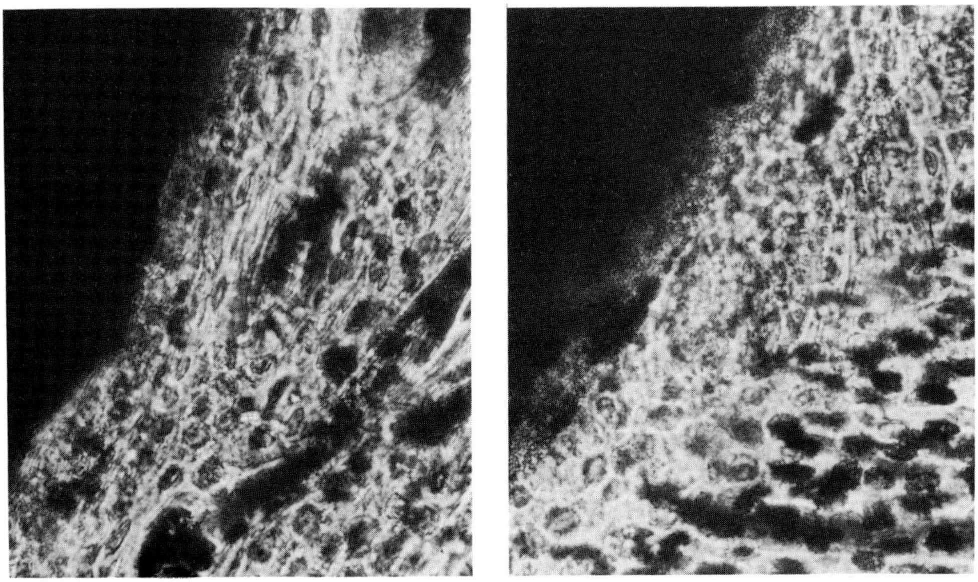

Figure 2 - Interface between bone and P1 composite (a) and bone and sintered hydroxyapatite (b) after 2 week implantation (x250).

Preliminary *"in vivo"* studies in rabbits showed a close contact between bone and implants after a 2 week implantation period for all composites, as may be seen in Figure 2a). New bone was formed around pins suggesting high compatibility and osteointegration with living tissues. A well-marked bioactive behaviour was observed and it seemed that no fibrous tissue formation occurred. Very similar results were also found for sintered hydroxyapatite after the same period of implantation, as presented in Figure 2b). Further studies are now underway to fully characterise the bone-bonding mechanism of these composites.

CONCLUSIONS

Glass reinforced hydroxyapatite composites showed to have higher fracture toughness than sintered hydroxyapatite. They proved to exhibit bioactivity *"in vitro"* suggesting that they have potential for bone ongrowth. Preliminary *"in vivo"* studies in rabbits seemed to confirm this hypothesis.

REFERENCES

1 Kangasniemi, K., De Groot, K., Wolke, J. and Yli-Urpo, A. J. Mater. Sci.: Materials in Medicine (1991) 2 133-137

2 Kangasniemi, I., De Groot, K., Becht, J. and Yli-Urpo, A. J. Biom. Mater. Res. (1992) 26, 663-674

3 Rawlings, R. D., Rogers, P. S. and Stokes, P. M. In: High Tech Ceramics, P. Vincenzini eds. Elsevier Science Publishers, Ammsterdam, (1987) 73-82

4 Holand, W., Vogel, W., Scubert, T. and Gummel, J. In: Bioceramics 2, Heidelberg, Germany, Japan (1989) 105-112

5 Kokubo, T., Biomaterials (1991) 12 155-163

6 Kitsugi, T., Yamamuro, T., Nakamura, T. and Kokubo, T. J. Biom. Mater. Res. (1989) 23 631-648

7 Amir, D. Gross, U. and Sela, J. Biomaterials, (1989) 10 585-589

8 Yoshi, S., Kakutani, Y., Yamamuro, T., Kokubo, T. and Takagi, M. J. Biomed. Mater. Res. (1988) 22 327-338

9 Santos, J. D., Knowles, J. C., Hastings, G. W. and Bonfield, W. British patent Nº 9213774.4, filed on 29th June 1993

10 Santos, J. D., Reis R. L., Knowles J. C., Hastings, G. W. and Monteiro, F. J. Biomaterials (1994) 15 5-10

11 Santos, J. D., Knowles, J. C., Monteiro, F. J. and Hastings, G. W. In: Bioceramics 5, Yamamuro T, Kokubo T, Nakamura T eds. Kokunshi Kankokai, Kyoto, Japan (1992) 35-41

12 Chantikul, P., Ansti, G. R., Lawn, B. R., Marshall, D. B. J. Am. Ceram. Soc. (1981) 64 539-543

13 Santos, J. D., Reis R. L., Knowles J. C., Hastings, G. W. and Monteiro, F. J. J. Mater. Sci.: Materials in Medicine (in press)

14 Bonfield, W. and Tully, A. E. J. Biomed. Engng. (1982) 4 23-27

15 Hobatho, M. C. and Ashman, R. B. J. Biomechanics (1989) 22 895-900

16 Kokubo, T., Kushitani,H., Sakka, S. and Yamamuro, T. J. Biomed Mater. Res. (1990) 24 721-734

17 Li, P., Yang, Q., Zhang, F and Kokubo, T. J. Mater. Sci. (1992) 3 452-456

Bioceramics, Volume 7 Edited by Ö.H. Andersson and A. Yli-Urpo
(*Proceedings of the 7th International Symposium on Ceramics in Medicine, Turku, Finland, July 1994*)

Characterization and Stability of Bioactive HA-G-Ti Composite Materials and Bonding to Bone

S. Maruno,[1] K. Imamura,[1] T. Hanaichi,[1] S. Ban,[2] H. Iwata[3] and H. Itoh[4]
[1] Nagoya Institute of Technology, Showa-ku, Nagoya 466, Japan
[2] School of Dentistry, Aichi-Gakuin Univ. Nagoya 464, Japan
[3] Dept. Orthopedic Surg., Nagoya Univ. Nagoya 466, Japan
[4] Tokyo Kouseinenkin Hospital, Tokyo 162, Japan

ABSTRACT

The construction details and features of the hydroxyapatite containing glass (HA-G) ceramic coating layer with bioactive surface, not only strongly adhered to the metal substrate through an intermediate glass layer but also bonded firmly to bone, and then *in vivo* stability of the glass coated implants are presented. The apposition of bone to the HA-G-Ti composite implants and mineralization in soft tissue surrounding the implant have been studied by high resolution FE-SEM, FT-IR and X-ray microdiffraction.

INTRODUCTION

The bioactive HA-G-Ti composite materials consisting of titanium alloy (Ti-6Al-4V) substrate coated with hydroxyapatite containing glass (HA-G) layer have an excellent biomaterial characteristics applicable to a bioreplacement such as a cementless artificial joint and end osseous dental implant [1~3]. The composite could be prepared without any structural and composition changes of HA crystalline particles, just as shown in plasma sprayed HA coating process [4], by a conventional firing method such as enameling. It is strongly dependent on the structure of the composite and the properties of the glass used for preparation of the composite whether the composites are suitable biocompatible materials for implant application over a prolonged period of time. It can be understood, therefore, that the biocompatibility and the implant stability of the HA-G-Ti composites are closely connected with the physicochemical properties of the glass, as well as the bioactivity potential due to the HA (involving in dissolution into living body). It was already described that the glass used had the excellent properties *in vitro* experiments including mechanical strength [5~7].

In the present work, the following matters are described: 1) the constitutional features and the bioactive surface formation of the HA-G ceramic layer which comprises a plurality of sub-layers with different HA contents, 2) the biological stability of the glass in body environment, 3) the bonding behavior of the bone to the composite and 4) the formation of bone mineral in tissue surrounding the implants. It has been concluded that the HA-G-Ti composites have desirable biomaterial characteristics for the bonding to bone, as well as the excellent biocompatibillty and the biochemical stability, with respect to long term fixation of the implant to bone.

EXPERIMENTAL

The method and procedure for preparing the HA-G-Ti functionally gradient composites, the physical properties of the HA containing glass with different concentration of crystalline HA particles, as well as the adhesion between the HA-G layer and the titanium substrate and then the requisite properties for the glass frit used, were reported elsewhere [7]. The HA-G-Ti composite rods (3.1 mm in diameter) with bioactive surface and glass only coated titanium (G-Ti) rods were implanted in dog femur and removed together with the surrounding bone after an elapse of a given period of time (1~38 month). The glass rods with the same composition as the glass frit were also implanted in compliance with necessity for comparison. Specimens for optical microscope observation of cross-section of the various HA-G-Ti composites were polished at an angle of 5° obliquely to the surface, using standard metallurgical techniques. The sample preparation for SEM observation with EDX analyzer, FT-IR and X-ray diffraction measurements was the same as the method described previously [1]. Especially, X-ray microdiffractometer with the curved position sensitive proportional counter (PSPC) which simultaneously detects diffracted X-ray intensity was used for measuring an irradiated area as small as 30 mm in diameter.

RESULTS AND DISCUSSION

The etching action of the HA-G-Ti composite, which mainly acts to remove to the surface layer covered with glass phase, is essentially important to produce a bioactive and biocompatible surface layer of the composite. The etched surface portion is in a roughened state having many cracks or ditches in order of microns in width and voids and having the fine HA particles exposed (Figure 1). The glass phase still remains in (b) and somewhat the over-etched surface structure is seen in (d). The most suitable surface for the bioactive composite implant is considered to be (c), wherein a great number of fine HA particles are stuck together with glass phase. Moreover, the surface irregularities with a bioactive characteristic as well as voids and cracks ensure bone ingrowth into the surface portion of the composite.

The composite structure of a substrate / intermediate glass layer / HA-glass layer is markedly observed (Figure 2). The HA-glass layer comprising a continuous glass phase with dispersed HA particles is well adhered to the substrate through an intermediate glass layer and constructed so as increase gradually the HA content from the innermost sub-layer toward the outermost sub-layer. The greater number of voids is prone to be developed in the vicinity of the intermediate of the HA-G layer. That is, the constitutional features of the HA-G ceramic layer inducing the specific gradient structure, the interfacial reaction between HA and glass and their wettability are greatly influenced by the properties characteristics of particulate crystalline HA and glass frit.

The fine ultrastructural appearance on the surfaces of the as glass-coated samples and implanted ones (a) showed no specific structure, having only a continuous smooth surface (Figure 3). The results of EDX analyses for both the samples suggested only slight differences on qualitative level, i.e., the decrease in several percents of alkaline elements, Na and K, for 38 month implant. There is

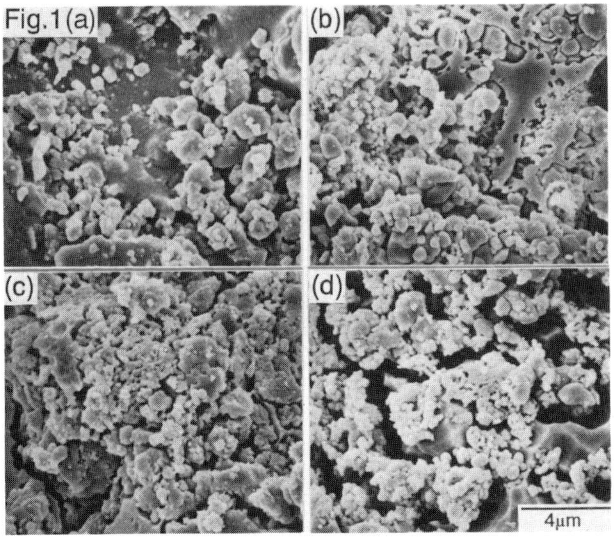

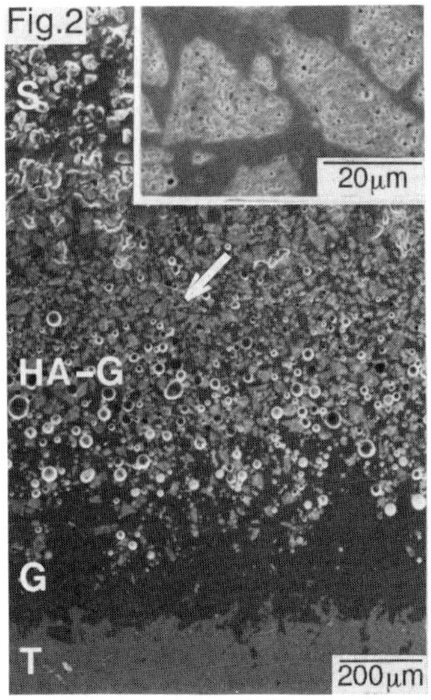

Figure 1 SEM photographs showing the change in the etched surface of the HA-G-Ti composites with the outermost sub-layer comprising 70wt%HA containing glass. The samples were etched in mixture solution of 3%HF and 5%HNO_3 for various times. (a) as coated, (b) 30sec, (c) 120sec, (d) 300sec.

Figure 2 SEM photograph of cross-section of the HA-G-Ti composite polished at an angle of 5° obliquely to the surface, wherein the surface (top of photo: s) is not yet etched chemically. The rectangular frame is the higher magnification of the area shown by arrow . T, G and HA-G indicate Ti-6Al-4V substrate intermediate glass layer and HA-glass ceramic layer, respectively. The measurements were carried out accelerating voltage less than 5kV for reducing electron damage.

not significant difference in the surface microstructure between (a) and (c) though the mottled look of the surface is observed in (c), where (c') is more evident. In addition, there are no clearly signs of the differences between (A) and (C), and then the detected elements in (A) and (C) surely correspond to the main elements of the composition of glass except carbon, because the glass does not contain inorganic bone substance (CaO, P_2O_5, i.e., apatite) and consists of alumino-borosilicate glass with silica content more than 67 wt%. Thus, the mottled pattern in (c) is considered to be due to ultrafine deposits of some organic matter (less than 10 nm in diameter). The connective tissue, formed at the interface between the G-Ti rod implant and bone during 38 month implantation, has a relatively smooth surface with fine ripple-mark (b), The result (B), wherein sulfur element was strongly detected and silicon does not exist, suggests that the connective tissue is mainly composed of chondroitin sulfuric acid which contains the bone mineral substance such as apatite. However, the ratio of calcium component to phosphorus one was extremely large as compared with that of natural bone. No inflammation reaction was recognized. The these results show that the glass used for the HA-G-Ti composite is stable for a prolong period of time.

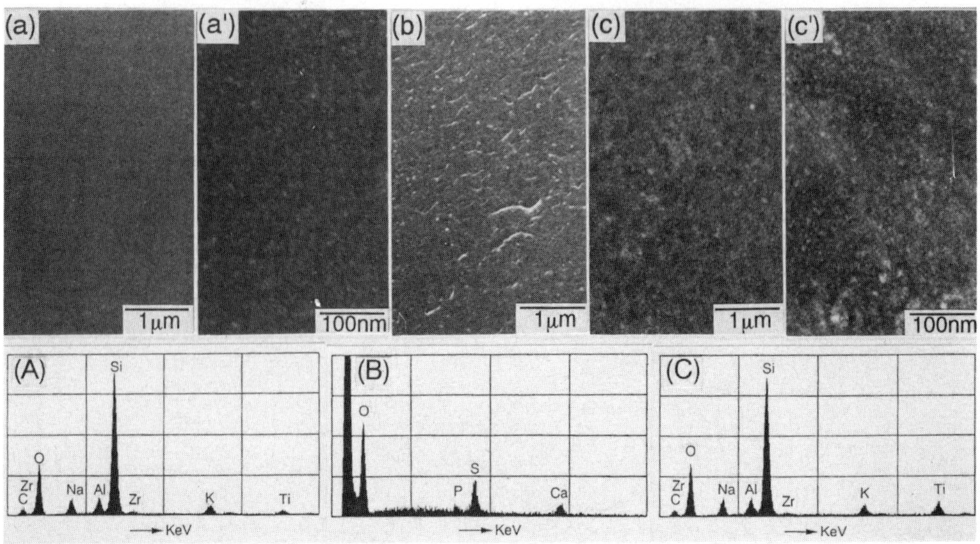

Figure 3 High resolution FE-SEM photographs showing the surface (a and c) of glass coated titanium rod implant, having been in contacted with bone (a) and perimysium (subcutis) (c) in dog and pulled out after 38 month implantation, wherein the conduction carbon were coated to prevent electrostatic changing. (b) is the surface of connective tissue film covered the bone facing the rod implant. (a') and (c') are higher magnification of (a) and (c), respectively. (A), (B) and (C) indicate the results of EDX analyses for (a), (b) and (c), respectively. The measurements were carried out accelerating voltage less than 5kV for reducing electron damage.

The extracted implant surface is covered with bone in apposition to the HA-G layer of the composite (Figure 4). The appearance of shear failure of bone indicates that the HA-G-Ti composite has been firmly bonded to the surrounding bone tissue; in (A) the bundles and / or thread of collagen fibers as well as bone mineral such as apatite deposited in a fibrotic tissue and in (B) amoebiform and shell-like bone tissues are observed. The failure mode was found to be predominantly at the portion of the surrounding bone (newly formed bone), not at the interface between the HA-G ceramic layer and the substrate.

The HA-G-Ti composite implant pulled out after 3 month was firmly attached to the surrounding bone and the average pull-out strength equaled 14.8±4.5MPa. A newly mineralized tissue is clearly observed around the extracted implant (Figure 5-b). The X-ray microdiffraction patterns from regions 1, 2 and 3 in (a), in close agreement with that from cortical bone 4, were identified as that of apatite with low crystallinity. In particular the region 1, had been closely contacted with the implant, demonstrates a higher degree of crystallinity of apatite, as compared with other regions. The micro FT-IR measurements, on the other hand, indicated that the apatite is biological apatite which contains carbonate component. These results show that the local mineralization in the cancellous bone tissue surrounding the

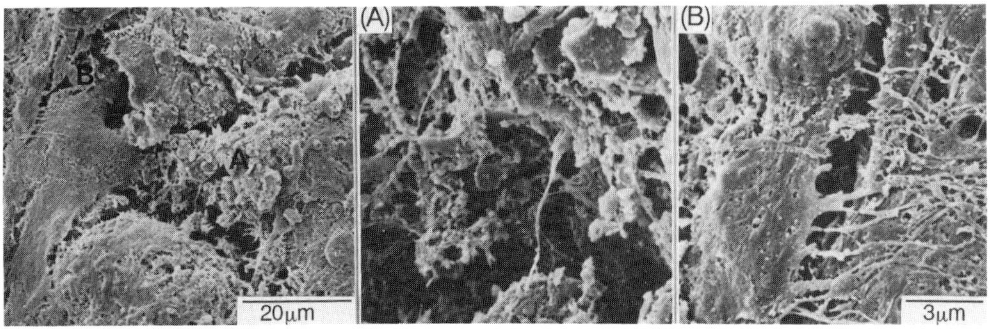

Figure 4 SEM photographs showing the surface of the HA-G-Ti composite implant, having been in contact with cortical bone in dog femur and pulled out after three month implantation. (A) and (B) are higher magnification of rectangular frame areas.

extracted implant occurs, resulting in the formation of the calcified tissue containing crystalline apatite. Moreover, the stages of local mineralization are considered to proceed in the sequence of 1, 2 and 3.

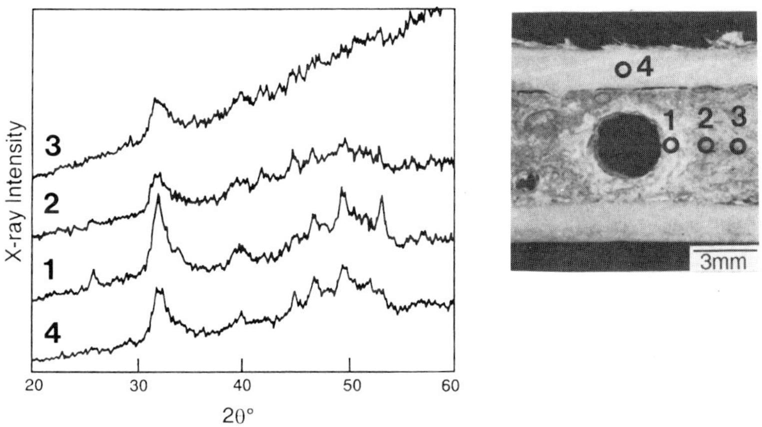

Figure 5 X-ray diffraction patterns (a) from the selected micro-areas (30 μm in diameter) marked by No. 1, 2, 3 and 4 in (b). The optical micrograph (b) shows the transverse section (perperdicular to the implant direction) of bone tissue surrounding the HA-G-Ti composite implant, pulled out after three month implantation and cortical bone (No. 4). The measurements were performed under 50kV-300mA for Cu-Kα radiation.

The bone ingrowth into the outermost sub-layer of HA-G ceramic layer, as well as the deposition of new bone, can be confirmed in the microscopic observation

(Figure 6). The bundles of mineralized collagen fibers and the ingrowing fibrotic tissue, with filamentous structure (80~120 nm in diameter), are clearly observed. Although it could not discern from (b) whether the interconnected and ingrown rod filaments indicate a periodic structure of hard tissue collagen fiber containing bone minerals, they are identified as well integrated to the HA-G ceramic. The precipitation and deposition of bone mineral crystallites on glass phase were scarcely recognized as shown in Figure 6. It has been found, moreover, that the dissolution / precipitation phenomena [8] of the HA particles exposed on and dispersed in the HA-G ceramic layer of the composite are concerned closely with the formation of the bond between the HA-G ceramic and bone.

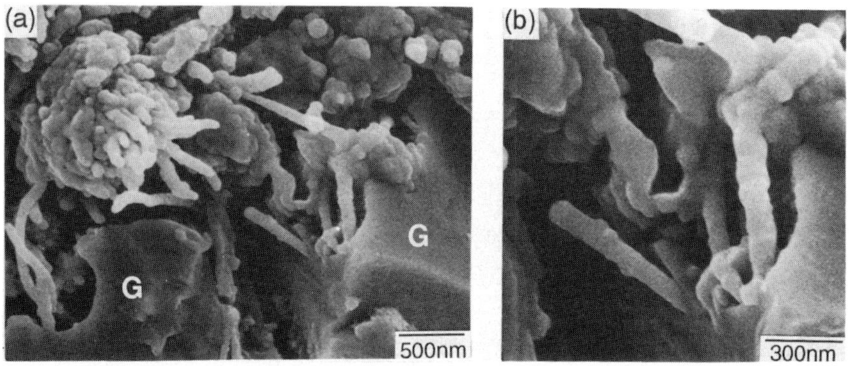

Figure 6 High resolution FE-SEM photographs showing the cross-section of the subsurface of HA-G coated layer of the implant, embedded in dog femur for 38 months. The crack 200~250nm in width seems to be originated in cutting impact by diamond cutter blade. G is the part of the glass matrix. (b) is higher magnification of (a).

REFERENCES

1. Maruno, S., Itoh, H., Ban, S., Iwata, H. and Ishikawa, T. <u>Biomaterials</u> (1991) 12 225-230
2. Maruno, S., Imamura, K., Iwata, H., Ban, S. and Itoh, H. In <u>Bioceramics Vol. 5</u>, Kobunshi kankokai (1992) 165-172
3. Ishikawa, T., Itoh, H., Maruno, S., Iwata, H. and Ban,S. <u>Clinical Mater.</u> (1993) 14 277-285
4. Radin, S. R. and Ducheyne, P. <u>J. Biomater. Sci.:Mater. in Med.</u> (1992) 3 33-42
5. Ban, S., Maruno, S. and Hasegawa, J. <u>Jpn. J. Appl. Phys.</u> (1990) 29 L.1523-26
6. Ban, S., Hasegawa, J. and Maruno, S. <u>Biomaterials</u>, (1992) 12 205-209
7. Maruno, S., Ban, S., Wang, Y. F., Iwata, H. and Itoh, H. <u>J. Ceram. Soc. Japan</u> (1992) 4 362-367
8. Bagambisa, F. B., Joos, U. and Schilli, W. <u>J.Biomed. Mater. Res.</u> (1993) 27 1047-1055

Bioceramics, Volume 7 Edited by Ö.H. Andersson and A. Yli-Urpo
(Proceedings of the 7th International Symposium on Ceramics in Medicine,
Turku, Finland, July 1994)

Development of a Bioactive Glass-Ceramic with the Incorporation of Titanium Particles

B.A. Taylor, R.D. Rawlings and P.S. Rogers

Department of Materials, Imperial College, Prince Consort Road, South Kensington, London SW7 2BP, UK

ABSTRACT

A bioactive glass-ceramic known as Apoceram, has been reinforced using particulate titanium. The composites were fabricated by hot pressing the powder mixtures of the Apoceram parent glass and titanium in the range 10-40vol.%. The hot pressing temperature was adjusted so that crystallisation of the matrix occurred during pressing, thus alleviating the need for any post pressing heat treatment. The fracture toughness and strength were measured and correlated with the microstructure and density. After optimisation of the processing parameters (time, temperature and pressure), both the strength and toughness of the composites were improved as compared with the monolithic material, in spite of a reaction between the particles and matrix. The reaction layer was identified as a compound of titanium and silicon.

INTRODUCTION

Bioactive glasses and glass-ceramics are implant materials which are surface reactive, ie controlled tissue-implant responses result in a chemical bond across the implant/tissue interfaces. Their common characteristic is the formation of a biologically active apatite layer which provides the bonding interface. The advantage of these materials is the removal of the necessity and associated problems for fixation, eg implants fixed by mechanical interlocking and cements are liable to loosening over a long period of time, but the bioactive materials generally have poor mechanical properties especially their low toughness. About a decade ago it was discovered that glass-ceramics containing apatite [A] and wollastonite [W] crystals have good mechanical strengths and are also capable of bonding to bone[1,2,3,4]. Although these have better mechanical properties than other monolithic bioactive materials, such as sintered hydroxyapatite [5], the brittleness displayed still limits their potential use for medical applications. It may be possible to improve the mechanical performance of bioactive glass-ceramics by the incorporation of a reinforcement phase, such as particles of a ductile metal. These ductile reinforcements may increase the toughness of the monolithic materials by offering additional dissipating mechanisms such as plastic deformation and crack bridging. An improved fracture resistance has been observed in a number of ductile reinforced systems including Bioglass + 316L stainless steel [6], Al_2O_3 + Al [7], and Al_2O_3 + Ni [8].

Apoceram is an A-W containing glass-ceramic in the Na_2O-CaO-Al_2O_3-SiO_2 system, originally developed as a cast material at Imperial College.[3,4] Early work revealed that the bioactivity was good and the mechanical properties promising. By improving fabrication procedures using the powder route and hot pressing, the monolithic material displayed better strength and toughness values, but the toughness remained still inadequate for many applications[9].This paper reports the results of an investigation to enhance the mechanical properties by introducing ductile metal particles in an Apoceram matrix. Titanium was chosen for the particulate reinforcement as it has low toxicity and is well tolerated by the body.

EXPERIMENTAL PROCEDURE

The Apoceram parent glass, type CP1, was prepared by melting raw batch powders in platinum crucibles [1500°C] and subsequently quenching in iced water. Its composition was [weight %] SiO_2-50.65, CaO-28.62, P_2O_5-7.04, Al_2O_3-6.45, Na_2O-4.47, CaF_2-2.77. The glass frit was milled and the powder sieved with a -38μm size aperture.The titanium was an irregular sponge and particles lay in the size range 35-70μm. Mixtures of CP1 and titanium were weighed, thoroughly mixed and hot pressed under vacuum. The heating rate was 7°C/min and the samples were held at 1000°C for 1 hour using pressures between 4.5MPa and 21.5MPa. For comparison purposes monolithic CP1 was also pressed in conjunction with each fabrication profile used for the composites. Initially the composites were studied as a function of titanium content. Once their properties were established, the most promising composition was studied further with view to optimising its processing parameters (pressure, temperature and time).

Mechanical properties [strength and toughness] were measured by means of a three point bend test. The specimen configuration was a rectangular beam with a nominal size of 2.5x5x30mm. The tests were carried out using an Instron test machine at room temperature. The crosshead speed was 1mm/min and the load cell was 1kN.

The bulk and theoretical densities of the samples were determined by an immersion method based on Archimedes' principle. Detailed microstructural and compositional assessment was made using a scanning electron microscope [SEM] and energy dispersive X-ray analysis [EDAX].

The effects of the titanium reinforcement upon the sintering densification and crystallisation were evaluated using dilatometry and differential thermal analysis [DTA]. They were conducted at 10°C/min, from room temperature to 1000°C. All samples were cold pressed and tested in an inert [argon] atmosphere in order to prevent oxidation of the titanium from the atmosphere.

RESULTS AND DISCUSSION

Mechanical Properties

CP1 fabricated with the following hot pressing profile (4.5MPa, 1000°C and 1 hour) has strength and toughness values of 132 MPa and 1.81 $MPam^{1/2}$ respectively. For the same fabrication parameters the composites showed a severe deterioration in strength compared with the monolithic material [Fig. 1a]. The strength peaked at 108 MPa with 20vol% titanium content and fell away as the content increased. However the toughness of the composite was equal to or slightly improved upon the glass-ceramic values, the highest value of 2.02 $MPam^{1/2}$ being for the 30 vol% composite [Fig. 1b].

The density of the samples was lower than that calculated from the rule of mixtures [Ti - 4.5 g/cm^3 and CP1 - 2.7 g/cm^3] being between 92-97% of theoretical. It is considered that the deviation from the rule of mixtures is due to porosity. Nevertheless it should be borne in mind that changes in the matrix structure and any reaction layers would also affect the rule of mixtures correlation.

SEM examination showed a homogeneous distribution of particles and no evidence of agglomeration. Typical micrographs of 20 vol % titanium and monolithic CP1 are shown in figures 2 and 3 respectively. The matrix of the composite contained apatite and wollastonite phases, as was previously reported for the monolithic material. The matrix became very coarse on addition of titanium, except immediately around the particles where the microstructure was finer. These small crystals radiated outwards from the particles and gradually merged into the coarser matrix. Wollastonite is surface nucleated and this may account for these fine crystals surrounding the titanium particles. A reaction layer, typically 1-2μm thick, was found at the particle/titanium interface. The reaction layer was identified as a compound of titanium and silicon, with a small amount of phosphorus. This layer debonded, particularly at the reaction layer/particle interface, creating flaws. In extreme cases the titanium particles became almost completely debonded from the matrix and may therefore act as pseudo pores. Voids are strength limiting, and the reduction in strength as compared with the monolithic material correlated with the porosity of the composites. As the Ti content increased the particle contact became more likely and the reaction layers ran around the outside of the contact points, joining the layer of one particle with another. This feature may explain the reduction in strength observed with composites containing higher titanium contents.

A suitable composition was chosen for further development. The 20vol% composite displayed the highest strength of the composites and a fracture toughness better than that exhibited by the monolithic material. As a consequence the 20 vol% composite was selected for development. The main features which could be improved were the particle/matrix reaction layer and the associated debonding. It was demonstrated that increasing the pressure of fabrication [time and temperature constant at 1 hour and 1000^oC respectively] reduced the areas of debonding and formation of voids by improving the powder packing and densification. The strength of monolithic CP1 was unchanged with increasing pressure, but there was a reduction in toughness. In contrast increasing the pressure was generally beneficial for the composite. Tripling the pressure to 13MPa yielded the best overall results, producing strength and toughness values of 150MPa and 2.45$MPam^{1/2}$ respectively. These properties were substantially better than those of the monolithic materials and there was also much less scatter in the composite results, indicating an improvement in reliability. This composite is shown in figure 4; there is now good bonding between the particle and reaction layer.

Table 1. Dilatometry results.

Sample	Sintering [^oC]		
	Onset	End	Maximum Rate
CP1	**700**	**920**	**815**
CP1 + 20vol% Ti	**715**	**885**	**815**

Table 2. DTA results.

Sample	[A]^oC			[W]^oC		
	Onset	End	Peak	Onset	End	Peak
CP1	834	884	865	924	956	943
CP1 + 20vol% Ti	827	871	861	893	938	932

Dilatometry and DTA

Typical dilatometry results for compacted CP1 glass and the 20vol% Ti composite powders are shown in table 1, which gives the temperatures at which the onset, end and maximum rate of sintering occur. A reduction in length for CP1 and the composite of approximately 18% and 14% respectively, was observed. It can be seen that although the temperature at which the maximum rate occurs is unchanged, the temperature range of sintering has been reduced significantly [50^oC] by the addition of titanium. It is thus seen that the addition of titanium particles hinders the sinterability of the glass-ceramic matrix. This may be due to the physical barrier for sintering glass powders created by the reinforcement, impeding the viscous flow of the glass. Particles will induce differential sintering rates which in turn will produce transient stresses during sintering. These stresses can locally retard the densification rates of the composite and generate flaws. As mentioned earlier wollastonite is surface nucleated, this nucleation upon the titanium particles may have accelerated the crystallisation processes which in turn would increase the viscosity of the glass inhibiting full densification.

The apatite and wollastonite crystallisation temperatures are given in table 2. The onset temperature for apatite has been little affected by the titanium particles, however the wollastonite onset temperature has been lowered significantly, by approximately 30^oC. It is proposed that this reduction in the onset temperature of wollastonite contributed to the reduction in the sintering range, as discussed above.

SUMMARY

Titanium particles were successfully employed as a reinforcement agent in a hot pressed bioactive glass-ceramic called Apoceram. A reaction layer between the particles and matrix was observed and found to be a compound of titanium and silicon. However the extent of this reaction was minimised by control of the process parameters. For a composite containing 20 vol% Ti optimisation of the pressure produced mechanical properties substantially better than those observed for the monolithic material; strength and toughness values comparable with those for human bone were obtained.

The crystallisation temperatures of apatite and particularly wollastonite were lowered on addition of titanium. Wollastonite is surface nucleated and the titanium particles were acting as sites for nucleation. The acceleration in crystallisation hindered the sintering of the composites and inhibited full densification.

REFERENCES

1. Kokubo T. et al. Bull. Inst. Chem. Research (1982) 60 260-268.

2. Kokubo T. et al J. Mat. Sci. (1986) 21 536-540.

3. Carpenter P.R., Campbell M.,Rawlings R.D. and Rogers P. S. J. Mat. Sci. Letters (1986) 21 1309-1312.

4. Ponton C.B. SERC Report (1990).

5. De Width G. Vandijk H. A. J., Hattu N. and Priss K. J. Mat. Sci. (1981) 16 1592-1598.

6. Ducheyne P. and Hench L. L. J. Mat. Sci. (1982) 17 592-606.

7. Andersson C. A. and Aghajanian M. K. Ceramic Engineering Proceedings (1988) 9 621-626.

8. Sun X. and Yeomans J.A. Special Ceramics 9, British Ceramics Proceedings 49, Institute of Ceramics (1992) 297-308.

9. Alanyali H. PhD Thesis (1992) Imperial College, South Kensington, London.

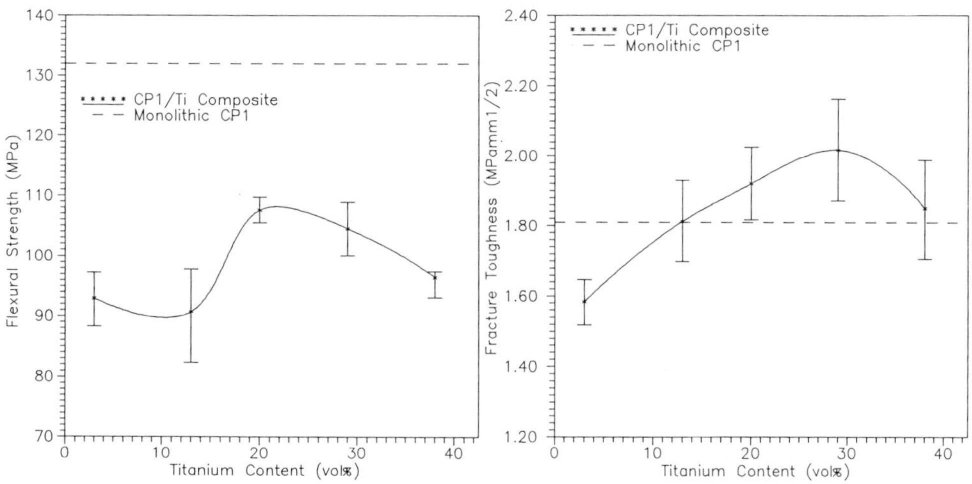

Figures 1a and1b illustrate the variation of strength and toughness respectively with increasing volume percent titanium particles in CP1 glass-ceramic.

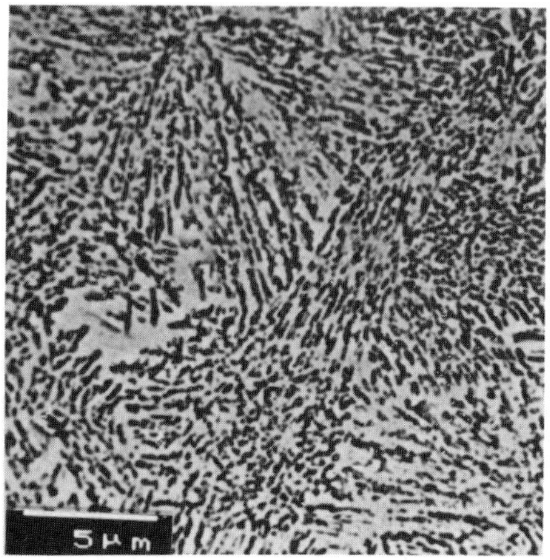

Figure 2: Monolithic CP1
[1000°C, 4.5 MPa, 1 hour]

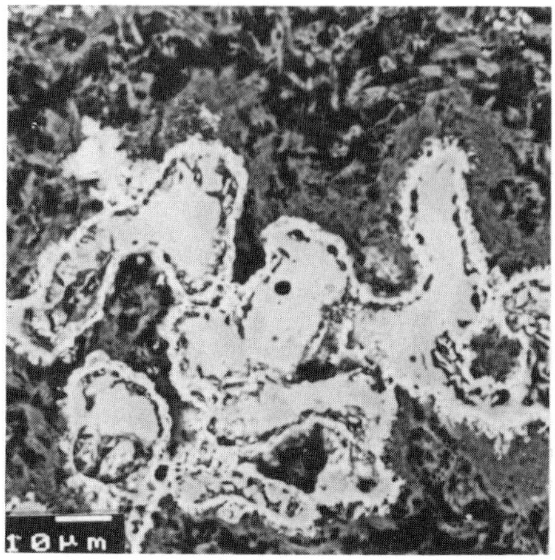

Figure 3: CP1 + 20 vol% Ti
[1000°C, 4.5MPa, 1 hour]

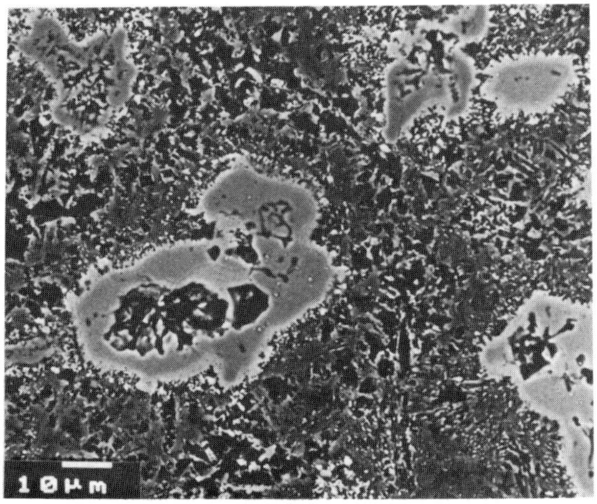

Figure 4: CP1 + 20 vol% Ti, [1000°C, 13MPa, 1 hour]

Bioceramics, Volume 7 Edited by Ö.H. Andersson and A. Yli-Urpo
(*Proceedings of the 7th International Symposium on Ceramics in Medicine, Turku, Finland, July 1994*)

Effect of Electrochemical Deposition of Calcium Phosphate on Bonding of the HA-G-Ti Composite and Titanium to Bone

S. Ban, S. Maruno,* N. Arimoto, A. Harada, M. Matsuura and J. Hasegawa
Department of Dental Material Science, School of Dentistry, Aichi-Gakuin University, 1-100, Kusumoto-cho, Chikusa-ku, Nagoya 464, Japan
* Department of Electrical and Computer Engineering, Nagoya Institute of Technology, Gokiso-cho, Showa-ku, Nagoya 466, Japan

ABSTRACT

Calcium phosphates were deposited on the hydroxyapatite-glass-titanium (HA-G-Ti) functionally gradient composite bars as well as pure titanium bars by electrochemical process. After implantation in femur of rabbits, the bonding strengths of the four groups of the specimens to the bone were determined by the pull-out method. At 3 weeks after implantation, the bonding strengths of the composite and of the titanium with the electrochemical calcium phosphate coating were significantly higher than those without it, respectively ($p<0.05$). However, at 9 weeks after implantation, the bonding strengths with the coating did not significantly differ from those without it, respectively ($p>0.05$). These results demonstrate that the electrochemical calcium phosphate coating increases the bonding strength of the implant to bone in the early stages of implantation.

INTRODUCTION

Electrochemical deposition is one of the methods frequently used for the calcium phosphate coating on metal substrate [1-5]. The great advantage of this method is its ability to produce homogeneous coatings on the metal substrate with complicated shape. However, the adhesion of the deposition to the metal substrate is weak [2]. On the other hand, Maruno et al. developed a hydroxyapatite-glass-titanium (HA-G-Ti) functionally gradient composite consisting of a substrate coated with a glass-HA ceramic layer, wherein the ceramic layer comprises a continuous glass phase with dispersed HA particles and the surface having a roughened state is prepared [6]. Animal study showed that the bonding strength of the composite-implants to dog femora were significantly higher than the titanium-implant without the HA-G coating up to 3 months after implantation [7]. However, it is desirable to achieve the high bonding strength as quickly as possible after implantation. It was also found that needle-like carbonate apatites with low crystallinity were formed on the surface of the composite for 3 months *in vivo* and for 6 months *in vitro* [8]. It was also found that the needle-like calcium phosphates could be formed on the titanium for less than 20 min by an electrochemical process [9]. In the present study, we applied the electrochemical deposition of calcium phosphate on the surface of the HA-G-Ti composite in order to obtain a higher bonding strength of them to bone as fast as possible after implantation.

MATERIALS AND METHOD

The preparation method for the HA-G-Ti composite is as follows. Slips of the mixture of the glass and HA powder with different contents of HA to the glass were coated on pure titanium bars, 2 mm in diameter, with sandblasted surface and fired at 800-900°C. The HA-glass layer about 100 μm in thickness were produced on the titanium bars by repeating the coating and firing. Finally, the surface of the composite was etched in an aqueous solution of 3% HF and 5% HNO_3 to remove the surface layer covered with glass and to obtain the bioactive surface layer with HA fine particles exposed. The details of the preparation method for the HA-G-Ti composite were described previously [6].

Calcium phosphates were electrochemically deposited on the HA-G-Ti composite bars, 2.2 mm in diameter and 11 mm in length, and pure titanium bars, 2 mm in diameter and 11-12 mm in length, in the cell of three-electrode configuration. The counter electrode was a platinum plate, 10 mm in width, 20 mm in length, and 0.1 mm in thickness. The electrode potentials were measured versus a saturated calomel electrode (S.C.E.) and the current was maintained to be constant by a galvanostat with a function generator. The electrolyte was prepared by 137.8 mM of NaCl, 2.5 mM of $CaCl_2 \cdot 2H_2O$ and 1.6 mM of K_2HPO_4 into distilled water and buffered at a pH of 7.2 with trishydroxyaminomethane and HCl. The specimens clamped to the working electrode were polarized by a cathodic loading of 100 mA for 20 minutes at 62°C in the electrolyte.

To determine the effect of the electrochemical calcium phosphate coatings on the bonding, four groups of the specimens were implanted into the rabbit femora: the titanium bar with (*Ted*) and without (*Tas*) the coating and the HA-G-Ti composite bar with (*Hed*), and without (*Has*) it, as shown in Figure 1. The holes for implantation were made transversely through the lateral cortex of the femur by using a drill, 2.3 mm in diameter. Randomly-selected three or four specimens were implanted into femur of rabbits. The wound was irrigated with saline solution after the implantation was completed. The periosteum, the fascia, and the skins were closed with nylon sutures, respectively. The rabbits were sacrificed at 3 weeks and 9 weeks after implantation and the femora were taken out from the

Figure 1 Specimens used in this study. From the left, *Has*, *Hed*, *Tas*, and *Ted*.

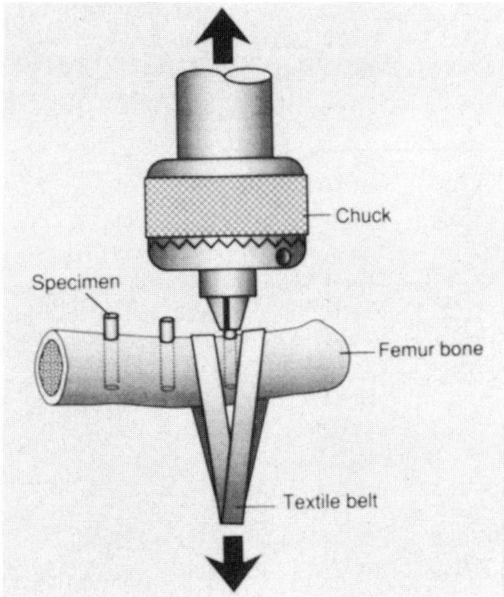

Figure 2 Schematic illustration of the pull-out test.

bodies. Immediately, the femora were used for radiographic observation, and the bonding strengths of these four groups of the specimens to the bone were determined by the pull-out method using Instron testing machine. The specimens were fixed to a chuck and the bone were pulled out by textile belts to obtain homogeneous shear stresses surrounding the specimen (Figure 2). The cross-head speed was 5 mm/min. The pull-out bonding strength, σ, was calculated by dividing the pull-out force by the contact area which was derived from the diameter of the specimen and the average thickness of the cortical bone. The specimens extracted from the bone were dehydrated with serially graded alcohol after formalin fixation and transferred to amyl acetate, following by drying with a critical point dryer. The specimens after drying were examined by field emission type scanning electron microscopy (FE-SEM) and Fourier transform infrared absorption spectroscopy (FTIR).

RESULTS AND DISCUSSION

FE-SEM photographs of the surface of the HA-G-Ti composite and Ti bars with and without the electrochemical coating are shown in Figure 3. The surface of Ti without the electrochemical coating (*Tas*) shows a typical sandblasted metal texture, whereas the surface of Ti with the coating (*Ted*) shows the precipitation of needle-like crystals. It was previously confirmed that these precipitates were identified to be a calcium deficient carbonate apatite with low crystallinity such as living bone [5]. The surface of the HA-G-Ti composite without the electrochemical coating (*Has*) is covered with glass and HA crystals, whereas the surface of the composite with the coating (*Hed*) is covered with needle-like precipitates which is

Figure 3 FE-SEM photographs of the surface of the HA-G-Ti composite and titanium bars with and without the electrochemical coating before implantation.

nearly the same morphology with those of the titanium with the coating (*Ted*). On the other hand, these needle-like crystals were also observed on the surface of the HA-G-Ti composite when it was soaked in the simulated body fluid for 6 months *in vitro* and was implanted in the dog femur for 3 months *in vivo* [8]. It was also found that these crystals were identified to be a calcium deficient carbonate apatite elongated to *c*-axis direction.

The results of the pull-out tests are presented in Table 1. The bonding strength of each group at 9 weeks after implantation was definitely higher than that at 3 weeks after implantation, respectively. Then, statistical comparisons between test groups at 3 weeks and 9 weeks after implantation were respectively performed by using two-way analysis of variance (ANOVA). It was found that the bonding strengths of the HA-G-Ti composite to the bone were significantly higher than those between the titanium and bone ($p<0.01$), and the electrochemical coating significantly increased the bonding strengths of them to the bone at 3 weeks after implantation ($p<0.05$). However, there were no statistically significant differences between the groups with and without the electrochemical coating at 9 weeks after implantation ($p>0.05$), although the bonding strengths of the HA-G-Ti composite to the bone were significantly higher than those of the titanium to the bone (p<0.01). These results demonstrate that the electrochemical calcium phosphate coating increases the bonding strength of the implant to the bone in the early stages of implantation.

Table 1 Results of pull-out bonding test

Group code	Material	Treatment	3 weeks No of test	3 weeks σ (MPa) mean	3 weeks σ (MPa) sd	9 weeks No of test	9 weeks σ (MPa) mean	9 weeks σ (MPa) sd
Tas	Ti	as sandblasted	8	4.4	2.3	8	18.5	3.6
Ted	Ti	electrochem.coating	5	8.2	2.9	8	18.8	4.3
Has	HA-G-Ti	as etched	8	9.3	4.1	6	23.7	4.1
Hed	HA-G-Ti	electrochem.coating	8	13.2	5.0	5	24.4	5.5

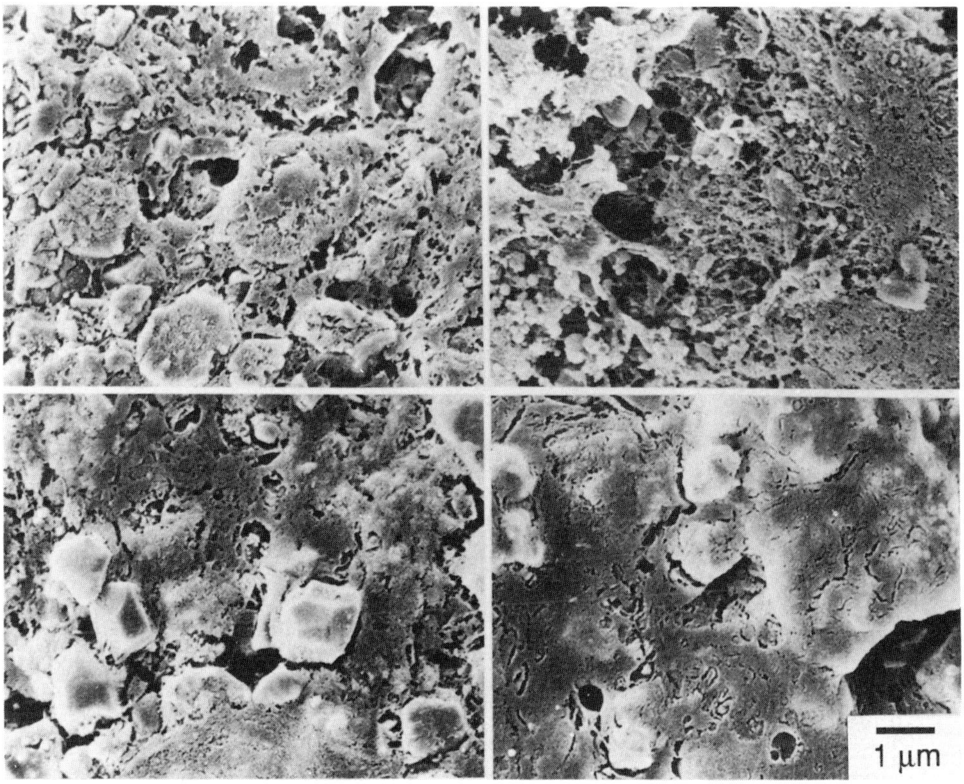

Has *Hed*

Figure 4 FE-SEM photographs of the surface of the HA-G-Ti composite with (*Hed*) and without (*Has*) the electrochemical coating extracted from the rabbit femur at 3 weeks after implantation. Upper photographs are the portion which have been bound to the cortical bone and lower ones are the portion in contact with bone marrow.

FE-SEM photographs of the surface of the HA-G-Ti composite after pull-out test are shown in Figure 4. The collagen bundles and fractured bone tips are observed on the surface of the composite with and without the coating (*Hed* and *Has*) in contact with the cortical bone. This suggests that both the composite with and without the coating have been tightly bound to bone, and the failure is located at the new bone portion. In the portion in contact with bone marrow, it can be seen that the surface of the composite with the electrochemical coating (*Hed*) is covered with homogeneous precipitates, i.e., new bone, whereas some HA particles are still observed on the surface of the composite without the coating (*Has*). It was also found that all the specimens were covered with new bone and there was no gap between the specimens and bone at 9 weeks after implantation. It can be assumed that the bonding of the implants to the bone preliminary depended upon the mechanical-interlocking force between them, because the bonding strength of the specimens with and without the coating showed statistically the same value when the gap between the implant and bone were completely filled with new bone.

On the other hand, the needle-like precipitates were occasionally formed on the surface of the composite *in vivo* and *in vitro* as mentioned previously [8]. These results demonstrated that calcium phosphate coatings deposited by the electrochemical method promotes the new bone formation between the implant and bone, because the needle-like precipitates can act as nucleation sites for apatite formation. Therefore, it is concluded that, although the adhesion of the coating to the substrate is relatively weak, the electrochemical calcium phosphate coating on HA-G-Ti composite and titanium increase the bonding strength due to the mechanical interlocking force in the early stages of the implantation, because new bone can fill the gap between the implant and bone in earlier than that without the coating.

ACKNOWLEDGMENT
The authors thank Mr. K. Imamura, Nagoya Institute of Technology, for technical assistance in preparation of the HA-G-Ti composite. This research was supported by a Grant-in-Aid for General Scientific Research (No. 05671645) from the Ministry of Education, Science, and Culture of Japan.

REFERENCES

1. Shirkhanzadeh, M. J. Mater. Sci. Lett. (1991) 10 1415-1417
2. Royer, P. and Rey, C. Surf. Coat. Tech. (1991) 45 171-177
3. Monma, H. J. Ceram. Soc. Jpn. (1993) 101 737-739
4. Ban, S. and Maruno, S. In Bioceramics Vol.5 Kobunshi Kankokai, Kyoto, Japan (1992) 49-56
5. Ban, S. and Maruno, S. Jpn. J. Appl. Phys. (1993) 32 L1577-L1580
6. Maruno, S., Ban, S., Wang, Y-F., Iwata, H. and Itoh, H. J. Ceram. Soc. Jpn. (1992) 100 362-367
7. Ishikawa, T., Itoh, H., Maruno, S., Iwata, H. and Ban, S. Clin. Mater. (1993) 4 277-285
8. Ban, S. and Maruno, S., Iwata, H. and Itoh, H. J. Biomed. Mater. Res. (1994) 28 65-71
9. Ban, S. and Maruno, S. Abst. Ann Meet. Ceram. Soc. Jpn., Nagoya, Japan (1994) Abst. No. 3F04

Bioceramics, Volume 7 Edited by Ö.H. Andersson and A. Yli-Urpo
(Proceedings of the 7th International Symposium on Ceramics in Medicine, Turku, Finland, July 1994)

Mechanical Behaviour of a Hydroxyapatite-Zirconia Particles Composite

J. Caetano-Zurita,[1] O. Bermúdez,[2] I. López-Valero,[3] E.B. Stucchi,[4] J.A. Varela,[4] J.A. Planell[5] and S. Martínez[1]

[1] Dep. Cristal.lografia, Mineralogia i Dip. Minerals – Fac. Geologia – Universitat de Barcelona – C/ Martí Franquès, S/N°, 08028 – Barcelona – Spain
[2] Departamento de Física, Universidad del Cauca, Popayán (Colombia)
[3] Osteosynth, S.L. – C/ Major, 65 – 080629 – Torrelles de Lobregat – Barcelona – Spain
[4] UNESP – Instituto de Química de Araraquara, R. Prof. Francisco Degni, S/N°, 14800 Araraquara, Sao Paulo – Brazil
[5] Dep. Ciència dels Materials i Enginyeria Metal.lúrgia – UPC – Avda. Diagonal 647, 08028, Barcelona – Spain

ABSTRACT

Fibre and particle hydroxyapatite-zircona composites have been prepared by uniaxial pressing and sintering at different temperatures, and their mechanical properties have been measured and related with their texture and mineral dynamics. The products obtained are porous, with densities ranging from 46% to 78% of the theoretical density.
A good correlation has been found between the increase in density and the K_{IC} increase at sintering temperatures below 1200°C. At higher temperatures, the densification process of the material continues, but K_{IC} decreases. This decrease is related with the reaction of the zircona material, with the nucleation and growth of $CaZrO_3$ crystals and with the formation of defects or weakness points in the composite.

INTRODUCTION

Hydroxyapatite is the mineral component of natural bone and, as such, it has been studied extensively as a candidate biomaterial for its use in prosthetic applications (1). The main weakness of these materials lays in their poor strength and fracture toughness which make them unsuitable for load-bearing biomaterials (2,3). Composites of polymers reinforced with hydroxyapatite for improved mechanical properties have been studied, but they are not always adequate for load-bearing applications.
Fibre reinforcement is a well known method for the improvement of the fracture toughness of brittle materials. Hydroxyapatite matrices reinforced either with ceramic fibres, such as silicon carbide, carbon, alumina or with metals, such as titanium and alloys have been reported (4). Partially stabilized zirconia has been widely studied due to its high strength, high fracture toughness and good biocompatibility (5). It has been thought that the addition of zirconia, particles or fibres, to a hydroxyapatite matrix may lead to an improvement of the mechanical properties of the composite(6) and will not affect its biocompatibility (7,8).
The present study evaluates the fracture toughness of such composites by means of SENB specimens in terms of the processing parameters, and performs their fractographic analysis.

EXPERIMENTAL PROCEDURE

The hydroxyapatite powder was supplied by OSTEOSYNTH, S.L., Barcelona - Spain. The powder was calcined at 800°C for 3h and desaggregated in a ball-mill. The zircona powder from Merck and Y_2O_3 - stabilized circona fiber from Zircar, was used as starting material. The circona powder was milled in a McCrone-Mill for 20 min. The hydroxiapatite powder was mixed with 10%, 20% and 30% (by weight) zirconia powder and with 30% of zircona fiber in an acetone medium in a ball-mill. The resulting powder was dried, de-agglomerated and mixed with an ethylene bis-stearoylamide type lubricating wax (Hoeshst), in a rotatory mixer for two hours. Bars 6 mm x 6 mm x 50 mm were made by uniaxial pressing in a steel die, at a pressure of 2000 $Kg.cm^{-2}$, and sintered at 1100, 1150, 1200 and 1250°C. The products were characterized by means of X-ray diffraction (Siemens D-500) and scanning electron microscopy (Jeol 6400). The density of the compacts was measured by the Archimed's method using mercury.

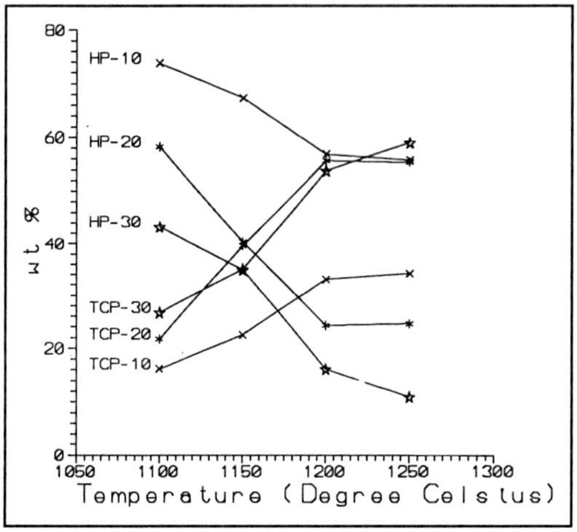

Figure 1. Relative evolution of hydroxyapatite in relation to TCP for the different temperatures.

For the measurement of fracture toughness, the samples were polished (300 grit) on their intended support and load faces, and a 3 mm notch was machined with a diamond saw. The samples were then tested in four point bending at a crosshead speed of 0.5 mm/min in an electromechanical Instron 4507 testing machine, equipped with a load cell of 1000 N. The support span (S_1) = 2 x load span (S_2) = 40 mm. The fracture toughness, K_{IC} was calculated from the following equation:

$$K_{IC} = Y\,S\,\sqrt{a} \qquad S = \frac{3\,P\,l}{B\,W^2}$$
$$Y = 1.99 - 2.47\left(\frac{a}{W}\right) + 12.97\left(\frac{a}{W}\right)^2 - 23.17\left(\frac{a}{W}\right)^3 + 24.8\left(\frac{a}{W}\right)^4 \qquad (1)$$

l = (S_1 - S_2)/2, **W** is the depth of the specimen and B its width; **a** = length of the notch. The compressive strength was measured using samples of dimensions 5 mm x 7 mm x 10 mm, at a crosshead speed of 1 mm/min, and the Young's modulus was estimated from the slope of the stress-strain curve.

RESULTS AND DISCUSSION.

In a previous work (9)(10), the sintering process of composite materials based in hydroxyapatite and zirconia particles was studied. It was shown that a chemical reaction between ZrO_2 and hydroxyapatite takes place which decomposes hydroxyapatite into TCP, CaO and H_2O and where CaO reacts with ZrO_2 to produce $CaZrO_3$ and to stabilize the cubic phase. The relative evolution of hydroxyapatite in relation to TCP for the different temperatures is shown in Fig. 1 for the different studied materials. This decomposition produces CaO whilst cubic stabilized zircona is formed. Their reaction leads to the formation of $CaZrO_3$, as shown in Fig. 2. The diffusion of Ca inside the zirconia particles takes place through a $CaZrO_3$ reaction layer which slows down the diffusion process of CaO and consequently reduces the decomposition rate of hydroxyapatite (10). This mechanism is shown in Figs. 1 and 2 where the stabilization of the decomposition of hydroxyapatite and of the formation of $CaZrO_3$ for specimens with 10 and 20% ZrO_2 can be observed. For samples with 30% ZrO_2, the transformation is not stabilized yet due to the higher ZrO_2 content.

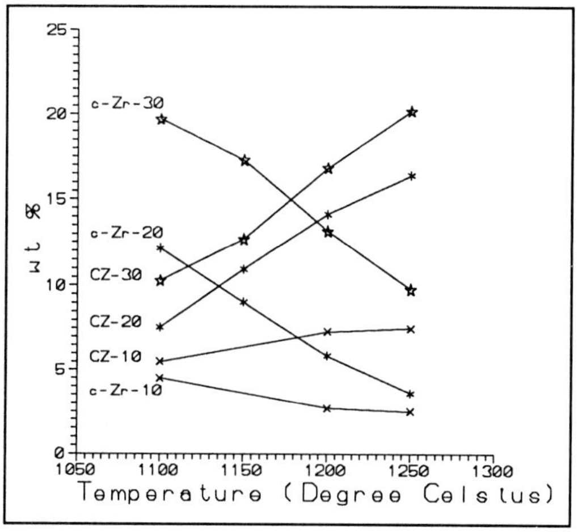

Figure 2. Evolution of cubic stabilized zircona to $CaZrO_3$ for the different temperatures.

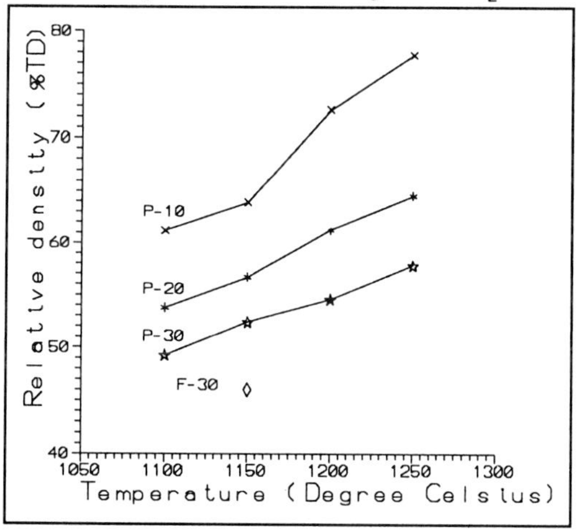

Figure 3. Density of hydroxyapatite-zircona composite with sintering temperature.

As this transformation takes place, the apparent density of the specimens decrease since H_2O is lost and the zirconia density increases because the CaO diffusion stabilizes its cubic form. The sintering of the material counteracts this effect since density increases as shown in Fig. 3.

The evolution of fracture toughness with ZrO_2 and temperature has been represented in Fig. 4. It can be noticed that between 1100 and 1200 °C fracture toughness increases with apparent density, whislt over this temperature the effect is reversed and fracture toughness decreases with increasind apparent density. Koji Ioku (11), used ZrO_2 with 3% Y_2O_3 dispersed in hydroxyapatite with a ratio hydroxyapatite:3YZ of 80:20. After isostatic press preparation a decrease in the values of K_{IC} are obtained when apparent density decreases for specimens sintered at temperatures higher than 1200 °C. These authors relate this decrease with the transformation of hydroxypatite and β-TCP into α-TCP. When SEM pictures of materials containing 30% zircona fibres stabilized with Y_2O_3 are observed, Fig. 5, and they are compared with the SEM pictures of the original fibres, Fig. 6, it can be noticed that the sintering of hydroxyapatite is correct, since particles have joined together, but due to the compactation method open porosity is present. It can be also clearly observed the decomposition of the zircona fibres, since crystals perpendicular to the fiber length have nucleated, which will enhance the possibility of fracture across the grain boundaries of such crystals.

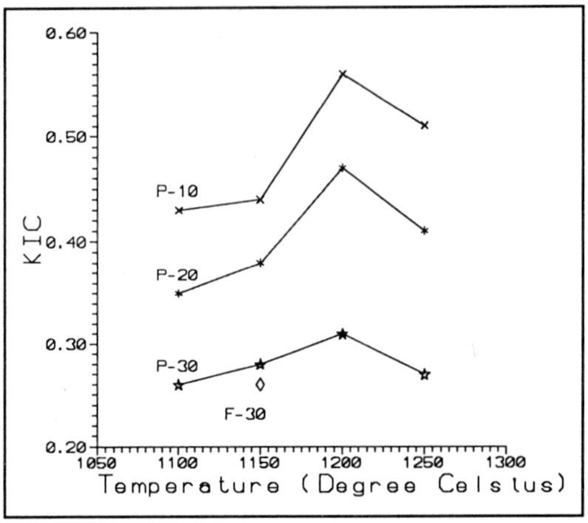

Figure 4. Evolution of fracture toughness with ZrO_2 contend and sintering temperature.

REFERENCES

1. De Groot, K., Proceedings of Seventh CIMTEC-World Ceramics Congress, 1990.
2. Jarcho, M.; Bolen, C.H.; Thomas, M.B.; Bobick, J.; Kay, J.F.; Doremus, R.H., J. Mater. Sci. **11**, 1976, 2027-2035.
3. De Groot, K., in Bioceramics of Calcium Phosphate, CRC, Florida, 1983, 102.
4. Ruys, A.J.; Zeigler, K.A.; Brandwood, A.; Milthorpe, B.K.; Morrey, S. and Sorrel, C.C., Proceedings of the 4th International Symposium on Ceramics in Medicine, **4**, 1991, 281-286.
5. Melendo, J.M. y Rodriguez,A.D., Bol.Soc.Esp.Ceram. Vidr., **30**, 3, 1991, 157-164.
6. Soltész, U., Bioceramics: Material characteristics versus in vivo behaviour, Annals of the New York Academy of Sciences, Vol.**523**, 1988, 137-156.
7. Fujita, M.; Oda, M.; Harikoshi, M.; Enomoto, S. and Sato, A., Proceedings of Fourth World Biomaterials Congress, 1992, 385.
8. Satoh, Y. and Niwa, S., Proceedings of the 3rd International Bioceramic Symposium 3, 1990, 101-108.
9. Caetano, J.; R.M. Manent; Valero, I.L.; Stucchi, E.B.; J.A. Varela; S. Martínez. in: Elaboration, Thermomechanical and Physicochemical Properties of Ceramics. S.Martinez & N. Clavaguera Ed. ISBN: 84-475-0496-4. Barcelona 1993. pp.19-25.
10. Caetano, J.; R.M. Manent; Valero, I.L.; Stucchi, E.B.; J.A. Varela; S. Martínez. European Society for Biomaterials, Davos 1993.
11. Koji Ioku, M. Yoshimura & S. Somiya. Biomaterials 1990, vol. 11, pp.57-61.

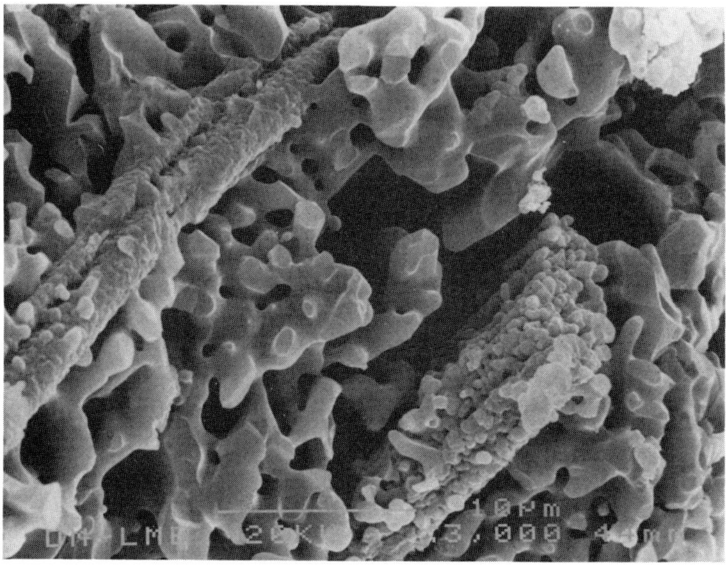

Figure 5. SEM micrograph of materials containing 30% of ZrO_2 fiber sintered at 1150°C.

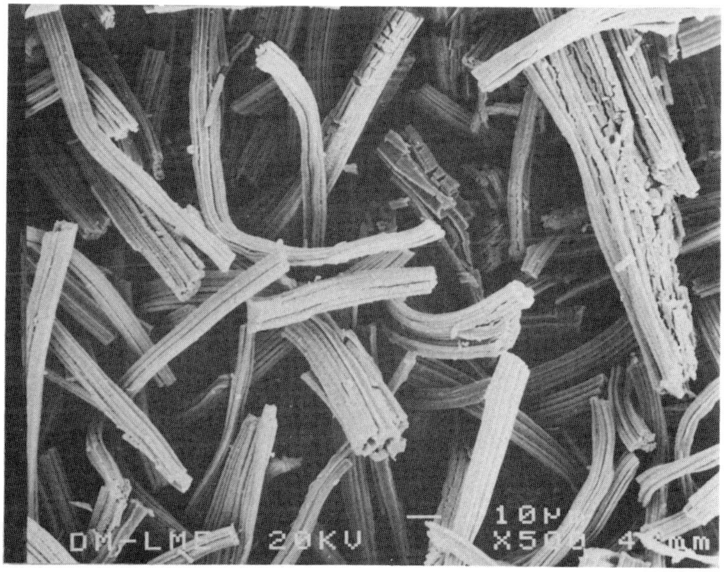

Figure 6. SEM micrograph of the original zircona fiber.

Bioceramics, Volume 7 Edited by Ö.H. Andersson and A. Yli-Urpo
(*Proceedings of the 7th International Symposium on Ceramics in Medicine, Turku, Finland, July 1994*)

Compliance of a Calcium Phosphate Cement with Some Short-term Clinical Requirements

M.P. Ginebra, E. Fernández, M.G. Boltong, J.A. Planell, O. Bermúdez* and F.C.M. Driessens

Dept. of Material Sciences and Metallurgy, Universitat Politècnica de Catalunya, Avda. Diagonal 647, 08028 Barcelona, Spain

* At present: Dep. de Física, Universidad del Cauca, Popayán, Colombia

ABSTRACT

Short-term clinical requirements proposed for calcium phosphate cements are: (1) initial setting time between 4 and 8 min, (2) final setting time between 10 and 15 min, (3) maximum attainable strength, and (4) attainment of this maximum strength within the shortest possible time. From previous studies it was known that the setting times of a calcium phosphate cement made of α-TCP could be controlled by using solutions of Na_2HPO_4 as cement liquid and that the strength could be improved by controlling the milling. Combinations of these factors were used to find out whether the cement derived from α-TCP could be brought to complete compliance. It could be concluded: (a) that compliance to requirements 1 and 2 simultaneously was possible, (b) that this caused a drop in the maximum compressive strength from 70 to 45 MPa, and (c) that at body temperature the maximum strength is achieved much faster than at room temperature.

INTRODUCTION

From the extended clinical experience with acrylic bone cements [1], dental cements, dental gypsum and dental stones [2] it can be concluded that the clinically desired range for the initial setting time I of a calcium phosphate cement is about

$$4 \leq I \leq 8 \text{ min} \qquad (1)$$

an that for the final setting time F about

$$10 \leq F \leq 15 \text{ min} \qquad (2)$$

As such most calcium phosphate cements have longer setting times [3-5], but in most cases the so-called common ion effect can be used to reduce the setting times [6].

Other short term clinical requirements are: (3) maximum attainable strength and (4) attainment of this maximum strength within the shortest possible time. This time should not be longer than the duration of the period of anesthesia, because then the patient may start to move again.From a previous study [6] it was known that a solution of Na_2HPO_4 could be used as a cement liquid in order to accelerate the setting of a cement prepared from α-TCP (alpha-tertiary calcium phosphate) as cement powder. The maximum compressive strength attained with this cement [7] was 60 to 70 MPa when 2% of precipitated hydroxyapatite was used as a nucleator in the cement powder. The purpose of the present study was to find out whether compliance could be attained with the requirements (1) and (2) concerning the setting times and which effect this would have on the maximum attainable strength. Also, the time necessary to reach this maximum strength had to be determined in order to obtain a survey of the overall compliance whith these requirements.

MATERIALS AND METHODS

α-TCP was prepared by heating an appropriated mixture of $CaHPO_4$ and $CaCO_3$ in an electric furnace at 1300 ºC in air during 6 hrs and subsequent quenching in air. Milling was done 1 hr with an agate ball mill; 2% precipitated hydroxyapatite was added and homogeneization of the cement powder was carried out by further milling during 10 min.

As cement liquids aqueous solutions containing 0, 1, 2, 2½, 3 or 4 % Na_2HPO_4 were used. Liquid/powder ratios were varied over the range $0.30 \leq L/P \leq 0.38$ g/g. Cement pastes were prepared by mixing liquid and powder in a mortar during 1 min. Brass rings with an inner diameter of 9.5 mm and a height of 5 mm were filled with the cement paste, while resting on a glass plate. Then the initial and the final setting time of the pastes were carried out with Gilmore needles at room temperature [3]. Determinations were done in triplicate for each combination of L/P ratio and Na_2HPO_4 solution. In this way compliance with requirements (1) and (2) was checked.

Those combinations of L/P ratio and Na_2HPO_4 solution, for which this compliance was found, were used to prepare cement cylinders with a height of 12 mm and a diameter of 6 mm. These were soaked during 1 day at 37 ºC in Ringer's solution and then used to determine the maximum attainable compressive strength (n=10) with an Instron Universal Testing Machine Type 4507. In a similar way the dependence of the compressive strength on the time of soaking was determined in order to find the time period necessary to reach the maximum attainable compressive strength. For this purpose in one series soaking at 37 ºC was started immediately after complete setting of the cement. However, in a second series soaking at 37 ºC was started 2½ hrs after complete setting and maintenance at room temperature. The latter was done in order to see what the additional effect of the lower temperature would be (compare [8]).

RESULTS

The results for the measurements of I and F are given in Table 1.

They show that compliance with conditions (1) and (2) is paid for by a reduction in the maximun attainable strength with 33 to 40 %. In addition, Table 2 shows that wehether or not the sample was kept first for 2½ hrs at room temperature before bringing it at 37 ºC had no effect on the maximum attainable compressive strength.

From table 3 it is apparent that the increase in strength is much faster when the temperature of the cement is brought to 37ºC inmediately after setting: within 5 hrs it reaches a compressive strength of practically 20 MPa.

Table 3. Compressive strength at L/P=0.32 and 3% Na_2HPO_4 in the cement liquid as a function of the age of the samples (n=10).

Series 1: start of soaking at 37 ºC inmediately after setting
Series 2: start of soaking at 37ºC 2½ hrs after setting

Age of sample (hrs)	Series 1	Series 2
3	12 (1)	5 (1)
5	19 (2)	11 (2)
7	23 (2)	17 (2)
9	23 (2)	21 (2)
11	25 (2)	26 (3)
14	n.d.*	29 (3)
16	n.d	32 (3)
18	n.d	35 (5)
24	36 (3)	35 (5)

*n.d. = not determined

In this respect it behaves like the commercial product Norian Fracture Grout [9], but the final compressive strength of our cement is nearly double that value (36 MPa) whereas the commercial product reaches a level of only 25 MPa. Yet, this commercial product gave a better mechanical quality of fixation of intramedullary rods than a commercial low-viscosity PMMA bone cement [10]. Hence, it might be worthwhile to compare calcium phosphate cements and PMMA bone cement also for other orthopaedic conditions like the fixation of metallic endoprostheses in hip and knees.

ACKNOWLEDGEMENTS

This study was supported by a grant from the Dirección Científica y Técnica of Spain. The authors would also like to thank the CICYT for funding this work through project MAT90-0755.

REFERENCES

1. Standard specification for acrylic bone cement F451-86, Annual book of ASTM Standards, ASTM, Philadelphia 1990, p. 82-88

2. J.W. Stanford, Guide to dental materials and devices, 4th Ed. Americal Dental Association, Chicago 1968

3. F.C.M. Driessens, M.G. Boltong, O. Bermúdez and J.A. Planell, J. Mater. Sci. Mat. Med. 4, 503-508 (1993)

4. Y. Takezawa et al., J. Japan Soc. Dent. Mat. Devices 6, 426-431 (1987)

5. A. Mirtchi, J. Lemaitre and E. Munting, Biomaterials 10, 634-638 (1989)

6. E. Fernández, M.G. Boltong, M.P. Ginebra, O. Bermúdez, F.C.M. Driessens and J.A. Planell, Clinical Materials, in press (1994)

7. F.C.M. Driessens, M.G. Boltong, J.A. Planell, O. Bermúdez, M.P. Ginebra and E. Fernández, in: P. Ducheyne and D. Christiansen (Eds) Bioceramics, vol. 6, Butterworth-Heinemann Ltd., London, 469-473 (1993)

8. M.T. Fulmer and P.W. Brown, J. Mater. Res. 8, 1687-1696 (1993)

9. D.C. More, E.P. Frankenburg, J.A. Goulet, G.P. Graziano, R.S. Maitra, L.A. Farjo and S.A. Goldstein, 10th Eur. Conf. Biomaterials, Davos, 1993, p.94

10. R.D. Coutts, S.A. Hacker, B.R. Constantz, J.M. Kerina and V.D.N. Nguyen, 10th Eur. Conf. Biomaterials, Davos 1993, p. 24

Bioceramics, Volume 7 Edited by Ö.H. Andersson and A. Yli-Urpo
(*Proceedings of the 7th International Symposium on Ceramics in Medicine, Turku, Finland, July 1994*)

Comparative Study of Two Types of Bioactive Bone Cements Containing $MgO-CaO-SiO_2-P_2O_5-CaF_2$ Glass and Glass-Ceramic Powder

J. Tamura, K. Kawanabe, M. Kobayashi, T. Yamamuro, T. Nakamura,*
T. Kokubo,** S. Yoshihara‡ and T. Shibuya‡

Department of Orthopaedic Surgery, Faculty of Medicine, Kyoto University, Kyoto, Japan
* Research Center for Biomedical Engineering, Kyoto University, Japan
** Division of Material Chemistry, Faculty of Engineering, Kyoto University, Japan
‡ Nippon Electric Glass Co. Ltd., Otsu, Japan

ABSTRACT

Two types of bioactive bone cement containing either $MgO-CaO-SiO_2-P_2O_5-CaF_2$ glass or glass-ceramic powder were made to evaluate the influence of crystal phase of the bioactive bone cement on its mechanical and biological properties. The former was produced from glass powder and Bis-GMA resin and the latter from glass-ceramic powder containing apatite and wollastonite crystals and Bis-GMA resin. Histological examination showed direct bonding between the bone and the cement for each bioactive cement. The bioactive bone cements bonded to the bone through reactive layer. In specimens with the bioactive bone cement containing glass powder the reactive layer was composed of two layers: a radiopaque outer layer (Ca-P-rich layer); and a relatively radiolucent inner layer (Si layer). For those with the bioactive bone cement containing glass-ceramic powder the Ca-P-rich layer was formed, however, the Si layer was absent.

INTRODUCTION

Kawanabe et al. reported a bioactive bone cement consisting of $CaO-SiO_2-P_2O_5-CaF_2$ glass powder and bisphenol-a-glycidyl methacrylate (Bis-GMA) resin (1). This cement forms a chemical bond with living bone through a Ca-P-rich layer and has a high mechanical strength. In this study, the influence of crystal phase of the bioactive bone cement on the mechanical and biological properties was determined by mechanical and *in-vivo* test on bioactive bone cement containing either $MgO-CaO-SiO_2-P_2O_5-CaF_2$ glass or glass-ceramic powder.

MATERIALS AND METHODS

Cement Preparation

Glass with a nominal composition of CaO 44.9, MgO 4.6, SiO_2 34.2, P_2O_5 16.3, and CaF_2 0.5 by weight ratio was first melted in a SiC furnace at 1500°C for

4 h. The melt was quenched between water-cooled steel rollers and formed into ribbons that were pulverized in an alumina ball mill to produce the glass powder. Glass-ceramic with the same composition containing apatite and wollastonite (AW-glass-ceramic) was synthetized by the method previously reported (2). The glass-ceramic plate was pulverized in an alumina ball mill to produce the glass-ceramic powder. The average particle size of the glass and glass-ceramic powder was 5 μm. The powder was silane-treated with γ-methacryloxy propyl trimethoxy silane (Shinetsu Chemical Industry, Tokyo Japan) and benzoyl peroxide, 0.4% per unit weight ratio, was added. The Bis-GMA resin was prepared from equal weights of Bis-GMA and triethylene-glycol dimethacrylate (TEGDMA) (Shin-nakamura Chemical Industry, Wakayama, Japan) and N.N-dimethyl-p-toluidine, 0.2% per unit weight of resin, was dissolved into the mixture(1).

For each type of bioactive bone cement, two kinds of cements, 70 and 80% by weight glass/glass-ceramic powder, were evaluated. Each cement was prepared by mixing the powder into the Bis-GMA resin and stirring for 1 min. Polymerization occured within 4 to 6 min.

Compression Test

Specimens for mechanical testing were prepared by inserting the mixed, but unpolymerized cement into steel mold (6 mm diameter, 12 mm long). After hardening, the specimens were removed and soaked in simulated body fluid (3) at 37°C. After soaking for 24 hs the specimen was subjected to compression testing using an Instron-type testing machine (Autograph model AG-10TB, Shimadzu, Kyoto, Japan) at a cross-head speed of 0.5 mm/min.

Animal Experiment

For evaluation of bioactivity a cortical bone defect was made on the medial aspect of the proximal metaphysis of both tibiae of male Wistar rats weighing 2 50-300 g (10 weeks old) and filled with one of the four bone cements. Each cement was implanted in three animals under general anesthesia with Nembutal (40 mg/kg body weight) and polymerized *in situ*. The animals were then sacrificed at 4 weeks post surgery (12 rats in total).

Histological Examination

The cement containing portion of the tibiae containing the cements was excised and fixed in 10% phosphate-buffered formalin solution. The specimens were dehydrated in serial concentrations of ethanol and then embedded in polyester resin. Thin sections (500 μm thick) were cut perpendicular to the axis of the tibia with a band saw (EXAKT BS-3000, EXAKT Apparatebau, 2000 Norderstedt, Germany), and were ground to a thickness of 120 μm using a diamond lap disk (Maruto Ltd., Tokyo, Japan) for evaluation by contact

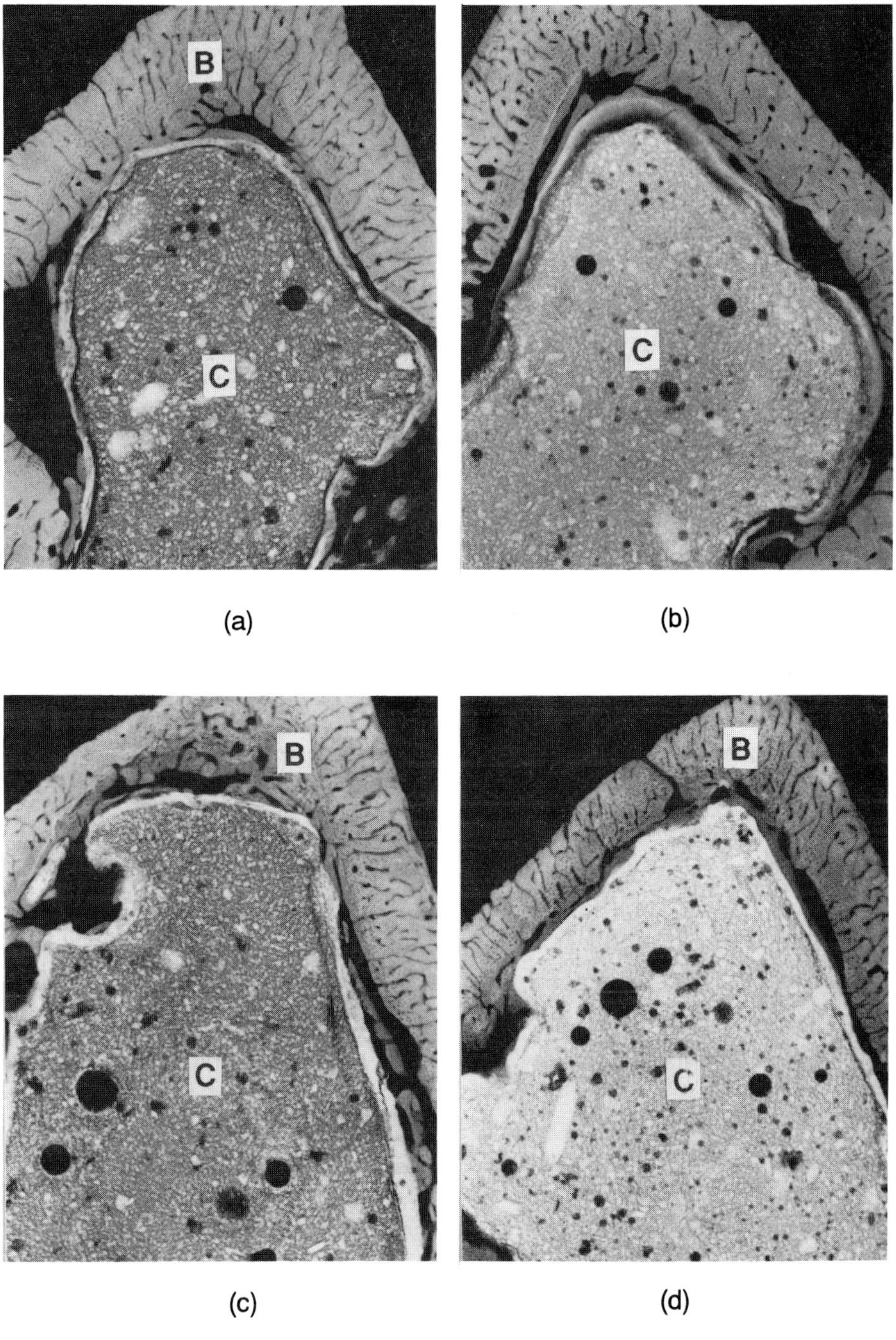

Figure 1 Contact microradiogram of bioactive bone cement in rat tibiae 4 weeks after implantation.
Cement containing (a)70% and (b)80% glass powder.
Cement containing (c)70% and (d)80% glass-ceramic powder.
C; cenent, B;Bone. (original magnification X40)

microradiography (CMR). Additional sections (500 μm thick) were polished with diamond paper for analysis of the interface between the bone and the bioactive bone cement using a scanning electron microscope (Hitachi S-800, Tokyo, Japan) with an energy-dispersive X-ray microanalyzer (Horiba EMAX-3000, Tokyo, Japan). X-ray intensities of silicon, calcium and phosphorus were analyzed on the cement/bone interface.

RESULTS

Mechanical Properties

Structural tests on the cement containing 70 and 80% glass powder showed compressive strength of 165 and 180 MPa, respectively, whereas those of the cement containing 70 and 80% glass-ceramic powder were 183 and 194 MPa, respectively.

Evaluation of Bioactivitiy

Analysis of the implanted specimens showed that around each type of cement, new bone formation was present and each was directly bonded to the bone without intervening fibrous tissue. A reactive layer was seen between the cement and the new bone. Cement with a high percentage of glass showed a thick reactive layer around its perimeter. In specimens with the bioactive bone cement containing glass powder, the reactive layer was composed of two layers: a radiopaque outer layer; and relatively radiolucent inner layer. In specimens with the bioactive bone cement containing glass-ceramic powder the reactive layer was radiopaque, however the radiolucent layer was absent (Figure 1).

Evaluation of the Cement/Bone Interface

Scanning electron microscopy and electron-probe microanalysis of the interface in each cement revealed high levels of calcium and phosphorus. This suggests that these cements bond to bone through a so-called 'Ca-P-rich layer'. In specimens with cement containing 80% glass powder, a layer having a low calcium level was seen inside the Ca-P-rich layer. In this layer the silicon level was slightly lower than in the cement, but was still relatively high (Si layer) (Figure 2). This region may correspond to the radiolucent layer observed by CMR. This layer was narrower in specimens with cement containing 70% glass powder. However, in specimens with the bioactive bone cement containing glass-ceramic powder, the Ca-P-rich layer was formed but the Si layer was absent.

DISCUSSION

Recently, new bone cements have been developed to overcome the problems

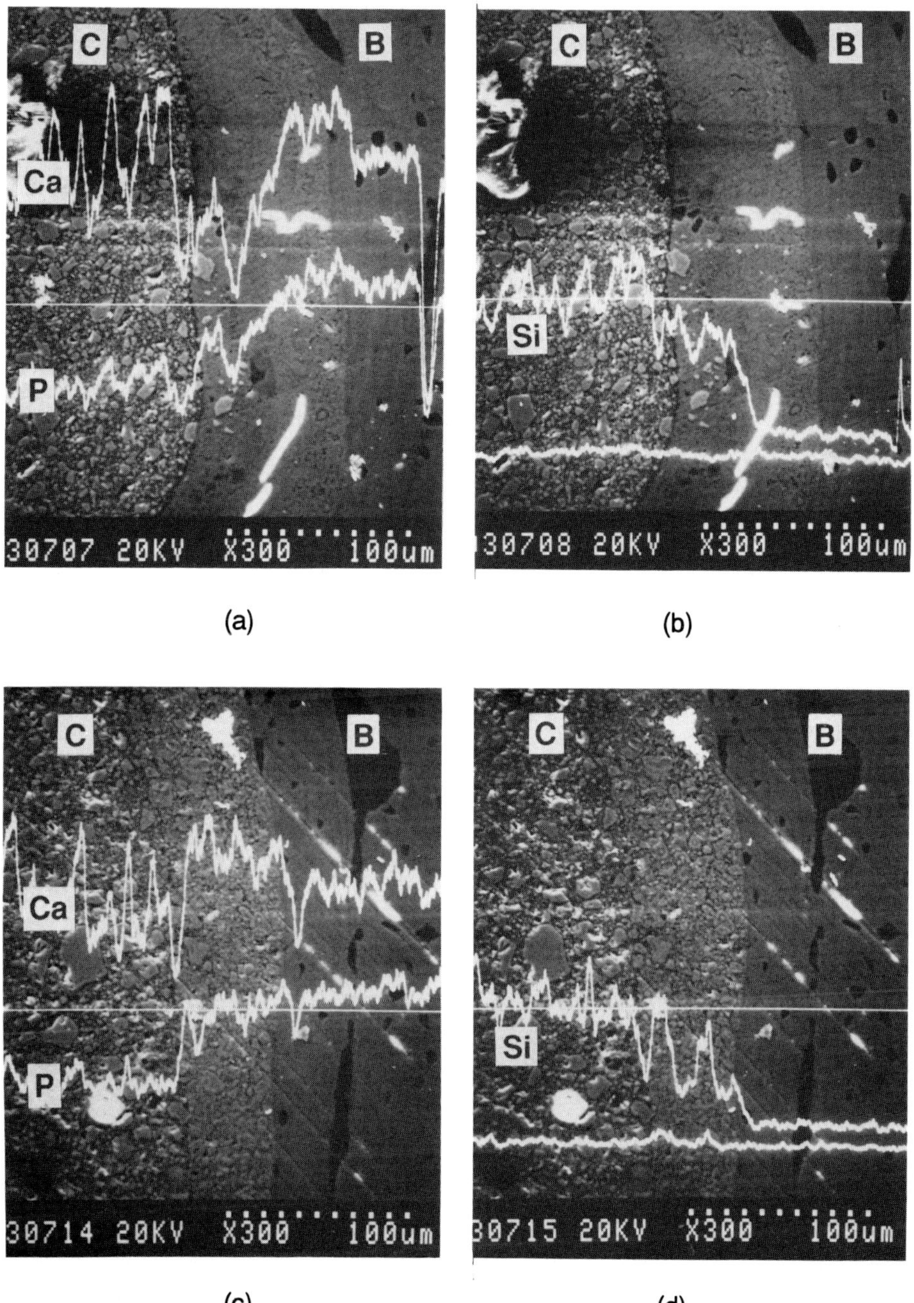

Figure 2 Scanning electron micrograph and electron-probe microanalysis of the interface between bioactive bone cement and bone. (a), (b) cement containing 80% glass powder, (c),(d) cement containing 80% glass-ceramic powder. C; cement, B; bone.

associated with PMMA bone cement. We have been trying to make bioactive bone cements using bioactive glass powder (4,5,6). Previously, it was reported that a combination of silane-treated $CaO-SiO_2-P_2O_5-CaF_2$ glass powder and Bis-GMA resin achieved bioactivity and high mechanical strength (1). In this study, $MgO-CaO-SiO_2-P_2O_5-CaF_2$ glass and glass-ceramic powder were used to investigate the influence of crystal phase in the powder on the mechanical properties and bioactivity. The compressive strength of each type of bioactive bone cement was more than twice that of PMMA cement (68MPa) (1), and high bioactivity was observed. A radiolucent layer was present inside the Ca-P-rich layer in specimens with cements containing glass powder. In this layer, the calcium level was decreased, whereas the silicon was relatively high (Si layer). It was thought that calcium had dissolved outside to make this layer radiolucent. A similar reactive layer was also observed in specimens with cements containing $CaO-SiO_2-P_2O_5-CaF_2$ glass (1). The Si layer was absent in specimens with cements containing glass-ceramic powder.
In our previous report, $MgO-CaO-SiO_2-P_2O_5-CaF_2$ glass-ceramic (A-W-glass-ceramic) in solid form bonded to bone through a Ca-P-rich layer and a Si layer was not observed, whereas $MgO-CaO-SiO_2-P_2O_5-CaF_2$ glass (mother glass of A-W-glass-ceramic) bonded to bone in the presence of Ca-P-rich and Si layers (7). The results of the present study with regard to the presence of the Ca-P-rich and Si layers agree with our previous findings for these materials implanted in solid form.

This work was supported in part by a Grant-in-Aid for Scientific Research (No.05558112) from the Ministry of Education, Science and Culture, Japan.

REFERENCES

1. Kawanabe, K., Tamura, J., Yamamuro, T., Nakamura, T., Kokubo, T. and Yoshihara, S., J. Appl. Biomater., (1993) 4 135-141
2. Kokubo, T., Ito, S., Sakka, S. and Yamamuro, T., J. Mater. Sci., (1985) 20 2001-2004
3. Kokubo, T., Kushitani, H., Sakka, S., Kitsugi, T. and Yamamuro, T., J. Biomed. Mater. Res., (1990) 24 721-734
4. Kokubo, T., Yoshihara, S., Nishimura, N., Yamamuro, T. and Nakamura, T., J. Am. Ceram. Soc., (1991) 74 1739-1741
5. Taguchi, Y. Yamamuro, T., Nakamura, T., Nishimura, N., Kokubo, T., Takahata, E. and Yoshihara, S., J. Appl. Biomater., (1990) 1 217-223
6. Tamura, J., Kawanabe, K., Yamamuro, T., Nakamura, T., Kokubo, T., Yoshihara, S. and Shibuya, T., Bioceramics Vol.6, Butterworth-Heinemann Ltd, (1993) 463-468
7. Kitsugi, T., Yamamuro, T., Nakamura, T. and Kokubo, T., J. Biomed. Mater. Res., (1989) 23 631-648

Bioceramics, Volume 7 Edited by Ö.H. Andersson and A. Yli-Urpo
(*Proceedings of the 7th International Symposium on Ceramics in Medicine, Turku, Finland, July 1994*)

Safety Assessment of Hydroxyapatite Composite Resin, a New Bioactive Bone Cement

A. Maruoka, M. Saito, H. Sakai, N. Sugano and K. Hino
Department of Orthopaedic Surgery, Osaka-Minami National Hospital, 677-2 Kido-cho, Kawachinagano-shi, Osaka, 586, Japan

ABSTRACT

We developed hydroxyapatite composite resin(CAP), a new bioactive bone cement composed of 80 % w/w hydroxyapatite granules and a bis-phenol-A glycidyl methacrylate (BIS-GMA)-based resin. CAP achieves direct contact with bone because complete polymerization at the surface of the resin is prevented by interaction with oxygen. The toxicity of the CAP polymer and acrylic monomers was examined by determining (1) the cytotoxicity of polymerized CAP, (2) the cardiovascular effects of the CAP monomers, and (3) the release of free CAP monomers during polymerization. The results indicated that the toxicity of CAP provides no barrier to its use as a bioactive bone cement.

INTRODUCTION

The use of polymethylmethacrylate (PMMA) cement for joint arthroplasty is an established method. PMMA cement for joint arthroplasty has been an established method. However, PMMA cement has some biological disadvantages, such as the growth of fibrous tissue at the PMMA cement-bone interface and thermal damage to the surrounding bone during polymerization. To overcome these disadvantages, we developed hydroxyapatite composite resin (CAP) as a new type of bioactive bone cement. CAP is composed of 80 % w/w hydroxyapatite granules and a bis-phenol-A glycidyl methacrylate (BIS-GMA)-based resin. The acrylic monomers used were BIS-GMA, 2,2'-bis-(4-methacryloylethoxyphenyl) propane (BIS-MEPP) and triethyleneglycol dimethacrylate (TEG-DMA) at a ratio of 35/35/30. Polymerization was initiated by mixing approximately equal volumes of two pastes, one of which contained HA granules plus the acrylic monomers and benzoyl peroxide (the initiator), while the other contained HA granules plus the acrylic monomers and diethanol-p-toluidine (the accelerator). It has been reported that CAP has a superior mechanical strength to PMMA cement and the ability to make direct contract with bone [1]. This property is demonstrated because complete polymerization of the surface of the resin is prevented by interaction with oxygen. Thus, HA granules are exposed on the surface of the polymerized CAP and allow direct bonding with bone. Since the obstruction of complete polymerization of the acrylic monomers plays an important role in the biocompatibility of CAP, the effects of the three monomer components need to be investigated. In the present study, the toxicity of CAP was examined by determining (1) the cytotoxicity of polymerized CAP, (2) the cardiovascular effects of the CAP monomers (BIS-GMA, BIS-MEPP, and TEG-DMA) in comparison with methylmethacrylate monomer, and (3) the release of free CAP monomers during polymerization in comparison with the release of methylmethacrylate monomer.

MATERIALS AND METHODS

1. Cytotoxicity of polymerized CAP

The toxicity of the polymerized CAP was assessed by a cell culture technique. Semicircular disks made of polymerized CAP and glass (the control) were prepared and sterilized by heating at 180 ℃ for 1 h. Then pairs of disks were placed into plastic Petri dishes 60 mm in diameter. Free CAP monomers were almost completely washed out during this preparation process. A suspension of trypsinized subcultured V79 cells was diluted to provide 100 cells per dish. After 1 week of incubation, colony formation was assessed by counting according to the method of Kotoura [2].

2.Cardiovascular effects of the CAP monomers

The cardiovascular effects of the three monomers (BIS-GMA, BIS-MEPP, and TEG-DMA) were compared with those of methylmethacrylate monomer. Rabbits were anesthetized with intravenous pentobarbitone (25 mg/kg) and intramuscular xylazine (2 mg/kg). Tracheal intubation was performed after tracheotomy and mechanical ventilation was provided by a rebreathing circuit with 100 % oxygen. After the induction of anesthesia, a femoral artery catheter was inserted and the arterial blood pressure was measured using a calibrated electromanometers, an amplifier, and a chart recorder (Nippon Colin Ltd.) The heart rate was also obtained from the femoral artery trace. A peripheral venous catheter was inserted via an ear vein to allow injection of the monomers. A total of 14 rabbits weighing 2.4-3.5 kg were used. Baseline readings were obtained before intravenous injection of the monomers at a dose of 0.1 g/kg. MMA monomer was administered to four rabbits, BIS-GMA monomer to two, BIS-MEPP monomer to three, and TEG-DMA monomer to three. MMA monomer and TEG-DMA monomer were also given to 1 rabbit each at 1.0 g/kg, BIS-GMA nor BIS-MEPP could be injected at this dose because of their high viscosity (Table 1).

3. Release of free CAP monomers during polymerization

The CAP pastes were mixed at room temperature and for 90 sec with a spatula, after which the mixture was poured into a plastic Petri dish (20 mm deep and 36 mm in diameter). Four minutes after the start of mixing, the dish containing CAP in the dough phase was immersed in a known volume of 37 ℃ physiological

Table 1. The items and doses of administration of injections to 14 rabbits

	Dose of each monomer (g/kg) 0.1	1.0	Rabbit
MMA	4	1	(NO. 1 - 5)
BIS-GMA	2	0	(NO. 6, 7)
BIS-MEPP	3	0	(NO. 8 - 10)
TEG-DMA	3	1	(NO. 11-14)

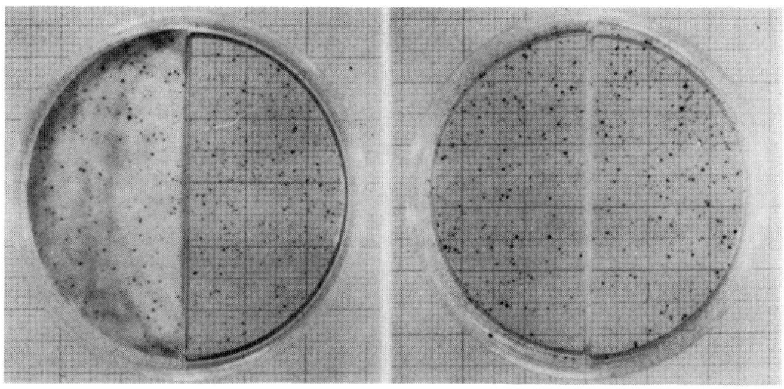

CAP control

Fig1. Cytotoxicity of polymerized CAP

saline. Thirty minutes after the start of mixing, the saline was analyzed by gas chromatography and the total amount of each monomer was measured. Then the amount was divided by the surface area of the dish. The same procedure was also performed for methylmethacrylate monomer and monomer release during polymerization was determined.

RESULTS

1. Cytotoxity of polymerized CAP

Figure 1 shows colony formation on the semicircular disks. These was no statistical difference between the CAP and control disks.

2. Cardiovascular monitoring

Table 2 shows serial mean arterial blood pressure reading when 0.1 g/kg of MMA monomer was injected into 4 rabbits. One of the animals died 4 minutes

Table 2. Changes in arterial pressure of 4 rabbits injected 0.1 g/kg of MMA monomer. # Death

Time from injection (minute)	Arterial pressure (mmHg) NO. 1		NO. 2		NO. 3		NO.4	
0'	117	54	156	97	81	55	158	104
0.5'	124	70	38	22	19	5	97	34
1'	148	85	103	53	35	9	39	25
2'	147	79	127	62	79	36	38	21
3'	143	73	-	-	63	35	32	17
4'	144	73	131	70	68	36	20	17
5'	110	57	132	77	85	40	#	
10'	141	68	142	90	85	50		

after the injection, 2 others developed transient hypotension within 30 sec after injection that recovered after 2-3 minutes, and the remaining animal showed no changes. Heart rate did not show any significant changes. When 10 times the dose of MMA monomer (1.0 g/kg) was injected, the rabbit died immediately. Figure 2 shows the changes of mean arterial pressure after the injection of MMA monomer. The broken line shows the rabbit which was injected with 1.0 g/kg of the monomer, while the solid lines show the rabbits injected with 0.1 g/kg. Figure 3 shows the changes of mean arterial pressure after the injection of CAP monomers. Neither BIS-GMA monomer nor BIS-MEPP monomer had any influence on the arterial pressure or the heart rate. 3 rabbits injected with TEG-DMA monomer (0.1 g/kg) showed no changes. This monomer dose had no

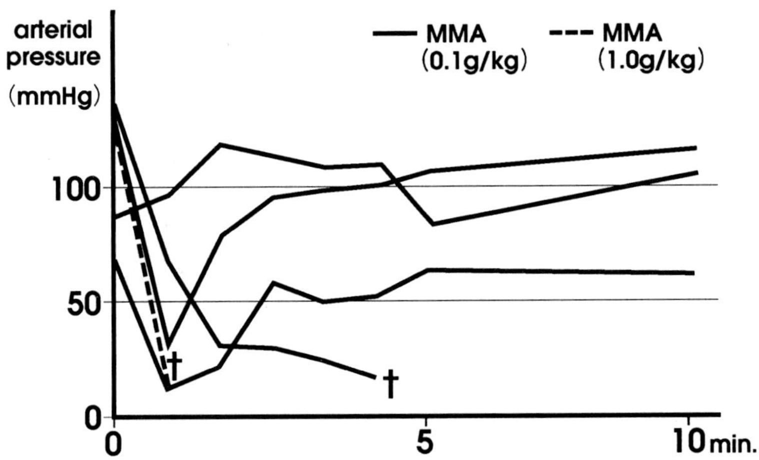

Fig 2. Changes in the mean arterial pressure following injections of MMA monomer.

† Cardiac arrest

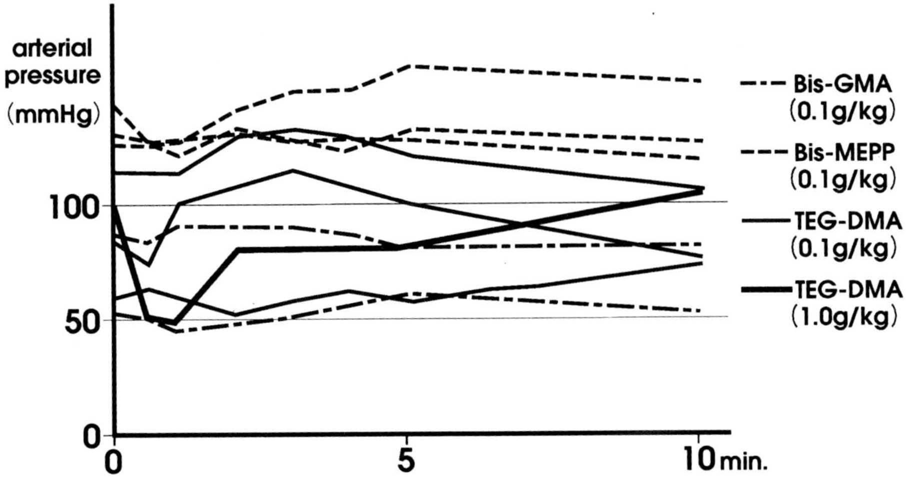

Fig 3. Changes in the mean arterial pressure following injections of monomers.

influence on the arterial pressure or heart rate. The bold solid line (Figure 3) shows the 1 rabbit injected with 1.0 g/kg of TEG-DMA monomer. Even this high dose of TEG-DMA monomer produced only transient hypotension and bradycardia. After 2 minutes, both the arterial blood pressure and the heart rate returned to baseline spontaneously. These date demonstrated that the CAP monomers had less cardiovascular effect than MMA monomer.

3. Release of free CAP monomers during polymerization

The amount of each CAP monomer released within 30 minutes was 1.4 μ g/cm^2 for BIS-GMA, 0.3 μ g/cm^2 for BIS-MEPP, and 380 μ g/cm^2 for TEG-DMA. The amount of MMA monomer released within 30 minutes after the start mixing was seventh times greater than that of TEG-DMA, while BIS-GMA and BIS-MEPP showed for less release than MMA monomer and TEG-DMA.

DISCUSSION AND CONCLUSIONS

Since the report of Bowen, BIS-GMA has been widely used in dentistry [3], and chronic pulpitis has only rarely occurred in association with this resin. This type of inflammation is mainly considered to be associated with the microleakage of bacteria [4],[5]. In previous in vitro studies, a polymerized BIS-GMA-based resin did not show any cytotoxity [6]. However, the biological effects of the free monomers released from this resin during polymerization have not been investigated, because the amount used for dental purposes was small and because polymerization occurred on the surface of the teeth. In contrast, orthopaedic surgery requires for more cement than dental use and complete polymerization of the resin in CAP is prevented by oxygen allowing free

Table 3. The solution rate of MMA monomer and CAP monomers (37 °C saline)

MMA	1.2 %
BIS-GMA	15 ppm
BIS-MEPP	3 ppm
TEG-DMA	4000 ppm

monomers to be released. Accordingly, we determined the toxicity of CAP monomers in this study. We found that CAP monomers were safer than MMA monomer with regard to cardiovascular effects. The high dose of TEG-DMA (1.0 g/kg) produced only transient hypotension and bradycardia, quite unlike the same dose of MMA monomer. These results indicate that the toxicity of CAP is no barrier to its use as a bone cement. The solution rates of MMA monomer and CAP monomers are shown in Table 3. The difference shown in the table indicates that MMA monomer is more likely to enter the blood than CAP monomers. In this study, CAP monomers were injected intravenously. However, this procedure was less than optimal, especially, for BIS-GMA and BIS-MEPP. In the future, the CAP monomer concentrations in blood and the major organs should be investigated after CAP is inserted into a long bone such as the femur. The metabolism of CAP monomers also needs to be clarified.

REFERENCES

1. Saito M, Maruoka A, Mori T, Sugano N, Hino K. Experimental studies on a new bioactive bone cement: hydroxyapatite resin. Biomaterials 1994; 15:156-160.
2. Kotoura Y, Yamamuro T, Shikata J,Kakutani Y,Kitsugi T. A method for toxicological evaluation of biomaterials based on colony formation of V 79 cells. Arch Orthop Trauma Surg 1985; 104:15-19.
3. Bowen RL. Properties of a silica-reinforced polymer for dental restorations. Am Dent Ass J 1963;66:57-64.
4. Brannstrom M, Nyborg H. Cavity treatment with a microbial fluoride solution: growth of bacteria and effect on the pulp. J Prosthet Dent 1973; 30: 303-310.
5. Bergenholtz G, Cox CF, Loesche WJ, Syed SA. Bacterial leakage around dental restorations: its effect on the dental pulp. J Oral Pathol 1982; 11:439-450.
6. Rathbun MA, Craig RG, Hanks CT, Filisko FE. Cytotoxicity of a BIS-GMA dental composite before and after leaching in organic solvents. J Biomed Mater Res 1991; 25: 443-457.

Cells and Materials

Bioceramics, Volume 7 Edited by Ö.H. Andersson and A. Yli-Urpo
(*Proceedings of the 7th International Symposium on Ceramics in Medicine, Turku, Finland, July 1994*)

Osteoblast and Osteoclast Responses to Calcium Phosphates

J.D. de Bruijn, Y.P. Bovell and C.A. van Blitterswijk
Laboratory for Otobiology and Biocompatibility, Biomaterials Research Group, Bldg. 55, University of Leiden, Rijnsburgerweg 10, 2333 AA Leiden, The Netherlands

ABSTRACT

Utilizing several *in vitro* and *in vivo* studies, a variety of calcium phosphates were examined for their bone-bonding and biodegradation properties. For the *in vitro* studies, use was made of osteoblast and/or osteoclast enriched rat bone marrow cultures that were grown on fluorapatite, tricalcium phosphate and various types of hydroxyapatite that differed in degree of crystallinity. For the *in vivo* studies, the same range of hydroxyapatite was used and implanted either in femora of young adult male Wistar rats, or in the acetabulum of goats.

The results of these studies show that more extensive dissolution/reprecipitation occurs on degrading and low crystalline calcium phosphate surfaces. As a consequence, bone tissue is formed more rapidly on these dynamic material surfaces. Irrespective of the type of material, osteoblasts first form afibrillar mineralized globular accretions, that subsequently fuse to form a cementing substance in which collagen fibres become incorporated. Osteoclasts only appear to have the ability to resorb low crystalline hydroxyapatite and tricalcium phosphate. In culture, osteoclastic resorption of these calcium phosphates occurred only after osteoblasts had been grown on the various materials. This suggests that osteoblast-derived substances initiate osteoclastic resorption activity.

INTRODUCTION

Much research is directed towards investigating the bone-bonding and biodegradation properties of calcium phosphate bioceramics. From a materials' point of view, bone-bonding is related to the formation of a surface carbonate apatite layer, while biodegradation or dissolution is mainly governed by the crystal structure, neck geometry and crystallinity of the material. Little is known, however, of the cellular processes that underlie bone formation onto biomaterials, and bone-bonding and biodegradation of calcium phosphates. Therefore, experiments are necessary to determine (i) why osteogenic cells show bonding-osteogenesis with bioactive materials but contact-osteogenesis with bioinert materials (Osborn and Newesely 1980), (ii) how this newly formed bone is produced at the biomaterial surface, (iii) how the interface structures observed between bone and biomaterial surfaces should be interpreted, and what their potential importance is, and (iv) which cell types can actively degrade calcium phosphates, and via what mechanism. By answering these questions, we could eventually develop new, custom-made biomaterials, to comply with the required conditions for use in specific applications.

Herein, we will present results from consecutive *in vitro* and *in vivo* studies in which both osteoblast and osteoclast responses to a range of calcium phosphates were examined.

MATERIALS AND METHODS

Hydroxyapatite varying in degree of crystallinity

Plasma-sprayed layers

Three types of hydroxyapatite (HA), varying in crystallinity, were plasma-sprayed onto either Thermanox^{tm} coverslips (for cell culture experiments), 1 mm diameter titanium rods (for implantation in rat femora), or acetabular cups (for implantation in goats). Low and semi-crystalline hydroxyapatite coatings were prepared by varying the plasma-spray parameters, while the highly-crystalline hydroxyapatite coating was prepared by a thermal treatment of the low-crystalline coating. The resulting crystallinities were approximately 10%, 60% and 95% respectively. All samples were sterilized with ^{60}Cobalt gamma-irradiation (2.5MRad) prior to the experiments.

Pressed blocks

Hydroxyapatite blocks varying in crystallinity, were produced by compressing spray-dried hydroxyapatite powder (CAM-Implants bv, the Netherlands) under vacuum for 10 s at 100 MN/m^2. The blocks were subsequently calcined or sintered in air at 600°C (HA600), 900°C (HA900) or 1250°C (HA1250). The microporosities were 56%, 52% and <2% respectively. Small sections were cut, the surfaces were polished with 1 μm diamond paste and the samples were ultrasonically cleaned prior to sterilization.

Cell cultures

Osteoblast-enriched cell culture

Rat bone marrow cells were derived from femora of 100-120 gram male Wistar rats according to the method described by Maniatopoulos et al. (1988). The cells were grown in α-Minimal Essential Medium containing 100 μg/ml penicillin G, 50 μg/ml gentamicin, 0.3 μg/ml Fungizone, 15% foetal bovine serum, and fresh additions of 10 mM ß-glycerophosphate, 50 μg/ml ascorbic acid and 10^{-8} M dexamethasone. The primary cultures (200 μl droplets on 1 cm^2 samples) were refed 3 times weekly and maintained at 37°C in a humidified atmosphere of 95% air/5% CO_2 for 18 days.

Osteoclast-enriched cell culture

Rat bone marrow cells were removed and grown as described above, without dexamethasone added to the culture medium. Cells were cultured for 11 days, fixed in Karnovsky's fixative and stained for tartrate-resistant acid phosphatase (TRAP) as described elsewhere (de Bruijn et al. 1994a).

Two-stage culture

In the two-stage culture, an osteoblast-enriched cell population was grown on the various substrata for 18 days to form bone-like tissue (de Bruijn et al. 1994a). These cells were subsequently removed with trypsin and a new, osteoclast-enriched cell population was grown onto the various preconditioned substrata.

Implantation experiments
The hydroxyapatite coated titanium implants were either press-fit inserted into the femora of 200 gram male Wistar rats as described previously (de Bruijn et al. 1994b), or placed, as hydroxyapatite coated acetabular cups, in the acetabulum of goats. Post-operatively, the animals were allowed free weight bearing.

Scanning electron microscopy and light microscopy (SEM and LM)
All specimens were fixed in Karnovsky's fixative prior to dehydration and critical point drying. Half the femora were prepared for SEM by submerging them in liquid N_2 and fracturing the bone tissue from the implant surface. The specimens were subsequently gold coated and examined in a Philips S525 at 15kV. The remaining femora and all acetabular cups were embedded in methyl methacrylate and sectioned for LM (approximately 10 μm thick sections) with an innerlock diamond saw.

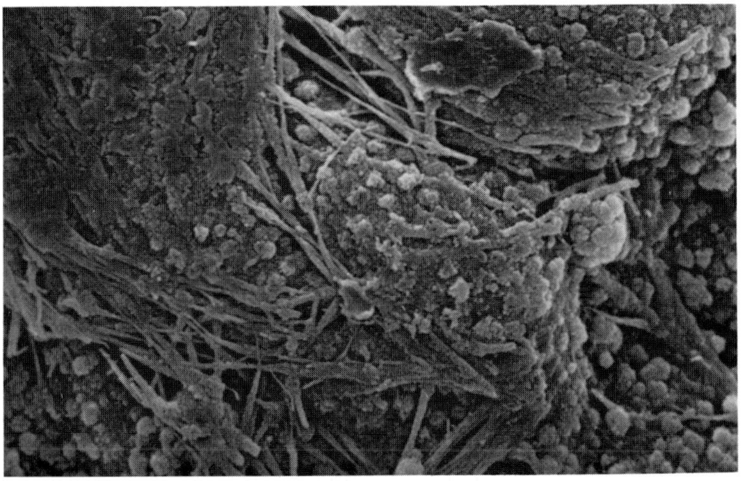

Figure 1. High power scanning electron micrograph showing individual globular accretions, in which collagen fibres are incorporated, on the surface of highly crystalline hydroxyapatite, in vitro. The supra-nodular cell layers have been partially removed by compressed air to visualize the interfacial composition between the material and the substratum. Field width = 15 μm

RESULTS AND DISCUSSION

Osteoblast responses
Both *in vitro* and *in vivo*, the initial matrix that was formed onto the various material by osteogenic cells, was a cementing substance that comprised afibrillar globular accretions (Fig. 1). These accretions subsequently fused to form a homogenous substance in which collagen fibres became incorporated. Occasionally, collagen fibres were also seen directly apposed to the implant surface. The surfaces of low-crystalline hydroxyapatite and other fast degrading materials, such as tricalcium phosphate,

showed a clear change in surface morphology (Fig. 2a), as compared to highly-crystalline hydroxyapatite and other slow-degrading materials, such as fluorapatite. This was the result of degradation and (re)precipitation which occurred more abundantly on the former materials. On highly crystalline hydroxyapatite only focal degradation was present while the general surface morphology of the plasma sprayed coating could still be observed (Fig. 2b). As a result, newly formed bone tissue was more closely integrated with the surfaces of fast degrading materials than with slow degrading materials.

In vivo, both in rat femora and in the acetabulum of goats, more bone tissue was in contact with low-crystalline hydroxyapatite plasma sprayed coatings as opposed to highly crystalline coatings. Although this has not been quantified as yet, these observations correspond well with results obtained by van Blitterswijk et al. (1994).

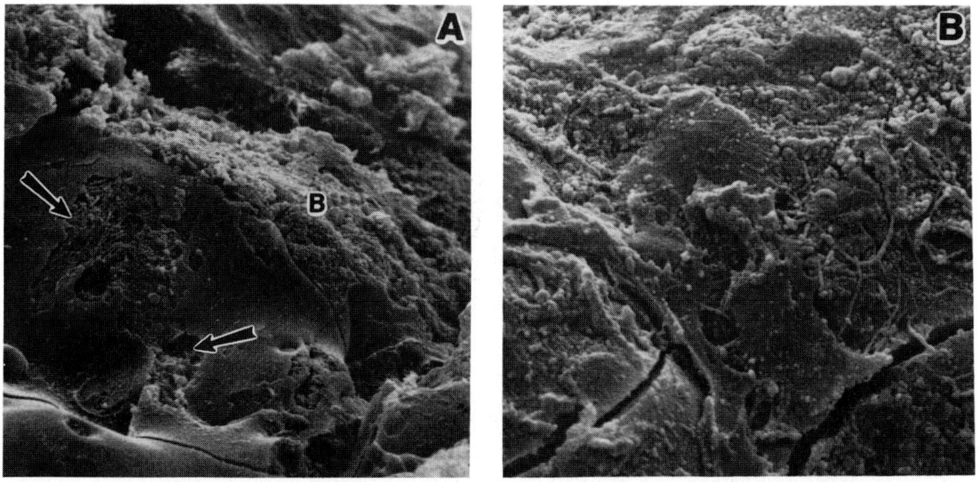

Figure 2. (A) Highly-crystalline hydroxyapatite coating implanted for 2 weeks in rat femora, on which bone tissue (B) has been formed. The general coating morphology is still visible, although focal degradation can be observed (arrows). **(B)** Low-crystalline hydroxyapatite coating also after 2 weeks of implantation in a rat femur. Note the distinctly altered surface on which extracellular matrix components have been deposited. Field widths: (A) 47 μm; (B) 23.3 μm

Osteoclast responses

No signs of osteoclastic resorption could be observed when the osteoclast enriched cell population was grown onto the various calcium phosphates. Only on pressed, low-temperature processed (calcined) hydroxyapatite, a roughening of the hydroxyapatite surface occurred in the vicinity of osteoclast-like cells. This was solely due to phagocytotic activity of these cells, a capacity that has also been described in the literature. When grown in the two-stage culture however, these cells were able to produce characteristic scalloped resorption lacunae in pressed hydroxyapatite discs that were calcined at either 600°C or 900 °C. Although the mineralized matrix produced by osteoblasts on the surface of densely sintered, highly-crystalline hydroxyapatite in the first stage of the culture was resorbed by osteoclastic cells, the hydroxyapatite itself was not susceptible to osteoclastic resorption. In the case of tricalcium phosphate, however, the osteoclastic cells were not only capable of resorbing the mineralized extracellular matrix, but also continued to resorb the underlying tricalcium phosphate surface (Fig. 3).

Hydroxyapatite coated samples that were implanted in rat femora did not show evidence of osteoclastic resorption. In contrast, however, acetabular cups that were plasma spray coated with a layer of low-crystalline hydroxyapatite and implanted in goats, showed distinct resorption in several areas. This was especially evident in areas of bone-remodelling around the implants (Fig. 4). Osteoclasts that had resorbed bone tissue, continued to resorb the underlying surfacial area of the low-crystalline hydroxyapatite coating. Interestingly, this surfacial coating layer also partially stained with the basic fuchsin/methylene blue stain employed, which suggests coating degradation and protein adsorption into this layer. The cement line, which is always present in areas of bone remodelling where new bone has been formed onto old bone, was in contact with the hydroxyapatite surface and was especially clearly visible in these resorption areas. Osteoclastic resorption of the plasma sprayed coatings was, however, only observed with the low-crystalline hydroxyapatite and not with the highly-crystalline hydroxyapatite coatings. These *in vivo* observation correspond well with the results that we have obtained *in vitro*.

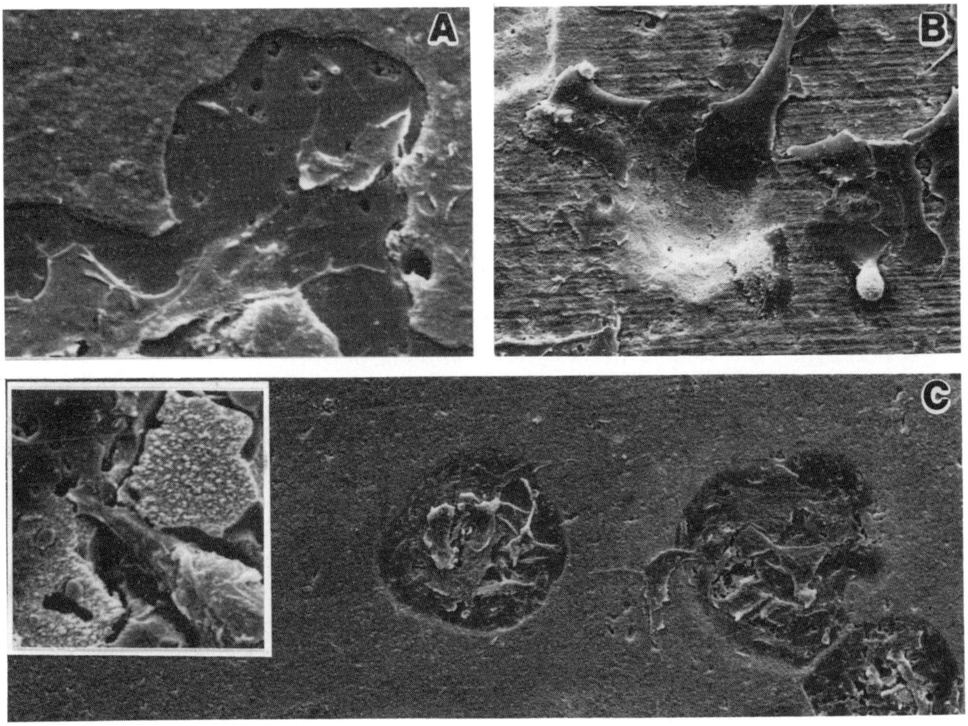

Figure 3A-C. Two-stage culture on **(A)** highly-crystalline hydroxyapatite (HA), **(B)** low-crystalline hydroxyapatite and **(C)** tricalcium phosphate (TCP). Note the distinct resorption the low-crystalline HA and TCP as apposed to highly crystalline HA. Inset in (C) shows degrading grains of the tricalcium phosphate. Bar: (A) 141 μm; (B) 330 μm; (C) 357 μm, inset 14 μm

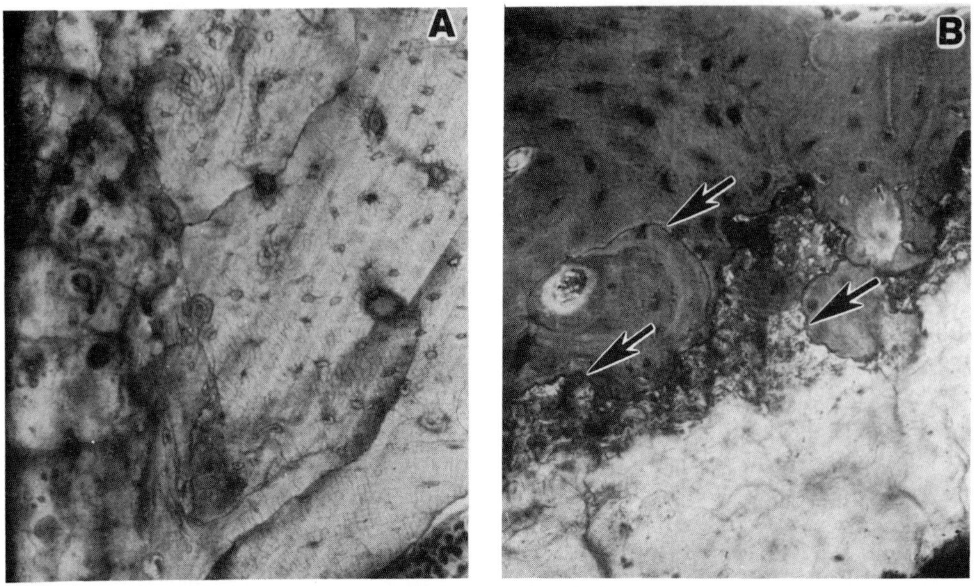

Figure 4. Light micrograph of an highly-crystalline **(A)** and low-crystalline **(B)** hydroxyapatite coating implanted the acetabulum of goats for 25 weeks. Note the clear resorption of the low-crystalline coating as opposed to the highly-crystalline hydroxyapatite. The cement-line, that is present in areas where bone remodelling has occurred, can both been seen at the material surface and in the Howships resorption lacunae (arrows).

CONCLUSIONS

Low-crystalline hydroxyapatite coatings are beneficial for early and rapid bone formation at their surfaces since they generate a surface apatite layer by dissolution/(re)precipitation. Furthermore, this material is susceptible to osteoclastic resorption, in contrast to highly-crystalline hydroxyapatite. Experiments are currently being performed to examine the long-term implant performance with these types of hydroxyapatite coatings.

REFERENCES

1. de Bruijn JD, Davies JE, Flach JS, Klein CPAT, de Groot K and van Blitterswijk CA, *Cells & Mater.* **3**, 115-127 (1993)
2. de Bruijn JD, Bovell YP, Davies JE and van Blitterswijk CA, *J. Biomed. Mater. Res.* **28**, 105-112 (1994a)
3. de Bruijn JD, Bovell YP, Klein CPAT, de Groot K and van Blitterswijk CA, *Biomaterials* (1994b, in press)
4. Maniatopoulos C, Sodek J and Melcher AH, *Cell Tissue Res.* **254**, 317-330 (1988)
5. Osborn JF and Newesely H, In: Dental Implants, Heimke G (ed), Carl Hansen Verlag, Munich, Germany, pp. 111-123 (1980)
6. van Blitterswijk CA (1994) *Pers. comm.*

Bioceramics, Volume 7 Edited by Ö.H. Andersson and A. Yli-Urpo
(*Proceedings of the 7th International Symposium on Ceramics in Medicine, Turku, Finland, July 1994*)

In Vitro Osteoclastic Resorption of Bone-like Apatite Formed on Bioactive Ceramic in Simulated Body Fluid or on Plates using Biomimetic Method

S. Yamada,[1] T. Nakamura,[1,2] M. Tanahashi,[3] T. Kokubo,[3] M. Oka[2] and T. Yamamuro[1]

[1] Department of Orthopedic Surgery, Faculty of Medicine, Kyoto University, 54 Kawara-cho, Shogoin, Sakyo-ku, Kyoto 606-01, Japan
[2] Research Center for Biomedical Engineering, Kyoto University, Kyoto, Japan
[3] Division of Material Chemistry, Faculty of Engineering, Kyoto University, Kyoto, Japan

ABSTRACT

A bone-like apatite layer was formed on apatite- and wollastonite-containing glass-ceramic (A-W GC) immersed in a simulated body fluid (SBF). Similar apatite was formed on plastic or glass plates which were treated with O_2 plasma, placed on grass grains in SBF, and immersed in a solution with ion concentrations 1.5 times those of SBF. Neonatal rabbit bone cells were cultured on these apatite layers. We removed most of the cells except osteoclasts by pronase E and EDTA, then observed the specimens using light and scanning electron microscopies. Resorption lacunae with osteoclasts were clearly shown on these apatite layers. Resorption was started within 10 hours.

INTRODUCTION

Surface-active ceramic bonds to bone chemically through the bone-like apatite layer formed on its surface *in vivo* (1). Long term observation revealed that thickness of this apatite layer decreased (2). However, the mechanism of its degradation is not known clearly. Since osteoclasts resorb bone during bone remodelling, they may degrade the apatite layer. To determine whether they can resorb the apatite layer, we cultured them on the bone-like apatite layer which was formed on A-W GC in an SBF (3). Recently, the method to form the bone-like apatite layer on various kinds of polymers and other materials was developed by Kokubo *et al.* (4). They called this the biomimetic method. We also cultured osteoclasts on the apatite layer formed on plastic or glass plates by this method. We observed the specimens by phase contrast, stereoscopic, and scanning electron microscopy.

MATERIALS AND METHODS

Preparation of Solutions

An SBF with ion concentrations (Na^{+} 142.0, K^{+} 5.0, Ca^{2+} 2.5, Mg^{2+} 1.5, HCO_3^{-} 4.2, Cl^{-} 148.0, HPO_4^{2-} 1.0, SO_4^{2-} 0.5 mM) nearly equal to those of human blood plasma and an aqueous solution (1.5 SBF) with ion concentrations 1.5 times those of SBF were prepared using the method of Kokubo *et al.* (4). The solution was buffered to pH 7.25 at 36.5 °C with 50 mM tris-hydroxymethyl aminomethane [$(CH_2OH)_3CNH_2$] (Sigma, St Louis, MO, USA) and 45 mM hydrochloric acid.

Apatite Formation on A-W GC

Mirror-polished A-W GC disks (15 mm in diameter, and 1 mm thick, Nippon Electric Glass Co., Ltd., Japan) were prepared by the method reported previously (5). Each disk was immersed separately in 30 ml SBF contained in a polystyrene bottle at 36.5°C for 5 days.

Apatite Formation on Plastic or Glass Plates

Using the method of Kokubo *et al.* (4), the bone-like apatite was formed on plastic and glass disks (14 mm in diameter, and 0.15 mm thick, purchased from Wako Pure Chemical Industries Ltd., Japan, and Matsunami Glass Co., Ltd., Japan). Briefly, the specimens were treated with O_2 plasma to strengthen the adhesion of the apatite layer to their surface, placed on CaO-SiO_2-based glass grains (150-300 μm in size) in 30 ml SBF for 1 day to form the apatite nuclei on their surface, and then immersed in 30 ml of 1.5 SBF at 36.5 °C for 3 days.

Cell Culture

Bone cell suspension containing osteoclasts was prepared using the method of Tezuka *et al.* (6). The long bones obtained from a 10-day old Japanese white rabbit were minced using scissors. The bone fragments were agitated in alpha-minimal essencial medium (α-MEM, Gibco, Grand Island, NY, USA) by vortex for 30 s. After sedimentation for 2 min, the cell suspension was harvested (repeated twice). The cells were washed in a centrifuge (400 rpm, 3 min), and suspended in 40 ml α-MEM with 10% foetal bovine serum (FBS, Whittaker M.A. Bioproducts, Walkersville, MD, USA) and antibiotics (100 U/ml penicillin G, 100 μg/ml streptomycin, Meiji, Japan). One milliliter of the cell suspension, containing 2 x 10^7 cells, was seeded into each well of 12-well multidishes (Corning, Corning, NY, USA). Each well contained one of the specimens. Incubation was maintained at 37°C in 5% CO_2 in air.

Microscopic Observation

The specimens were removed after 10, 24, 48, or 96 h, washed in phosphate-buffered saline, and treated with 0.001% pronase E (Sigma, St Louis, MO, USA) plus 0.02% EDTA for 10 min to remove the cells except osteoclasts (6). Some of the specimens were fixed in 80% ethanol for 1 min, stained for tartrate-resistant acid phosphatase (TRAP), and examined by stereoscopic microscopy for A-W GC or phase contrast microscopy for plastic and glass. The others were fixed in 2% glutaraldehyde in 0.1 M phosphate buffer at 4 °C overnight, dehydrated in an ascending series of ethanol, critical point-dried in CO_2, sputter-coated with platinum, and observed by SEM (S-800, Hitachi, Japan).

RESULTS

Resorption of the Bone-like Apatite Formed on A-W GC

Many TRAP-positive cells were found on the bone-like apatite layer formed on A-W GC using stereoscopic microscopy. SEM examination revealed many osteoclasts which had numerous dorsal microvilli and fine dense filopodia in peripheral areas. This appearance indicated that they were actively resorbing and migrating osteoclasts. Resorption lacunae were observed even after 10 h of culture, and they were more extensive after 24, 48, and 96 h of cultures. They had a track-like appearance (Fig. 1). At higher magnification, broken and fused crystals observed in resorbed areas were distinguished from the uniform needle-like crystals of the intact apatite.

Resorption of the Bone-like Apatite Formed on Plastic or Glass Plates

Many osteoclasts (TRAP positive) with resorption lacunae were observed on the bone-like apatite layer formed on plastic or glass plates using phase contrast microscopy and SEM (Fig. 2). The original surface of the plastic or glass plates appeared and no apatite remained in the resorbed areas.

DISCUSSION

The present study showed *in vitro* osteoclastic resorption of the bone-like apatite layer formed on A-W GC in SBF. The SBF can reproduce the process of apatite formation on surface-active ceramics occurring *in vivo* (3). The results suggest that the apatite layer formed on surface-active ceramics implanted in bone can be resorbed by osteoclasts and subjected to bone remodelling.

Many investigators have reported osteoclastic resorption of non-biological and artificial substrates *in vitro*. However, the track-like appearance of resorption lacunae were more clearly shown on the bone-like apatite layer in the present study than on the substrates of the previous studies. Osteoclasts were actively resorbing and migrating on the bone-like apatite layer. The apatite may be easily resorbed by osteoclasts and it may maintain the resorption and migration

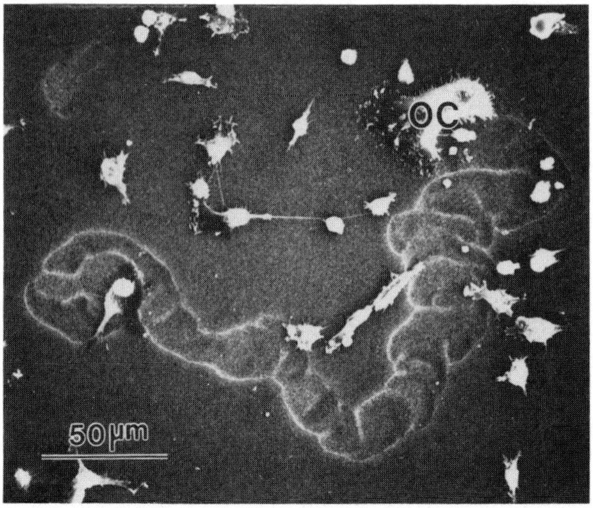

Figure 1 SEM photograph of an osteoclast and its track-like resorption lacuna on the bone-like apatite formed on A-W GC of 48 h culture. (OC=osteoclast. Original magnification x400.)

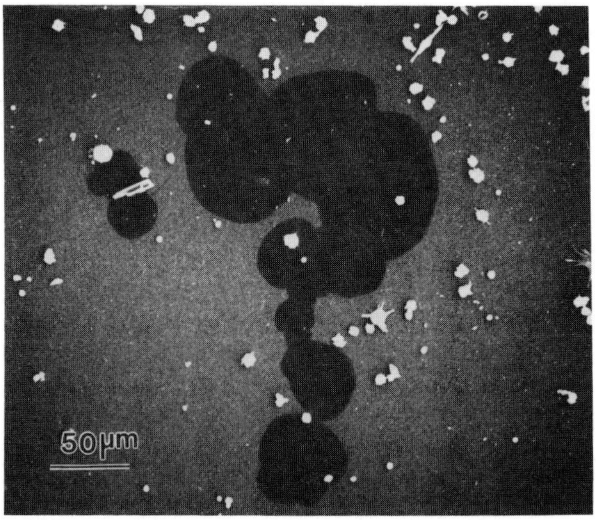

Figure 2 SEM photograph of a resorption lacuna on the apatite formed on glass plate of 96 h culture. The osteoclast was lost during preparation for SEM. (Original magnification x250.)

activity of osteoclasts. It was reported that the apatite was identified as hydroxyapatite with carbonate incorporated into the apatite lattice in a previous study using thin-film X-ray diffraction and Fourier transform infra-red reflection spectroscopy (3). These characteristics are similar to those of natural bone, and they may be suitable for osteoclasts. Proteins contained in FBS or produced by bone cells might be adsorbed onto the surface of the apatite, and they may also influence osteoclast activity. Further studies will be required to determine this. Additionally we think these materials may be used to investigate osteoclast resorption activity.

Acknowledgments. We wish to thank K. Okunaga and T. Shibuya (Nippon Electric Glass Co., Ltd.) for supplying the materials. This work was supported in part by a Grant-in-Aid for Scientific Research (No. 05454409) from the Ministry of Education, Science and Culture, Japan.

REFERENCES

1. Neo, M., Nakamura, T., Ohtsuki, C., Kokubo,T. and Yamamuro, T. J. Biomed. Mater. Res. (1993) 27 999-1006.

2. Neo, M., Nakamura, T., Ohtsuki, C., Kasai, R., Kokubo, T. and Yamamuro, T. J. Biomed. mater. Res., (1994) 28 365-372.

3. Kokubo, T., Ito, S., Huang, Z.T., Hayashi, T., Sakka, S., Kitsugi, T. and Yamamuro, T. J. Biomed. Mater. Res. (1990) 24 331-343.

4. Kokubo, T., Tanahashi, M., Yao, T., Minoda, M., Miyamoto, T., Nakamura, T. and Yamamuro, T. Bioceramics Vol. 6, ed by P. Ducheyne and D. Christiansen, Butterworth-Heinemann, Oxford, UK (1993) 327-332.

5. Kokubo, T., Ito, S., Sakka, S. and Yamamuro, T. J. Mater. Sci. (1986) 21 536-540.

6. Tezuka, K., Sato, T., Kamioka, H., Nijweide, P.J., Tanaka, K., Matsuo,T., Ohta, M., Kurihara, N., Hakeda, Y. and Kumegawa, M. Biochem. Biophys. Res. Commun. (1992) 186 911-917.

Bioceramics, Volume 7 Edited by Ö.H. Andersson and A. Yli-Urpo
(Proceedings of the 7th International Symposium on Ceramics in Medicine, Turku, Finland, July 1994)

Stimulatory Effects of Ceramic Particles on the Production of Bone Resorbing Mediators

Y. Nakashima, T. Shuto, K. Hayashi, T. Hotokebuchi, T. Inadome, T. Hara and Y. Sugioka
Department of Orthopaedic Surgery, Faculty of Medicine, Kyushu University, 3-1-1 Maidashi, Higashi-ku, Fukuoka, Japan 812

ABSTRACT

We investigated the effects of size and dose of various ceramic particles on the release of bone-resorbing mediators from macrophages. Mouse macrophage cell line, J-774 was co-cultured with either alumina or hydroxyapatite particles. Uniform suspensions of alumina or hydroxyapatite particles, ranging in 1-1000 μm were prepared, and amount of 0.002 and 0.02gram of the particles were used in each group. After forty-eight hours of exposure, each medium was harvested and tumor necrosis factor-α (TNF-α), interleukin-6 (IL-6) and interleukin-1 α (IL-1α) levels were assayed by enzyme-linked immunosorbent assay kit. The release of TNF-α from macrophages was increased by smaller size of both particles in a dose-dependent fashion. IL-1α and IL-6 production were enhanced in a similar manner. Hydroxyapatite particles, especially in small size (1μm), caused greater release of mediators than alumina particles. These results suggested that the particles small enough, even if it was biocompatible, had ability to stimulate bone resorption by increasing the production of bone-resorbing mediators.

INTRODUCTION

Recently osteolysis or bone resorption around prosthesis is a well recognized complication of total joint replacement. Histological studies of interfacial membranes present at the bone-implant interface demonstrate an aggressive granulomatous tissue containing histiocytes, fibroblasts, foreign body giant cells and a large number of intra- and extracellular particles from foreign materials (1.2). The importance of release of inflammatory mediators, including tumor necrosis factor (TNF), interleukin-1 (IL-1), interleukin-6 (IL-6) and prostaglandin E_2 (PGE_2), have been identified in human periprosthetic tissues and are released from cells *in vitro* in response to particles of prosthetic materials (3.4). However, many problems in the interpretation of the previous studies have been raised. Particulary, it is still controversial what kind of particles cause bone resorption. Moreover, it is unknown whether, or not, particles of newly applied materials, such as hydroxyapatite, cause bone resorption.
In this current study, we tested the hypothesis that any kinds of particles small enough, even if biocompatible, would basically cause the release of

inflammatory mediators involved in bone resorption. We used *in vitro* assays, which allow quantitative measurement of the effects of the particles; we also used particles of bioinert ceramic (alumina) and bioactive ceramic (hydroxyapatite) that were of comparable size and concentration.

MATERIALS AND METHODS

Particle Preparation

Alumina ceramic and hydroxyapatite ceramic particles were used in this study. These particles, prepared by grinding and sieving, were obtained from Kyocera Co. Ltd (Kyoto, Japan). The size range of particles measured by Microtrac Paricle size analyzer 9220 (Nikkiso Tokyo, Japan) were shown on Table 1.

	Particle size (μm)	
	hydroxyapatite	alumina
1μm group	1.22 ± 0.8	1.12 ± 0.8
100 μm group	102 ± 14	99 ± 9
1000μm group	1021 ± 56	1006 ± 89

Table 1. Size of particle in each group

Cell culture

The J774 cells (Riken cell bank, Tukuba, Japan) were maintained in RPMI-1640 (Gibco, Grand Island, New York) with 15% fetal calf serum. To passage the cells, they were treated with 0.25% trypsin and 0.2% EDTA at 37°C for 10 minutes, and then harvested by vigorous pipetting. The cells were counted and placed in twenty-four-well plates (1 X 10^6 cells per well), and allowed to adhere for twenty-four hours. Then, the medium was aspirated and fresh medium containing test substrate was added to each well. The amount of 0.002 and 0.02 gram of particles were added into each well. After 48 hours, supernatants from each well was harvested, centrifuged (5000 rpm, 10 min) and stored at -80°C until use.

TNF-α, IL-6 and IL-1α in the culture supernatant were assayed using an enzyme-linked immunosorbent assay kit for murine TNF-α and IL-6 (Endogen Inc., Boston, MA) and IL-1α (Genzyme, Cambridge, MA).

RESULTS

Macrophages that had been exposed to both of particles produced higher levels of TNF-α than control group in a size and dose dependent manner except 1000 μm-size group. The production of IL-6 and IL-1α were enhanced in a similar fashion. There were no significant difference between 1000 μm-size group and control group in all measurement. Especially, exposure to 1μm-size particle of both alumina and hydroxyapatite particles always stimulated release of significant levels of these factors than other groups. Moreover, hydroxyapatite particles stimulated 2-3 times greater production of these factors from macrophages when compared to alumina particles. (Figure 1-3). Phase-contrast microscopy demonstrated that both types of particles were readily phagocytosed by macrophages. These were seen in only 1μm-size groups.

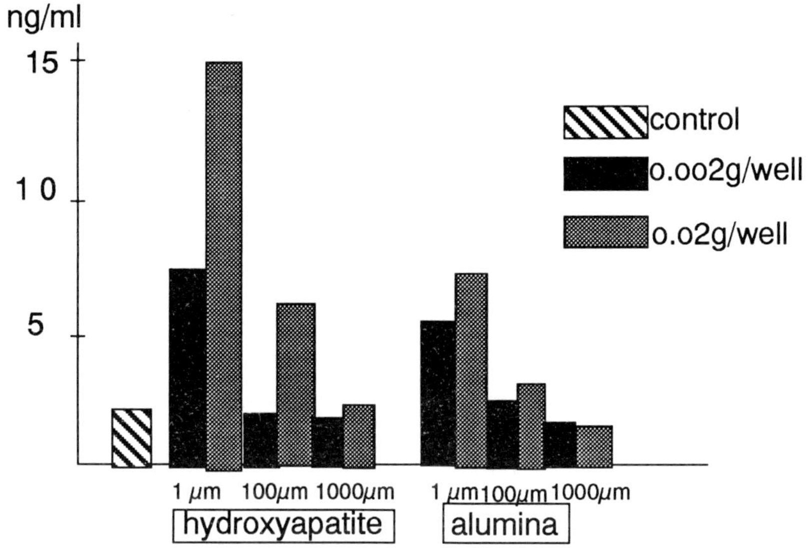

Fig. 1 TNF production from macrophages activated by ceramic particles

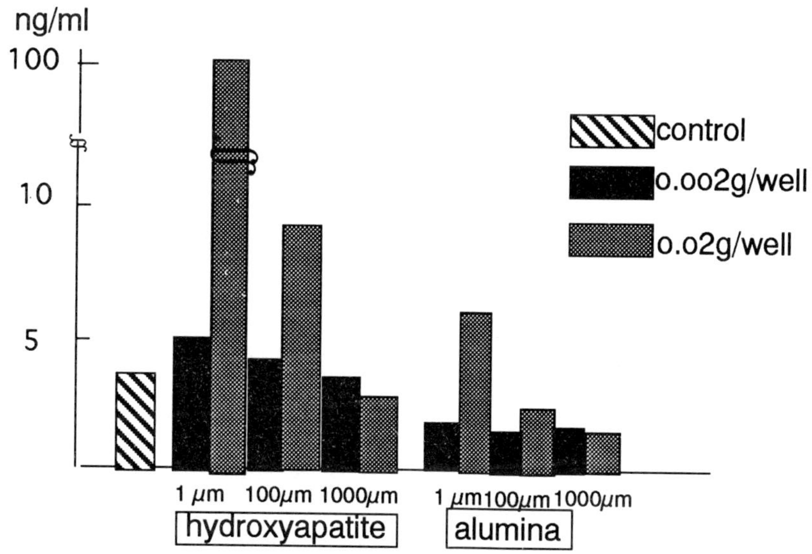

Fig. 2 IL-6 production from macrophages activated by ceramic particles

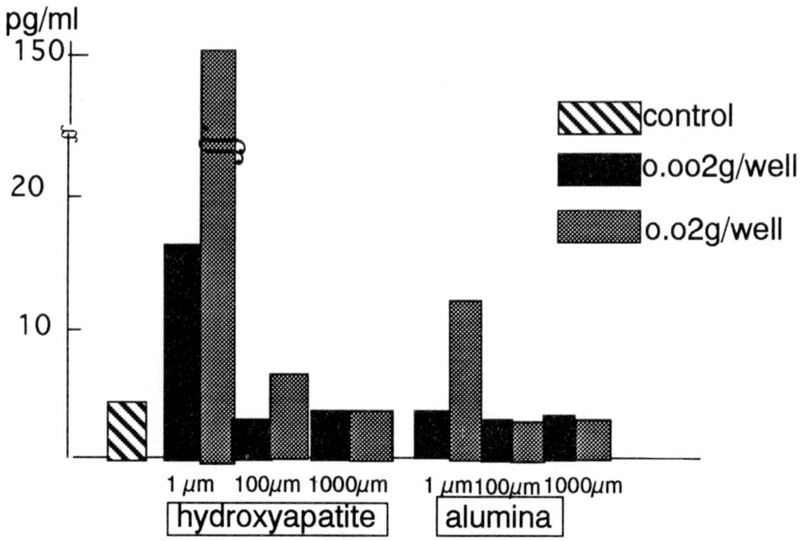

Fig. 3 IL-1 production from macrophages activated by ceramic particles

DISCUSSION

Althogh alumina is biocompatible, bioinert and stable in the biologic environment, its particle stimulated the release of inflammatory mediators in a size and dose dependent manner in this study. Clinically, there are several reports of osteolysis after alumina on alumina cementless total joint replacement (5.6). The occurrence of osteolysis in the absence of bone cement and polyethlene indicated that the particle from foreign material, regardless of it's type, played a role of cell activation process and also supported our results.

Interestingly, hydroxyapatite particles stimulated the release of inflammatory mediators several times greater than alumina particles in each mesurement. Kim et al. reported similar results that HA particles induced higher levels of IL-1 and IL-6 secretion from human bone marrow cells compared with titanium, cobalt-chrome and polyethlene particles (7). The reason for greater responce to hydroxyapatite particles was still unclear, but hydroxyapatite has relatively bioresorbable and biodegradative characteristics, which may contribute to this phenomenon.

On the other hand, the interest in new fixation methods using hydroxyapatite coating has considerably grown in recent years. Althogh short-term clinical results of hydroxyapatite coated hip prosthesis are sucessful, debonding of hydroxyapatite coating from substrate and the subsquent release of hydroxyapatite particles are major concerns in terms of implant fixation and bone resorption (8). We previously reported that hydroxyapatite coating was apt to debond from smooth-surfaced implants after direct shear loading, and roughing the surface of the implant prevented it (9.10). On the basis of the results in the previous and current study, methods must be improved to protect the relatively weak hydroxyapatite-substrate interface.

REFERENCES

1. M. Tanzer, W. J. Maloney, M. Jasty, W. H. Harris In J. Bone and Joint Surg. 1992, 74-A, 404-410

2. A. Tsao, L. Mintz, C. R. Mcrae, D. Stulberg and T. Wright In J. Bone and Joint Surg. 1993, 75-A, 19-26

3. W. A. Jiranek, M. Machado, M. Jasty, D. Jevsevar, H. J. Wolfe, S. R. Goldring, M. J. Goldberg and W. H. Harris In J. Bone and Joint Surg. 1993, 75-A, 863-879

4. S. M. Horowitz, S. B. Doty, J. M. Lane and A. H. Burstein In J. Bone and Joint Surg. 1993, 75-A, 802-813

5. B. Borssen, J. Karrholm and F. Snorrason In Acta. Orthop. Scand. 1991, 62, 73-75

6. M. Boehler, K. Knahr, H. Plenk Jr., A. Walter, M. Salzer, and V. Schreiber In J. Bone and Joint Surg. 1994, 76-B, 53-59

7. K. J. Kim, K. Sato, S. kotake, Y. Katoh and T. Itoh In Bioceramics 6 Butterworth-Heinemann, UK, 1993, 365-370

8. K. Soballe, S. Toksvig-Larsen, J. Gelineck, S. Fruensgaard, E. S. Hansen, L. Ryd, U. Lucht and C. Bunger In J. Bone and Joint Surg. 1993, 75-B, 681-687

9. Y. Nakashima, K. Hayashi, T. Inadome and Y. Sugioka In Bioceramics 6 Butterworth-Heinemann, UK, 1993, 449-455

10. K. Hayashi, T. Inadome, T. Mashima and Y. Sugioka In J. Biomed. Mater. Res. 1993, 27, 557-563

Bioceramics, Volume 7 Edited by Ö.H. Andersson and A. Yli-Urpo
(*Proceedings of the 7th International Symposium on Ceramics in Medicine, Turku, Finland, July 1994*)

Proteolytic Potential of Matrix Metalloproteinases and Serine Proteinases in the Host Response to THR Endoprostheses

M. Takagi,[1,4] S. Santavirta,[3] Y.T. Konttinen,[1] O. Lindy,[1] T. Sorsa,[2] H. Kurvinen,[1] M. Yamakawa[4] and A. Suda[4]

[1] Molecular Biology Laboratory, Department of Anatomy, University of Helsinki, Finland
[2] Institute of Dentistry, University of Helsinki, Finland
[3] Invalid Foundation Orthopedic Hospital, Helsinki, Finland
[4] Yamagata University School of Medicine, Yamagata, Japan

ABSTRACT

Proteolytic potential of matrix metalloproteinases (interstitial collagenase, gelatinase A and gelatinase B) and serin proteinases (cathepsin G and PMN elastase) were examined in 15 interface tissues between bone and implants in loose total hip replacement prostheses. The results were compared to those of 7 samples of non-inflammatory control synovial tissues. Functional enzyme activities of interstitial collagenase, gelatinase A, cathepsin G and PMN elastase were higher in the interface tissues than those in the control tissues. Induction of gelatinase B was found in the interface tissues. These results suggest that elevated proteolytic potential in the periprosthetic tissues causes extracellular matrix degradation around total hip replacement prostheses, which leads to the loosening of these implants.

INTRODUCTION

Progress of biomaterials and artificial joints has a great boon to patients suffering from joint diseases, whereas loosening of total hip replacement (THR) prostheses has became one of the inevitable problems (1,2,3). It appears evident that loosening of technically well inserted THR prostheses is, in part, due to cellular host responses in the periprosthetic tissues (2,3,4). We have formulated a hypothesis on loosening of THR prosthesis; loosening is usually a result of cyclic mechanical loading combined with biological responses in the periprosthetic tissues where proteolytic enzymes play an important role.

Extracellular matrix degradation *in vivo* occurs in the extracellular space at neutral pH. Therefore, the extracellular matrix metalloproteinases (MMPs) and serin proteinases, which are active at physiological pH, are considered important in the cascade of connective tissue degradation in both normal tissues and pathological conditions (5). We have recently demonstrated the tissue localization of MMPs, i.e., interstitial collagenase (MMP-1 type), gelatinase A (MMP-2, 72 kDa, fibroblast type) and gelatinase B (MMP-9, 92 kDa, neutrophil type), and that of serine proteinases, i.e. cathepsin G and polymorphonuclear neutrophil (PMN) elastase in the periprosthetic tissues of THR loosening (6,7,8). In this study, proteolytic potential of these enzymes were examined to clarify one important step in the cascade of matrix degradation as the result of cellular host response to the implants in THR loosening.

MATERIALS AND METHODS

Patients and samples. Fifteen samples of the interface tissues between bone and implants were obtained in revisions performed for aseptic THR loosening. Six patients were women and nine were men, and their mean age was 61.5 years (range 46 - 81 years). Thirteen THRs were originally performed due to primary osteoarthritis, and two to treat secondary osteoarthritis caused by congenital dislocation of the hip joint. Cemented and uncemented fixations were performed in five and ten cases, respectively. Cobalt-Chromium-Molybdenum (CoCrMo) and Titanium-Aluminum-Vanadium (TiAlV) alloys were used in ten and five cases, respectively. Seven samples of synovial capsular tissues were obtained at knee arthroscopy to examine meniscus injury, which patients did not have osteoarthritic changes or traumatic hemoarthrosis, and were used as a control group.

Tissue extraction and dialysis. Tissue samples were minced into small pieces and homogenized in the ice bath after addition of neutral salt extraction buffer (50 mM Tris-HCl, 10 mM $CaCl_2$, 2 M KCl, pH 7.5; 1 g tissue / 1 ml buffer). After incubation for one hour at +4°C, homogenates were centrifuged at 7.5×10^4 G for 60 minutes at +4°C. The supernatants for MMPs were dialyzed against 50 mM Tris-HCl, 5 mM $CaCl_2$, 10 mM $ZnCl_2$, 200 mM NaCl, pH 7.5, and that for serine proteinases were against 0.5M acetate buffer, pH 5.3 for 24 hours at +4°C, respectively.

Interstitial collagenase assay: To study the activity of endogenously activated interstitial collagenases, the tissue extract samples were incubated with type I collagen for 7 days. Type I collagen and the degradation products were separated by SDS-PAGE and quantified densitometrically (9).

Gelatinase assay: Human embryonic skin fibroblasts were cultured and the supernatant was collected after the stimulation of IL-1α as gelatinase A control (10). Leukocytes were purified from healthy human blood. The supernatant obtained from cell homogenates was subjected to gel filtration according to the protocol of Stricklin et al.(11). The gelatinase B containing fraction was used as control. Gelatinolytic activities of the interface tissue samples, control tissue samples, and gelatinase controls were examined using zymographic method (12). The activity of gelatinases was verified densitometrically and monitored by on-line computer analyzing system. Relative gelatinolytic activities of the samples were expressed in comparison with those of gelatinase A and B control, respectively.

Cathepsin G and PMN elastase assay: The measurement of cathepsin G and PMN elastase activities were performed by N-Suc-Ala-Ala-Pro-Phe-p-nitroanilide and N-Suc-Ala-Ala-Val-p-nitroanilide as substrates (13), respectively. The degradation of substrates were monitored spectrophotometrically as previously reported (7).

Statistical analysis. Difference between two groups were analyzed by unpaired Student's t-test. Results were expressed as mean ± the standard error of mean (Mean ± SEM). The level of significance was set at $P < 0.05$ and $P < 0.01$ (*).

RESULTS

Interstitial collagenase: Marked Type I collagenolytic potential was observed in the interface tissue sample obtained from loose THR prostheses. α-chains of native Type I collagen were degraded into the characteristic αA 3/4- and αB 1/4-

degradation products. No further degradation of αA chains could be observed in the studied samples (Figure 1). Interstitial collagenase activity was higher in the interface tissue samples ($26.1 \pm 5.4 \times 10^{-6}$ IU/L) than that of control tissue samples ($0.5 \pm 0.2 \times 10^{-6}$ IU/L, P<0.01, Table 1).

Gelatinase A and B: Marked gelatinolytic potentials corresponding to gelatinase A and B were observed in the interface tissue sample, whereas weak gelatinolytic potential corresponding to only gelatinase A was found in the control sample (Figure 2). Relative activities of gelatinase A and B were highly elevated in the interface tissue samples (gelatinase A: 10.0 ± 1.2, gelatinase B: 6.3 ± 0.6) when compared to those of control tissue samples (gelatinase A: 2.1 ± 0.3, P<0.01, gelatinase B: 0 ± 0, P<0.01, Table 1).

Cathepsin G and PMN elastase: Cathepsin G and PMN elastase values of the interface tissue samples (Cathepsin G: 26.0 ± 5.3 IU/L, PMN elastase: 65.2 ± 8.9 IU/L) were elevated when compared to those of the control tissue samples (Cathepsin G: 2.2 ± 1.5 IU/L, P<0.01, PMN elastase: 3.5 ± 1.5 IU/L, P<0.01, Table 1).

Statistical differences in the activities of interstitial collagenase, gelatinase A, gelatinase B, cathepsin G and PMN elastase were not found among different types of fixation (cemented versus uncemented) and types of alloy (CoCrMo versus TiAlV) (Table 1).

DISCUSSION

In chronic local implant reactions, the cellular host response is likely to consist of monocyte/macrophages, which express CD68, CD11b, non-specific esterase and MHC locus II coded Ia antigen, and fibroblasts, which reveal the immunoreactivities of proline 4-hydroxylase and carboxy-terminal of type I procollagen (2). The local cell reaction also includes scattered perivascular T lymphocyte infiltrations mostly consisting of the CD4-positive T lymphocyte subset (2,4). These immunopathological studies have lead to the analysis of molecular mechanism on weakening of periprosthetic tissues. In the periprosthetic tissues, immunoreactivity of MMP-1 type interstitial collagenase was observed in monocyte/macrophages, fibroblasts and vascular endothelial cells, that of gelatinase A was in fibroblasts and monocytes/macrophages, that of gelatinase B was in monocyte/macrophages, fibroblasts, vascular endothelial cells and occasional PMNs (6). Cathepsin G-immunoreactive cells were found in monocyte/macrophages, fibroblasts and PMNs, and PMN elastase-immunoreactive cells were observed in mononuclear cells and occasional PMNs (7,8). These immunohistochemical profiles suggested the participation of neutral proteinases, MMPs and serin proteinases, in the weakening of periprosthetic tissues around loose THR prostheses.

This study demonstrates elevated functional activities of interstitial collagenase, gelatinase A, cathepsin G and PMN elastase, and induction of gelatinase B activity in the tissue extract samples from the interface tissue between bone and implants, when compared to those of non-inflammatory synovial control tissues. These functional activities were observed in different types of fixation (cemented versus uncemented) and types of alloy (CoCrMo versus TiAlV). Once a cellular host reaction occurs in periprosthetic tissues and causes loosening, it appears to induce proteinases regardless of type of fixation or prosthetic materials.

Detection of proteolytic potentials, when combined with immunohistochemical localizations of the proteinases in the loosening of THR components (6,7,8), indicates the contribution of proteolytic enzymes to the weakening of periprosthetic tissues in loose THR prostheses and implies proteinase-inhibitor imbalance *in situ* (5). Deeper understanding of proteolytic cascade in loosening may in future lead to prevention of prosthetic loosening, because MMPs and serine proteinases can be chemically inhibited (5,14,15) and it may become possible to modulate the unfavorable proteinase activity by synthetic and recombinant reagents in pathological conditions.

ACKNOWLEDGEMENTS

This work was supported in part by grants from Finnish Academy, the Paulo Foundation, Finland, and Japan Society for the Promotion of Science, Yamagata Orthopedic Society for the Promotion of Science, Japan.

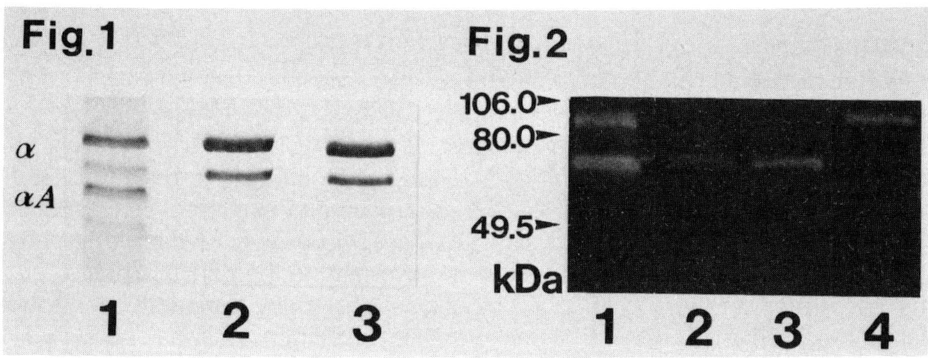

LEGENDS

Figure 1. Type I collagen degradation by interstitial collagenase. α: α-chains of native Type I collagen, αA: αA 3/4-degradation products, lane 1: interface tissue sample, lane 2: control tissue sample, lane 3: absence of sample.

Figure 2. Gelatin zymography. lane 1: interface tissue sample, lane 2: control tissue sample, lane 4: gelatinase A control, lane 4: gelatinase B control.

Table 1. Proteolytic potential in the interface tissue samples.

Interstitial collagenase (x10^{-6} IU/L)

Interface, total	(n=15)	26.1 ± 5.4*
cemented	(n= 5)	27.3 ± 8.9*
uncemented	(n=10)	25.5 ± 6.7*
TiAlV	(n= 5)	26.1 ± 9.1*
CoCrMo	(n=10)	26.0 ± 6.7*
Control	(n= 7)	0.5 ± 0.2*

Gelatinase A (relative activity)		Gelatinase B (relative activity)	
Interface, total	10.0 ± 1.2*	Interface, total	6.3 ± 0.6*
cemented	11.3 ± 1.7*	cemented	5.7 ± 0.7*
uncemented	9.3 ± 1.5*	uncemented	6.5 ± 0.8*
TiAlV	8.9 ± 1.8*	TiAlV	7.6 ± 1.0*
CoCrMo	10.5 ± 1.5*	CoCrMo	5.6 ± 0.7*
Control	2.1 ± 0.3	Control	0 ± 0

Cathepsin G (IU/L)		PMN elastase (IU/L)	
Interface, total	26.0 ± 5.3*	Interface, total	65.2 ± 8.9*
cemented	29.2 ± 7.1*	cemented	67.6 ± 7.4*
uncemented	24.4 ± 7.1*	uncemented	64.1 ± 12.2*
TiAlV	20.2 ± 7.2*	TiAlV	61.2 ± 19.5*
CoCrMo	28.9 ± 7.0*	CoCrMo	67.3 ± 7.6*
Control	2.2 ± 1.5	Control	3.5 ± 1.5

* : P<0.01 for interface tissue samples to control.

REFERENCES

1. Harris WH, Schiller AL, Scholler JM, Freiberg RA, Scott R. In J Bone Joint Surg (Am) 58:612-618, 1976.

2. Santavirta S, Konttinen YT, Bergroth V, Eskola A, Talloth K, Lindholm TS. In J Bone Joint Surg (Am) 72:252-258, 1990.

3. Santavirta S, Gristina A, Konttinen YT. In Acta Orthop Scand 63:225-232, 1992.

4. Santavirta S, Nordström D, Metsärinne K, Konttinen YT. In Clin Orthop 297:100-110, 1993.

5. Woessner JF. In FASEB Journal 5:2145-2154, 1991.

6. Takagi M, Konttinen YT, Santavirta YT, Sorsa T, Eizen AZ, Norldsletten L, Suda A. In Acta Orthop Scand (in press).

7. Takagi M, Konttinen YT, Santavirta S, Kangaspunta P, Sorsa T, Suda A, Rokkanen P. In Finn J Orthop Traumatol 16:91-97, 1993.

8. Takagi M, Konttinen YT, Santavirta S, Sorsa T, Suda A, Rokkanen P. In Progress in Rheumatology Edited by Machtey I. Petach Tikva: Golda Medical Center, 5:26-31, 1993.

9. Sorsa T, Suomalainen K, Turto H, Lindy S. In Med Biol 63:66-72, 1985.

10. Konttinen YT, Lindy O, Suomalainen K, Ritchlin C, Saari H, Vauhkonen M, Lauhio A, Santavirta S, Sorsa T. In Matrix 11:395-403, 1991.

11. Stricklin GP, Jeffrey JJ, Roswit WT, Eisen AZ. In Biochem 22:61-68, 1983.

12. Ingman T, Sirsa T, Lindy O, Koski H, Konttinen YT. In J Clin Periodontol 21:26-31, 1994.

13. Nakajima MC, Powers JC, Ashe BM, Zimmerman M. In J Biol Chem 254:4027-4032, 1979.

14. Lauhio A, Sorsa T, Lindy O, Suomalainen K, Saari H, Golub LM, Konttinen YT. In Arthritis Rheum 36:1335-1336, 1993.

15. Wewers MD, Casolaro MA, Sellers SE, Swayze SC, McPhaul KM, Wittes JT, Crystal RG. In N Engl J Med 316:1055-1062, 1987.

Bioceramics, Volume 7 Edited by Ö.H. Andersson and A. Yli-Urpo
(*Proceedings of the 7th International Symposium on Ceramics in Medicine, Turku, Finland, July 1994*)

The Effects of the Microstructure of Two Hydroxyapatite Ceramics on Cell Morphology and Proliferation

S. Best,[†] D. Lee,[*] J. Shaw[†] and W. Bonfield[†]

[†] IRC in Biomedical Materials, Queen Mary and Westfield College, Mile End Road, London E1 4NJ, England

[*] IRC in Biomedical Materials, Institute of Orthopaedics, University College and Middlesex Schools of Medicine, Brockley Hill, Stanmore, Middlesex HA7 4LP, England

ABSTRACT

Two hydroxyapatite powders, supplied by Plasma Biotal Ltd., were physically and chemically characterised before being die pressed and sintered over a range of temperatures. The resulting samples ranged in density from approximately 60% of the theoretical for hydroxyapatite to near full density. Cells from a osteogenic cell line (HOS TE85) were plated on to the ceramics and observed using scanning electron microscopy. Cells grown on the higher density samples were found to be polygonal with extensive fine cell processes. On the lower density surfaces, the cells tended to be elongated but lacked the array of fine processes. In addition, cell density after 24 hours was greater on the high density samples. [^{3}H]-thymidine incorporation, used to assess cell division indicated a higher level of cell proliferation on the higher density hydroxyapatite samples.

1. INTRODUCTION

There has long been interest in the use of hydroxyapatite as a bone replacement material and both *in-vitro* and *in-vivo* work has confirmed its osteogenic nature (1). The profound effect of hydroxyapatite powder morphology on the sintering behaviour of the material has been reported previously (2). However, despite the importance of hydroxyapatite in the bioceramics field, the effects of the microstructure of the ceramic on the response of cells, *in-vitro*, has not been widely reported.

This paper documents the preliminary results in trials with two hydroxyapatite powders of contrasting morphology. The powders were pressed and sintered under similar conditions to produce a series of hydroxyapatite ceramics with a variety of microstructures. Cell proliferation and metabolism is known to be influenced by the morphology of the growth substrate due to altered attachment characteristics. The morphology and proliferation of osteogenic cells on hydroxyapatite substrates of varying densities has, therefore been investigated.

The work reported in this paper forms part of a larger programme to study the *in-vitro* response of human osteoblast-like cells to ceramics of contrasting microstructure and composition.

2. METHODS

2.1 Powder Characterisation.

Two synthetic hydroxyapatite powders (designated A and B) were supplied by Plasma Biotal Ltd. X-ray diffraction was performed on powder compacts with a Siemens D5000 goniometer, 710 X-ray generator using $Cuk\alpha$ radiation at 30mA, 40KV. Scans were performed between 2θ values of 5° and 65° with a step size of 0.03° at 2 seconds per step. Infra-red spectroscopy was performed using a Nicolet *730sx FTIR* instrument operating in the absorbance mode. Samples were analysed in the as-received condition and after heat treatment at 1300°C for 1 hour. The calcium to phosphorus ratio and impurity content of the powders were measured by Ceram Research Ltd. using X-ray fluorescence.

The morphology of powders was studied using a JEOL 6300 scanning electron microscope with an accelerating voltage of 20KV. A Micromeretics High Speed Surface Analyser was used to determine particle surface area with the BET method. Particle size measurements were performed with a Malvern Particle Size Analyser using the 'Mastersizer' software facility. Samples were treated in an ultrasonic bath for one minute in order to break up any loosely bound agglomerates prior to measurement.

2.2 Specimen Preparation and Characterisation

Disc specimens (30mm diameter x 1.5 mm in thickness) were produced using a single action die. A pressure of 170MPa was exerted using a hydraulic press. Samples were then sintered at temperatures ranging from 1100°C to 1320°C with heating and cooling rates of 4°C/minute and a dwell of 1 hour at the maximum temperature. The density of sintered samples was assessed using Archimedes' principle. The grain sizes of the specimens sintered at the various temperatures were from micrographs obtained using a JEOL 6300F SEM.

2.3 *In-vitro* Testing

Hydroxyapatite samples sintered at 1120°C, 1180°C, 1260°C and 1300°C were selected for *in-vitro* studies, to provide a series of samples with a range of densities and grain sizes. The samples were washed and sterilised by autoclaving prior to use. Cells from an osteogenic cell line (HOS TE85) were used throughout the study (3). The cells were plated at a density of 2×10^4 $cells.cm^{-2}$ in Dulbecco's minimal essential medium (DMEM) with 10% foetal calf serum (FCS) and 1% non-essential amino acids (NEAA) at 37°C and 5%CO_2. After culture for 24 hours the samples were fixed in 2% glutaraldehyde in 0.1M phosphate buffer pH 7.4 and were prepared for electron microscopy using a Balzers CPD030 critical point dryer.

The morphology of the cells was studied using a JEOL 6300F scanning electron microscope with an accelerating voltage of 5kV.

2.4 Tritiated Thymidine ($[^3H]$-TdR) Incorporation

HOS TE85 cells were plated on to hydroxyapatite samples as above and cultured for 24 hours to allow the cells to attach. The medium was removed and replaced with fresh DMEM + 10% FCS + 1% NEAA + $1\mu Ci.ml^{-1}$ $[^3H]$-TdR and cultured for a further 24 hours. The cells were washed thoroughly with DMEM + 10% FCS + 1% NEAA + $50\mu g.ml^{-1}$ non-radioactive thymidine and digested in a solution containing $1\mu l$ papain in 1ml phosphate buffered saline + 5mM cysteine

HCl + 5mM EDTA at 60°C for 4 hours. Samples of the cell digest were assayed for [^{3}H]-TdR on a Perkin Elmer scintillation counter. Total DNA was assayed using the DAPI method (4).

3. RESULTS

3.1 Powder Characterisation

The X-ray diffraction traces and infrared analyses for the powders both indicated that the materials contained a small amount of carbonate in their as-received condition. However, both powders exhibited similar peak widths and heights and contained no other extraneous phases either before or after heat treatment.

XRF indicated that both powders had similar concentrations of impurities. Most of the trace elements contained were at levels of less than 0.1 percent, but in particular, the powders were both found to contain higher traces of silica (0.49% and 0.34% for powder A and powder B respectively) and magnesia (0.28% and 0.27% for powder A and powder B respectively). The calcium to phosphorus ratios for the two powders were 1.69 and 1.70 for powders powder A and powder B respectively.

Scanning electron microscopy of the two powders indicated that the morphology of the powders was similar in that they were both composed of agglomerates of sub-micron acicular crystallites. However, close observation revealed that powder B appeared to be composed of slightly smaller particles with more clearly defined acicular crystallites

The surface area and particle size analysis results (median particle size, ($D_{0.5}$) and the sizes below which 90% and 10% of the particles fell ($D_{0.9}$ and $D_{0.1}$)) are displayed in table 1.

Powder	Surface Area (m^2/g)	$D_{0.5}$(μm)	$D_{0.1}$(μm)	$D_{0.9}$(μm)
A	8.3	4.53	1.38	14.66
B	13.4	3.29	0.92	10.11

Table 1. Surface Area and Particle Size Analysis Results

The results indicate that the surface area of powder B was higher than that of powder A. The particle size distributions of both powders were approximately monomodal. However, the values of $D_{0.1}$, $D_{0.5}$ and $D_{0.9}$ for powder B were all lower than those of powder A.

3.2 Sintering Results

Figure 1 shows the variation in density with sintering temperature for the two powders. Samples of powder B appeared to show enhanced sintering characteristics over powder A at all sintering temperatures (see overleaf).

Observations made using the scanning electron microscope indicate that the specimens showed no abnormal grain growth over the range of sintering temperatures tested. However, after sintering at 1300°C, specimens of powder B had a slightly coarsened mean grain size of 3μm rather than 0.5-1μm for both powders sintered at lower temperatures.

3.3. *In-vitro* Morphology

Scanning electron micrographs of HOS TE85 cells grown on hydroxyapatite samples of varying density are presented in figure 2. Cells grown on the ceramics sintered at higher temperatures (i.e. the higher density samples) were found to be polygonal with extensive fine processes (figures 2a, b, e). Cells grown on the lower density surfaces were markedly more elongated but lacked the array of fine processes (figures 2c, d, f.). In addition, cell density after 24 hours was significantly greater on the higher density samples. The trend was the same for both powders although comparison between the powders indicated that the density and not the sintering temperature was the controlling factor.

3.4 Tritiated Thymidine ([^{3}H]-TdR) Incorporation

[^{3}H]-TdR incorporation results are presented in figure 3. Incorporation rates were lower in all hydroxyapatite samples than for Thermanox cover slips (an optimised growth surface). However, trends within the samples were noted. For both powders, increasing ceramic density resulted in an increase in proliferation as assessed by [^{3}H]-TdR incorporation. These results are in accordance with the increased density of cells grown on higher density samples reported above.

4. DISCUSSION

A number of groups have reported methods of producing and sintering hydroxyapatite (1), but very few document in detail the physical characteristics of their material. The results indicate that sintering temperature, with its associated alteration in density and surface morphology can affect cell growth on the materials. An increase in density elicited an increase in spreading of the cells on the material and an increase in proliferation. Our findings are in agreement with previous studies, which have reported an association between cell spreading and cell division (5).

5. CONCLUSIONS

a. High density samples promoted the spreading of HOS TE85 cells.
b. A positive relationship between sample density and cell proliferation was observed.
c. The morphology and proliferation effects were determined by sample density and not sintering temperature.

ACKNOWLEDGEMENT - Funding for this work was supplied by the Science and Engineering Research Council.

REFERENCES

1. LeGeros, R. Z., LeGeros, J. P., in An Introduction to Bioceramics, World Scientific Publ. Co. Pte. Ltd. 1993, 139-180.

2. Best, S., Bonfield, W., Doyle, C., in Bioceramics 2: Proceedings of the 2nd International Symposium on Ceramics in Medicine, .Publ. Deutsche Keramische Gesellschaft e.V. 1990, 57-64.

3. McAllister, R. M., Gardner, M. B., Green, A. E., Brandt, C., Nichols, B. H. Cancer, 1971 27 397-402.

4. Kapuscinski, J., Skoczylas, B., Annal. Biochem. 1977 83 252-257

5. Lee, D. A., Bentley, G., Archer, C. W. Osteoarthritis & Cartilage, 1993 1 137-142

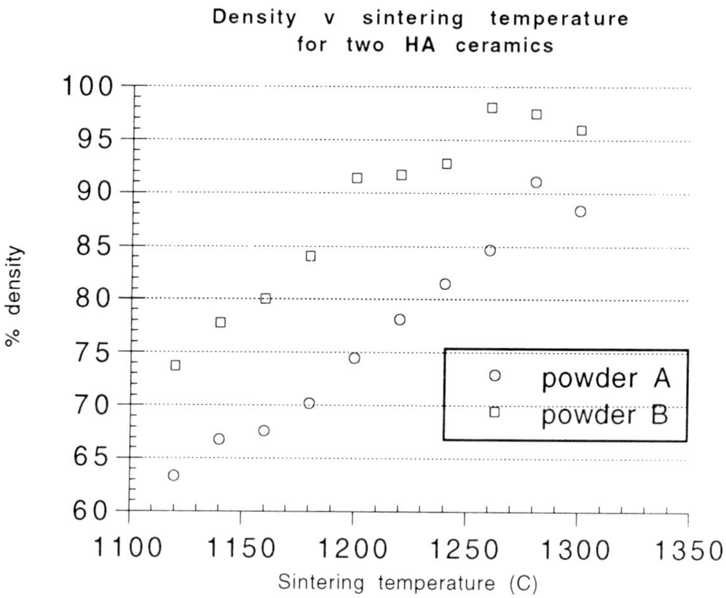

Figure 1. Density versus Sintering Temperature for Powders A and B

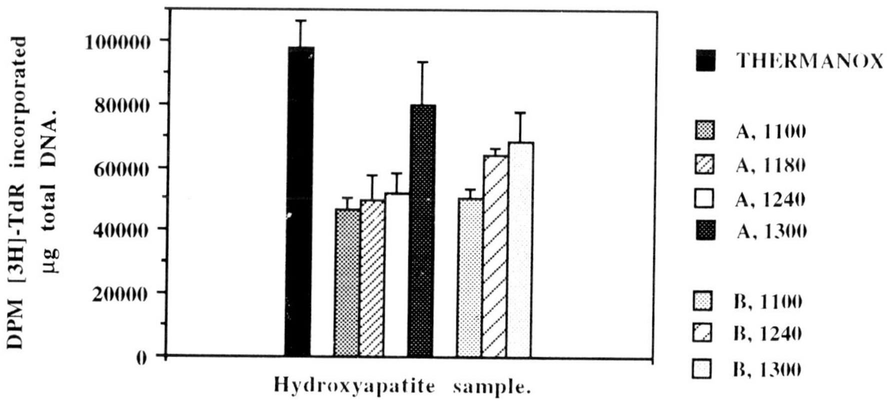

Figure 3. [^{3}H]-TdR Incorporation by HOS TE85 Cells Grown on Hydroxyapatite and Thermanox.
(Each point represents the mean and standard error of four replicates).

Figure 2. Scanning Electron Micrographs of HOS TE85 Cells Grown for 24 Hours on Hydroxyapatite Surfaces Sintering at Various Temperatures.

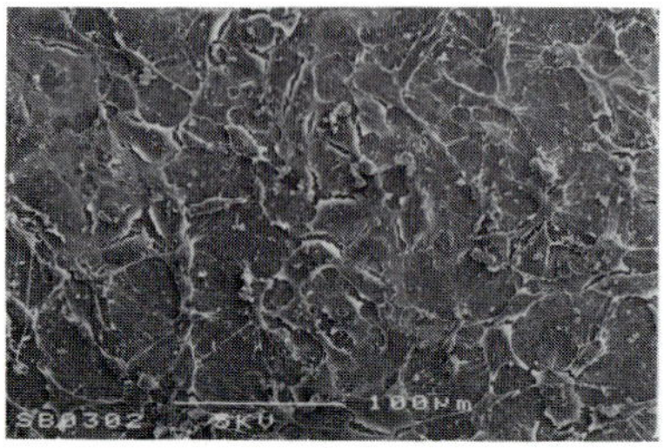

a. Powder B sintered at 1300°C

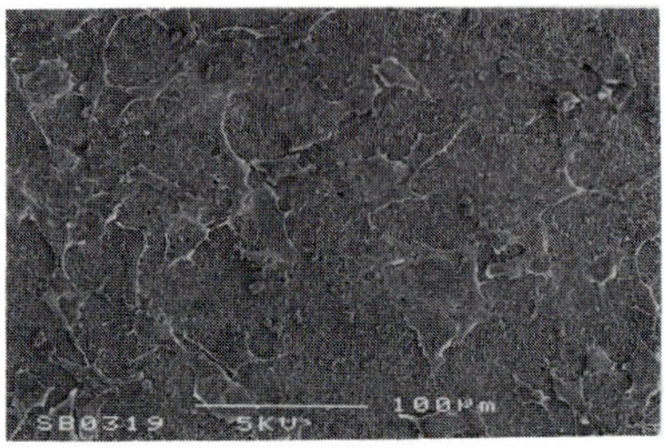

b. Powder A sintered at 1300°C

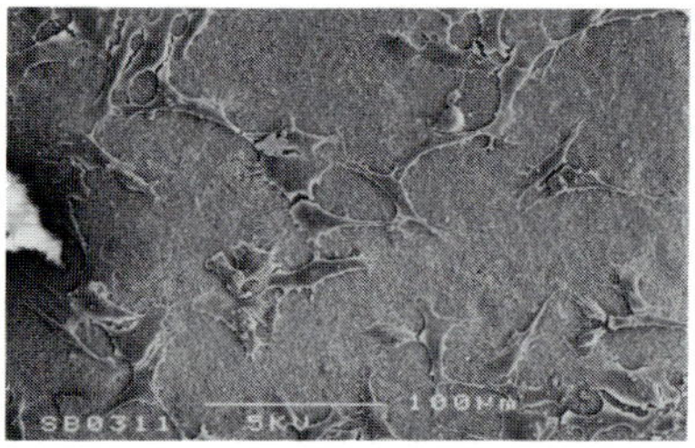

c. Powder B sintered at 1180°C

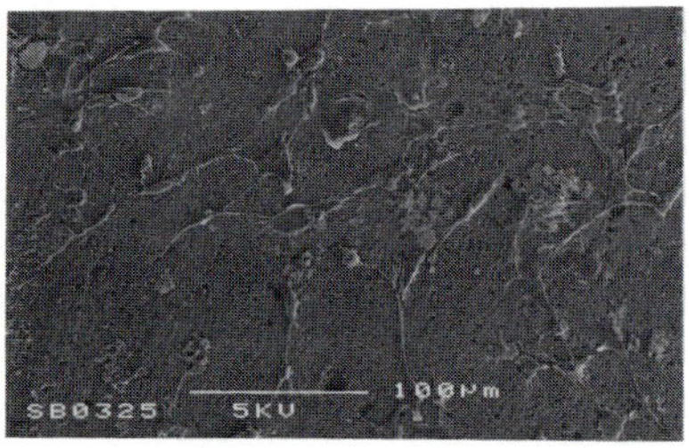

d. Powder A sintered at 1180°C

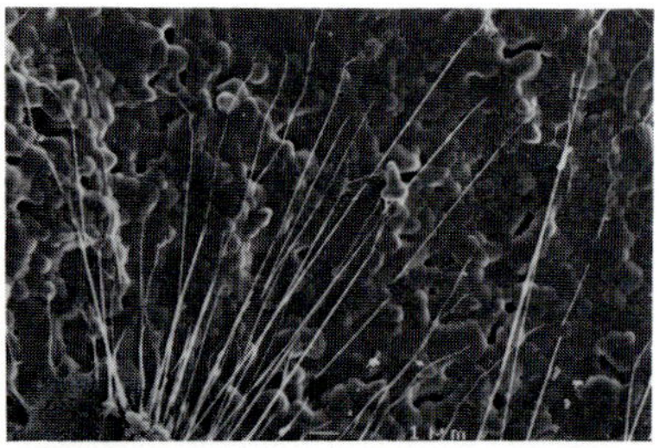

e. Powder B sintered at 1200°C

f. Powder A sintered at 1180°C

Bioceramics, Volume 7 Edited by Ö.H. Andersson and A. Yli-Urpo
(*Proceedings of the 7th International Symposium on Ceramics in Medicine, Turku, Finland, July 1994*)

Effect of Matrigel Coating on Bone Cell/Hydroxyapatite-coated Implant Interface

S.H. Maxian, M.C. Melican, M.L. Tiku and J.P. Zawadsky
Orthopaedic Research Laboratory, UMDNJ-Robert Wood Johnson Medical School, New Brunswick, New Jersey 08903, USA

ABSTRACT

The effect of coating hydroxyapatite (HA) coated rough titanium (Ti) implants with Matrigel (reconstituted basement membrane proteins) was investigated *in vitro*. Postnatal rat calvarial cells were seeded on tissue culture plates, low crystallinity (30-40% crystalline) HA coated and uncoated rough Ti flats under standard tissue culture conditions for 2, 7 and 14 days. Calcium dissolution was measured for the HA coated surfaces and showed 35-40ppm Ca ions in solution by 14 days. Presence of HA enhanced alkaline phosphatase activity/cell (APA/cell) production as compared to uncoated rough Ti implants without Matrigel coating. In the presence of Matrigel no differences in APA/cell production were seen between HA coated and uncoated rough titanium implants.

INTRODUCTION

Hydroxyapatite (HA) coatings of varying crystallinities have been shown to influence cell morphology, growth, matrix formation and mineralization *in vitro* and *in vivo* (1-4). Concerns over particulate debris from the coating have been addressed, while chemical dissolution is desirable and may be a prerequisite for enhanced cellular activity at the surface of the implants (4,5). It can be argued that the presence of a two interface system encountered with non-resorbable HA coatings can be problematic in the long term (1,6). Current research suggests that lower crystallinity HA coatings show the most resorption and may thus enhance mineralization. We have shown that resorbable HA coatings equate non-resorbable HA coated surfaces in bone attachment strengths, however decreased bone apposition was seen on resorbable coatings (1). We have shown greater cellular activity at the surface of resorbable than nonresorbable HA coatings in cell culture studies (7). The presence of a biologically active protein may help alleviate problems of decreased bone apposition as seen around our resorbable HA coatings. Pre-osteoblasts cultured on Matrigel (reconstituted basement membrane proteins) coated tissue culture plates showed evidence of formation of cytoplasmic processes similar to osteocytic canalicular networks (8). Thus, the effect of Matrigel coating on bone cell growth, osteoblastic phenotype expression and morphology was investigated on the rough titanium implants and low crystallinity resorbable HA coated titanium implants at 2, 7 and 14 days. This study investigated the hypothesis that Matrigel coating can enhance bone progenitor cell growth and osteoblastic phenotype expression on orthopaedic implants.

MATERIALS AND METHODS

Greater than 95% pure HA (Ca/P ratio of 1.67) was plasma sprayed onto Ti-6Al-4V alloy powder sprayed titanium (Ti) sheets at a thickness of 50-60μm by APS Materials Inc. (Dayton, OH). Ti and HA sprayed sheets were punched to obtain disks of 13mm in diameter which fit in the 24 well tissue culture plates. HA coating crystallinity was determined by X-Ray diffraction (XRD) on a Siemens D500 diffractometer (Figure 1). Implant surface roughness of HA coated and uncoated rough titanium implants was compared by profilometry at a speed of 1.0scan/μm for a scan distance of 2000μm. These two types of implants, HA coated (HA-rTi) and uncoated rough Ti (rTi), were sterilized by ethylene oxide. All cultures were maintained at 37°C in a humidified 5% CO_2 atmosphere, under standard sterile conditions, in high glucose DMEM media containing nonessential amino acids, 2mM Glutamine, 10% fetal calf serum, 1% penicillin/streptomycin B/amphotericin B as fungizone, and 50μg/ml Gentamycin solution.

Post-natal rat calvaria were dissected and cleaned of all periosteal and endosteal tissues and washed thoroughly with Hank's buffer. Calvaria were digested by incubation in collagenase for 20 mins and the first digestion solution discarded. Cells obtained from subsequent digestions were plated in complete media and grown to confluence. Osteoblast phenotype was confirmed by alkaline phosphatase (AP) cytochemical staining (Sigma kit#86). These pre-osteoblast cells were seeded at a concentration of $5x10^5$ cells per well in 1ml of media. Each experiment was run in triplicate at three different times for 2, 7 and 14 days. Tissue culture plates (24 well) without implants, with HA coated and uncoated rTi implants were investigated with and without Matrigel coating. Matrigel (10mg/ml; Collaborative Biomedical Products, Bedford, MA) was applied as a 100μl/well coating on the three types of wells and dried at room temperature until a gel was formed.

Flame atomic absorption analyses (Quantitative Technologies Inc., Whitehouse, NJ) of the incubation media at 2, 7 and 14 days for the HA coated implants were run in triplicate for the non-Matrigel coated implants. Alkaline phosphatase activity (APA) was determined by colorimetric assay with p-nitrophenylphosphate read at 410nm following trypsin removal of cells from the wells and implants. Cell growth was determined by an MTT assay (Sigma kit#CGD-1) following incubation for 3-4hr of attached cells. Cell number was determined by regression analysis from a predetermined MTT assay standard curve of known cell numbers. XRD of the HA coated implants with and without Matrigel after 2, 7, and 14 days in culture was performed. Statistical significance ($p<0.05$) was determined by ANOVA and Student's t-Tests.

RESULTS AND DISCUSSION

Ca ion in solution following incubation of the HA coated implants in complete media without cells showed 14.67 ± 1.86ppm at 2 days, 24.33 ± 3.28ppm at 7 days and 35.67 ± 4.84ppm at 14 days (Figure 2). Ca ion concentration in solution was significantly greater ($p=0.01$) at 14 days than at 2 days but not at 2 vs. 7 days or 7 vs. 14 days. Others have shown dissolutions of 20ppm by 4 weeks (10). This HA coating therefore shows rapid dissolution, which may be beneficial for cellular activity as we have previously shown on completely resorbable HA coatings.

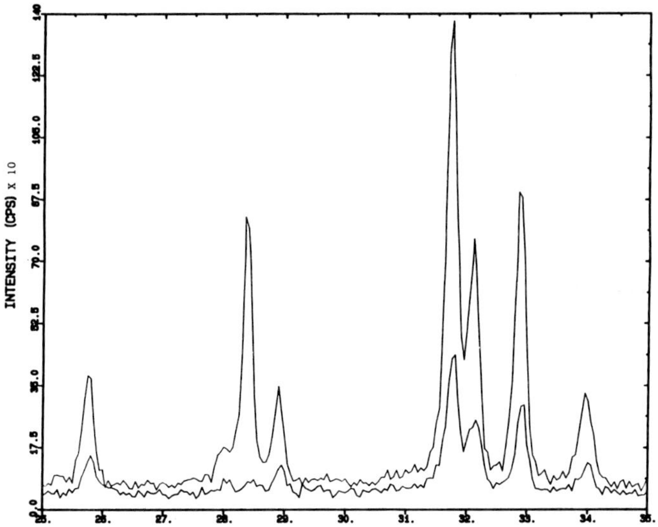

Figure 1: X-Ray diffraction patterns at two-theta degrees of 30-40% crystalline HA coating as compared to 89% crystalline HA coating.

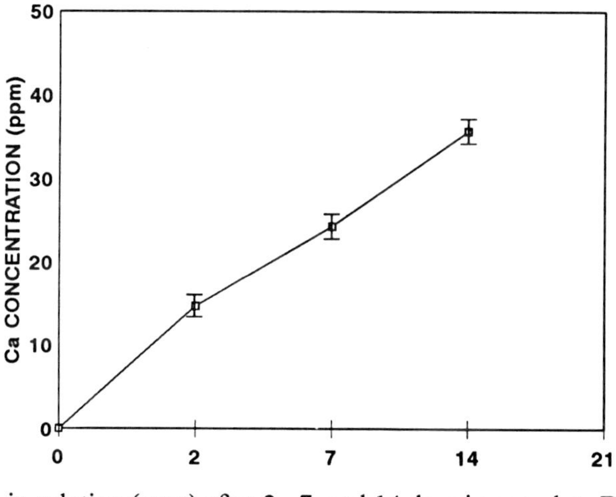

Figure 2: Ca ion in solution (ppm) after 2, ,7, and 14 days in complete DMEM media.

Surface roughness profiles of the HA coated and uncoated rTi implant showed respective average roughnesses (*Ra*) of about 26.4μm and 35.8μm and respective change in maximum and minimum profile heights of 120.1μm and 157.8μm. These roughnesses indicate sufficient openings for tissue ingrowth which have been suggested to be around 50-150μm diameter pores. Our hypothesis further stipulates that coating these implants with a biologically active protein will enhance cell attachment and growth into the pores of the coating.

XRD analysis of the surface of the HA coated implants with and without Matrigel showed no noticeable differences in patterns at 2, 7, or 14 days. This HA coating was 30-40% crystalline as determined by regression analysis of our most and least crystalline HA coatings (Figure 1).

Significantly greater APA (Figure 3) was seen on culture plates with and without Matrigel than on the two types of implants at all time periods. Without Matrigel significantly greater amount of APA was seen on the HA coated implants than the uncoated rTi implants at 2 and 14 days. With Matrigel significantly greater APA was seen on HA coated than uncoated rTi implants at 14 days only. Presence of Matrigel showed greater APA on the rough Ti at 2, 7 and 14 days, whereas for the HA coated implants APA was greater at 7 and 14 days in the presence of Matrigel. Effect of time was a significant variable for wells without implants and wells with HA coated and uncoated rTi implants without Matrigel at all time periods and with Matrigel at all time periods except 2 vs. 7 days for the two types of implants.

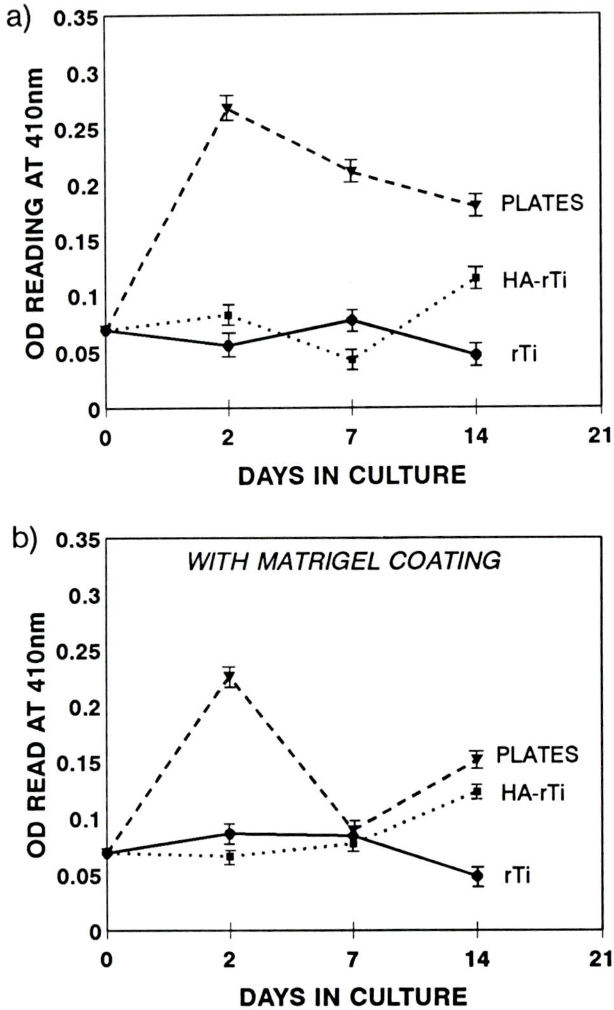

Figure 3: Alkaline phosphatase activity of cells shown as OD readings at 410nm for culture plates, HA coated rough Ti (HA-rTi) and rough Ti (rTi) implants (a) without Matrigel and (b) with Matrigel coating after 2, 7 and 14 days in culture.

Cell growth (Figure 4) was significantly greater in wells without implants only without Matrigel coating at all time periods and with Matrigel coating at 14 days only. Presence of HA coating showed significantly lower cell growth than on the uncoated rTi by 7 and 14 days without Matrigel coating. No significant differences were seen between these two implants at any time period with Matrigel coating. Matrigel coating significantly decreased cell growth on the tissue culture plates at 2 and 7 days. Presence of Matrigel showed significantly lower cell growth on the rTi implants, whereas no differences were seen on the HA coated implants with the addition of Matrigel. Effect of time was a significant variable for rTi implants with and without Matrigel with increases in cell numbers with time up to 7 days and for wells without implants at all time periods. HA coated implants showed no significant differences in cell numbers between the three time periods with and without Matrigel.

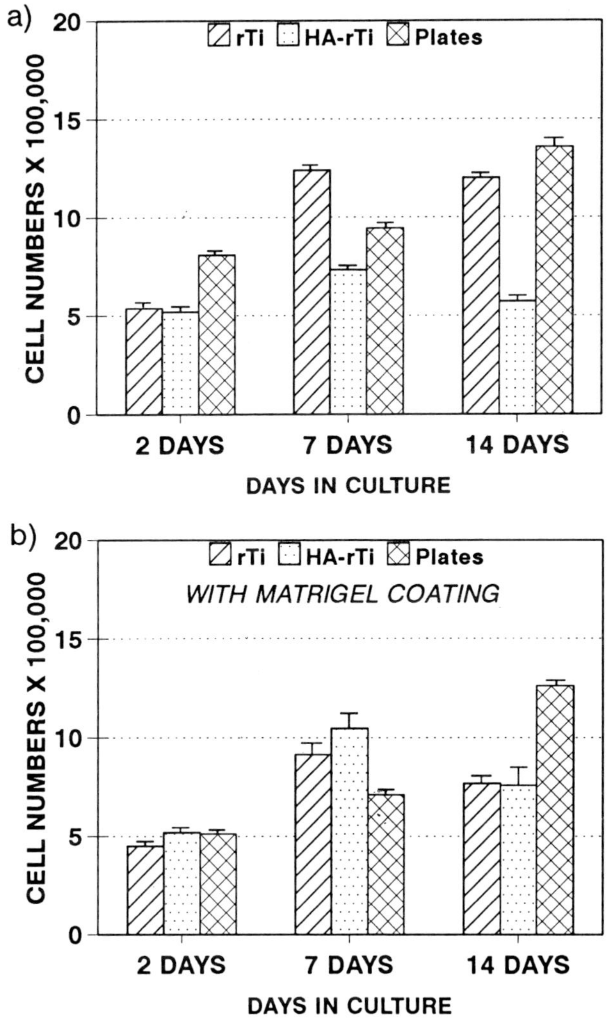

Figure 4: Cell growth from MTT assay for culture plates, HA coated rough Ti (HA-rTi) and rough Ti (rTi) implants (a) without Matrigel and (b) with Matrigel coating.

CONCLUSIONS

Discussing the APA/cell ratio is a better indication of the surface reactivity of the various implants and their ability to enhance osteoblastic phenotype. On the tissue culture plates and on the rTi implants a decline in APA/cell was seen from 2 to 7 to 14 days with and without Matrigel coating. On the other hand, APA/cell on HA coated implants with and without Matrigel showed an initial decline from 2 to 7 days, and then increased from 7 to 14 days to double values for uncoated rTi at 14 days.

In summary, these results suggest that the formation of an interface between bone cells and the various orthopaedic implants used is influenced by the chemical structure of the surface and is affected by the crystallinity and surface resorption of the HA coating. The presence of Matrigel did not show enhanced bone progenitor cell growth and osteoblastic phenotype expression on orthopaedic implants. However, presence of this protein coating did not have a detrimental effect on cell growth and alkaline phosphatase activity.

REFERENCES

1. Maxian, S.H., Zawadsky, J.P. and Dunn, M.G., J. Biomed. Mater. Res., (1993) 27, 717-728.
2. Maxian, S.H., Zawadsky, J.P. and Dunn, M.G, J. Biomed. Mater. Res., (1993) 27, 111-117.
3. Maxian, S.H., Zawadsky, J.P. and Dunn, M.G., in Bioceramics, Proceedings of the 6th International Symposium on Ceramics in Medicine, Philadelphia, USA, (1993) **6**, 135-139
4. de Bruijn, J.D., Flach, J.S., de Groot, K., van Blitterswijk, C.A. and Davies, J.E., Cells and Materials (1993) 3(2), 115-127.
5. Campbell, P., McKellop, H., Park, S.H. and Malcolm, A., Trans. 39th. Orthopaedic Research Society Meeting, San Fransisco, CA, (1993) 224.
6. van Blitterswijk, C.A., Leenders, H., Brink, J.v.d., Bovell, Y., Flach, J., de Bruijn, J. and de Groot, K., Trans. 19th. Society for Biomaterials Meeting, Birmingham, AL, (1993) 337.
7. Ducheyne, P. and Cuckler, J.M., Clin. Orthop. Rel. Res., (1992) 276, 102-114.
8. Maxian, S.H., Liesch, J.B., Tiku, M.L., Zawadsky, J.P. and Dunn, M.G., Trans. 19th. Society for Biomaterials Meeting, Birmingham, AL, (1993) 336.
9. Vukicevic, S., Luyten, F.P., Kleinman, H. and Ressi, A.H., Cell, (1990) 63, 437-445.
10. Whitehead, R.Y., Lucas, L.C., and Lacefield, W., Clinical Materials, (1993) 12, 31-39.

Bone Induction

Bioceramics, Volume 7 Edited by Ö.H. Andersson and A. Yli-Urpo
(*Proceedings of the 7th International Symposium on Ceramics in Medicine, Turku, Finland, July 1994*)

Coordinate Osteoblastic Differentiation on the Surface of Bioactive Materials

H. Ohgushi, M. Okumura, T. Takaoka, K. Inoue and S. Tamai
Department of Orthopedic Surgery, Nara Medical University, Kashihara City, Nara 634, Japan

ABSTRACT

To analyze the cascade of initial bone formation on bioactive and non-bioactive materials, we implanted various porous materials combined with rat bone marrow cells at subcutaneous sites. De novo bone formation appeared at about 3 weeks post implantation. Osteoblasts fabricated the bone on the surface of bioactive materials, primarily as mature bone (lamellar type). This was evidenced by the finding of polarized microscopy which showed parallel arrangement of bone collagen bundles. Thus, three-dimensional cellular orientation by coordinating osteoblasts on the materials' surface initiated the bone formation. However, de novo bone formation started away from the surface of non-bioactive materials and polarized microscopy demonstrated random arrangement of bone collagen fiber indicating immature bone formation (woven type).

These results indicate the importance of bioactive materials' surface to support osteogenic differentiation resulting in appositional bone formation (lamellar bone formation). Therefore, the surface of bioactive materials can mimic characteristic feature of normal bone surface.

INTRODUCTION

Recently, various materials applicable for the use in clinical cases in orthopedic, dental and maxillofacial surgery have been developed. Among them, some specific materials such as calcium phosphate ceramics and glass ceramics are called "bioactive" due to the bone bonding property [1-3]. However, the materials themselves can not show bone forming ability. It is known that stromal stem cells which can differentiate into osteogenic cells reside in the marrow cell population [4]. We have reported that neither the marrow cells nor the materials themselves showed bone formation at ectopic(subcutaneous or intramuscular) sites, whereas the marrow cells loaded porous materials resulted in bone formation [5-7]. When we use bioactive materials as carrier of marrow cells, the bone formation started directly on the material surface and thus formed bone bonded to the surface [7]. From histological, biochemical and gene expression studies, we hypothesized that attachment of osteogenic stem cells to the surface of the material is necessary for firm bone bonding through the cascade of osteogenic differentiation [8]. In contrast, osteogenic cells can not attach to the surface of non-bioactive (bioinert and biotolerant) materials such as alumina ceramics resulting in intervening fibrous tissue formation between the materials and bone tissue [9]. In present study, we analyzed the initial bone formation on various bioactive materials and compared with bone formation on non-bioactive materials. The histological sections were analyzed under light, fluorescence and polarized microscopy.

MATERIALS AND METHODS

Bioactive materials: porous hydroxyapatite [7], porous β-tricalcium phosphate[6] and porous calcium carbonate[10]. The average pore size of these materials are 230 um in diameter and void volume are 50 -60%. The shapes are disk (5mm in diameter and 2mm in thickness; Interpore International, Irvine, CA)
Non-bioacitive materials: porous alumina ceramics [9] (polycrystaline alumina ceramics, pore size 100 - 600um and void volume 70%, Kyocera Co., Kyoto, Japan) and calf derived cancellous bone graft substitute (Kiel bone, Braun Melsungen, Germany)[11].

These porous materials were combined with rat marrow cells and implanted at subcutaneous sites. Detailed procedure for marrow cell preparation has already been reported[5-7]. Briefly, marrow from the femoral and tibial diaphysis of Fisher rats were hydrostatically forced into centrifuge tube containing heparinized phosphate buffered saline (PBS). The marrow was disaggregated and concentrated by centrifugation resulting in 5×10^8 nucleated cells/ml. Porous discs were soaked in the marrow cell suspension then implanted at subcutaneous sites. Implants were harvested at 4 and 6 weeks post implantation. Histological sections were analyzed under fluorescence and polarized microscopy.

RESULTS AND DISCUSSION

As shown in Fig. 1, de novo bone formation occurred in the pore regions of hydroxyapatite(HA) ceramics combined with bone marrow cells at subcutaneous sites. The ectopic bone formation reminisces embryonic bone formation; this type of bone materializes as woven bone and subsequently changes to lamellar type bone. However, analysis of the initial bone formation in HA revealed that the process was not embryonic bone formation but rather mature appositional bone formation which initiated on the HA surface. As shown in Fig.1, in the composites of HA and marrow cells, orderly alignment of osteoblasts formed bone on the HA surface. Polarized microscopy demonstrated that the initial bone was primarily formed as lamellar type, evidenced by the parallel arrangement of collagen bundles. In physiological condition, lamellar bone is usually formed in apposition to a preexisting bone surface. Thus, like bone surface, in vivo implanted HA surface elicited this osteoblastic apposition resulting in lamellar bone formation. And importantly, the bone formation always started on the surface of materials. The lamellar bone formation initiated on the materials' surface were also seen in β-tricalcium phosphate and calcium carbonate (Fig. 2).

Kiel bone is bovine cancellous bone derived bone substitute and the materials can be categorized as biotolerant material. As shown in Fig. 3, when the Kiel bone was combined with marrow cells, de novo bone formation occurred away from the surface resulting in dense fibrous tissue interposition between the de novo bone and Kiel bone surface. Polarized microscopy demonstrated random arrangement of collagen fiber of the de novo bone appeared in the pore region of Kiel bone. The random arrangement of collagen fiber contrasts the lamellar architecture of the Kiel bone itself. This finding of woven bone formation on the material differs from the above findings of bioactive materials which showed lamellar bone formation occurred primarily on the surface. When we use porous alumina ceramics as an another type of non-bioactive materials, polarized microscopy demonstrated woven bone formation in pore regions of alumina ceramics and the bone formation always started away from the alumina surface resulting in initial fibrous tissue interposition between the bone and alumina surface.

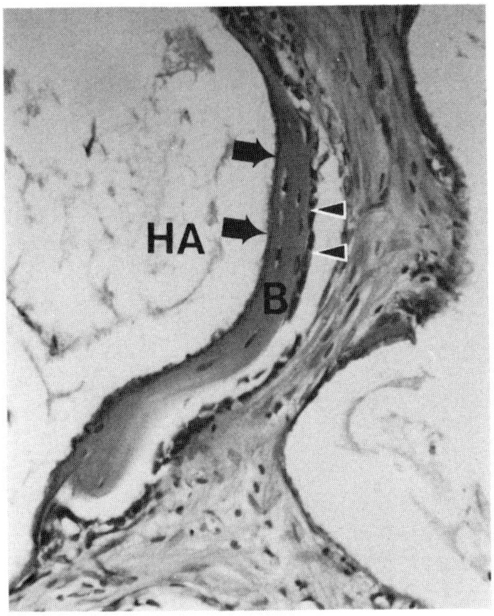

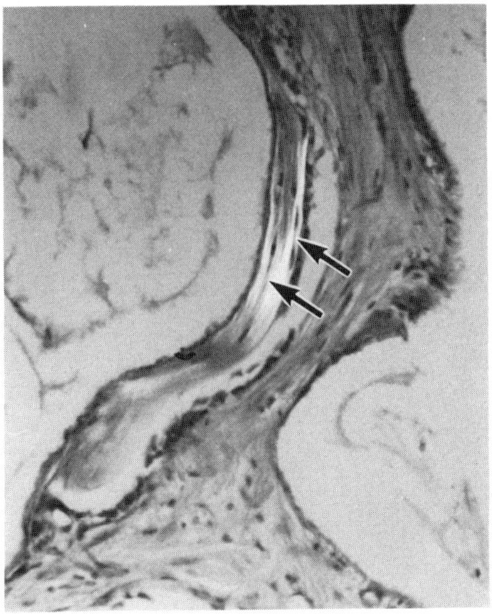

Fig. 1a (left)
Photomicrograph (x 200) of the composite of porous hydroxyapatite (HA) and marrow cells 4 weeks post subcutaneous implantation. Active osteoblasts (arrow heads) initiate bone formation (B) on the surface of HA. Arrows indicate the bone/HA interface.
Fig. 1b (right)
Photomicrograph of the same section under polarized light. Parallel orientation of brilliant lines (arrows) indicate the orderly arrangement of collagen fiber seen in lamellar bone.

Alkaline phosphatase (ALP) is known to reflect osteoblastic activity and bone Gla protein (BGP) is exclusively bone specific protein. We already reported that in the composite of the marrow cells and porous hydroxyapatite, ALP activity began to appear at 2 weeks post implantation and BGP could be detected post 3 weeks coinciding with the appearance of bone [12]. In addition, gene expression analysis showed that the ALP and BGP mRNAs were not detectable by Northern blot analysis at 1 week post implantation of the composite. At 2 weeks post implantation, both the ALP and BGP mRNAs began to appear and increased in their expression till 4 weeks post implantation. In contrast to these results, in HA implantation alone (without marrow cells), bone formation could not occur and these mRNAs were not detectable at any time [10]. These previous findings and present results indicate the importance of the bioactive materials surface for the differentiation of marrow stromal stem cells which results in firm bone bonding. However, the differentiation can not occur on the surface of non-bioactive materials.

Importance of the surface activity of the bioactive materials to evoke gene expression of progenitor cells is that the cell differentiation occurred in a highly organized manner evidenced by the orderly alignments of the differentiated cells (osteoblasts lining). This implicates that the surface activity can coordinate the osteoblastic function. Same observable facts occurred in porous tricalcium phosphate and calcium carbonate. These facts again confirm our previous assumption that tricalcium phosphate and calcium carbonate can be categorized as bioactive materials. This cellular event is schematically drawn in Fig. 4. On the surface of bioactive materials, babies (undifferentiatited cells) can sleep,

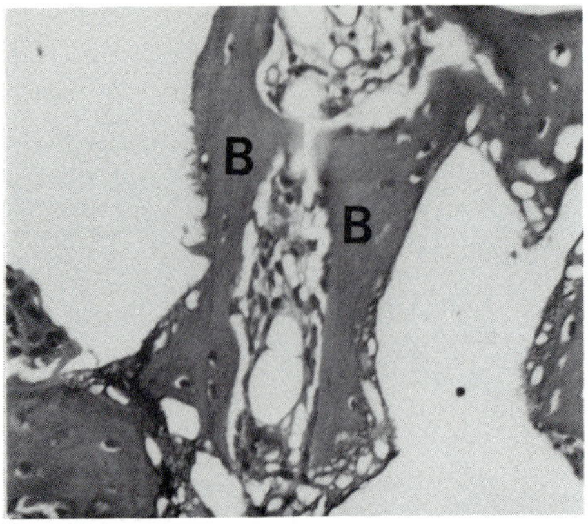

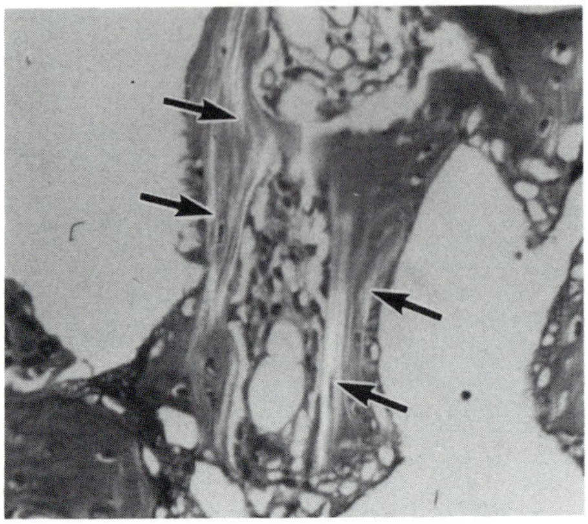

Fig. 2a (left)
Photomicrograph (x 200) of the composite of porous calcium carbonate and marrow cells 4 weeks post subcutaneous implantation. White area shows the ghost of calcium carbonate produced by decalcification procedures. There is no intervening fibrous tissue between de novo bone (B) and the surface of calcium carbonate.
Fig. 2b (right)
Photomicrograph of the same section under polarized light. Parallel orientation of brilliant lines (arrows) indicate the orderly arrangement of collagen fiber seen in lamellar bone.

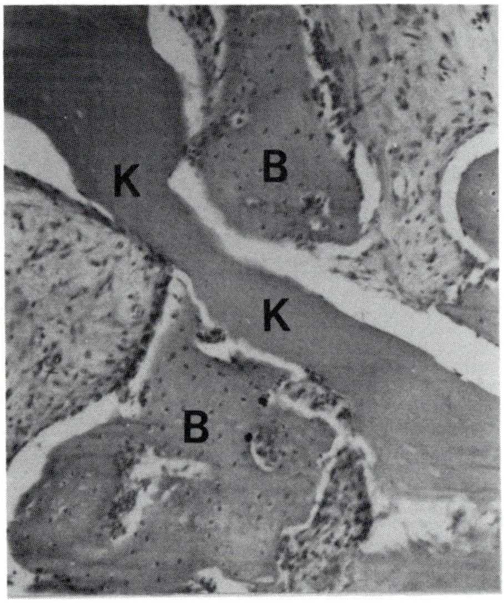

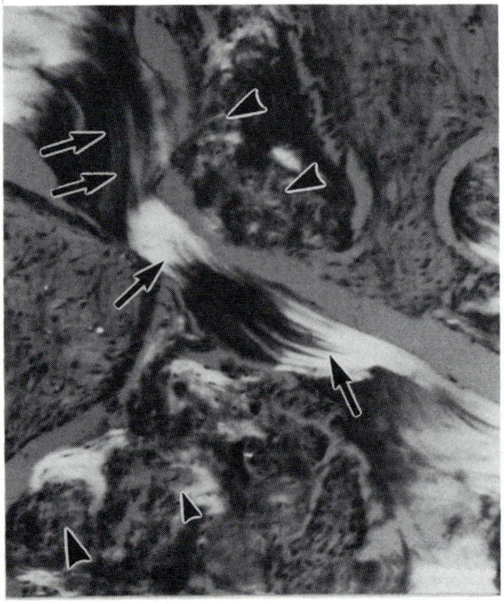

Fig. 3a (left)
Photomicrograph (x150) of the composite of Kiel bone and marrow cells 4 weeks post subcutaneous implantation. Dense fibrous tissue are seen between de novo bone (B) and Kiel bone (K). Kiel bone is identified by the presence of empty lacunae.
Fig. 3b (right)
Photomicrograph of the same section under polarized light. Random arrangement of brilliant lines (arrow heads) indicate the disorganized collagen fiber seen in woven type of de novo bone. In contrast, parallel orientation of brilliant lines (arrows) indicate the orderly arrangement of collagen fiber seen in lamellar type of Kiel bone itself.

Fig. 4
Schematic bone formation on bioactive and non-bioactive materials (from Ceramics Japan vol., 29, p66,1994, with permission).

grow very comfortably and finally become many healthy persons who cooperate to build bone matrix on the surface. The cooperation leads to making fine architecture of bone (lamellar bone) which bonds to the material's surface. However, surface of non-bioactive materials can not provide comfortable environment for babies and therefore, grown up persons can not cooperate to work resulting in random formation of bone matrix (woven bone) and importantly the bone can not bond to the surface of non-bioactive materials.

All of these results demonstrate that the surface of bioactive materials supports osteoblastic differentiation. The surface-dependent cell differentiation takes place in a highly organized manner and is followed by fabricating mature (lamellar) bone which bonds to the material. We hypothesize that the bioactive material has the capability to support cell differentiation and causes the expression of its specific phenotype on the material surface. Such cellular events finally result in the integration of the material with surrounding new tissue.

ACKNOWLEDGEMENTS

We wish to express our gratitude to Dr. Edwin C. Shors (Interpore International Corporation) and Kyocera Company for supplying various porous materials.

REFERENCES

1 L.L. Hench, Ann. NY Acad. Sci.(1988)523 54-71 (1988).

2 U.M.Gross, H.J. Schmitz, and V. Strunz,(1988) Ann. NY Acad. Sci,523 211-226

3 M. Jarcho, Clin. Orthop. Rel. Res.(1981)157 259-278

4 M. Owen, In: Bone and Mineral Research, Elsevier Science Pub., New York, (1985) Vol 3,pp 1-25.

5 H. Ohgushi, V. M. Goldberg, and A. I. Caplan, J. Orthop. Res.(1989) 7 568-578

6 H. Ohgushi, M. Okumura, S. Tamai, and E.C. Shors, J. Biomed. Mat. Res., (1990)24 1563-1570.

7 M. Okumura, H. Ohgushi, and S. Tamai, Biomaterials,(1991)12 411-416

8 H. Ohgushi, Y. Dohi, S. Tamai and S. Tabata, J. Biomed Mat. Res.(1993) 27 274-280

9 M. Okumura, H. Ohgushi, Y. Takakura, C.A. van Blitterswijk and K. Koerten, Bio-Medical Mat. & Engineering (1992) 2 191-201

10 H. Ohgushi, M. Okumura, T. Yoshikawa, K. Inoue, N. Senpuku and S. Tamai J. Biomed Mat. Res.(1992) 26 885-895

11 M. Okumura, H. Ohgushi, T. Yoshikawa and S. Tamai,In Biceramics Vol.2 German Ceramic Society, Cologne (1990) 86-93

12 T. Yoshikawa, H. Ohgushi, M. Okumura, S. Tamai, Y. Dohi and T. Moriyama, Calcif. Tissue Int. (1992) 50 184-188

Bioceramics, Volume 7 Edited by Ö.H. Andersson and A. Yli-Urpo
(*Proceedings of the 7th International Symposium on Ceramics in Medicine, Turku, Finland, July 1994*)

Osteogenic Response of Bone Marrow Cells in Porous A–W Glass Ceramic

M. Okumura, H. Ohgushi, K. Inoue, S. Tamai, S. Dohi,* T. Shibuya** and S. Komatsudani**

* Department of Orthopedic Surgery and Public Health, Nara Medical University, Nara, Japan
** Nippon Electric Glass Co. Ltd., Otsu, Japan

ABSTRACT

Porous A-W glass ceramics (apatite and wollastonite containing glass ceramics;A-W GC) alone and combined with rat bone marrow cells were inserted subcutaneously in the back of syngeneic Fischer rats. Fluorochrome-labelling was performed in the course of the experiment and the ceramics were harvested at 3, 4, 6, 8, 12 weeks post implantation. Undecalcified sections of the implants were examined by light and fluorescence microscopy. A-W GC without marrow did not show any bone formation, whereas marrow cells loaded A-W GC showed new bone formation in the pore areas. Fluorochrome analysis showed that osteogenesis generally started from A-W GC surface and proceeded in a centripetal direction towards the center of the pores. Scanning electron microscopical observations of the bone/A-W GC interface also revealed direct bone bonding to A-W GC. Biochemical analysis demonstrated high alkaline phosphatase activity and bone Gla protein contents of A-W GC with marrow cells. These results suggest that porous A-W GC can support osteogenic differentiation of bone marrow cells.

INTRODUCTION

Previously, we reported that a composite graft of rat bone marrow cells and porous calcium phosphate ceramics showed consistent de novo bone formation in the pore regions at subcutaneous sites(1-4). This method is suitable for the observation of primary bone formation on the materials surface without any influences from preexisting host bone. In this regards, bioactive materials such as hydroxyapatite ceramics showed direct bone apposition on the surface of the pore wall by the osteoblastic cell differentiation of marrow stromal stem cells. In addition, we analyzed Bone Gla protein (BGP) contents and alkaline phosphatase (ALP) activity in the de novo bone formed in porous hydroxyapatite ceramics and found that the BGP contents reflected the amount of bone area and ALP activity well correlated with the number of active osteoblasts(5). Therefore, these two biochemical parameters can be used for the assessment of bone forming capacity of porous materials. In the current experiment, we focused on the osteogenic response of marrow cells in porous A-W glass ceramics.

MATERIALS AND METHODS

Implant materials

Apatite-Wollastonite containing glass ceramics (Nippon Electric Glass Co. Ltd., Otsu, Japan) were used. The material had a porous structure with a mean void volume of 70 and 90% and well interconnected pores measuring 300 μm in diameter. The implants of 5 x 2.5 x 2 mm rectangular shaped A-W glass ceramics were prepared.

Marrow cell preparation

The detailed procedures were previously reported(2-4). Briefly, femora and tibiae of 5 Fischer strain 344 rats (7 week-old males) were removed and placed in saline. Both ends of the femur and tibia were cut away from the epiphysis. The marrow plug from the diaphyseal portion was then hydrostatically forced into a test tube containing heparinized phosphate buffered saline(PBS). The marrow in PBS was disaggregated by sequential passage through 18G and then 20 G needles to obtain a cell suspension. The cell suspension was centrifuged at 250G for 5 minutes and 200 µl of the supernatant was mixed with the cell layer by a vortex mixer. Following these procedures, the number of nucleated cells was about $5x10^8$/ml. To make the composite graft, the materials were soaked in this disaggregated marrow cell suspension at room temperature for about 10 minutes.

Surgical procedure

Syngeneic Fischer rats were anesthetized by intraperitoneal injection of Nembutal (2.5 mg/100g B.W.). Subcutaneous pouches were created by blunt dissection in the back of the rat following 8mm longitudinal incisions. A-W glass ceramics alone and combined with marrow cells were implanted in subcutaneous sites. The implants were harvested at 3, 4, 6, 8 and 12 weeks post implantation. During the postoperative time, multiple fluorochrome labelings were performed to observe the dynamics of the newly formed bone in the pores. The rats were given one dose each of tetracycline (50 mg/kg subcutaneously) and calcein (15 mg/kg intravenously).

Histological evaluation

The detailed sectioning procedure for undecalcified sections has been described in previous reports(3,4). Each implant was fixed in 70% ethyl alcohol, stained with Villanueva bone stain, dehydrated in graded series of ethyl alcohol and acetone and then embedded in methyl methacrylate. After cutting sections(200 µm in thickness) by using a milling machine, they were ground with a diamond lap disc (Speed Lap, Maruto Ltd, Japan). The prepared specimens were examined by light or fluorescence microscopy. For the scanning electron microscopical study, the flat surface of the implant was prepared after grinding the implant surface embedded in methyl methacrylate using a diamond compound. Surface of the implant was coated with a layer of carbon, about 10 nm thick. The ceramic/bone interface was analyzed with a scanning electron microscope connected to an energy dispersive spectrometer. The implant surface was first observed in the backscattered electron(BSE) imaging mode. In addition, line analysis of three elements (calcium, phosphorus and silicon) was performed along a line on the bone/ceramic interface.

Biochemical analysis

The implants for biochemical analysis were immediately crushed, homogenized in 0.2% Nonidet P40 containing 1mM $MgCl_2$ and centrifuged at 13,000 rpm for 10min at 4°C. The supernatant was assayed for ALP using-p-nitrophenyl phosphate as a substrate. ALP activity represented umol of p-nitrophenol released per one implants after 30 minutes of incubation at 37°C. BGP was extracted from the

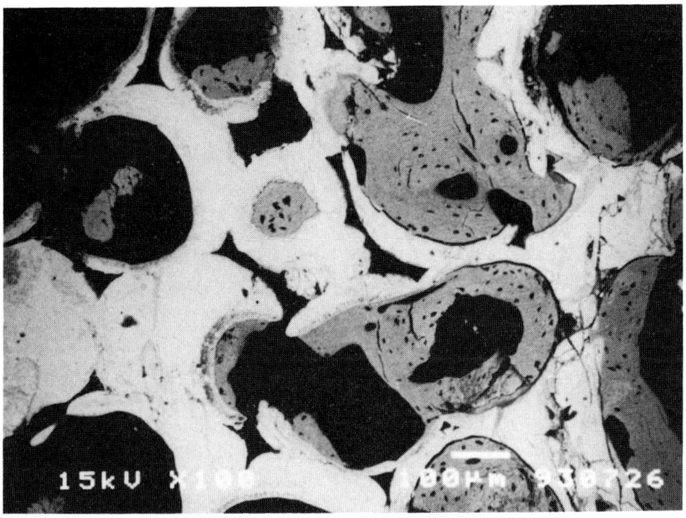

Figure 1.
Backscattered electron image photograph of A-W glass ceramics with marrow cells 8 weeks after implantation. White areas indicate A-W glass ceramics and gray areas indicate bone. Bar;100 μm

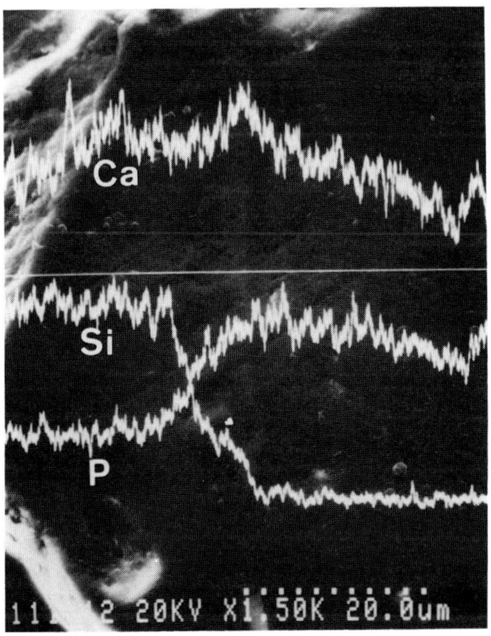

Figure 2.
Scanning electron micrograph of the interface between bone (right side) and A-W glass ceramics (left side). Line analysis of Ca, P and Si are shown as wave lines. Pathway of the electron beam is indicated by the center white line.

sediment after extraction of 0.2%Nonidet P40 by shaking in 1 ml of 20% formic acid at 4°C for 2 weeks. An aliquot (500μl) of the formic acid extract was then applied to the column of Sephadex G-25(fine) and eluted with 10% formic acid. Protein fraction were collected, lyophilized and used for radioimmunoassay of BGP as previously described(6).

RESULTS AND DISCUSSION

Apatite-Wollastonite containing glass ceramics (A-W GC) without marrow did not show any bone formation, whereas marrow cells loaded A-W GC began to appear new bone formation at 3 weeks after implantation in the pore areas of implants. After 4 weeks, all A-W GC with marrow cells showed bone formation. Fluorochrome analysis showed that osteogenesis generally started from the ceramic surface and proceeded in a centripetal direction towards the center of the pores. Scanning electron microscopical observations of the bone/A-W GC interface also revealed direct bone bonding to A-W GC. The histological findings showed the bioactive nature of A-W GC and osteogenic capacity of A-W GC in the presence of marrow cells. The osteogenic capacity was confirmed by biochemical analysis. The analysis demonstrated that ALP activities (μmol/30min/implant) of A-W GC with marrow at 4 and 8 weeks post implantation were 4.29 ± 1.20 and 3.57 ± 0.75 respectively. BGP contents (ng/implant) of A-W GC with marrow were 291 ± 123 and 930 ± 117 respectively. Only traces of these were detected in control implants of A-W glass ceramics without marrow cells.
Previous studies on dense type of A-W GC showed the material's bioactive characteristic features and high mechanical property even under load-bearing conditions in a body environment(7-10). In the present study, porous A-W GC combined with rat marrow cells implanted in subcutaneous site revealed extensive bone formation and the intimate bonding between the de novo bone and A-W GC surface on SEM-EPMA analysis. These results suggest that porous A-W GC can support osteogenic differentiation of marrow stromal cells on its surface.

REFERENCES

1. Ohgushi,H., Goldberg,V.M. and Caplan,A.I. J. Orthop. Res.(1989) 7, 568-578
2. Ohgushi,H., Okumura,M., Tamai,S. and Shors,E.C. J. Biomed. Mater. Res. (1990) 24, 563-1570
3. Okumura,M., Ohgushi,H. and Tamai,S. Biomaterials.(1991) 12, 411-416
4. Okumura,M., Ohgushi,H., Takakura,Y., Van Blitterswijk, C.A. and Koerten, H.K. Bio-Medical Mat. and Eng. (1992) 2, 191-201
5. Yoshikawa, T., Ohgushi, H., Okumura, M., Tamai, S., Dohi, Y. and Moriyama, T. Calcif. Tissue Int. (1992) 50, 184-188
6. Sugimoto,K., Dohi, Y., Dohi, K., Yonemasu, K., Fijimoto, J., Kanauchi, M., Yamanaka, F., Moriyama.,T.,and Ishikawa, H. Mineral Electrolyte Metab. (1987), 13, 152
7. Nakamura, T., Yamamuro, T., Higashi, S., Kokubo, T. and, Itoo, S., J. Biomed. Mater. Res. (1985) 19, 685-698
8. Kitugi, T., Yamamuro, T., Nakamura, T., Higashi, S., Kakutani, Y., Hyakuna, K., Itoo, S., Kokubo, T., Takagi, M. and Shibuya, T. J. Biomed. Mater. Res. (1986) 20, 1295-1307
9. Kitugi, T., Yamamuro, T. and Kokubo, T. J. bone and Joint Surg. (1989) 71-A. 264-272
10. Kokubo, T., Ito, S., Huang, Z.T., Sakka, S., Kitsugi, T. and Yamamuro, T J. Biomed. Mater. Res. (1990) 24, 331-343

Bioceramics, Volume 7 Edited by Ö.H. Andersson and A. Yli-Urpo
(*Proceedings of the 7th International Symposium on Ceramics in Medicine, Turku, Finland, July 1994*)

Osteogenic Capacity of Vascularized Periosteum Supported by Porous Hydroxyapatite

K. Inoue, H. Ohgushi, H. Yajima, M. Okumura, H. Ishida and S. Tamai
Department of Orthopedic Surgery, Nara Medical University, Kashihara City, Nara 634, Japan

ABSTRACT

We established a unique experimental model of the porous hydroxyapatite (HA) wrapped with a vascularized periosteal flap. In present study, we used this model to analyze the bone forming capacity of the vascularized periosteum. Using Fischer rats, the periosteum of tibia nourished by the saphenous vessels was elevated as a vascularized periosteal flap on a vascular pedicle. The HA was wrapped with the vascularized periosteal flap and transferred to the subcutaneous space in the inguinal region (Vascularized Periosteal Graft:VPG group). In controlateral side, the HA was wrapped with a fascial flap nourished by the saphenous vessels and transferred to the subcutaneous space in the inguinal region (Vascularized Fascia Graft:VFG group). At 4 and 8 weeks postoperatively, the implants were harvested and examined histologically.

In VPG group, the bone formation was observed not only on the surface area of HA (location of the periosteum), but also in the pore of HA. However, in VFG group, the bone formation was hardly detected. These results indicate that HA ceramics wrapped with the vascularized periosteal flap has significant bone forming ability.

INTRODUCTION

Vascularized bone graft has vigorous bone forming capacity compared with traditional free bone graft[1][2]. However, major problem of this method is that size of graft bone has its limitation. And the results obtained from many experimental studies on free periosteal flaps were inconsistent. We established a unique experimental model of porous HA wrapped with vascularized periosteal flap as this solution of above problem. The advantage of this model is that osteogenic property of periosteum can be available without coexistent host bone. Therefore, in clinical situation, there is no need to take bone from the patients.

MATERIALS AND METHODS

Materials

Implants of porous hydroxyapatite(HA) were made by converting the calcium carbonate skeleton of marine coral(genus Porites) to hydroxyapatite through a hydrothermal exchange reaction (Interpore 200 porous hydroxyapatite, Interpore International, Irvine, California). The implants chosen for this study have a pore volume of 50-60% and, fully interconnected pores, measuring 190-230 um in diameter. The HA showed good osteoconductive property[3].

The materials were cut into rectangular shape (20 mg).

Implantation of ceramics

We elevated a vascularized tibial periosteal flap (3 X 15 mm) nourished by a saphenous artery and vein from tibial surface. The HA ceramics were wrapped with this vascularized periosteal flap and implanted subcutaneously in the inguinal region (VPG group) of syngeneic rats(8 weeks, Fischer 344). In controlateral side, the HA was wrapped with fascial flap with saphenous vessels and implanted subcutaneously in the inguinal region (VFG group).

Histological analysis

Rats were sacrificed at 4 and 8 weeks after surgery. Harvested HA ceramics were used for histological evaluation. The ceramics were fixed in 10% buffered formalin, and then demineralized in formic acid. The specimen were stained with hematoxylin and eosin.

RESULTS AND DISCUSSION

In VPG group at 4 and 8 weeks after surgery, the bone formation was observed at the location of vascularized periosteal flap (near the surface area of HA), but also in the pore areas of HA (Fig.1). However, in VFG group, bone formation was not detected(Fig.2). These results indicate that HA ceramics wrapped with the vascularized periosteal flap has significant bone forming ability (Table 1).
The osteogenic capacity of periosteum has been recognized by numerous investigators, but is the subject of continuing controversy. Discrepant results in different studies have been attributed chiefly to differences in the viability of grafted periosteum. Bearing this in mind, Finley et al. in 1978 conducted an experimental study on vascularized periosteal flaps and observed excellent bone formation[4]. In contrast to the free vascularized periosteal flap, vascularized bone graft (living bone surrounded with periosteal tissue) is a technique to show intensive new bone formation. However, major problem of this method is that size of graft bone has its limitation and morbidity is clear for the patient. We established a unique experimental model of porous hydroxyapatite wrapped with vascularized periosteal flap. And this model showed extensive bone forming capacity of vascularized periosteal flap. Interestingly, the bone formation was also detected inside of the ceramic pore area(Fig.1), therefore the composite of the periosteal flap and porous ceramics is comparable to the vascularized bone graft.

Table 1

Bone formation following implantation

Weeks	VPG group	VFG group
4	8/8	0/8
8	8/8	0/8

At 4 and 8 weeks after implantation, all 8 ceramics showed bone formation in VPG group, whereas none of 8 ceramics showed bone formation in VFG group.

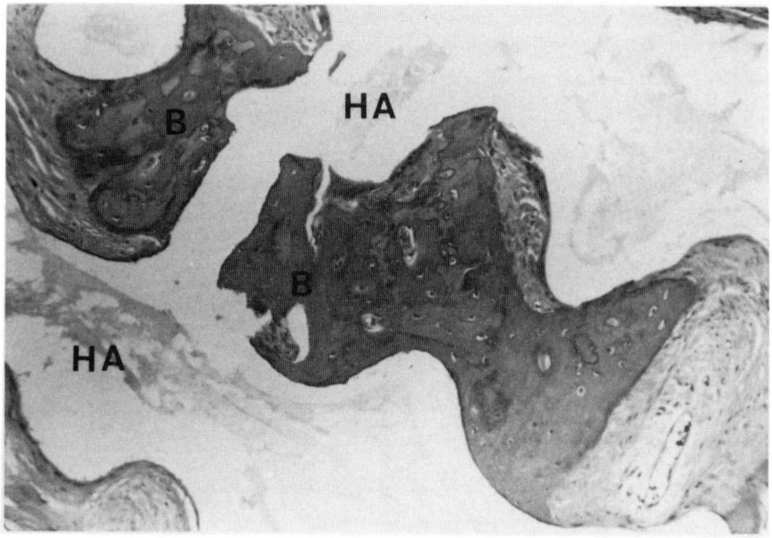

Fig.1
Eight weeks after implantation of HA wrapped with vascularized periosteal flap. The bone formation was observed in the pore area.(X 300)

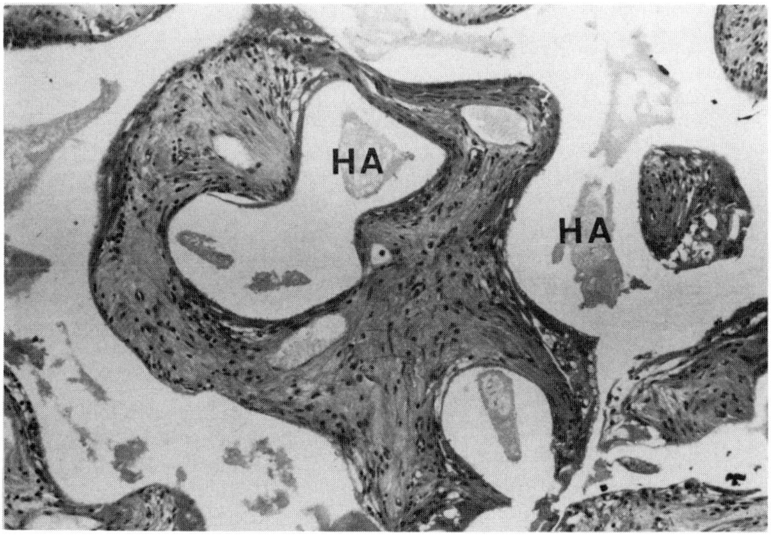

Fig.2
Four weeks after implantation of HA wrapped with vascularized fascial flap. The bone formation was not detected, only fibrous tissue in the pore area was observed.(X 300)

However, important question is that the osteogenic capacity of the HA wrapped with non-vascularized periosteal flap. In this regard, our recent results indicate the higher bone formation capacity of the ceramics wrapped with vascularized periosteal flap than that with non-vascularized periosteal flap. The method we used to measure the bone forming ability was biochemical analysis[5].
Therefore, the composite of vascularized periosteal flap and HA ceramics can be useful for alternative method of the vascularized bone graft.

REFERENCES

1)H. Yajima, et.al.:Vascularized fibular grafts for reconstruction of the femur. J. Bone Joint Surg. 75B:123-128, 1993.

2)H. Yajima, et.al.:Vascularized fibular grafts in the treatment of osteomyelitis and infected nonunion. Clin. Orthop., 293:256-264, 1993.

3)H.Ohgushi, et.al.: Bone formation process in porous calcium carbonate and hydroxyapatite. J.Biomed. Mater. Res., 26:885-895, 1992.

4)Finley, J.M. et.al.: Revascularized periosteal grafts : A new method to produce functional new bone without bone grafting. Plast. Reconstr. Surg., 61:1-6, 1978.

5)H. Ishida, et.al.: manuscripts in preparation.

Bioceramics, Volume 7 Edited by Ö.H. Andersson and A. Yli-Urpo
(*Proceedings of the 7th International Symposium on Ceramics in Medicine, Turku, Finland, July 1994*)

Aptitude of Ceramic Mineral Bone Substitutes Depending on Quality of the Host Tissue Bed – Animal Study

W. Mittelmeier, G. Harbauer, H. Mittelmeier and R. Gradinger
Orthopädische Klinik, Medizin. Universität zu Lübeck, Ratzeburger Allee 160, 23538 Lübeck, Germany
Orthopädische Klinik & Institut f. Experimentelle Chirurgie, Univ. des Saarlandes, 66424 Homburg, Germany

Abstract:

Animal tests for intraindividual investigation of implants in different host tissues are presented. The same, natural derived ceramic material (Pyrost) is implanted in standard bone defects, soft tissues (muscle) and combined tissue (iliac crest: muscle/bone surface).
The implant stimulates the bone formation significantly (osteostimulation), but it`s not osteoinductive. The velocity of bone formation depends on the contact surface with the intact bone bed. In the muscular site bone appears only after combination of the implant with autogenic marrow ("marrow inoculation"). The bone areas are expanding depending on time. Marrow is developing inside of the bone areas after 4 weeks.

Introduction:

Since more than 100 years there are efforts to search for the ideal bone substitute. Usually chemical and immunological properties of the respective implants were emphasized.
Lexer (3) differentiated the quality of the host tissue bed depending on the contact to the periosteum resp. to an intact host tissue. Schweiberer (5) classified the host tissue according to the vascularity and local necrosis.
Different regenerative bone substitutes consisting of synthetic calcium-phosphate were tested in the last 20 years. In 1983 we first presented a natural derived ceramic bone substitute with osteoconductive and osteostimulative activity (2; 4) . But we didn´t know the meaning of the host tissue bed when using these materials.

Materials:

Bovine bone is deproteinized and treated by high temperature ceramic procedure. The material, called Pyrost, has a porosity of 70 %, a pressure stability of 5.3 N/mm^2, it consists mostly of hydroxiapatite (93 % hydroxiapatite, 7 % ß-TCP) and is slowly resorbable.

Method:

We created the "4-sites-test" (or "4-point-test"), which allows the intraindividual and comparing investigation of implants in muscle pouches and femoral kondyle defects - each at both sides.
In another test we used an iliac crest defect instead of the left femoral condyle bore holes.
Based on the results of Burwell (1) which showed the meaning of bone marrow in procedures of bone regeneration, we tested combined grafts of the implant with autogenic marrow for accelerating or producing ectopic bone formation. In this technique of "marrow inoculation" we transferred only 0.2 ml marrow per implant.

The comparing tests were performed in adult rabbits (age >12 months). We implanted the same material (Pyrost) in the same size of 5x5x8mm in different tissues:
1. Bore hole defect in the right femoral condyle (6mm diameter,10mm length; n=28+16) without any implant.
2. Pyrost in the left femoral condyle defect of the same animal (n=28).
3. P. without autogenic marrow in the right quadriceps muscle (n=28+16);
4. P. with autogenic marrow in the left quadriceps muscle (n=28).
5. P. in iliac crest defect (5x8 mm, n=16), where 31 % of the implant surface is covered by bone and 69 % by muscular tissue.
6. In a complete further series we implanted P. in granulated form (diameter= 150 - 200 μm; n=16 in each site).
Histology was performed with undecalcified bone preparation after 2, 4, 6, 8, 12 and 24 weeks in the large series, after 4 and 6 weeks in small serieses. In each series at 6 weeks we obtained at least 10 cases. Each specimen was inspected in 10 cross section slides and investigated by automated image-analysis system.

Results:

1. The regeneration of the *empty femoral condyle* defect without the implant is slow and incomplete.
2. With Pyrost in the *femoral condyle* occurs a complete and fast reconstruction of the defect by bone and marrow within 4 weeks. In this situation a thick pattern of vascularization and new formed bone tissue inside of and around the implant can be observed.
3. By implantation of P. *in quadriceps muscle without marrow* the tissue and the implant was unable to form bone, marrow or cartilage. The implant is filled up by vascularized connective tissue.
4. With P. and *autogenic marrow* in quadriceps muscle a slow, but intensive and complete bone formation within the implant volume up to 24 weeks can be detected. Marrow is developing inside of the growing bone areas after 4 weeks. At the same time the grade of marrow inside of the bone areas is increasing to >70%.
5. When *granulated* P. was implanted with autogenic marrow in the quadriceps muscle, only small bone areas developed in early phases (up to 4weeks). But these bone areas don`t persist because of instability of implant and host tissue.

After 6 weeks only fibrous connective tissue exists between and around the granules.
6. After implantation of Pyrost in the iliac crest we observed a slower bony reconstruction: It only occurs by invasive bone formation from the adjacent bone tissue. Reconstruction is complete after 6 weeks.
Altogether there is a significant difference between the different test situations with regard to area resp. velocity of bone formation.

Discussion:

Only the combination of the cancellous stable ceramic implant with autogenic marrow was able to produce permanent ectopic bone formation - but not the granulated form of the same material.
The implant stimulates bone formation significantly (osteostimulation), but it`s not osteoinductive. It needs bone contact or marrow inoculation for bone formation. The velocity of bone formation depends on the contact surface to the intact bone bed and on the osteostimulation by the implant. The bone areas are expanding inside of the cancellous implant depending on time. Marrow volumes arise only within the new formed bone areas. Marrow is functioning as the impulse, Pyrost as a suitable casing for bone formation.

Osteoconduction/ osteostimulation/ osteoinduction, suitable cancellous structure and sufficient biomechanical stability are essential presuppositions of a regenerative bone substitute, but the result of bone formation depends likewise on a fitting host tissue.

Based on these investigations the method of marrow inoculation with Pyrost is meanwhile introduced in clinical application.

References:

(1) Burwell, R.G.: Osteogenesis in cancellous bone grafts. Clin Orthop 1965, 40:35
(2) Katthagen, B.-D.: Bone regeneration with bone substitutes. Springer, Berlin Heidelberg New York, 1986
(3) Lexer, E.: Die freien Transplantationen, Neue Deutsche Chirurgie 1924, 26:15
(4) Mittelmeier, W.: Knochenneubildung im ersatzschwachen Lager. In Osteoplastiken und artifizielle Knochenregeneration bei der Osteogenese, Demeter, München-Gräfelfing, 1986
(5) Schweiberer , L.: Die Bedeutung der autologen Spongiosatransplantation sowie Fragen der Revascularisation von Transplantaten. Nova Acta Leopoldina 1976, 228:371

Fig. 1

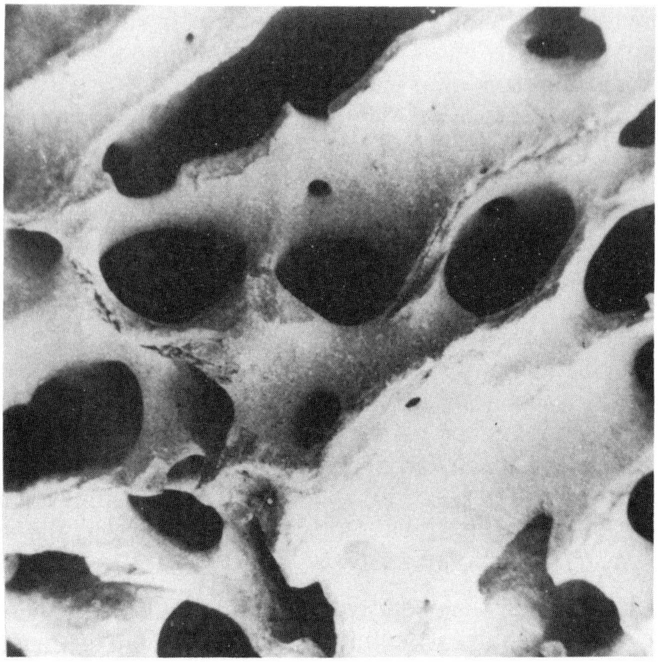

Fig.1: Microstructure of the ceramic bone substitute Pyrost. Cancellous structure with large pore sizes of 300 - 1500 μm and wide connecting passages.

Fig. 2

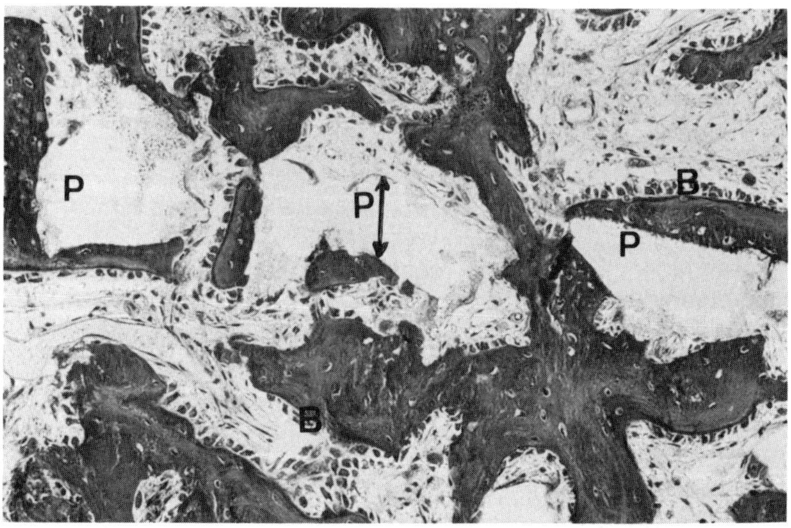

Fig. 2: Pyrost (P) in bore hole defect of the left femoral condyle after 2 weeks (150 x). Dense bony growth (B) inside of the implant and filling up the defect. Intimate contact between new formed bone and implant.

Fig. 3

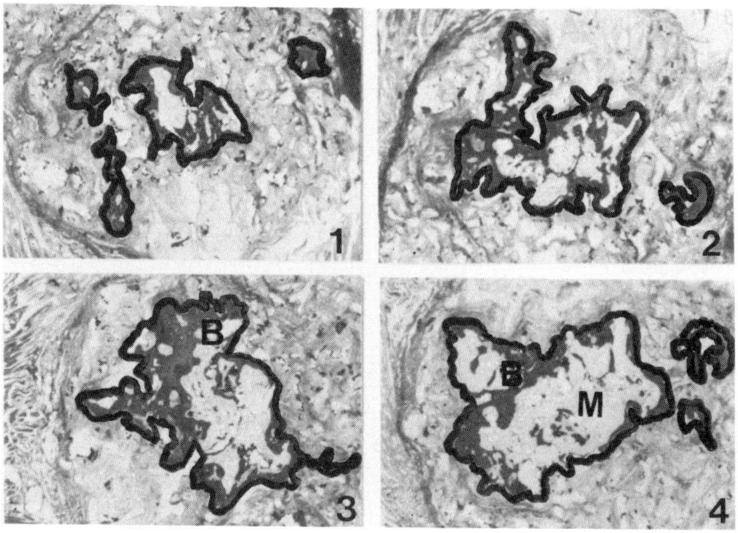

Fig.3: Implant with autogenic marrow in the quadriceps muscle (12 weeks, 8x). 4 cross section slides of the same specimen show the 3-dimensional structure of new bone area. Bone formation within the implant volume is slower, but also intensive and complete. Marrow (M) developing inside of the growing bone areas.

Fig. 4

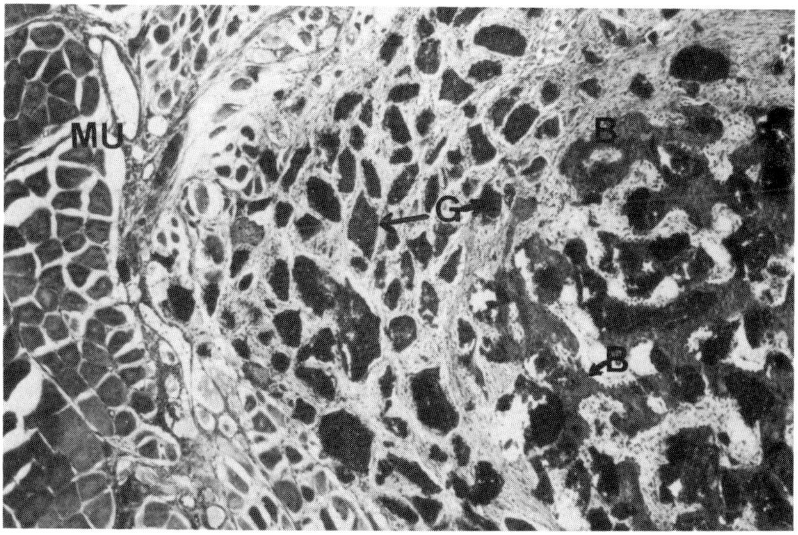

Fig.4.: Pyrost in granulated form (G) in quadriceps muscle (4 weeks): Bone areas only developed in early phases (up to 4weeks). But these bone areas don`t persist because of instability of implant and host tissue (muscle=MU).

Bioceramics, Volume 7 Edited by Ö.H. Andersson and A. Yli-Urpo
(Proceedings of the 7th International Symposium on Ceramics in Medicine, Turku, Finland, July 1994)

Bone Reconstruction and Revascularization of Total Diaphysis Defects by Natural Derived Ceramic Bone Substitutes – Animal Study

W. Mittelmeier, T. Braun, G. Giebel and R. Gradinger

Orthopädische Klinik, Medizin. Universität zu Lübeck, Ratzeburger Allee 160, 23538 Lübeck, Germany

Traumatologie & Institut f. Experimentelle Chirurgie, Univ. des Saarlandes, 66424 Homburg, Germany

Abstract:

Two different natural derived bone substitutes with a cancellous structure (*Pyrost, Bio-oss*) were implanted in complete diaphyseal defects of 36 rabbit femora and 12 rat femora after combination with autogenic marrow.
Stabilization was performed by plate osteosynthesis.
The complete reconstruction of the bone defects occured after 6 - 10 weeks (average 8 weeks) The implants therefore were completely histologically and biomechanically integrated in the bony microstructure.

Introduction:

In many clinical situations we are forced to reconstruct major bone defects, when osteogenetic potency of the tissue bed is limited and damaged.
Autogenic bone often is not available in sufficient quantity, particularly during childhood. For reasons of transplantation immunology and revascularization the effectiveness of allogenic transplants is limited.

Different studies in regenerative bone substitutes tested only a osteostimulative or osteoconductive capacity of different materials in suitable tissue beds. Consequently it is a demand for valid data about the applicability and apptitude of regenerative bone substitutes in situations of high weight bearing in the diaphysis of long bones.

Therefore different bone substitutes were introduced during the last years.

For some bone substitues as Pyrost and Bio-oss an osteostimulative effect in orthotopic bed was shown in animal experiments. But the reconstruction of segmental diaphysial defects with bone substitutes is not established until now.

Materials:

Pyrost and Bio-oss. Both materials are derived from bovine bone; both have a cancellous structure as natural bone with wide pores and wide interconnecting channel system; both are consisting chiefly of calcium-phosphates. Pyrost is completely deproteinized and sintered by a burning procedure, it consists of ceramic material, chiefly of hydroxiapatite (93 % HA, 7 % ß-TCP) *(1).* Bio-oss contains non-ceramic calcium-phosphate (55 %, mostly HA) and a residue of proteines (<135 ppm) *(4);* it has >3 x less compression stability. In first animal experiments both materials showed a stimulation of bone ingrowth within the orthotopic bone bed *(1,3).*
By means of combination with autogenic marrow bone regeneration could be achieved and advanced also in the ectopic areas *(3).*

Method:

Central complete diaphysis areas were resected including periost and marrow in 36 adult rabbit femora (1.2 cm defect length) and 12 adult rat femora (0.6 cm d.l.). The bone substitutes were inoculated by autogenic marrow and inserted in the defect area. The autogenic marrow was prepared from the resected bone area and inoculated on the implant material after dilution by Ringer´s solution (1:2). Stabilization was performed by plate osteosynthesis (6 screws) on lateral side.
Histology was performed with decalcified bone preparation after 2/ 8/ 24 and 48 weeks in 70 μm sections. Each specimen was inspected in 10 cross section slides and investigated by automated image-analysis system.

Results:

The histological examination and microangiograms showed an early and nearly complete vascularization after 2 weeks inside of both implants in rats and rabbits. A complete reconstruction of the bone defects could be observed after 8 weeks, if no luxation (2 cases/ Pyrost: 1 in rabbit, 1 in rat) or fracture of the implant (5 cases/ Bio-oss: 3 in rabbits, 2 in rats) occured. The implants therefore were histologically and biomechanically completely integrated in the bony macrostructure and microstructure: "functional integration":

In short time results a thick pattern of new formed bone tissue inside and around the implant was developping. New mineral bone trabeculae could be detected directly on the implant trabeculae without any separating interface layer. In early phases of regeneration the ingrowth and the orientation (direction of trabecular structures, density) of the new formed bone were not directed by biomechanical pattern.

After 8 up to 48 weeks the bony tissue inside of and around the implant was characterized by biomechanical aspects (histomorphometric analysis). Broad bony pillars were built to transfer the axial and bending forces between implant surface and cortical bone at the medial side. These stabilizing pillars extended mostly through the medial part of the implants with multiple direct contacts to the implant trabeculae. Under the plate (lateral) bone regeneration was performed in long time results by cancellous pattern with relatively small density of trabeculae - like patterns of fracture healing under a plate.

Discussion:

By use of the presented materials it is possible to reconstruct a diaphyseal bone defect in a good vascularized orthotopic tissue bed quickly and completely (osteoconduction, osteostimulation, biocompatibility). The cancellous structure of the implants with a wide interconnecting pore system and an own stability advance obviously the fast ingrowth of bony tissue and integration. There was no significant difference in bone formation rate comparing the both materials.

Bone regeneration with both implants chiefly was performed by aspects of high biomechanical strength. As shown in these animal experiments with early full weight bearing, the application of suitable bone substitutes in complete large bone defects has its limits in biomechanical aspects, but not in the tissue integration. Necessary is a complete and fast ingrowth of bony tissue into the wide interconnecting pores of the implant, but also an initial high stability against compression and bending forces combined with a suitable degradation rate. By ceramic procedure Pyrost obtained a higher compression stability, but also a slower degradation activity.

References:

(1) Katthagen, B.-D.: Bone regeneration with bone substitutes. Springer, Berlin Heidelberg New York, 1986

(2) Mittelmeier, H. et al.: Neue Wege des Knochenersatzes. Ortop Praxis, 1984, 20, 389f.

(3) Mittelmeier, W.: Knochenneubildung im ersatzschwachen Lager. Demeter, München-Gräfelfing, 1992.

(4) Schlickewei, W.: Experimentelle Untersuchungen zum Knochenersatz mit bovinem Apatit. H Unfallheilkd 1991, 216, 59-68.

Fig. 1

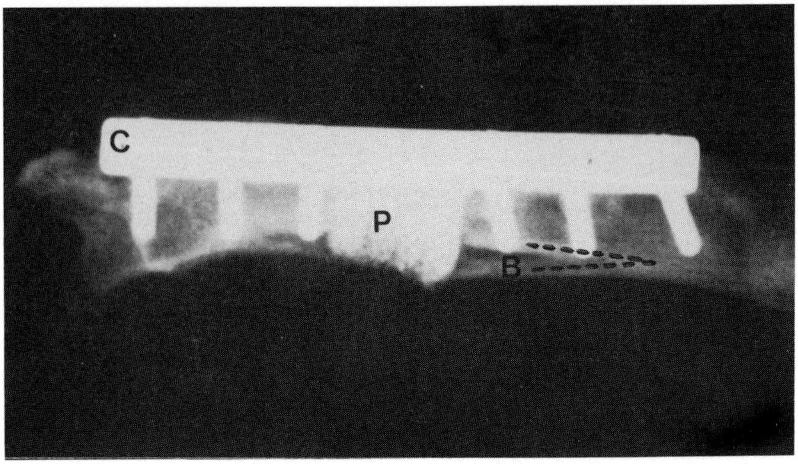

X-ray after 8 weeks postop. in a complete defect of femoral diaphysis (rat, Pyrost): Bone formation and integration inside of and around the implant (P) by biomechanical criteria. Bone formation (B) as cranial and distal osseous bridges and pillars to the implant surface.

Fig. 2

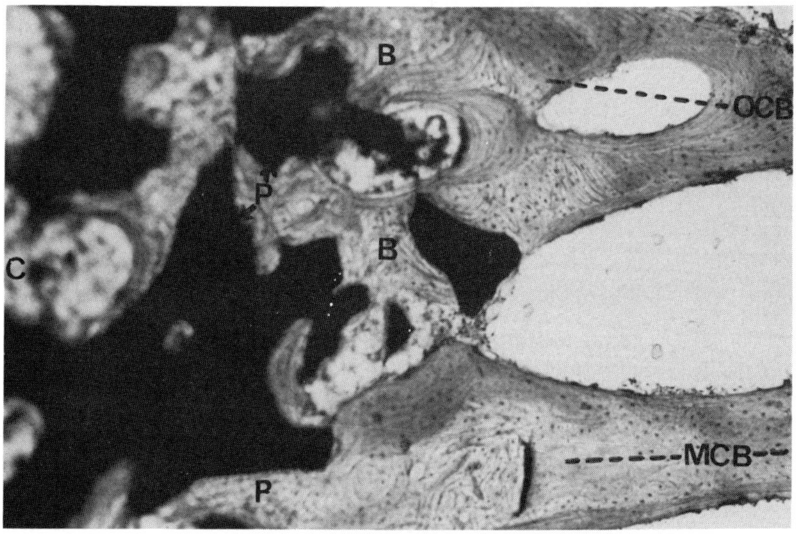

Histology: rabbit, 24 weeks (Pyrost) : Medial-distal area of the implant. Original cortical bone (OCB) in continuity to the implant surface (P) with direct contact. Orientation of modified cortical bone (MCB) like a pillar (y-shape) after remodeling by biomechanical pattern. Deep ingrowth of new bone (B).

FIG. 3

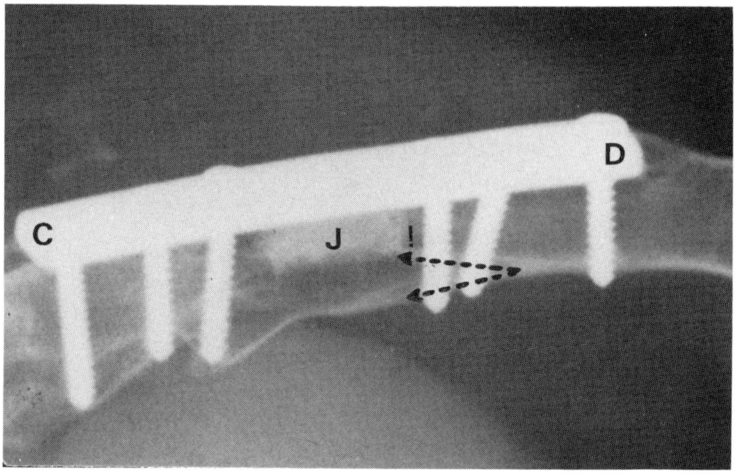

FIG. 4

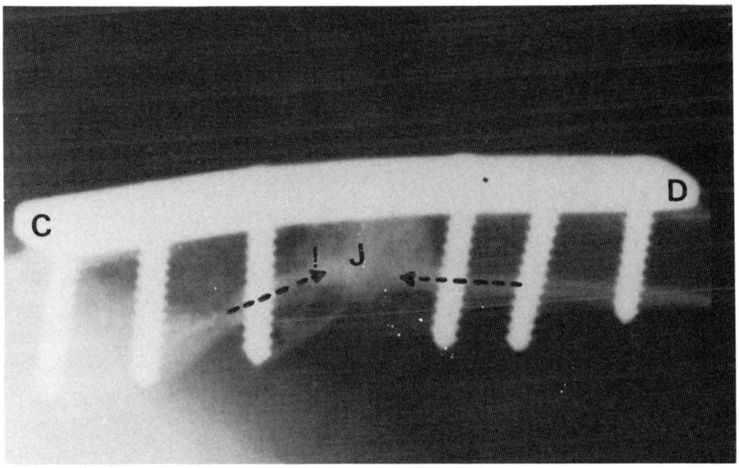

After 24 and 48 weeks postop. (rabbit, Bio-oss): The continuity of the medial cortical bone is reconstructed by a dense bone belt through the medial side of the implant (I). After 48 weeks complete integration in osseous host tissue. (C=cranial/proximal, D=distal).

Bioinert and Wear-Resistant Materials

Bioceramics, Volume 7 Edited by Ö.H. Andersson and A. Yli-Urpo
(Proceedings of the 7th International Symposium on Ceramics in Medicine, Turku, Finland, July 1994)

Medical-Grade Alumina During the Past Two Decades

G. Willmann
Cerasiv GmbH, Innovatives Keramik-Engineering, Medical Products Division, Fabrikstrasse 23-29, D-73207 Plochingen, Germany

ABSTRACT

Medical-grade alumina has been used in various clinical applications for the past two decades. The primary applications have been in orthopedics for the femoral ball head for hip endoprostheses. Up to 1994 more than 2 million heads have been used worldwide. The following topics will be discussed: Properties and design of ceramic femoral ball heads for total hip replacement, safety factors, standards and government regulations, and future development.

The development of aluminium oxide (alumina) ceramics which form part of the range of modern engineering ceramics started in Germany in 1974 with a project promoted by the German BMFT. Participating in this project were industrial companies, physicians and institutes. At that time, the BMFT had insisted on the interdisciplinary design of the project, and hence material scientists, veterinary surgeons, orthopaedists and the manufacturers cooperated in the development of alumina ceramics from the start. The first objective were components for artificial hip joints, e.g. the ball-and-socket joint of hip joint prostheses. First, the material was selected and then the constructional design of the ball-and-socket joint elaborated. From the start, the people participating in the project believed that any artificial ball-and-socket joint consisting of a ball head in alumina and a socket should be completely made of alumina ceramics due to the basic properties of alumina ceramics. The ceramic companies participating in the project were: Feldmuehle, Rosenthal-Technik and Friedrichsfeld. The most important pioneers in the field of ceramic materials were Erhard Dörre and Günter Heimke. Today, alumina ceramics as specified by ISO 6474 are generally approved as biomaterials for use in endoprosthetics. After alumina ceramics had stood the test for use in hip joint prostheses, people thought about using alumina ceramics in other articular endoprostheses (1-4).

Table 1 Important properties of medical-grade alumina according to ISO 6474

Basic property	important because of
high resistance to corrosion	bioinertness
excellent hardness and good surface finish	high wear resistance
high Young's modulus and high compression strength	stiffness, no deformation
high mechanical strength	good fatigue properties, safety, reliability
high purity material	long-term stability

The applications specified below are known or have been investigated: ball heads of hip joint endoprostheses, sockets of hip joint endoprostheses, socket inlays of hip joint endoprostheses, femoral or tibial components of knee prostheses, shoulder prostheses and finger prostheses.

The use of alumina ceramics in endoprosthetics is based on the **resistance to corrosion** of alumina (5) (see table 1). Resulting from this is the material's **bioinertness** which is most important in medical applications (3).

The material's **hardness** is extremely high. Resulting from this is the material's **low wear** which is most important in medical applications. This property is particularly significant when it comes to ball heads of hip joint endoprostheses.

The **elasticity modulus** is 380 GPa (much higher than steel). As a result, the material provides for **rigid and non-deformable** implants to be produced for use in endoprosthetics and dental implantology.

Alumina has **high mechanical strength.** Implants are always designed as load-accepting implants. The implants remain in the body over many years, and therefore must not be susceptible to possible reductions in strength, which could affect their load-carrying function.

The **high resistance to compression** of alumina is very important. Especially in endoprosthetics, e.g. in hip joint prostheses, high loads must be accepted over long times. This means that the prosthetic components must not be susceptible to elastic or plastic deformations, since such deformations would cause the wear of the prosthetic components to increase immediately.

Implants are exposed to relatively strong corrosion influences. Normally, glass phases are found on the grain boundaries in ceramic materials, which are due to impurities imported through the raw materials. The glass phases will degrade in the body and cause the material to loose its mechanical strength. Following from this is that top priority must be attached to the chemical purity of the alumina used.

Experience has shown that there are two approaches possible in the production of ball-and-socket joints for hip joint endoprostheses (1, 2, 6):

a) Alumina heads running in sockets made from polyethylene (UHMW-PE).

b) Alumina heads running in alumina sockets or in alumina socket inlays.

Approach No. a) (using polyethylene sockets) provides for low material wear to be achieved, and enables the long-term use of alumina ball heads and polyethylene sockets in the body. Approach b) using a combination of alumina heads and alumina sockets or alumina socket inlays provides for minimum material wear to be achieved, and hence is to be preferred over approach No. a). From the physician's

point of view, handling artificial joints with sockets in polyethylene is much easier.
The alumina ceramic material used in medical applications consists of alumina (Al_2O_3) that has been doped with magnesium oxide (MgO) as specified by ISO 6474 in 1981 and presently is under revision (7).
Apart from the wearing behaviour, the material's mechanical strength is another highly significant parameter which must be optimized. There is a close correlation between the mechanical strength and the density, the median grain size and the glass phases available on the grain boundaries. The density must be maximized, the grain size must be kept small and the share of glass phases available on the grain boundaries must be minimized. This means that high-purity raw materials must be used, which today are available in the market, and that the production process used must provide for reproducable results (see table 2).
Apart from the material's long-term mechanical strength, the most important property consists in the minimum wear of the alumina ceramic parts or their counterparts, i.e. the alumina ceramic ball. Apart from the hardness of alumina and the surface quality of the ceramic ball heads (average peak-to-valley height of 0.01 μm), the capability of adsorbing polar molecules (e.g. water, body fluids) is of utmost significance. Owing to this capability, alumina ceramics in the body adsorb the polar molecules available there, thus creating the basis for the formation of a moistening liquid film which in the end provides the lubricant required between the ceramic ball head and the polyethylene socket, or between the ceramic ball head and the ceramic socket or socket inlay. If this liquid film is interrupted as a result of excessively high loads or insufficient tolerances, the material wear will increase.
Despite all the favourable properties of alumina it should be noted that alumina is brittle (5). When using the material's fracture toughness as a mesure of its brittleness, it is plain to see that the fracture toughness of alumina is smaller than the one of cast iron, which means that ceramics are more brittle than cast iron.

Table 2 Properties of medical-grade alumina

Property	Units	ISO 6474*	ISO 6474**	BIOLOX
Purity	% in weight	-	-	> 99.9
Apparent density	g/cm^3	≥ 3.90	≥ 3.94	> 3.95
Al_2O_3 contents	% in weight	≥ 99.5	≥ 99.5	> 99.7
SiO_2+ alkali oxides	% in weight	≤ 0.1	-	-
SiO_2	% in weight	-	< 0.01	< 0.01
Na_2O	% in weight	-	< 0.01	< 0.01
$SiO_2+Na_2O+CaO+Fe_2O_3$	% in weight	-	< 0.1	-
CaO	% in weight	-	-	< 0.01
Fe_2O_3	% in weight	-	-	< 0.015
Grain size	μm	≤ 7	≤ 4.5	3
Bending strength	MPa	≥ 400	≥ 450	> 550

* 1981 ** 2nd ed. to be edited

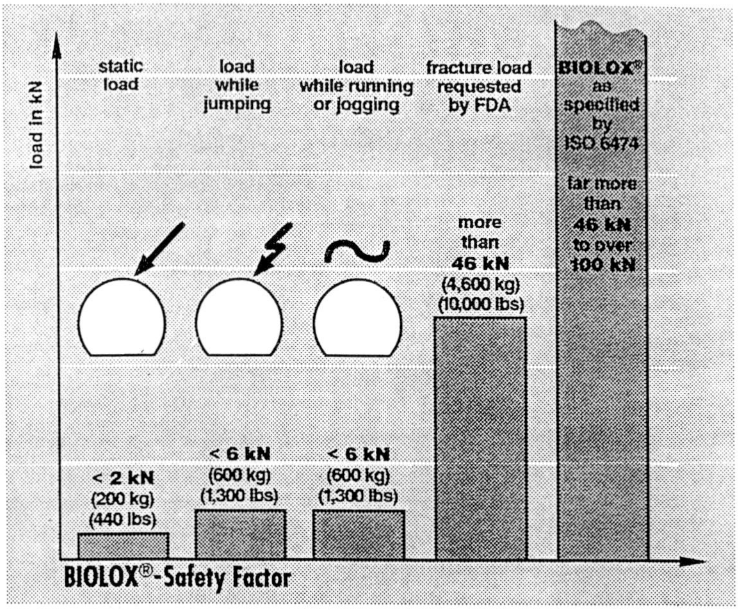

Figure 1 Physiological loads compared to fracture loads of ceramic ball heads

The modular structure of hip joint endoprostheses requires the ceramic ball head to be fixed onto a metal cone by way of taper fitting. This method of fitting the stem of the hip joint endoprosthesis to the articular ball head of the hip joint endoprosthesis was found to be particularly suited for metal components. It is generally requested by orthopaedic surgeons that ceramic ball heads too be fitted to the prosthetic stem in the same way. Using this method for ceramic ball heads also is possible by providing a conical bore on the ceramic ball. For this, the cone must be shaped at an angle of 1:10 in accordance with the German standard specifications DIN 254 and DIN ISO 3040. The cone angle resulting from this is 5° 42.5 minutes.

Such concept has proved to be successful and has been used since two decades. The failing loads of ceramic balls achieved with this concept correspond to a safety factor of 10 to 50 times the body weight (see Fig. 1).

In Europe, the socalled Euro-cone (12/14) has proved to be particularly successful. Basically speaking, it is possible for this reason to use articular ball heads of any manufacturer together with the prosthetic stems of different manufacturers.

For production of alumina ball head see (8).

Modern hip joint endoprostheses are designed according to the principle of separation of functions, which means that the best properties only are asked of the biomaterials used and known today. In respect of the hip joint endoprosthesis this means (see Figure 2).

a) The load-carrying functions and especially the functions subject to high alternating loads are assumed by titanium- or cobalt/chromium-alloys. This means that the stem of the endoprosthesis and the socket are made from such alloys.

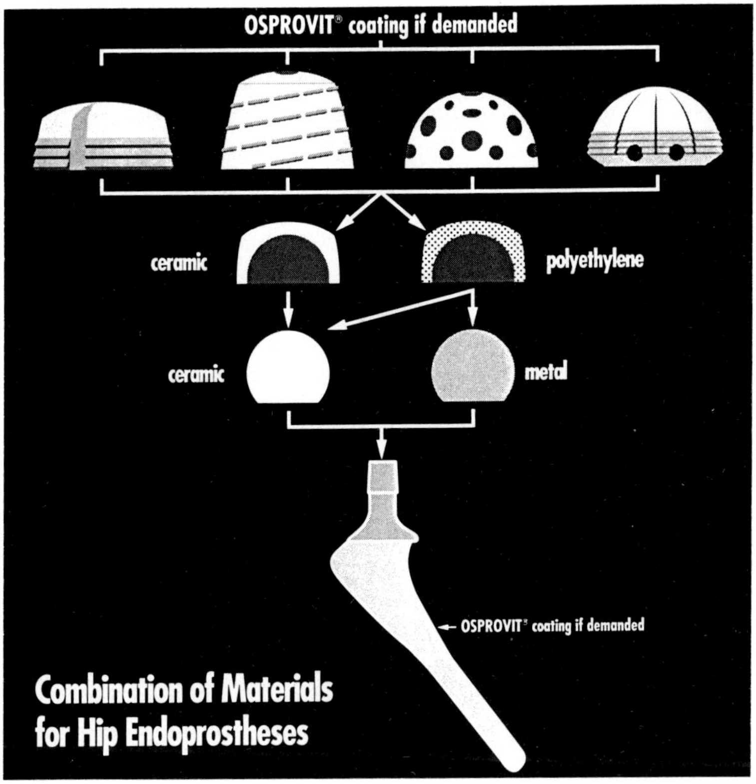

Figure 2 Optimization of hip-systems by designing according to the principle of separation of functions (see text) (OSPROVIT = Hydroxyapatite plasma coating)

b) The ball head is made either from alumina or from cobalt/chromium. The surgeon is free to choose the type of ball he wishes to use.
c) Today's sockets are modular sockets and consist of two components. The external span of the socket - which may be any design - is made from bioalloy. The external span will accept the inlay in alumina ceramics or in polyethylene.
d) The socket and the stem alike may be coated with bioactive hydroxyapatite (4) which due to its osseoconductive properties fosters the adhesion of the bone to the prosthetic components (see Fig. 2).
e) The following material combinations may be used for the configuration of articular ball heads and socket inlays: **cobalt/chromium ball head and polyethylene inlay,** which represents a good combination in respect of wear. **Ceramic ball head and polyethylene inlay,** which represents an even better combination in respect of wear. **Ceramic ball head and ceramic socket inlay,** which represents the best combination in respect of wear. The surgeon is completely free to choose the combination he prefers, as long as the prostheses are designed according to the constructional principle described above.
The product liability or the principles for its interpretation are based on the state of the art. Today's state of the art is documented by the total of accessible information (literature, books, congresses), state regulations (Federal Health Board, FDA,

MHW) and laws. Since business is done at a more and more international level today, the standard specifications referred to in particular are the German DIN standards, the British standards (BS) and the American ASTM standards. The production process used in industrial companies manufacturing components for the medical industry generally is based on the principles of Good Manufacturing Practise (GMP) and the ISO standard specification 9001.
Generally, it is ensured by the regulations of the FDA, the international standards and the GMP directives in combination with ISO 9001 that only safe and well-tested implant components are available in the market.
CERASIV (formerly Feldmuehle) have started to face these requirements at a very early stage, and exclusively supply components in alumina ceramics as specified by ISO 6474, which are manufactured in compliance with FDA recommendations and with ISO 9001. Since 1974, the company has produced and supplied more than 1.5 million (1994) BIOLOX heads which have been inserted successfully.
There is a growing market for ceramic implants, which from the point of view of industrial companies is quite attractive. Apart from components in alumina ceramics as specified by ISO 6474, components in zirconium oxide (ball heads almost exclusively) have been on offer for some time. From the point of view of material scientists, zirconium oxide ceramics are particularly interesting because they are new and because of their remarkable mechanical strength and their high fracture toughness which is higher than the one of alumina ceramics.
There are different types of zirconium oxide available in the market today: Mg-PSZ or Y-PSZ or Y-TZP. At present, it is still unknown which of the 3 material groups will be qualified in the end for use in medical applications (8).

REFERENCES

1. Ravaglioli, A. and Krajewski, A. Bioceramics-Materials, Properties, Applications, 1992

2. Heimke, G. In Concise Encyclopedia of Medical & Dental Materials Pergamon Press, Oxford, 1990, 28

3. Black, J. Biological Performance of Materials 1992

4. Willmann G. In Technology and Health Care Elsevier, Amsterdam, 1993, 141

5. Morrell, R. Handbook of Properties of Technical & Engineering Ceramics 1985

6. Hench L.L. In Amer. Ceramic Soc. Bull. 1993, 4, 93

7. Willmann, G., In Zahnärztl. Praxis, 1990, 41, 286

8. Clarke I.C., Willmann G. In Bone Implant Interface Mosby, St. Louis, 1994, 203-252

Bioceramics, Volume 7 Edited by Ö.H. Andersson and A. Yli-Urpo
(*Proceedings of the 7th International Symposium on Ceramics in Medicine, Turku, Finland, July 1994*)

Treatment of Malignant Bone Tumours by Replacement with Alumina Ceramic Prostheses. Long Term Follow-up of Two Cases

A. Suda, T. Sato, A. Ishikawa, M. Takagi and T. Osanai
Department of Orthopedic Surgery, School of Medicine, Yamagata University, Iida-Nishi 2-2-2, Yamagata City, 990-23, Japan

ABSTRACT

We have experienced 14 cases treated by replacement with alumina ceramic prostheses after wide resection of malignant bone tumors. Radiological and clinical studies were performed by us in two cases having had long-term process after operation (case 1:fibrosarcoma of the left femur, twelve years after operation, case 2:osteosarcoma of the right femur, nine years after operation). Radiograms of case 1 showed tapering and atrophy of the distal cortex of the preserved femur within one year after replacement with prosthesis of lower half length of the left femur, sinking of several mm length of the prosthesis, anterior movement of 1 mm length of the proximal end of the prosthesis, and overbridging by bone-formation on the contact region on the prosthesis and the femur. Radiograms of case 2 showed tapering and atrophy of the distal region of the femur within several months, but did not show sinking of the prosthesis. Bone atrophy disappeared six months after operation. Range of motion of the knee was zero to one hundred degrees in the case 1 and zero to one hundred and fifteen degrees in the case 2. Function of knee joints was satisfactory in two cases.

INTRODUCTION

Recently, we tend to choice the limb salvage for most of primary malignant bone tumors. Replacement with artificial bones were often used for the limb salvage after wide resection of malignant bone tumors. It is known that alumina ceramics have more biocompatibility than metal under static or dynamic condition[1]. We can not find a report of cases with a long-term follow-up after replacement with alumina ceramic implants for malignant bone tumors. Radiological and clinical studies were performed by us in two cases having had long-term process after replacement with alumina ceramic prostheses for the limb salvage by wide resection of malignant bone tumors.

CASES

Case 1:Y.K. 45 years old, male.
In April, 1981, he complained of pain of the left knee. In Nov. 1981, osteolytic lesion and partial destruction of the cortex was observed in the distal metaphysis of the left femur in the radiograms(Fig.1). The pathological diagnosis of fibrosarcoma was obtained under the light microscopic observations of specimens by biopsy from the distal metaphysis of the left femur. Preoperative chemotherapy by intra-arterial infusion with adriamycin, mitomycin, 5-FU, and vincristine was very effective. The distal femur of 21 cm in length including tumor was resected after preoperative chemotherapy, and the replacemnt of distal femur and knee joint with alumina ceramic prostheses, which had the shaft of 13 mm in diameter and 21 cm in length, was done on Dec. 24, 1981. The external fixation with plaster cast was done for one month after operation. Then, the active ROM exercise of his knee joint was began. He began weight bearing and walking with a knee brace six weeks after operation.
Radiograms of case 1 showed tapering and atrophy of the distal cortex of the preserved femur within one year after replacement with prostheses of lower half length of the left femur, sinking of several mm length of the prosthesis, anterior movement of 1 mm length of the proximal end of the prosthesis, and overbridging by bone-formation on the contact region on the prosthesis and the femur. Sinking of the prosthesis was not recognized since one year after operation.
Twelve years after operation, range of motion of the left knee is zero to one hundred degrees and functional evaluation by Enneking[2] is excellent without disease(Fig.2, Fig.3).

Case 2:Y.Y. 19 years old, female.
In Sept. 1984, she complained of pain and swelling of her right knee. Radiograms showed irregular sclerosing and partial destruction of the cortex in the distal metaphysis of the right femur(Fig.4). Since the pathological diagnosis of osteosarcoma was obtained by biopsy, preoperative chemotherapy by intra-arterial infusion with adriamycin and

vincristine, and by intra-venous infusion with high dose methotrexate was done. The replacement of distal femur and knee joint with alumina ceramic prostheses, which had the shaft of 11 mm in diameter and 10 cm in length, was done on May 9, 1985.

Radiograms showed tapering and atrophy of the distal region of the femur within several months after operation, but, did not show sinking of the prosthesis, Bone atrophy disappeared six months after operation. Nine years after operation, range of motion of the right knee is zero to one hundred and fifteen degrees and evaluation by Enneking[2] is excellent(Fig.5, Fig.6).

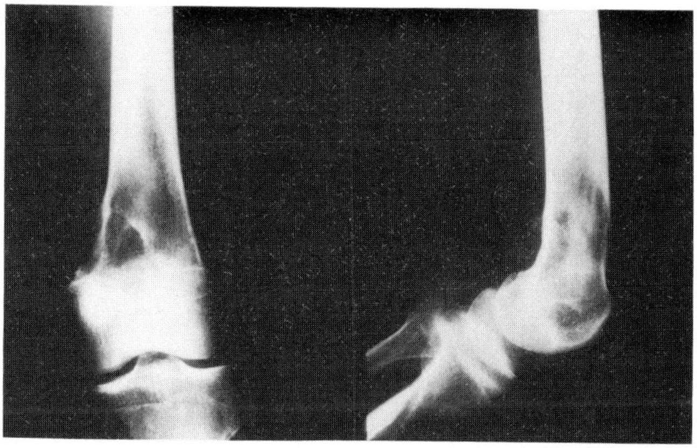

Fig.1 Preoperative radiograms (case 1)

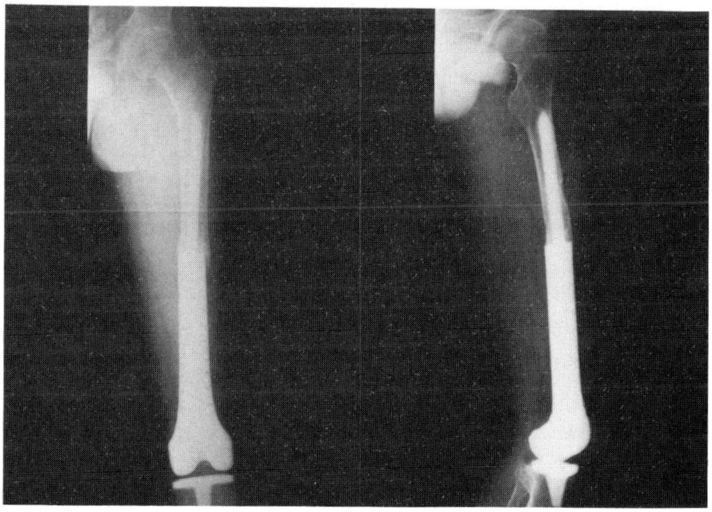

Fig.2 Postoperative radiograms (case 1)

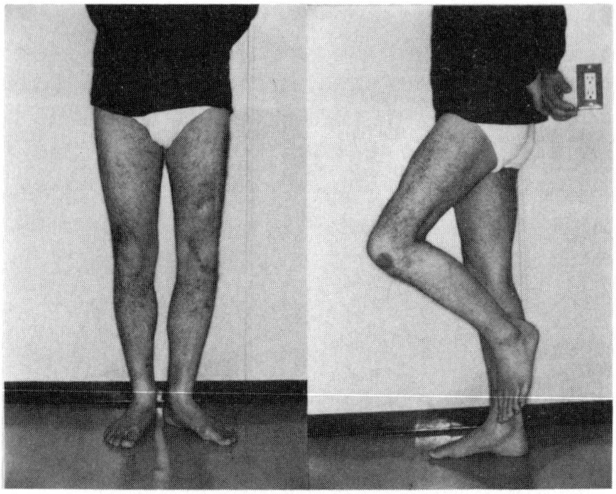

Fig.3 Twelve years after operation (case 1)

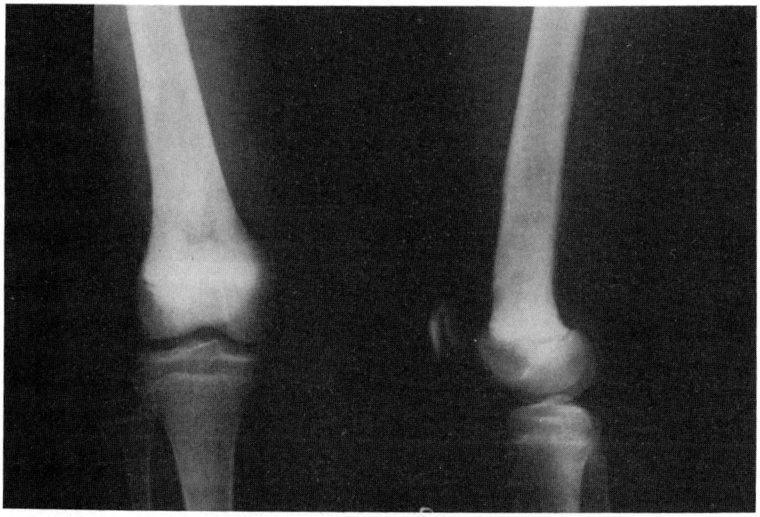

Fig.4 Preoperative radiograms (case 2)

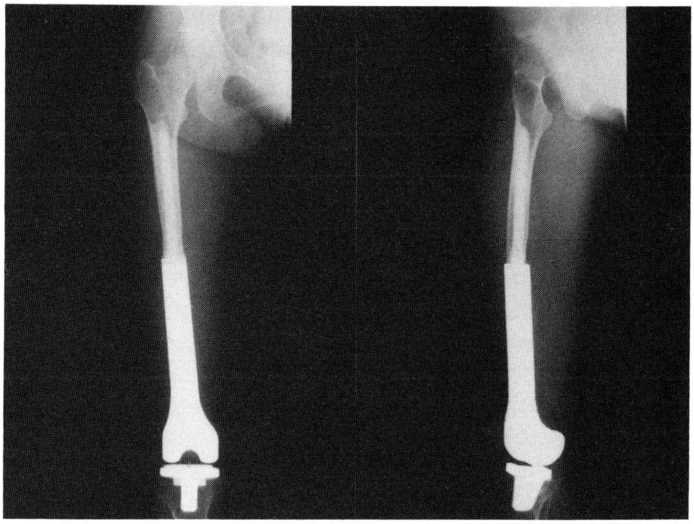

Fig.5 Postoperative radiograms (case 2)

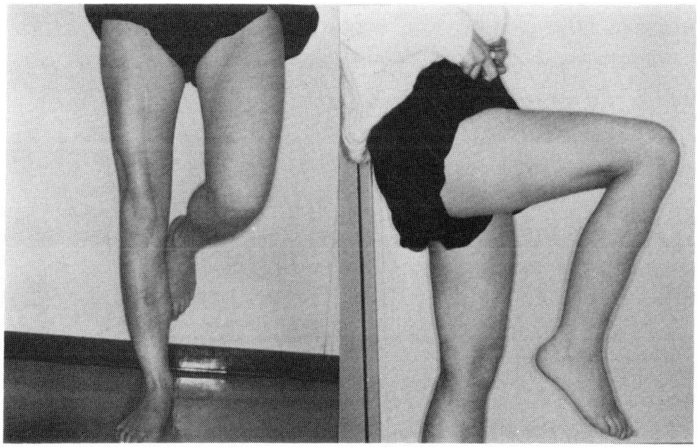

Fig.6 Nine years after operation (case 2)

DISCUSSION

It is known that alumina ceramics have more biocompatibility than metal under static or dynamic condition[1]. We have experienced 14 cases treated by replacement with alumina ceramic prostheses after wide resection of malignant bone tumors[3]. Radiological and clinical studies were performed by us in two cases having had long-term process after replacement with alumina ceramic prostheses for the limb-salvage operation. Radiograms showed tapering and atrophy of the distal cortex of the preserved femur, or sinking of the prosthesis in the early stage after operation, but these findings were not recognized since at least one year passed after operation. It seems that the damage of periosteum by operation caused tapering and atrophy of the distal cortex of the preserved femur at the early stage after operation, however, that some stimulation of the distal cortex by weight bearing led to the improvement of tapering and atrophy of bone, and to the overbridging by new bone-formation. Anyway, the external fixation by plaster cast at the early stage after operation will be important in order to get the good fixation of the prosthesis in the bone.
We have successfully developed new ceramic material, using a unique technique, which dispersed 50 vol% zirconia in the hydroxyapatite matrix. The new ceramic material was shown to have an excellent affinity for the bone and the excellent mechanical strength under static or dynamic condition[4,5]. We expect that our new bioactive ceramics will contribute to the development of the prostheses.

REFERENCES

1)Hattori,H.,Matsunaga,T.,Sato,S.,Hanai,K. and Akahoshi, Y.:Orthopaedic Ceramic Implants. 1982, 2, 9-13
2)Enneking,W.F.:The second International Workshop on Design and Application of Tumor Prosthesis for Bone and Joint Reconstruction:Vienna 1983
3)Suda,A.,Watanabe,Y. and Takahashi,N.:Orthopaedic Ceramic Implant. 1983, 3, 107-112
4)Suda,A., Sato,T., Takagi,M., Osanai,T. and Suzuki,O.:J. Jpn Orthop. Assoc. 1989, 64, 249-259
5)Suda,A.,Osanai,T.Ishikawa,A.,Takagi,M.,Tanabe,M.,Suzuki, O., Kakutani,K. and Harada,Y.:Bioceramics, 1992, 5, 191-196

Bioceramics, Volume 7 Edited by Ö.H. Andersson and A. Yli-Urpo
(*Proceedings of the 7th International Symposium on Ceramics in Medicine, Turku, Finland, July 1994*)

A Retrieval Study of Al_2O_3-Heads Running Against UHMWPE- and CFRP-Cups

H.J. Früh, R. Ascherl,* C. Kaddick, W. Siebels and G. Blümel
Institut für Experimentelle Chirurgie der Technischen Universität München
* Orthopädische Klinik der Medizinischen Universität Lübeck, Germany

ABSTRACT

The articulating surfaces of retrieved ceramic heads (n=37, Al_2O_3) implanted with cemented and non cemented UHMWPE-cups or cementless CFRP-cups (Carbon Fiber Reinforced Plastic) were compared using surface measurement techniques and Scanning Electron Microscopy (SEM). There were no surface changes regarding roughness, geometry (diameter and sphericity) and morphology of retrieved ceramic balls compared to virgin balls of the related product period. However this statement must be limited to the observed implantation times.

INTRODUCTION

Wear of artificial joint surfaces is one of the most important factors limiting the functional lifetime of total hip endoprostheses. To minimize wear debris Al_2O_3 according to ISO 6474 [11] is used since 1969 paired with Al_2O_3 [7,16], since 1975 paired with cups made from UHMWPE [7,15] and since 1988 articulating against cups made from CFRP [14]. The advantages of ceramic balls compared to metal balls were published in several studies [3,17]. Laboratory experiments [1,7,14] and in-vivo testings [2] showed the advantages of CFRP as a cup material due to minimized wear and mild biologic response to wear products. The purpose of this study was to determine the changes of ceramic balls running against different cup materials in total hip replacement.

MATERIALS AND METHODS

Thirty-seven retrieved heads (BIOLOX®, Al_2O_3, ∅ 32 mm) implanted with cemented and non cemented UHMWPE-cups (n=15) and with non cemented CFRP-cups (Carbon Fiber Reinforced Plastic, thermoset matrix with pitch based carbon fibers (CAPROMAN®), n=22) were investigated. No significant differences between the CFRP and the UHMWPE group concerning age, sex, side, body-height, body-weight and the reason for revision (stem loosening) were evident. The average time until revision was 22 (7 - 56) months for the CFRP group and 108 (15 - 192) months for the UHMWPE group. SEM was used to visualize the influence of the two cup materials on the surface morphology of the Al_2O_3 balls. Quantitative measurements were performed using a HOMMEL tester T4000 (Hommelwerke, Schwenningen, Germany) to determine the surface roughness (roughness parameters according to DIN 4768 and 4776 [4,5] rsp. ISO 3274, ISO 4287-1, and ISO 4288 [8,9,10] (cut off $\lambda_c = 0.08\mu m$): R_a, $R_{y,max}$, R_q, R_p, R_k, R_{pk}, R_{vk}, M_{r1}, M_{r2}). Using the parameters R_p, R_k, R_{pk} and R_{vk} it is possible to determine peaks and flutes not recognized by the CLA-value (R_a). The sphericity deviations were measured to compare the retrieved balls with their virgin size (δD: diameter deviation, δa_{30}: sphericity deviation in 30°, δa_{pole}: sphericity deviation at the pole).

RESULTS

In SEM of all surfaces smooth areas could be seen comparable to virgin balls (figures 1 and 3). Using a higher resolution (>1000x) polishing tracks caused by the manufacturing process equivalent to metal balls or balls of other ceramic manufacturers could be detected. The isolated holes or cavities visible in figures 2 and 4 are well known on ceramic surfaces. In the group with cemented cups (UHMWPE) no surface differences due to a possible interaction with bone cement particles (three body wear) were observed (figures 1 and 2). In this group slightly leveled boundaries on the surfaces of balls with longer implantation times could be discovered (figures 1 and 2).

The surface profiles of all balls showed similar appearance (figures 5 and 6). The roughness parameters R_a (CLA) and R_q and $R_{y,max}$ showed normal values of polished ceramic surfaces with smooth areas and isolated cavities (table 1). Sometimes the parameters were obviously higher than the acceptance ranges in ISO 4288 (range of cut off λ_c [10]). Due to cavities those values were excluded. The percentage of peak areas was evidently smaller than the percentage of flute areas: $R_{pk} < R_{vk}$, $R_{pk*} < R_{vk*}$. The parameter M_{r1} (percentage of peaks) and M_{r2} (percentage of cavities, $M_{r2*} = (100\% - M_{r2}) > M_{r1}$) support this finding (DIN 4776 [5] and ISO 4287-1 [9]).

The differences in diameter of the virgin balls in comparison to the balls after retrieval were within the reproducible measuring accuracy (table 1, δD). The results of the sphericity measurements showed high standard deviations due to small differences and no uniform tendency in the determined values (δa_{30}, δa_{pole}).

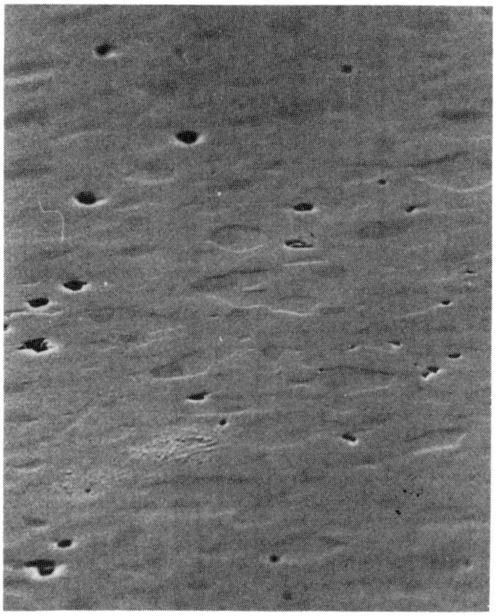

10µm

Figure 1 Surface of a retrieved ceramic ball (Al_2O_3) in SEM, running 168 months against UHMWPE (1000 x).

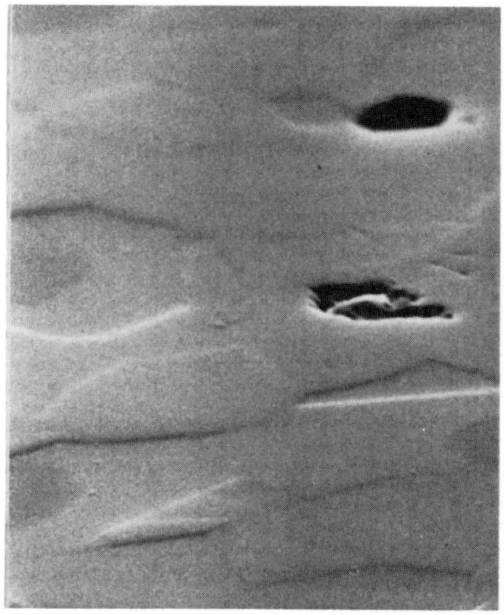

3µm

Figure 2 Surface of a retrieved ceramic ball (Al_2O_3) in SEM, running 168 months against UHMWPE (detail of figure 1, 5000 x).

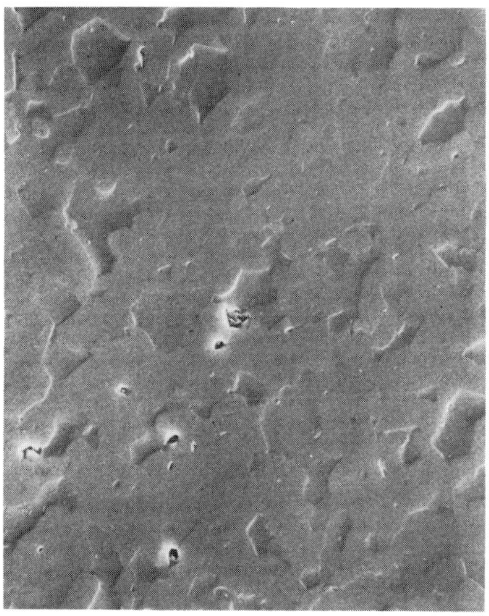

10µm

Figure 3 Surface of a retrieved ceramic ball (Al_2O_3) in SEM, running 28 months against CFRP (1000 x).

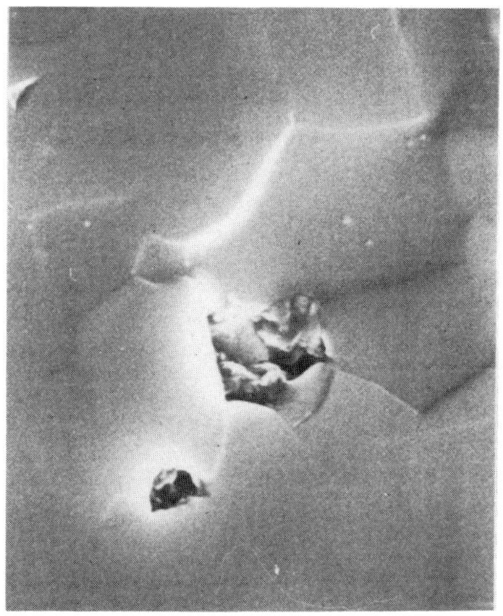

3µm

Figure 4 Surface of a retrieved ceramic ball (Al_2O_3) in SEM, running 28 months against CFRP (detail of figure 3, 5000 x).

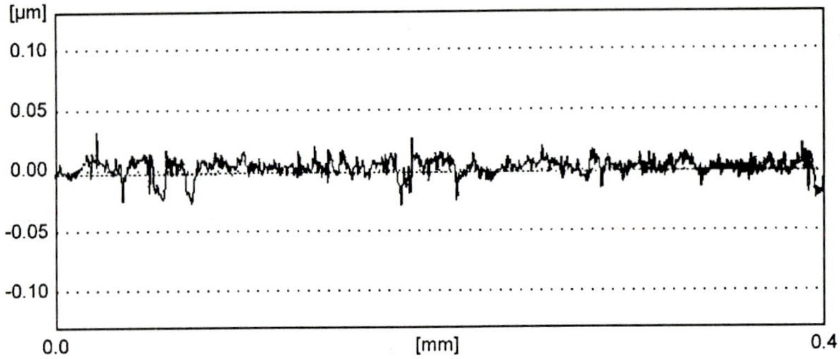

Figure 5 Surface roughness profile of a retrieved ceramic ball (168 months running against UHMWPE, $R_{y,max}=0.090\mu m$, $R_a=0.007\mu m$, figure 1).

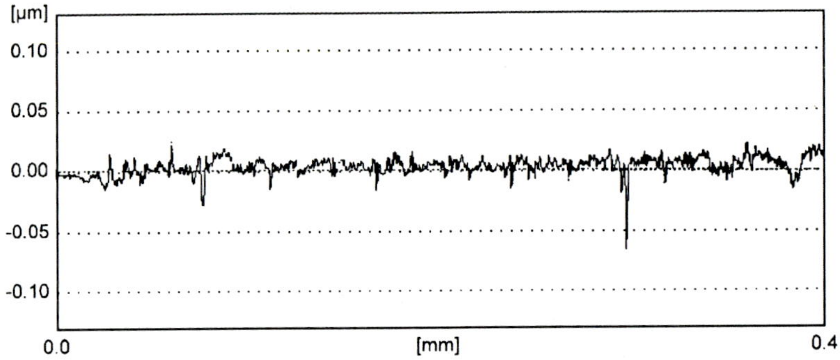

Figure 6 Surface roughness profile of a retrieved ceramic ball (28 months running against CFRP, $R_{y,max}=0.111\mu m$, $R_a=0.006\mu m$, figure 3).

Table 1 Roughness parameters corresponding to ISO 4287-1 and ISO 4288 and sphericity and geometry of the retrieved ceramic balls [MW ± SD].

Cup material	$R_{y,max}$ [µm]	R_a [µm]	R_q [µm]	R_p [µm]	M_{r1} [%]	R_{pk*} [µm]	R_{pk} [µm]
UHMWPE	0,172 ±0,088	0,010 ±0,005	0,015 ±0,007	0,067 ±0,039	6,6 ±2,2	0,044 ±0,036	0,016 ±0,015
CFRP	0,221 ±0,122	0,015 ±0,010	0,027 ±0,042	0,096 ±0,084	7,4 ±2,9	0,065 ±0,076	0,023 ±0,022

Cup material	R_k [µm]	R_{vk} [µm]	R_{vk*} [µm]	M_{r2} [%]	δD [mm]	δa_{30} [µm]	δa_{pole} [µm]
UHMWPE	0,026 ±0,010	0,041 ±0,015	0,130 ±0,075	84,9 ±4,1	0,002 ±0.005	0,1 ±1,3	0,8 ±2,3
CFRP	0,041 ±0,042	0,051 ±0,027	0,147 ±0,097	85,2 ±4,0	0.002 ±0.005	0,5 ±2,0	0,1 ±2,0

(δD: diameter deviation, δa_{30}: sphericity dev. (30°), δa_{pole}: sphericity dev. (pole))

DISCUSSION AND CONCLUSION

The roughness parameters showed smooth areas on the surface related to the findings in SEM. No surface differences due to three body wear by bone cement particles were observed comparing cemented and non cemented cups [3]. In the UHMWPE-group slightly leveled boundaries on the surfaces of balls with long implantation times could be detected. No differences of surface changes regarding the roughness of ceramic balls before and after implantation paired with UHMWPE or CFRP could be distinguished (table 1, $p>0.05$, t-test). The roughness parameters reflect typical ceramic profiles with smooth areas and isolated holes ($R_{pk}<R_{vk}$ [5]) contrary to metal surfaces with piled up profiles along the edges of scratches [3,12]. The differences of the sphericity between the virgin balls and the balls after retrieval were within the reproducible measuring accuracy ($p>0.05$, t-test). The sphericity deviations of all retrieved ceramic heads were within the demanded standard.
In conclusion there were no surface changes regarding the roughness and morphology of ceramic balls paired with UHMWPE or CFRP compared to virgin balls of the related product period. However this statement must be interpreted carefully due to the short implantation time of the CFRP-group.

Acknowledgements: The authors would like to thank the CERASIV company for providing the original geometric measurements of ceramic heads used for this study.

REFERENCES

1 BMFT-Abschlußbericht: Förderungskennzeichen 01 VG 274 (1991)

2 Boenisch, B., Ascherl, R., Liebendörfer, A., Geissdörfer, K., Schmeller, M.L., Kerschbaumer, S., Scherer, M.A., Erhardt, W., Scheer, W., Blümel, G.: Five-year functionality study of carbon fibre reinforced acetabular implants in dogs. In: Heimke, G., Soltész, U., Lee, A.J.C. Clinical implant materials. Advances in Biomaterials, Vol. 9, Elsevier, Amsterdam, NL (1990) 391-396

3 Cooper, J.R., Dowson, D., Fisher, J., Jobbins, B. Ceramic bearing surfaces in total artificial joints: resistance to third body wear damage from bone cement particles. Journal of Medical Engineering (1991) 15 63-67

4 DIN 4768: Ermittlung der Rauheitskenngrößen R_a, R_z, R_{max} mit elektrischen Tastschnittgeräten (1990)

5 DIN 4776: Rauheitsmessung: Kenngrößen R_k, R_{pk}, R_{vk}, M_{r1}, M_{r2} zur Beschreibung des Materialanteils im Rauheitsprofil (1990)

6 Dörre, E., Dawihl, W. Ceramic hip endoprostheses. In: Hastings, G.W., Williams, D.F. Mechanical properties of biomaterials. Wiley (1980) 113-127

7 Hinterberger, J., Ungethüm, M. Untersuchungen zur Tribologie und Festigkeit von Aluminiumoxidkeramik-Hüftendoprothesen. Zeitschrift für Orthopädie (1978) 116 294-303

8 ISO/CD 3274: Surface texture: Instruments for the assessment of surface texture; Profile method (1993)

9 ISO/CD 4287-1: Surface texture; Terms, definitions and parameters of surface texture (1993)

10 ISO/DIS 4288: Determination of surface roughness parameters using stylus instruments (1993)

11 ISO/DIS 6474: Implants for surgery - Ceramic materials based on high purity alumina (1992)

12 Kato, K. Tribology of ceramics. Wear (1990) 136 117-133

13 Nevelös, A.B., Evans, P.A., Harrison, P., Rainforth, M. Examination of alumina ceramic components from total hip arthroplasties. Journal of Engineering in Medicine (1993) 207 155-162

14 Plitz, W., Walter, A. Tribology of carbon fibre reinforced epoxy resin acetabularcups. 1st World Congress of Biomechanics, La Jolla, California, USA (1990) 345

15 Streicher, R.M., Semlitsch, M., Schön, R. Ceramic surfaces as wear partners for polyethylene. In: Bonefield, W., Hastings, G.W., Tanner, K.E. Bioceramics, Vol. 4, Butterworth-Heinemann (1991) 9-16

16 Walter, A. On the material and the tribology of alumina-alumina couplings for hip joint prostheses. Clinical Orthopaedics and Related Research (1992) 282 31-46

17 Zichner, L., Paschen, U., Starker, M. In vivo wear of the articulation surfaces made of Al_2O_3 ceramics and polyethylene in total hip endoprostheses. In: Willert, H.G., Buchhorn, G.H., Eyerer, P. Ultra high molecular weight polyethylene as biomaterial in orthopedic surgery. Hogrefe & Huber Publishers, Toronto, USA (1991) 148-151

Bioceramics, Volume 7 Edited by Ö.H. Andersson and A. Yli-Urpo
(*Proceedings of the 7th International Symposium on Ceramics in Medicine, Turku, Finland, July 1994*)

Investigation of Retrieved Femoral Biolox Heads

G. Willmann, U. Kemmer* and K. Zweymüller**

* Cerasiv GmbH, Medical Products Division, Fabrikstrasse 23-29, D-73207 Plochingen, Germany

** MA 17 – Orthopädisches Krankenhaus, Gersthof, Wielemansgasse 28, A-1180 Wien, Austria

ABSTRACT

Alumina femoral heads have been used for hip endoprostheses since 20 years. 3 retrieved BIOLOX heads - the oldest one was 13 years in vivo - have been investigated to find out if the articulating surface has changed due to wear in vivo. No change in diameter, sphericity, and surface roughness could be detected.
The bore in the BIOLOX heads was investigated, too. No change of the ceramic heads' bore surface and no fretting of the head could be detected. The heads have kept the specifications valid at the time of fabrication.

CERAMIC FEMORAL BALL HEADS FOR ARTIFICIAL HIP JOINTS

A modern endoprosthesis for total hip replacement (THR) is designed as a modular system, i.e. a stem out of a bioalloy, a femoral ball head out of cobalt-chrome or alumina ceramic, and an acetabular cup or cup insert out of polyethylene (= PE-UHMW) (1, 2). The articulating components, the cup and the head, cause wear debris. This wear debris causes osteolysis, loosening of the endoprosthesis, and in most cases a revision. The objective of the R & D work during the last decades was to minimize the wear debris and revision rate and to maximize the survival rate and the average time of implantation (1).
Since 1974 femoral ball heads out of ceramic, i.e. alumina (= aluminum oxide ceramic, Al_2O_3) are successfully used worldwide, mainly in Europe. The fundamental properties for this application are: the bioinertness and the outstanding hardness and wear resistance of alumina heads combined with PE cups or with alumina cups (2, 3, 4).
So far more than 2 million ceramic femoral ball heads have been successfully used in Australia, Europe, Japan, and North America. The most well known alumina is BIOLOX which was developed by Feldmühle (5). Up to 1994 about 1.5 million BIOLOX heads have been used.
There are various aluminas (6). Only an alumina according to the standards ISO 6474, ASTM F 603, and FDA recommendations has the properties specified for high quality femoral ball heads (2). An alumina material that fulfills these specifications may be called "medical-grade alumina". There are only very few companies in Europe and Japan, none in the USA, that are able to manufacture such ball heads

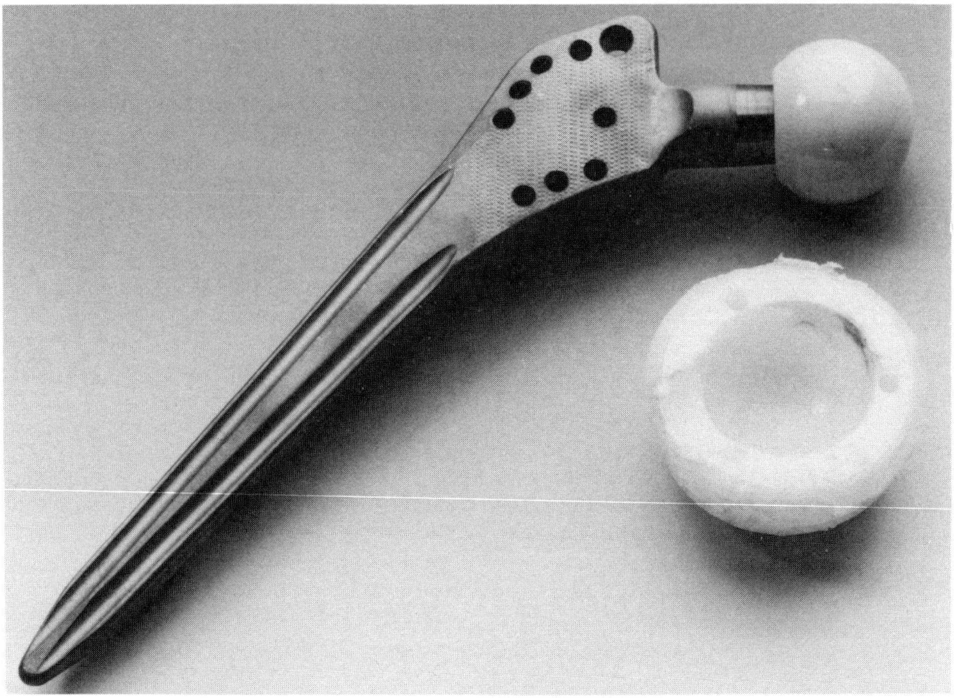

Figure 1 One of the retrieved hip endoprosthesis equipped with a BIOLOX head

Table 1
Explanted BIOLOX ball heads

Name	number	in vivo
Case M.B.	9305F8	13 years
Case M.H.	9305F6	9 years
Case E.S.	9305F7	8 years

Femoral ball heads out of alumina proved to be safe components for endoprostheses for THRs. The best proof is the investigation of retrieved components. There are very few reports about retrieved alumina ball heads. In this paper some cases of retrieved Zweymüller-prostheses will be discussed. These prostheses were equipped with BIOLOX heads (see figure 1 and table 1). The first implantation and the revision were performed in the hospital Gersthof in Vienna, Austria.

EXAMINATION OF THE ARTICULATING SURFACE

The objective is to find out if the articulating surface of the BIOLOX ball head has changed while being implanted.
Furthermore we want to find out if BIOLOX shows signs of fretting - a phenomenon violently discussed in the USA at present (7, 8).
What is most demanded of the articulating surface of a ball head is the guarantee of a low friction mobility during the whole time of implantation with minimized production of wear particles.

Table 2
Important basic characteristics for ball heads

Characteristics	based on
minimized wear	high hardness, form stability, stiffness corrosion resistance, surface quality
safety	high strength, corrosion resistance, homogeneity, production safety, quality control
scratch resistant	high hardness

Table 3
Diameter of the ball heads (in mm)

no.	in vivo	acc. to records	retr. heads	diff.
F8	13	31.954	31.946	0.008
F6	9	31.980	31.978	0.002
F7	8	31.990	31.989	0.001

Table 4
Surface roughness R_a in μm

no.	years in vivo	retrieved head		
F8	13	0.06	0.06	0.06
F6	9	0.08	0.07	0.05
F7	8	0.06	0.05	0.07

Prerequisites for that are (see table 2):

- minimized wear as a result of an excellent high hardness of about 2000 HV 0.1
- excellent form stiffness as a result of the high elastic modulus of 380 GPa and a compression strength of 4000 MPa,
- optimal surface quality.

Above in connection with the

- ideal wetting behavior (9), which is the basis for the formation of a lubricant. This has made BIOLOX an optimal counterpart for polyethylene.
  Basis for the comparison of the retrieved ball heads are data filed by our quality assurance when the ball heads were despatched, i.e.
- sphericity measured across the pole and under 30°
- diameter (at the equator)
- surface roughness R_a

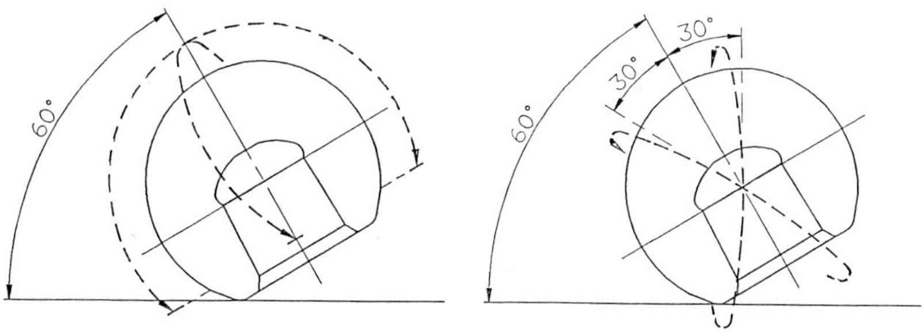

Figure 2 Measurement of the sphericity

Figure 2 shows that the sphericity is still within the tolerances of the dimensional accuracy of +/-2 μm. A certain tendency can be perceived, i.e. that the sphericity diminishes slightly with its duration if implantation. Estimated value could be **0.2 μm per year.** This statement is not proven statistically, however.
The diameter of the ball heads (table 3) is still within the tolerances of the dimensional accuracy. If you evaluate the tendency, the diameter changes from **considerably below 0.1 μm per year to only slightly below 0.1 μm per year.**
The surface quality, characterized by R_a, shows neither any change in comparison with the usual values nor any tendency as far as the duration of implantation (see table 4) is concerned.

EXAMINATION OF THE BORE

The taper fitting for modular hip endoprostheses has proved to be successful both for CoCr ball heads and for BIOLOX ball heads (1, 2).
Several authors in the USA report on fretting in connection with metal ball heads (2, 7, 8, 10).

Prerequisites for fretting are:
I. micro movement between male and female taper in the ball head.
II. galvanic corrosion as a result of the electro-chemical voltage series when different metal materials are in contact with each other.
BIOLOX is an electric insulator with a specific electric resistance of 10^{15} Ω cm and for that reason does not cause the effects mentioned under II.
The following analysis shall clarify if eventually appearing micromovements have caused a change at the male metal taper of the prosthesis stem and at the female taper in the BIOLOX ball head.
Table 5 shows the typical specifications for the angle, straightness, roundness, and roughness of the metal taper.
The analysis shows that the metal taper has not changed with regard to its specification.
In table 6 specification characteristics for the female taper in the ball head are compared with values measured at the explants. All characteristics are within the tolerances and therefore unchanged.
An examination with regard to scratches - this would mean a disruption of the microstructure in the BIOLOX ball heads - was made by means of dye penetrant and showed a negative result.
At the upper end of the female taper a deposit was identified which, by means of EDAX in a SEM was identified as titanium alloy and therefore as the print of the male taper. This print has to appear as a result of the taper fitting.

Table 5
Taper fitting - changes at the male taper

Characteristics	typical spec. (type 2)	F7	F8
straightness	< 12 μm	*)	11 μm
roundness	< 15 μm	5 μm	6 μm
roughness R_t	60-90 μm	86 μm	97 μm
angle	5°35'-5°42.5'	5°37.8'	5°38.6'

*) dirty

Table 6
Taper fitting - change of the conical bore (female taper) in the BIOLOX ball head

Characteristics	typical spec.	F7	F8
straightness	0-3 μm	*)	*)
roughness	0.6-2 μm	1.3 μm	1.0 μm
angle	5°43'-5°48.5' average 5°46'	5°46.5	5°46.3'

*) dirty

CONCLUSION

The femoral ball heads investigated here have been in good function for many years (8 to 13 years). Prerequisites for minimized wear debris and a safe component are that the following properties do not change while the femoral head is in function (in vivo): roundness of the ball head, surface finish (roughness) of the articulating surface (see table 1). Other very important things are: no scratches on the articulating surface (three body wear due to particles from bone cement or bone), no corrosion, no change in density and microstructure (corrosion of the grain boundaries).
The investigated BIOLOX femoral heads show no change when compared to the specification and standards valid at the time when manufactured. So the conclusion is that the BIOLOX femoral heads fulfill all specifications and standards even over periods of ten years and that these heads are safe components.

REFERENCES

1. Buchhorn, G.H. and Willert H.-G. Technical Principles, Design and Safety of Joint Implants 1994

2. Clarke, I.C. and Willmann G. In Bone Implant Interface" Mosby, St. Louis, 1994, 203-252

3. Williams, D. Concise Encyclopedia of Medical & Dental Materials 1990

4. Ravaglioli, A. and Krajewski A. Bioceramics 1992

5. Feldmühle, now CERASIV, Plochingen, Germany

6. Brook, R.J. Concise Encyclopedia of Ceramic Materials 1991

7. Workshop on Modularity of Orthopedic Implants, Miami, USA, 1992

8. Maximizing Stability while Addressing Fretting and Wear Debris, Company Publication Smith & Nephew Richards, Memphis, USA

9. Dawihl, W. and Dörre E. In Biological Performance of Materials - Fundamentals of Biocompatibility". Marcel Dekker, New York, 1992

Bioceramics, Volume 7 Edited by Ö.H. Andersson and A. Yli-Urpo
(Proceedings of the 7th International Symposium on Ceramics in Medicine, Turku, Finland, July 1994)

The Importance of the Raw Material Sources for Medical Grade Zirconia

W. Burger,* H.G. Richter,* C. Piconi,** A. Cittadini,*** D. Bossi*** and H. Boccalari****

* Cerasiv GmbH, P.O. Box 1149, Plochingen, Germany
** E.N.E.A., Area Energia e Innovazione, Via Anguillarese 301, Roma, Italy
*** Universita Cattolica del Sacro Cuore, Largo Francesco Vito, Roma, Italy
**** Fabbricazioni Nucleari, P.O. Box 16, Bosco Marengo (AL), Italy

ABSTRACT

There is interest in zirconia being used in biomedical applications as ball heads for total hip prostheses. Two potential types are under discussion: partially stabilized zirconia (PSZ) and tetragonal zirconia polycrystals (TZP) materials. Because of its enhanced material properties especially TZP stabilized with yttria is favourable. To eliminate high amounts of natural radioactive impurities, the precursors are purified. It depends on the kind of the precursor and the purification method to achieve powders having a very low impurity level. Thermodynamical and kinetic investigations on high purity coprecipitated and yttria coated zirconia powders show a different behaviour.

INTRODUCTION

Partially stabilized zirconia was introduced to the orthopedic community a few years ago because of its attractive mechanical properties. It seems, that partially stabilized zirconias are suitable and safe materials for femoral ball heads of THR-systems [1,2]. The so-called PSZ-grades, which consist of cubic matrix grains having tetragonal precipitates within those grains, are of less interest because of their relatively high porosity of about 4 %. Furthermore PSZ-ceramics are normally based on powders having significant amounts of radioactive and chemical imupurities [3]. The more attractive zirconia ceramics to being used in biomedical applications are TZP-ceramics. TZP ceramics are usually stabilized by yttria. The microstructure of Y-TZP is very fine grained and the densities reach nearly theoretical values. For the fabrication of Y-TZP ceramics, the powders must be very fine grained and have a very high chemical purity. In order to qualify the zirconia powders for the fabrication of hip joints, their radioactive impurities have to be minimized by the purification of the precursors, because the radioactive impurities form solid solutions with the zirconia and show a similar chemical behaviour as zirconia itself.

Typical purification methods are destillation, sublimation, or extraction. Especially chromatographic or ion-exchange extraction processes are of interest for the purification. It depends on the precursor, which of the purification methods is preferably applied. This means, that for the purification of $ZrCl_4$ a distillation process is preferably applied, because $ZrCl_4$ is a liquid and can be distilled off from impurities very easily. If $ZrOCl_2$ or $ZrO(NO_3)_2$ are used as precursors, probably a different purification method like chromatographic extraction has to be used.

After the purification process of the chemical zirconium salts is finished, normally a precipitation process follows and finally the precipitated powder is calcined. During the precipitation process the size of the precipitates is controlled in order to reach very fine grained crystallites of a size between 50 - 250 nm after calciniation.

For the production of Y-TZP ceramics, the yttria has to be distributed very homgeneously in the zirconia matrix. The most common method to realize this is a coprecipitation process: the stabilizing amount of yttria is added to the solution of the purified zirconium salt as an yttrium salt before the precipitation and calcination process is started [4]. But there exists another approach to introduce yttria: a coating of the already calcined zirconia particles [5]. Three different high purity powders have been investigated.

RESULTS

Table 1 shows the chemical analysis of three different powders. It is evident, that powder 3 is the one having the highest chemical purity level, followed by powder 1. Powder 1 has a lower amount of stabilizing yttria. In this powder the zirconia particles are coated by yttria, while powder 2 and 3 are based on a coprecipitation process. Significant amounts of chemical impurities are detected in powder 2 (Al_2O_3, SiO_2, TiO_2 and CaO).

Room temperature X-ray investigations showed a big difference in the amount of tetragonal phase in the starting powders. The amount of the tetragonal phase in the coprecipitated powders (2 and 3) is about 73 %, while in the coated powder (1) more than 90 % of the particles are still monoclinic.

impurity	unit	powder 1	powder 2	powder 3
Na2O	[wt-%]	<0,05	<0,05	<0,05
K2O	[wt-%]	0,01	0,01	0,01
MgO	[wt-%]	<0,1	<0,1	<0,1
CaO	[wt-%]	<0,01	0,03	<0,01
Al2O3	[wt-%]	0,1	0,28	<0,05
SiO2	[wt-%]	<0,03	0,11	<0,03
TiO2	[wt-%]	0,01	0,11	0,02
Fe2O3	[wt-%]	0,01	<0,01	<0,01
HfO2	[wt-%]	1,96	1,78	1,67
Y2O3	[wt-%]	4,8	5,2	5,2

Table 1: Chemical impurities of the Y-TZP powders

Since the starting powders showed a different amount of tetragonal phase, it was expected, that the monoclinic → tetragonal phase transformation is different in the powders. Figures 1 and 2 show the phase transformation behaviour of the coated and coprecipitated powders. In powder 3 the phase transformation is already completed at 800 °C. Further heating leads to a higher peak intensity, which is due to the crystallization effect. By cooling down after a treatment at 1400 °C, no monoclinic phase appears any more. The tetragonal phase remains stable. In the coprecipitated powder 2, the forming of the tetragonal phase is completed at about 1000 °C. Despite of this difference, the behaviour is like powder 3. While powder 3 has a specific surface area of 14 m^2/g, in powder 2 only 9 m^2/g were found.

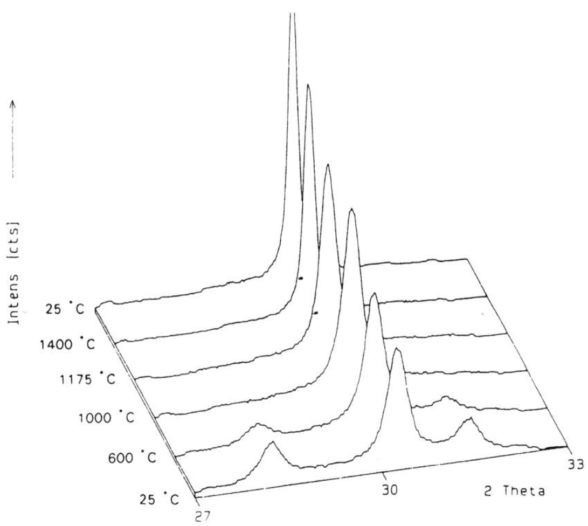

Figure 1: Phase transformation in the coprecipitated powder 3

The specific surface area of powder 1 was 13,5 m^2/g. But the behaviour concerning the phase transformation monoclinic → tetragonal was quite different from powder 3. Until the well known transformation temperature of about 1175 °C no remarkable intensity of the [111]-reflection of the tetragonal phase was formed. Starting up from that temperature, a remarkable increase coupled with a decrease of the intensity of the monoclinic phase was detected. Although the intensity of the monoclinic phase decreased steadily, still at 1300 °C a monoclinic peak could be detected. The phase transformation monoclinic → tetragonal was completed at 1400 °C. As in powder 2 and 3 already described, no retransformation tetragonal → monoclinic was detected.

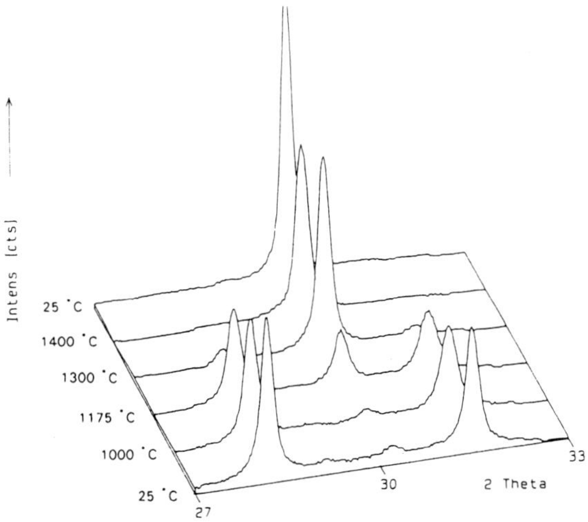

Figure 2: Phase transformation in the yttria coated powder 2

Sintering trials have demonstrated that the sintering activity of the coated powder is higher than that of the coprecipitated powders. Still at 1500 °C the achievable density of powders 2 and 3 is less than in powder 1, although the powders were prepared under identical conditions. This temperature lead to discontinous grain growth. At a sintering temperature of 1430 °C the microstructure of all materials is very homogeneous. The grain size is smaller than 0,5 μm. The microstructure of material 3 sintered at 1430 °C is shown in figure 3. Due to the high sintering activity the temperature for material 1 can be reduced further. This leads to an extremely fine grained microstructure as demonstrated in figure 4. The grain size is about 0,2 μm.

DISCUSSION

Although the powders were prepared under the same conditions, the sintered materials showed very different behaviour. While in material 1 a density ≥6 g/cm^3 can be realized very easily, such a density of parts having a very fine grained and homogeneous microstructure was not achieved in materials 2 and 3. Based on a theoretical density of 6,1 g/cm^3, a porosity of about 3 % remains in the sintered material 3. It seems reasonable to assume, that the pores remain during sintering because of a very high shrinkage rate, which is caused by the low green density, and high

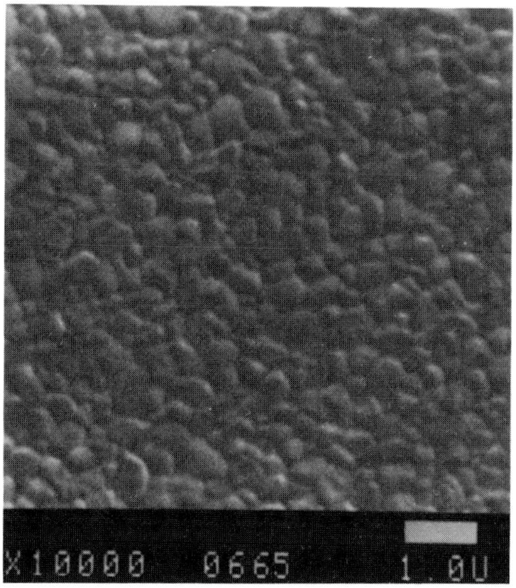

Figure 3: Microstructure of material 3 after sintering at 1430 °C

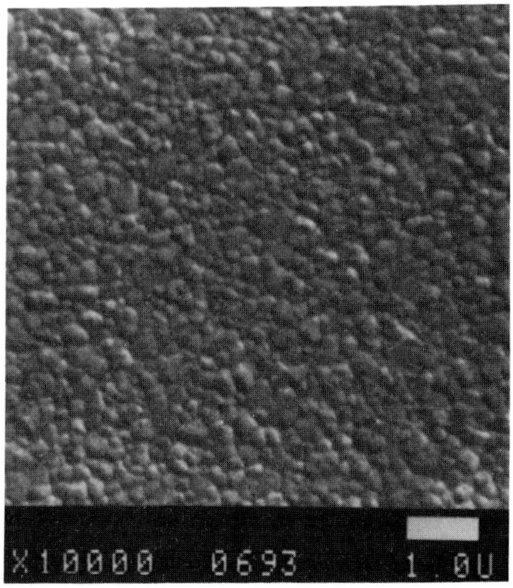

Figure 4: Microstructure of material 1 sintered at 1410 °C

sinteractivity of the powder, which is caused by the high specific surface area, small grain size and monoclinic → tetragonal phase transformation.

Material 2 has about the same density as material 3. Although the specific surface area of powder 2 is lower than in powder 3, the densification during the sintering process is assisted by the glass forming oxides like TiO_2, SiO_2, Al_2O_3 and CaO. It is assumed that a very thin amorphous grain boundary phase remains in the microstructure.

For material 2 and 3 it seems reasonable, that a pressure supported densifiction process would lead to higher densities and strengths. This goal can be reached by a HIP-process, which can follow a pre-sintering process at moderate temperatures in order to eliminate pores. For material 3 another approach is possible: the optimization of the binder system in respect to reach a higher green density and a different sintering behaviour.

In material 1 densities of more than 99 % are reached at moderate temperatures. The densification process is assisted by the specific surface area and the behaviour of the monoclinic → tetragonal phase transformation, which still takes place during the shrinkage process, leads to a very controlled densification process.

CONCLUSIONS

- The achievable sintered densities of ceramics based on yttria coated powders are very good; an extremely fine grained microstructure can be realized.
- For the stabilization of the tetragonal phase in coated powders lower amounts of yttria are needed than in coprecipitated powders.
- Y-TZP ceramics based on yttria coated zirconia powders are very favourable to be used in biomedical applications.

LITERATURE

[1] Garvie, R. C., Urbani, C., Kennedy, D. R. and McNeuner, J. C.; J. Mat. Sci., **19** (1984), 3224 - 3228

[2] Christel, P. et al.; J. Biomedical Mat. Research, **23** (1989), 45

[3] Burger, W. and Willmann, G.; In: Bioceramics, Vol. 6, Philadelphia, USA, (1993), 299 - 304

[4] Haberko, K., Ciesla, A. and Pron, A.; Ceramurgia International, **1** (1975), 111 - 116

[5] Dransfield, G. P. et al.; In: Euro-Ceramics, Vol. 1, Augsburg, D, (1989), 275 - 279

ACKNOWLEDGEMENT

This work was funded by the European Union by supporting a project which is actually running under the BriteEuRamII programme.

Bioceramics, Volume 7 Edited by Ö.H. Andersson and A. Yli-Urpo
(*Proceedings of the 7th International Symposium on Ceramics in Medicine, Turku, Finland, July 1994*)

Yttria-Stabilized Zirconia Ceramic for Improved Hip Joint Head

J.M. Drouin and B. Cales
Ceramiques Techniques Desmarquest, Z.I. No. 1 – 27025 – Evreux – France

ABSTRACT :

The high mechanical properties of Y-TZP allow to design a lot of zirconia hip joint heads. The influence of the geometrical characteristics of the head itself, the characteristics of the metallic trunnion and the conditions of press-fitting on the mechanical strength of the head is presented here.Several examples of design optimisation of zirconia ceramic heads using both finite element analysis and burst tests on metallic trunnions are presented. It has been shown that trunnion/head angle mismatch lower than 20' do not influence the femoral head fracture strength. Because of the lower Young's modulus of TA6V trunnions, they lead to higher fracture strength than observed with Cr-Co trunnion. However the fracture strength can be significantly improved by increasing the surface roughness of the trunnion.

INTRODUCTION

Yttria-stabilized zirconia ceramics (Y-TZP) were successfully introduced in orthopaedy for about 9 years. The improved mechanical properties of zirconia ceramic, combined with excellent biocompatibility and wear properties, make this material one of the best candidates for orthopaedic prosthesis of new generation. For this reason, zirconia ceramics are more and more widely used in orthopaedy in replacement of alumina ceramics and, to a certain extent, of metals.

The high fracture strength and toughness of Y-TZP ceramics allow to design a lot of reliable zirconia hip joint heads with geometrical parameters that can be varied to large extents. Each of these heads must be drawn and optimized to perfectly fit the metallic trunnions onto which they will be implanted. Thus, it is possible to manufacture small diameter heads, down to 22.22 mm, with a 8/10, 10/12 and even 12/14 mm cones, or heads with larger cones, like 28 mm diameter heads with 14/16 mm cones. Table I summarizes the various possible combinations and also indicates in each case the maximum length between short and long neck. Among all these combinations, only 22.22 mm heads with 14/16 cone cannot be manufactured for geometrical reasons. For other designs, in order to increase the reliability, the stress distribution inside the head under loading must be analyzed. Indeed, the knowledge of magnitude and location of high tensile stresses in hip joint head is of prime importance because they could affect the survival probability of these components.

TABLE 1

TYPE OF CONE	HEAD DIAMETER			
	32	28	26	22.22
	Maximum length between long and short neck			
14/16	7 mm	7 mm	4 mm	
12/14	10 mm	8 mm	5 mm	0 mm
10/12	10 mm	10 mm	6 mm	3.5 mm
8/10	10 mm	10 mm	7 mm	5 mm

It has been shown in previous publications [1,2,3], that two main tensile stress concentrations appear at the inner surface of the ceramic cone, when the femoral head is press-fit on the trunnion : -i) the first one is located at the top of the conical hole corresponding to bending deformation; -ii) the second one is located at the metallic trunnion/ceramic head interface as a consequence of hoop stresses. The magnitude of these stress concentrations depends on the location and extent of contact area between the ceramic head and the metallic trunnion. Therefore, the contact area greatly influences the fracture strength of the femoral heads [2,3], and an optimized design based on finite element analysis (F.E.A.) should be defined for each head geometry. Thus, to minimize the tensile stress concentrations it is necessary to increase -i) the distance, d, between the top of the metallic trunnion and the top of the ceramic cone to typically $\geq$ 2 mm; -ii) the head/trunnion contact area, provided it do not extend beyond limit at the bottom of the ceramic cone [2,3].

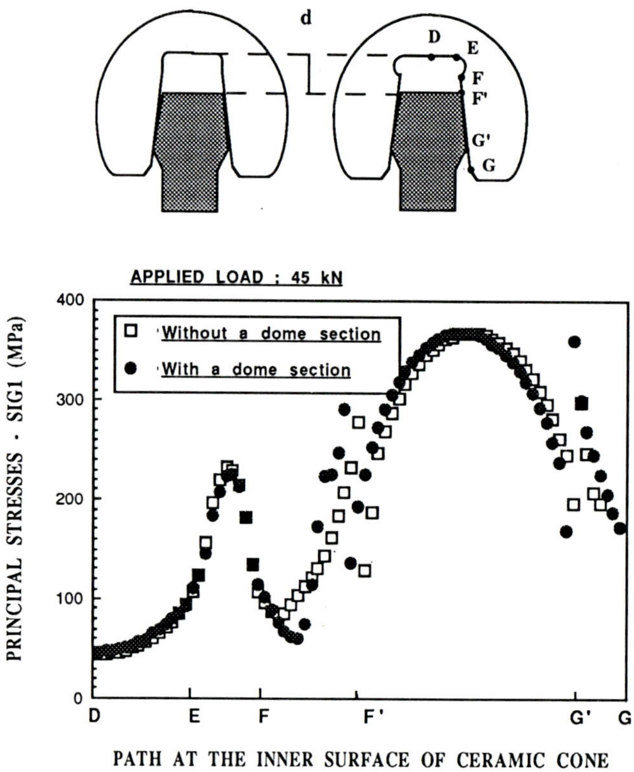

Figure 1 :

Head designs with or without a dome section at the top of the cone (a) and corresponding tensile stress profile (b).

Under these conditions, the introduction of a dome section at the top of the ceramic cone (Fig. 1a), which has previously been described as a more favourable design modification to reduce stress concentration [4], has in fact no direct influence on the stress distribution, but just can be associated to an increase of the distance d at the top of the ceramic cone. Indeed, two heads having the same design, one with a dome section and the other without, and characterized by the same distance d (Fig. 1a) will exhibit exactly the same stress concentrations, as shown in Figure 1b, and therefore the same fracture strength.

The real fracture strength of a femoral head does not only depend on the head design but also on the geometrical and physical characteristics of the metallic trunnion. The purpose of the present study was to investigate the influence of the metallic trunnion characteristics, i.e. angle mismatch, surface roughness, ..., on the head fracture strength. It has been done by using both F.E.A. with appropriate contact elements and standardized burst tests. F.E.A. method has been previously described in detail [2,3].

INFLUENCE OF HEAD/TRUNNION ANGLE MISMATCH ON HEAD FRACTURE STRENGTH

The maximum head/trunnion angle mismatch $\Delta\alpha$, currently used is about 10' and corresponds to a tolerance interval of 0'/-5' about for metallic trunnion and 0'/+5' for ceramic cone. This conventional value was selected with the objective of providing good reliability and efficiency of hip-joint prosthesis. Fessler and Fricker [5] studied the stress distribution in alumina femoral head from theoretical stress analysis and shown that $\Delta\alpha$ should be minimized in order to reduce the maximum hoop stress at the trunnion/head interface. These authors also established that, for alumina heads, $\Delta\alpha$ should be positive, i.e. initial contact between metallic trunnion and ceramic cone should occur at the top of the metallic trunnion and not at the bottom of the ceramic cone. On the other hand, because of small radial displacements of the metallic trunnion during press-fitting, they recommend to maintain $\Delta\alpha$ at a very small positive value [5].

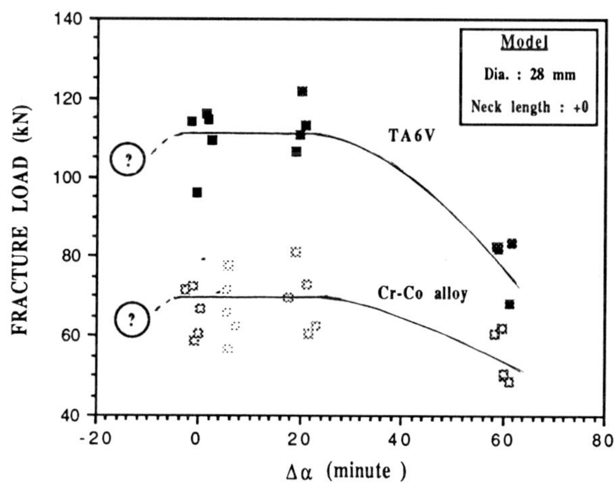

Figure 2 : Influence of the head/trunnion angle mismatch on the head burst strength on TA6V and Cr-Co trunnions.

In the present study, the influence of the head/trunnion angle mismatch on the fracture strength of zirconia heads (PROZYR®) was analyzed in order to define reliable limits for the angle mismatch.

Figure 2 shows the variation of zirconia head fracture strength versus $\Delta\alpha$, using TA6V and Cr-Co trunnions with a surface roughness R_{max}=30 μm. Surprisingly, it appears in both cases that for $\Delta\alpha$ between 0' and 20' there is no significant influence of angle mismatch on the fracture strength of zirconia heads. Only for angle mismatches significantly higher than 20' the fracture strength drastically decreases. The head/trunnion angle mismatch could thus be increased beyond the current limit of 10' without any degradation of head fracture strength, for burst testing.

These results could be explain on the basis of finite element analysis. Indeed, F.E.A. reveals that when the angle mismatch $\Delta\alpha$ increases, the location of the maximum tensile stress moves along the inner surface of the ceramic cone from the head/trunnion interface to the top of the conical hole. In the same time the level of these tensile stresses increases. Thus, up to a "threshold" value of $\Delta\alpha$, which is strongly correlated to the value of distance d, the maximum tensile stress remains located at the trunnion/head interface and could be considered as constant. Beyond this limit the maximum tensile stress drastically increases and moves to the top of the conical hole. The Figure 3 illustrates the variation of the maximum tensile stress deduced from F.E.A. versus $\Delta\alpha$, for femoral heads having the same design than those used for burst tests (Figure 2) and characterized by the same distance d of 3 mm. There is a fair agreement between theoretical and experimental results. This behaviour was also confirmed by analyzing the location of fracture origin on fractured heads after burst tests which moves from trunnion/head interface to the top of the conical hole when the angle mismatch $\Delta\alpha$ was increased.

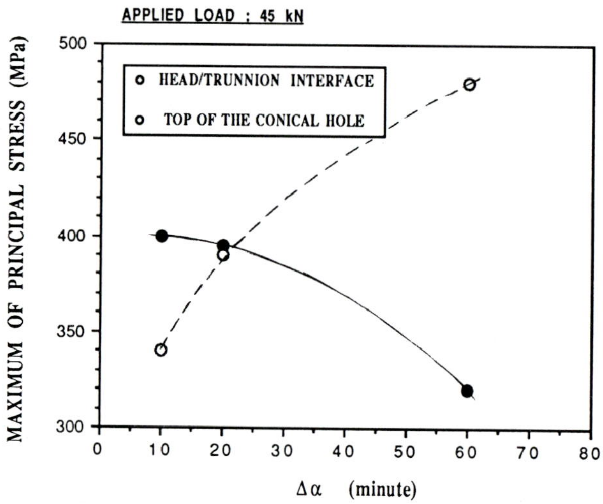

Figure 3 : Variation of the effective maximum principal stress versus head/trunnion angle mismatch, as a combination of principal stresses at head/trunnion interface and at the top of the cone.

INFLUENCE OF SURFACE FINISH AND COMPOSITION OF METALLIC TRUNNION ON HEAD FRACTURE STRENGTH

The values of fracture strength of various 28 mm and 22.22 mm zirconia heads (PROZYR®) on 10/12 TA6V metallic trunnions having different surface roughness are summarized in Table II. For each design, it may be noted that higher is the surface roughness of the metallic trunnion, higher is the ceramic head fracture strength. In the present case, an improvement of 30% about is observed for 28 mm heads. This behaviour has several origins.

28 mm ZIRCONIA HEADS

SURFACE FINISH	NECK LENGTH		
	-3.5 mm	0 mm	+3.5 mm
Rmax = 30 microns	110 kN	105 kN	85 kN
Rmax = 200 microns	140 kN	130 kN	110 kN

22.22 mm ZIRCONIA HEADS

SURFACE FINISH	NECK LENGTH 0 mm
Rmax = 3 microns	78 kN
Rmax = 30 microns	90 kN

TABLE II

First, under very low applied load, the head/trunnion contact area is very small and mainly concern the top of the metallic trunnion. Such a local plastic micro-deformation is enhanced by surface roughness. When the load is increased, the stress at the trunnion/head interface increases up to values that can be higher than the yield strength σ_y of the metal, thus leading to plastic deformation of the metallic trunnion at the interface. The resulting deformation of the metallic trunnion, which is enhanced by surface roughness, increases the contact area between femoral head and trunnion and hence the fracture strength. In addition, for a given load, lower is the yield strength σ_y of the metallic trunnion, higher is the trunnion/head contact area. For instance, TA6V trunnions with yield strength $\sigma_y \approx 795$ MPa, will exhibit a higher plastic deformation at trunnion/head interface than Co-Cr trunnions with yield strength $\sigma_y \approx 830$ MPa. This is the reason of the higher fracture strength observed for TA6V trunnion($\approx$ 50 %) when compared to Cr-Co trunnions (Figure 2).

The Young's modulus E must also be considered to explain the difference in burst values between TA6V and Cr-Co trunnions . The Young's modulus directly controls the forces transferred from the trunnion to the head at the trunnion/head interface and therefore the value of the maximum tensile stress in the head. Thus, the lower Young's modulus of TA6V trunnions also contribute to improve the femoral head fracture strength (Figure 2). This has been confirmed by F.E.A.. Figure 4 shows the variation of maximum principal stress σ_1 versus applied load for both TA6V and Cr-Co trunnions, with the same geometrical characteristics and surface roughness $R_{max.} = 30$ μm. Cr-Co trunnions lead to higher tensile stresses in the femoral head and consequently to a lower fracture strength. However, it should noted that, in some cases, the same fracture strength could be reached on Cr-Co trunnions as on TA6V trunnions, after improvement of the surface roughness.

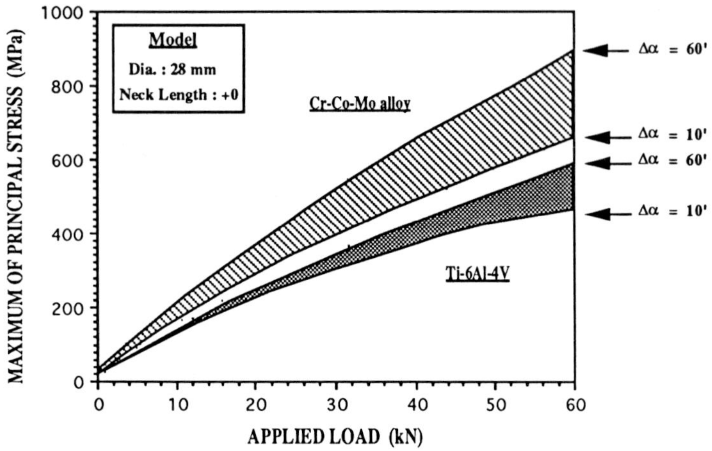

Figure 4 : Variation of the principal stress σ1 versus the applied load for both TA6V and Cr-Co trunnions.

It is therefore essential that the approval of a given zirconia hip joint head design should be submitted to a standardised burst test using metallic trunnions identical to that used for implanted prosthesis and having the same specifications (geometry, surface finish, chemical composition, ...). Any change of the metallic trunnion characteristics, even minor like diameter tolerances, surface roughness or metal composition, could have dramatic consequence on the fracture strength of the head. In this respect, standardization of metallic trunnions, as it has been recently proposed within ISO activity, should necessarily integrate all these characteristics.

CONCLUSION

The present study confirmed that the fracture strength of ceramic femoral head not only depends on the head design but also on the characteristics of the metallic trunnion and the conditions of press-fitting. It has been shown that trunnion/head angle mismatch lower than 20' do not influence the femoral head fracture strength. Because of the lower Young's modulus of TA6V trunnions, they lead to higher fracture strength than observed with Cr-Co trunnion. However the fracture strength can be significantly improved by increasing the surface roughness of the trunnion. The experimental results are well explained by F.E.A.

REFERENCE

[1] G. Kleer, D. Siegele and U. Soltész, *Biomechanics : Current Interdisciplinary Research*, Ed. : S.M.Perren, E. Schneider, 353 - 358 (1985).

[2] J.M. Drouin and B. Cales, *Actualités en Biomatériaux Volume II*, Ed. Romillat, 110 - 114 (1993).

[3] J.M. Drouin and B. Cales, *10th European Conference on Biomaterials*, Sept. 1993, Davos, Switzerland.

[4] Middleton, J., G.N. Pande, H.G. Richter and G. Willmann, *10th European Conference on Biomaterials*, Sept. 1993, Davos, Switzerland.

[5] H.Fessler and D.C. Fricker, *Proc. Instn. Mech. Engr,* 203 : 15 - 34 (1989).

Bioceramics, Volume 7 Edited by Ö.H. Andersson and A. Yli-Urpo
(*Proceedings of the 7th International Symposium on Ceramics in Medicine, Turku, Finland, July 1994*)

Long Term Stability of a Surgical Grade Zirconia Ceramic

B. Cales and Y. Stefani
Ceramiques Techniques Desmarquest Z.I. No.1 27035 – Evreux – France

ABSTRACT

Ceramic material are now currently used in Orthopaedy and especially for hip joint heads manufacturing. A long term stability of their mechanical properties is thus required to insure long implantation time. However, on the basis of the previously published results on the aging behaviour of these materials, it is not easy for ceramic users to estimate the time dependence of mechanical properties of such ceramics. The aging behaviour is in fact directly controlled by both microstructure and chemical composition and provided these two characteristics have been optimised very stable ceramic materials can be manufactured. As it has been shown in this paper, this applies for both alumina and zirconia ceramics and the aging behaviour of a surgical grade zirconia ceramic has been detailed.

INTRODUCTION

Of greatest significance for orthopaedic applications is the long term stability of implant materials. In the case of femoral prosthesis, the search of new friction combinations, with reduced wear rates, led to the development of ceramic bearing components. Alumina ceramics were first introduced (1) and clinical results demonstrated the benefit of lowered polyethylene wear rates (2). Zirconia was then introduced in orthopaedy for 9 years to overcome the brittleness problem of alumina. Indeed, yttria-stabilized zirconia ceramics (Y-TZP) are characterized by outstanding fracture strength and toughness, at least double that of dense aluminas. Such mechanical properties result from the transformability of tetragonal zirconia grains which leads to an enhancement of resistance to crack propagation. This mechanism has been described in detail in several papers (3). The long term stability of the mechanical properties of these ceramics, both alumina and zirconia, is required to provide long implantation times.

Several papers have been concerned by the aging of ceramic materials in biological conditions (4-10) and confusing results were often reported, making the aging behaviour of ceramic materials unclear for non-ceramists. Therefore, the objective of this paper is to give an overview of the aging behaviour of alumina and zirconia ceramics and to explain the time dependence of their mechanical properties on the basis of their main physico-chemical characteristics. The aging behaviour of a surgical grade zirconia ceramic will be then detailed.

INFLUENCE OF MICROSTRUCTURE AND CHEMICAL COMPOSITION ON THE AGING BEHAVIOUR OF CERAMICS FOR IMPLANTS

Alumina ceramics

Several papers on the in-vivo or the in-vitro aging of dense alumina ceramics have reported a significant fracture strength degradation, even after short time aging in physiological conditions (Figure 1). Dalgeish and Rawling (4) observed a 9-10% reduction in strength for three different aluminas. Their samples exhibited a small but significant porosity (2 to 6%) that could be at the origin of the strength degradation. Indeed, Krainess and Knapp (11) have indicated that a small permeation of the samples could generate a significant strength degradation, up to 30%. Similarly, the studies of Osterholm and Day (5) and Mandrino et al. (6) also concluded to a reduction in fracture strength of alumina ceramics after 6 to 12 months of in-vivo aging. They clearly established that the chemical composition and especially the calcium oxide content in alumina was responsible for the decrease of the fracture strength (5). This was confirmed by Sinharoy who shows the influence of calcium migration on the strength reduction of dense alumina aged in steam (12). It might be concluded from these results that alumina ceramics are subjected to strength degradation in simulated body environment. However, other experimental results, as well as clinical results, shown that under optimized conditions of microstructure (pore free material) and chemical composition (very low calcium oxide content) the mechanical properties of alumina ceramics are stable over long period of time. It is therefore not realistic to give general rules regarding the aging behaviour of alumina ceramics since it will directly depends on the physical characteristics of each ceramic (microstructure, chemical composition).

Zirconia ceramics

The situation is similar for Y-TZP zirconia ceramics, although only papers concerning in-vitro aging of laboratory made zirconia ceramics indicated a

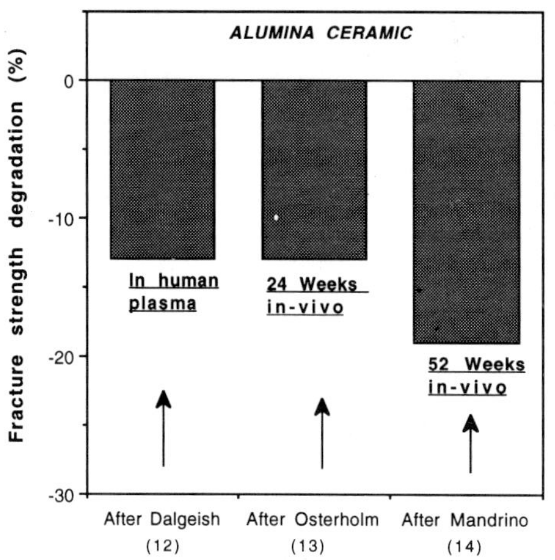

Figure 1 Fracture strength of dense alumina in simulated body environment.

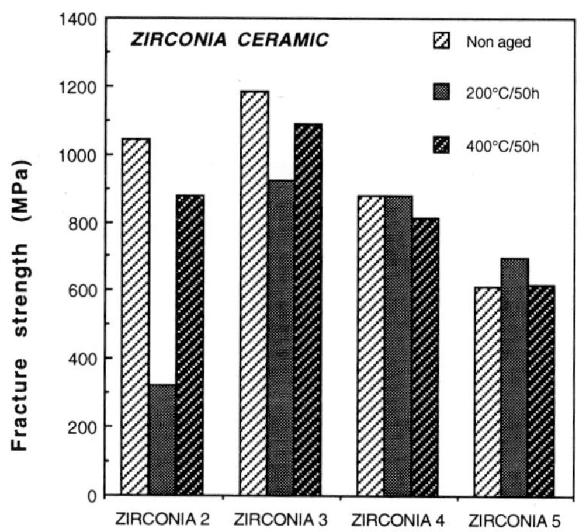

Figure 2 Influence of ageing in steam at 200°C and 400°C on the fracture strength of several Y-TZP ceramics (After Ref. 14).

strength decrease (7,8), while in-vivo experiments on commercial zirconia hip joint heads demonstrated a long term stability without any change of the fracture strength and toughness (9,10). It is not therefore quite easy for non-specialists to really appreciate the long term stability of a given Y-TZP zirconia ceramic on the basis on literature survey.

It has been previously reported that Y-TZP ceramics could be sensitive to aging in steam at 150-400 °C. Such treatment can lead to a more or less important tetragonal-to-monoclinic (T-M) transformation of zirconia grains and consequently to a significant decrease of fracture strength. Most of the authors considered such low temperature degradation as a general characteristic of all Y-TZP ceramics, irrespective to the physico-chemical characteristics of the material. Such a conclusion is not correct and it has been clearly shown by the theoretical studies of Lange (13) that the T-M transformation rate is directly dependent of physical characteristics of Y-TZP ceramics, and mainly grain size and chemical composition (yttria content). Therefore, it cannot be considered as a general behaviour for T-TZPs.

Indeed, aging behaviour in water vapour was reported by Swab (14) for several commercial Y-TZP ceramics having different yttrium oxide contents and mean grain sizes, as shown in figure 2. Zirconia 2 which is characterized by a quite low yttrium oxide content (2 mol %) exhibits a strong fracture strength decrease. Zirconias 3, 4 and 5 are characterized by a yttrium oxide content of 3 mol % and show much more stable behaviours. They also exhibit different levels of fracture strength which are expected to depend upon microstructure parameters, such as residual flaw sizes and grain sizes. Concerning grain size, it is observed that the lowest is the mean grain size, the higher the stability in steam at 150-400°C. This is the case for instance in Figure 2 for Zirconia 4 which is characterized by a mean grain size of 0.18 μm only. However, an extremely fine mean grain size is not a good objective because, as it has been shown by Theunissen (15), below 0.4 μm about the tetragonal grains become to be stable. They partly loss their metastable nature and T-M transformation no longer can be induced. A direct consequence is a decrease of mechanical properties before any aging, and especially of fracture toughness (15). The resistance to crack propagation is thus reduced. It is more convenient for Y-TZP ceramic to be characterized by mean grain sizes of the order of 0.5 - 0.7 μm, especially in the

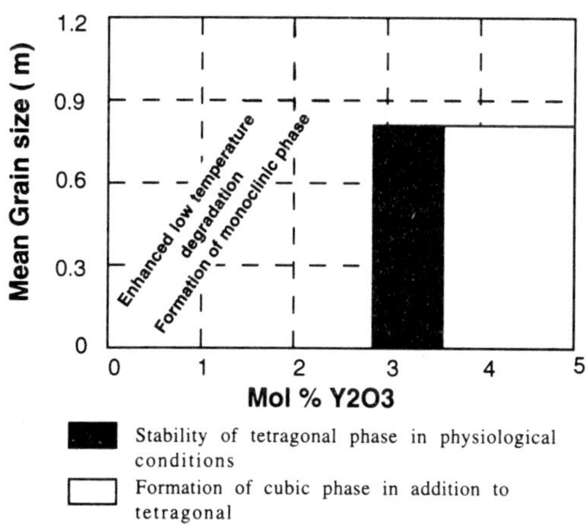

Figure 3 Microstructure-composition stability diagram for Y-TZP ceramics.

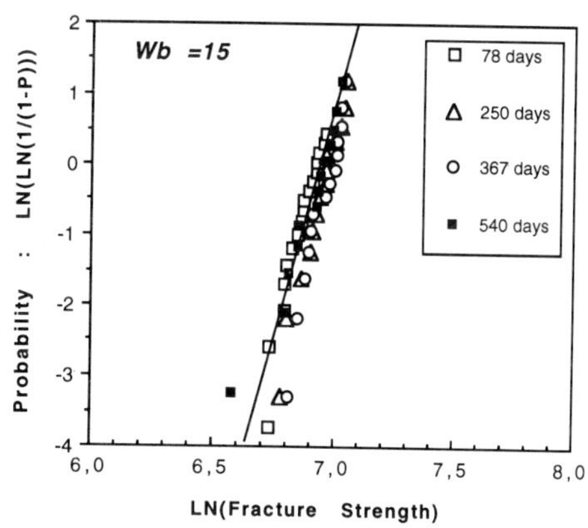

Figure 5 Weibull plots of aged PROZYR® zirconia ceramic.

case of biomedical application which correspond to much less severe aging conditions. Which such microstructures, the metastability of tetragonal grains is fully effective to enhance both fracture strength and toughness.

It is thus possible to draw a microstructure-chemical composition diagram as shown in Figure 3. Tetragonal Y-TZP ceramics with mean grain size and yttrium oxide content corresponding to the dark grey area will exhibit stable mechanical properties in physiological conditions. However, although they are stable, each of such Y-TZP ceramics will exhibit its own in-vivo aging behaviour which needs to be characterized specifically. For Y-TZP ceramics with yttrium contents lower than 2.8 mol% or large grain sizes (≥ 1 μm) rapid low temperature degradation can occur with strength decrease. On the other side, for excessive yttrium oxide content (i.e. higher than 3.5 mol %) cubic phase can be formed, which is also detrimental for the fracture strength, but have no drastic consequence on the aging behaviour. Therefore, it is an evidence that the aging behaviour of Y-TZP ceramics should be analyzed case by case, depending on the microstructure and chemical composition of the material and any general behaviour should be considered as incorrect.

LONG TERM STABILITY OF A SURGICAL GRADE ZIRCONIA CERAMIC

PROZYR® ceramic is a Y-TZP tetragonal ceramic which is characterized by a yttrium oxide content close to 3 mol % and a mean grain size of 0.5 - 0.6 μm about. According to the previous diagram (Figure 3), it is expected to have stable mechanical properties when aged in physiological conditions. This behaviour has been clearly confirmed by both in-vitro and in-vivo aging experiments and by the mechanical analysis of retrieved implant material. The in-vitro aging (Ringer's solution - 37°C) of this zirconia ceramics is depicted in Figure 4 where is reported the variations of flexural strength over a period of 2 years. It clearly appears that the fracture strength of this material remain unchanged. Furthermore, a statistical analysis of the fracture strength scattering, using the Weibull approach, confirms that there is absolutely no change in the mechanical properties after two years in Ringer's solution (Figure 5). The same Weibull's modulus (slope of the line) is observed for non-aged zirconia and after two years in-vitro. Phase change analysis on aged samples also indicate no tetragonal to monoclinic transformation during in-vitro aging (9).

The effect of in-vivo aging is also reported in Figure 4. Doted circles indicate the mean fracture strength of the PROZYR zirconia ceramic measured on small flexural bars directly cut into zirconia hip joint heads, either non-aged or retrieved after an implantation time of two years. No fracture strength degradation is observed and this result is particularly significant in so far as the

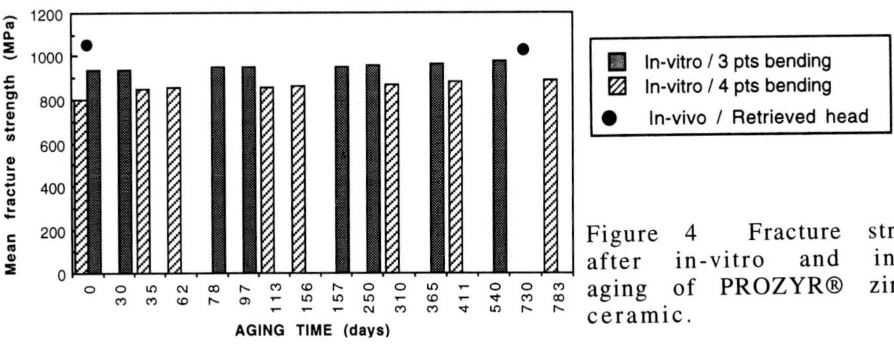

Figure 4 Fracture strength after in-vitro and in-vivo aging of PROZYR® zirconia ceramic.

zirconia ceramic has been subjected to both corrosion (human synovial liquid) and fatigue. Indeed, on assuming a 4 hours per day activity of the patient, the material supported between 5 and 10 millions of load cycles. The same stability of mechanical properties were observed after animals implantation (9).

Such a behaviour clearly demonstrate that this zirconia ceramic is perfectly stable and do not exhibit any low temperature degradation in physiological conditions. It is therefore expected to have constant mechanical properties for implant made with PROZYR® ceramic. This has been confirmed by the mechanical analysis (Standard burst tests) of retrieved heads. The fracture strengths are reported in figure 6 and compared to the mean value of fracture strength in-process measured on several lots of heads with the same design, using the same burst test procedure. The statistical higher and lower limits for the two head designs are also indicated. It appears that the burst strength values of the retrieved heads are higher that the mean fracture strength for the two designs and cannot be considered as affected by aging in human for 2 to 3 years.

CONCLUSIONS

The aging behaviour of surgical grade ceramics is directly related to their microstructure and chemical composition. This is case for dense alumina and Y-TZP zirconia ceramics. Depending on their microstructure and chemical composition either stable or time-dependent mechanical properties will be observed. It is therefore incorrect to give general behaviour for aging, irrespective to their physico-chemical characteristics. For instance, in the case of Y-TZP ceramics, provided the yttrium oxide content is in the range 2.8 - 3.4 wt. % and the mean grain size is lower than 0.8 µm, stable mechanical properties are observed after 2 years aging in physiological conditions. This is the case of the PROZYR® zirconia ceramic the mechanical properties if which remain perfectly stable after in-vitro, in-vivo aging and human implantation.

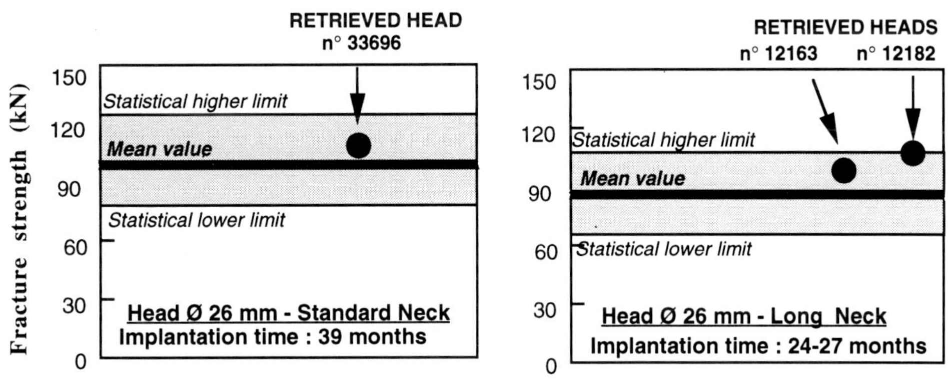

Figure 6 Fracture strength of retrieved heads compared to SPC fracture strengths before implantation.

REFERENCES

1. Boutin, P., *Rev. Chir. Ortop.*, **58**: 229-246 (1972).

2. Zichner, L. P. and H-G. Willert, *Clinical Orthop. Related Res.* , **292**: 86-94 (1992).

3. Garvie, R.C. and J.R. Hellman, *J. Mat. Sci.*, **21**: 1253-1257 (1986).

4. Dalgeish, B.J. and R.D. Rawling, *J. Biomed. Mat. Res.*, **15**: 527-542 (1981).

5. Osterholm, H.H. and D.E. Day, *J. Biomed. Mat. Res.*, **15**: 279-288 (1981).

6. Mandrino, A., R. Eloy, B. Moyen, J.L. Lerat and D. Tréheux, *J. Mat. Sci.*, **3** : 457-463 (1992).

7. Thompson, I. and R.D. Rawlings, *Biomaterials*, **11**: 505-508 (1990).

8. Drummond, J. L., *J. Am. Ceram. Soc.*, **72**: 675-676(1989).

9. Calès, B. and Y. Stéfani,*19th Annual Meeting Society for Biomaterials*, April-May 1993, Birmingham, AL, USA. To be published in J. Biomed. Mat. Res.

10. K. Shimizu et al., *J. Biomed. Mat. Res.*, **27:** 729-734 (1993).

11. Krainess, F.E. and W.J. Knapp, *J. Biomed. Mat. Res.*, **12**: 241-246 (1978).

12. Sinharoy, S., L.L. Levenson and D.E. Day, Am. Ceram. Soc. Bull., **58** : 464-466 (1979).

13. Lange, F. F., *J. Mat. Sci.*, **17**: 240-246(1982).

14. Swab, J. J., *J. Mat. Sci.*, **26**: 6706-6714 (1991).

15. Theunissen, G. S. A. M., J. S. Bouma, A. J. A. Winnbust and A. J. Burggraaf, *J. Mater. Sci.*, 27: 4429-4438, (1992).

Bioceramics, Volume 7 Edited by Ö.H. Andersson and A. Yli-Urpo
(*Proceedings of the 7th International Symposium on Ceramics in Medicine, Turku, Finland, July 1994*)

Zirconia for Medical Implants – The Role of Strength Properties

H.G. Richter, W. Burger and F. Osthues
Cerasiv GmbH, P.O. Box 1149, 73201 Plochingen, Germany

ABSTRACT

The strength of specimens made of different zirconia qualitities and of alumina were determined in dependence on hydrothermal treatment and on loading rate. The load bearing capacity of hip joint heads made of zirconia were measured using different taper qualities. It is shown that the strength properties measured under standard environmental conditions must not be regarded as the only decisive criterion for choosing an appropriate ball head materials.

INTRODUCTION

Ball heads made of alumina-ceramics have been used as parts of total hip joint prostheses successfully for 20 years. Zirconia became a material of interest for this application several years ago due to its promising strength properties (1, 2). Hip joint heads have to be able to sustain high loads for a long time and under the influence of aggressive environment. Therefore, the time dependent strength is investigated. It is known from literature (3, 4, 5) that in contrast to alumina, zirconia ceramics are susceptible to the combined influence of humid environments, temperature and stress. This influence is investigated at temperatures which are used for sterilization. The experiments were run on two zirconia qualitites and on alumina for comparison.

MATERIALS

From the different types of partially stabilized zirconia only two types are regarded to be candidate materials for hip joint heads, namely zirconia made from coprecipitated powder and zirconia made from yttria-coated zirconia powder. These two materials are based on two different powder processing routes by which yttria containing zirconia can be produced. In the coprecipitation route the stabilizing amount of yttria is added to the solution of the zirconia salt as an yttrium salt before the precipitation; in the coating route the already calcined zirconia particle are coated by yttria particles. In the coprecipitated zirconia the amount of yttria normally is between 5.0 and 5.4 wt.-%. In the case of yttria-coated zirconia powder the optimum amount of yttria is regarded to be lower, i.e. between 4.5 and 5 wt.%. For the present investigation commercially powders were used. The alumina-ceramic used for comparison is BIOLOX from CERASIV.

EXPERIMENTAL

From these materials 4-point-bending specimens with dimensions according to ASTM C-611 or DIN 51110 pt. 1., i.e. 45 mm x 4 mm x 3 mm (length x width x height) were prepared as well as hip joint heads of the type 28S.

The bending strength was determined in standard environment as well as after hydrothermal treatment. Hydrothermal treatment was perforemed at 140 °C for up to 120 hours. Burst tests on hip joint balls were run according to ISO 7206 part 5.

RESULTS AND DISCUSSION

1. Strength after hydrothermal treatment

In Table 1 the strength mean values of the materials investigated are compiled in dependence on the aging time. It can be seen, that the coprecipitated material has a higher strength at standard environmental conditions but lower strength after hydrothermal treatment of 120 hours when compared with the coated material. The strength of BIOLOX lies around 530 MPa and is independent of the hydrothermal treatment.

With regard to the aging mechanisms in zirocnia reference is made to the literature (e.g. 6).

Table 1: Strength of different zirconia qualities and of BIOLOX alumina in dependence on the aging time.

	Strength [MPa] after Aging time [h]			
Material	0	24	48	100
Zirconia - coprecipitated	1110	942	876	836
Zirconia - coated	986	923	922	874
Alumina (Biolox)	525	542	528	537

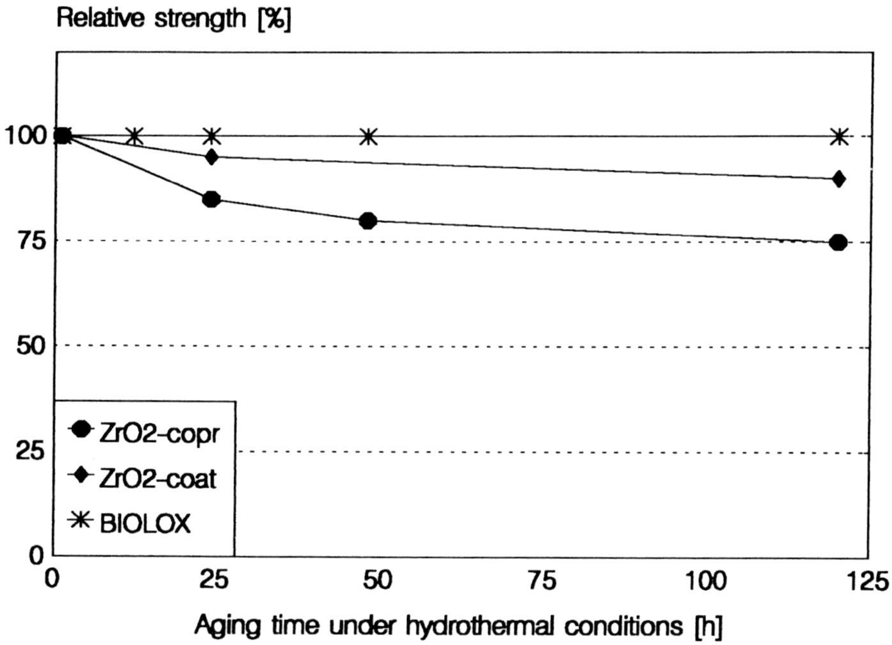

Fig. 1: Relative strength versus time of aging under hydrothermal conditions.

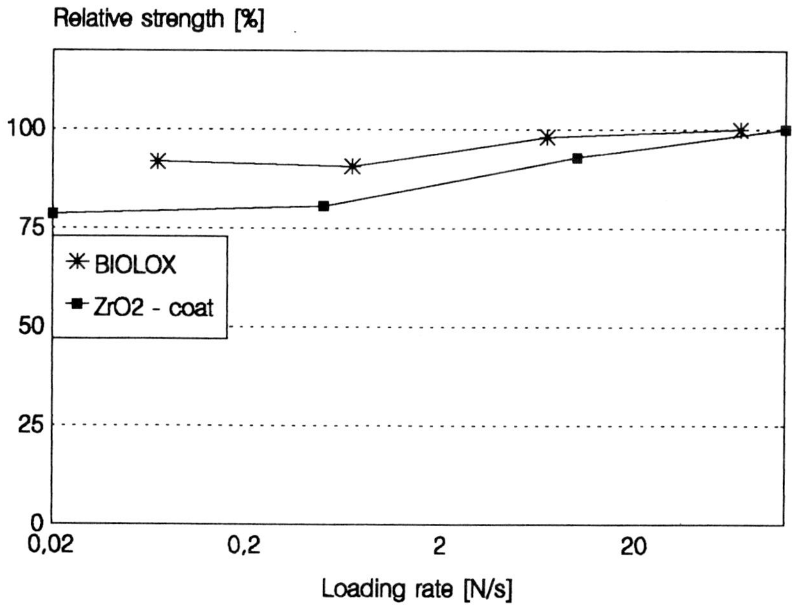

Fig. 2: Relative strength versus loading rate.

2. Fatigue

In Figure 2 the decrease of the mean strength with decreasing loading rate is plotted using relative strength values, i.e. the ratio of the strength to the strength at the highest loading rate used in these experiments is used. The strength at the highest loading rate is also called the "inert strength". It is evident that BIOLOX is less susceptible to fatigue than the zirconia qualities. From these data the static fatigue parameter n can be calculated. High values of n indicate low susceptibility to fatigue. The zirconia qualities have an n-value of around 40, BIOLOX an n-value of about 65.

3. Phase change

It was observed that the surface mainly of the coprecipitated zirconia has undergone a change. Therefore the phase structure and the thickness of this layer were investigated. Using X-ray-diffractometry it was found that the amount of monoclinic phase changed with the duration of the hydrothermal treatment. Figure 3 shows this dependence for the coprecipitated and the coated material. In the coprecipitated material an increase from about 5 % before hydrothermal treatment to more than 80 % monoclinic phase could be observed after a 24 h treatment, whereas in the coated zirconia material the increase of the monoclinic phase increases only gradually with the aging time and reaches only about 60 % after an aging time of 120 h. No change at all was found with alumina.

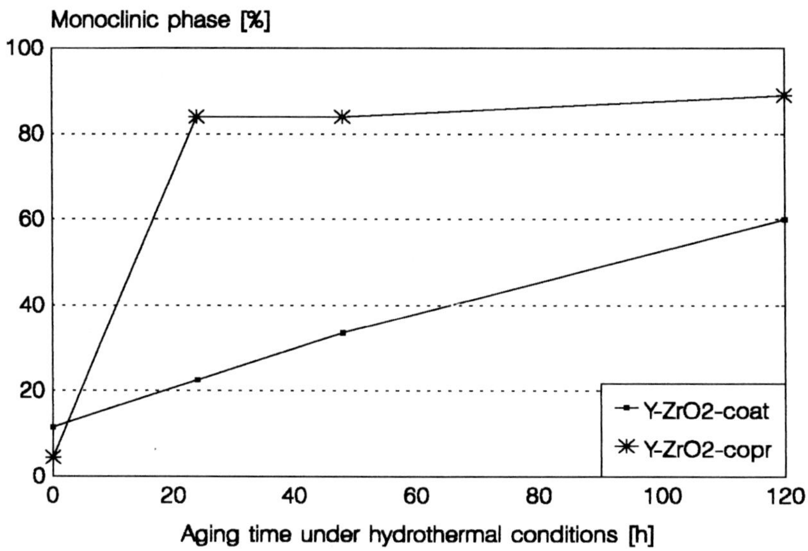

Fig. 3: Monoclinic volume fraction versus aging time.

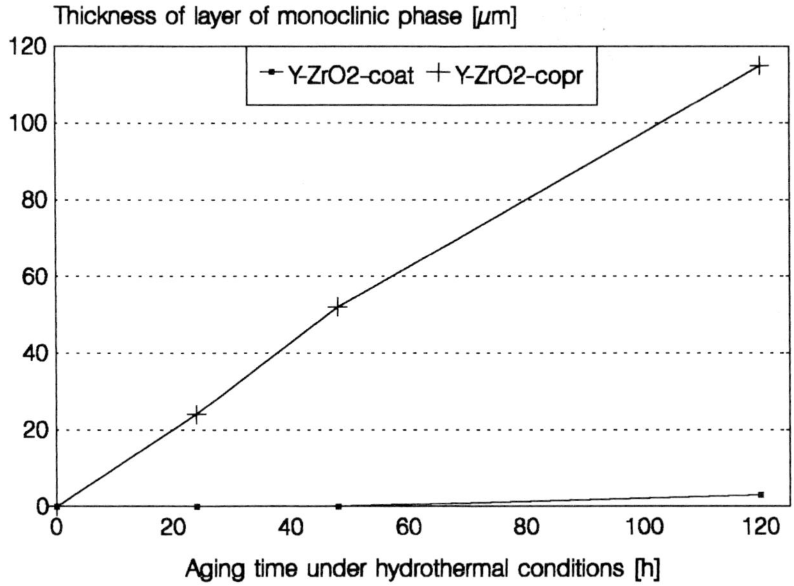

Fig. 4: Thickness of the monoclinic layer in dependence of the aging time.

The thickness of the monoclinic layer can easily be measured in dependence on the aging time under hydrothermal conditions. The results are illustrated in Fig. 4. A strong increase in the thickness of the monoclinic layer can be observed on the precipitated zirconia material, whereas no comparable thick layer could be found on the coated zirconia quality.

4. Burst strength of hip joint balls

It is of predominant interest to assess the performance of hip joint balls produced of the zirconia materials investigated. In a preliminary study especially the influence of the testing taper characteristics used for the burst test was investigated. Results are given in Table 2. It can be seen that the burst strength load strongly depends on properties like material, Young's modulus, hardness and surface structure. There is a siginificant tendency that tapers made of the harder material reduce the load bearing capacity of hip joint balls.

This means that for the load bearing capacity of hip joint balls the material strength is not the only determining factor. Further, it is evident, that the zirconia qualities show a slightly increased susceptibilty to fatigue.

Table 2: Burst strength of zirconia (copr.) hip joint balls (type 28L) in dependence on taper parameters

Taper Material	Hardness HV 10	Surface structure Ra [μm]	Burst strength [kN]
Ti-alloy	352	4,6	80
Ti-alloy	310	48,4	93
CoCr-alloy	435	2,7	44
CoCr-alloy	644	2,7	47
NiCrMo-alloy	310	60,3	108

CONCLUSIONS

As in the case of alumina, zirconia intended for use as implant material hase to be assured very carefully for time and environmental depending strength properties in order to make sure that only the optimum suited material is used. Zirconia made of yttria-coated zirconia powder has pronounced advantage over coprecipitated zirconia powder. The load bearing capacity of hip joint ball heads is almost as strongly influenced by the taper characteristics as by the ball material.

REFERENCES

1. Garvie, R.C., Urbani, C., Kennedy, D.R. et al J. Mater. Sci. 1984, 19, 3224-3228

2. Christel, P. et al J. Biomed. Mater. Res. 1989, 23, 45

3. Yoshimura, M. et al J. Mater. Sci. Lett 1987, 6, 465-467

4. Sato, T. and Shimada, M. J. Amer. Ceram. Soc. 1985, 68 356-359

5. Schmauder, S., and Schubert, H. J. Amer. Ceram. Soc. 1986, 69, 534-540

6. Lawson, S., Gill, C. et al In: Third Euro-Ceramics, Vol 3 (1993) 507-512

Clinical Results

Bioceramics, Volume 7 Edited by Ö.H. Andersson and A. Yli-Urpo
(*Proceedings of the 7th International Symposium on Ceramics in Medicine, Turku, Finland, July 1994*)

Obliteration of Frontal Sinuses with Bioactive Glass after Chronic Suppurative Sinusitis. One Year Follow-up

K. Aitasalo,[1] M. Peltola,[1] J. Suonpää,[1] A. Yli-Urpo,[3] Ö. Andersson,[4] M. Varpula[2] and R-P. Happonen[3]

Departments of Otolaryngology,[1] Radiology[2] and Dentistry,[3] University of Turku and Åbo Akademi University,[4] Finland

ABSTRACT

Bioactive glass S53P4 was employed as an obliteration material in 5 osteoplastic frontal sinus operations. Clinical examination, CT-scans, laboratory monitoring of inflammatory parameters as well as liver and kidney function and histopathological examination of two postoperative biopsies indicated that the material was stabile and well tolerated. No signs of reinfection were noted. Further analysis of the CT-scans with ROI-selection revealed a mean decrease of 15 percent in the density of the obliteration material. This is consistent with the drying of the wet glass granules in a phantom model. Bioactive glass is a promising and well tolerated alloplastic material for frontal sinus operation.

INTRODUCTION

The principle of the osteoplastic obliteration of the frontal sinus is that a sinus completely obliterated cannot give rise to suppuration. The method is reserved for those chronically infected sinuses which have failed to respond to other means of treatment. The same method is also used in the treatment of osteomata, trauma and mucoceles of frontal sinus region.

A careful stripping of all the sinus mucosa with removal of the inner cortical layer is an essential part of the operation. This is followed by the sinus obliteration, which has become an integral part of the frontal osteoplastic operation. However, the material with which this is best achieved is still unsolved (1).

Bergara and Itoiz (2) and Montgomery and Pierce (3) demonstrated that implanted autogenous fat accomplish total obliteration of the sinus cavity and is resistant to infection and osteoneogenesis if the mucous membrane and the inner cortical bony lining are removed completely. Mucous membrane also fail to regenerate in the obliteration cavity. McNeil (4) found that if the fat was traumatized or the cavity infected, fat transformed to fibrous tissue. He implanted also experimentally muscle, Gelfoam and blood clot, all of which failed to achieve obliteration.

Alloplastic materials have been employed to eliminate the second procedure required to obtain an autograft and because failure frequency has in many reports exceeded 25 percent (5). Failla (6) reported good results in obliterating the traumatized frontal sinus with acrylate. Barton (7) reported succesful obliteration with proplast in eight patients with chronic frontal sinusitis followed-up for 1 to 5 years. However, the radiographic appearance of the obliterated cavity resembled more closely fibrous tissue than replacement with bone.

In the ENT-Clinic, Turku University Hospital we have employed bioactive glass as an obliteration material in the osteoplastic frontal sinus operations since 1991. In the present study the stability of the obliteration material was controlled by CT-scans analyzed by the region of interest (ROI)-selection.

MATERIAL AND METHODS

Five patients, two men and three women (mean age 45.6 years), suffering from chronic suppurative frontal sinusitis were treated surgically by osteoplastic frontal sinus obliteration. The operation was performed through coronal incision approximately 3 cm behind the hairline. The anterior bony wall of the frontal sinus was raised and the diseased mucous membrane was stripped from the sinus and all its recesses. The inner table of the sinus was drilled down with a diamond burr. The cortical plate was thinned and any septa within the sinus or on the bone flap was removed. The upper orifice of the nasofrontal duct was obliterated with lyofilized dura. Granules and blocks of bioactive glass S53P4 (8) were placed in the frontal sinus cavity. Two sizes of granules (A. 0.9-1.0 mm and B. 0.63-0.8 mm) were mixed. The diameter of the blocks were more than 2 mm. The glass obliterated the sinus completely. After this, the bony flap was replaced and fixed with titanum microplates. The patients were given i.v. antibiotics for five postoperative days. Laboratory monitoring of the patients haematology and serum CRP, asparate transferase (SGOT) and creatinine was performed before the operation and three, six and twelve months after the operation. CT-scans of the frontal region were repeated one week, six and twelve months after the operation.

The density of the material used for obliteration was analyzed by the region of interest (ROI)-selection by the CT Pace TM Scanner System from nine separate control regions in each CT-image. The system is based on digital absorbtion coefficients of each pixels in the selected region transformed to Hounsfield (H-) values. For the valid comparision of these values at each of the nine 3 mm wide regions special attention was paid to equalize the head position at each scanning sessions.

External reference phantom consisted of glass granules with same sizes as used for obliteration (A, B, and C.). The granules were scanned as dry and moistened with saline (NaCl 9mg/ml). The density of the material were analyzed in ten regions with the ROI-selection. All parameters of the CT-scanning (thickness and the position of the granule mass, radiation energy etc) were equalized with the patient's scannings. Wet granules were used to simulate the situation at the time of the obliteration. A comparison was made between the measurement results of the patients and the phantom model.

In two patients the extraction of the miniplates offered a possibility to harvest samples from the osteoplastic flap and the obliteration material for histopathological study. These operations were performed 12 and 16 months after the primary operation.

The statistical analysis was carried out using analysis of variance for repeated measures and analysis of covariance for repeated measures, where the possible confounding factors were included in the model as covariates and their effect adjusted when needed.

RESULTS

Densitometry of the phantom models

The effect of the moistening of the mixture of the granules of three sizes on the H-values is presented in figure 1a. In the figure the H-values of the 30 selected regions are shown on the right side and corresponding H-values of the wet granules on the left side. The mean H-values of wet and dry granules were 1182 and 742, respectively (difference 37.2 %). The H-values recorded separately from three granule sizes are shown in table 1.

Table 1.
The results of the densitometry by ROI-selection in three granule sizes of biactive glass scanned as wet (granules in saline solution) and dry (only granules) and expressed with Hounsfield values (means of 10 regions).

Granule size	Hounsfield values Wet	Dry	Difference	%
(A) 0.9-1.0 mm	1175.4	771.4	404.2	34.4
(B) 0.63-0.8 mm	1159.7	653.3	506.1	43.7
(C) A + B	1212.7	804.1	408.6	33.7

Clinical study

All the five patients were almost symptomless at the follow-up examinations and no signs of inflammation or mucous membrane regeneration in the frontal sinus were seen in the postoperative CT-scans. The bioactive glass particles accomplished total obliteration of all the sinuses with only minor visible changes during one year follow-up period. In the postoperative laboratory controls all the parameters monitored fell within the reference values.

Figure 1. The densities (analyzed by ROI-selection and expressed as Hounsfied values) of wet and dry bioactive glass granules in the CT-scans of the phantom model (a) and the densities of the frontal sinuses in the CT-scans at the follow-ups (b). Nine regions of each five patients were analyzed.

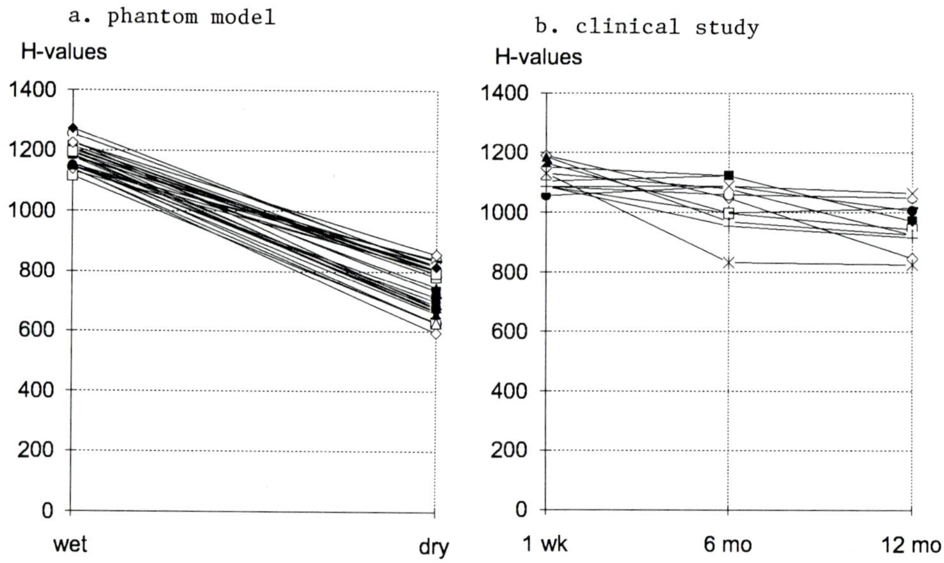

The results of the densitometric ROI-selection of the frontal sinus scans of the five patients immediately after the operation and after 6 and 12 months follow-ups are shown in Figure 1b. For the statistical analysis the H-values of three subregions are averaged in the figure. Accordingly, three density averages of each obliterated cavity are presented by one line in the figure. The mean H-value at the immediate postoperative control scans was 1130.7. At the 6 months follow-up examination it was 1029.5 and after 12 months 956.4. The average decrease was 101.2 (8.9 %) and 173.8

(15.4%) at 6 and 12 months, respectively. The interindividual differences were minor showing quite similar changes in the surface and inside the obliteration material. The decrease from the initial state was significant and comparable with findings in the phantom model.

Histopathological study

The histopathological examination of the bone and glass samples harvested at the removal of miniplates of two patients showed fibrosis around the glass granules. The surface of the bioactive glass was surrounded by a thick reaction layer. No new bone formation between the glass particles could be seen. No inflammatory changes or foreign body reactions were demonstrated.

DISCUSSION

The osteoplastic operation of the frontal sinus is restricted to patients with chronic suppuration and failure to respond to other means of treatment. In this study the patients represent a group with severe longlasting infection and great risk of complications and residual infections. The rather high failure rates in earlier studies and our own experiences have resulted in a need to seek for new stabile obliteration materials. Although the number of patients and their follow-up time is limited the present experience shows that the bioactive glass seems to obliterate the frontal sinus as well as fat and spongious bone. The second operation for harvesting the autogenous material can be avoided.

The stability of the bioactive glass was also shown in two biopsy samples showing no inflammation or foreign body reaction inside the sinus. The bioactive glass granules were well tolerated and accomplished a good obliteration at the anterior wall of the frontal cavity. The laboratory data also indicated the safety of this synthetic material.

The repeated CT-scans of the frontal sinuses during the follow-up period showed only minor visible changes with no signs of regeneration of mucous membrane or reinfection of the cavity or inside the osseous laminae of the sinuses.

The ROI-selection was used because this kind of densitometry reveals accurately minor chances e.g. microabcesses, mucous membrane regeneration etc. inside the obliteration material. No signs of such failures were seen. The decrease of the density by 15.4 percent during the follow-up seems to be due to fibrosis and drying of the obliteration material. This is supported by the results of the study with the phatom model. The scanned granules were either totally wet resembling the situation at the end of the surgical procedure or dry.In the patient's frontal sinus the real situation is somewhere between wet and dry.

We conclude that bioactive glass is a promising and well tolerated alloplastic material for frontal sinus operation. It is easy to handle and sterilize. Total accurate obliteration of the sinus is achieved with different sizes of granules and blocks. The obliteration is maintained by the stability of the material.

REFERENCES

1. Lawson,W. In: Surgery of the Paranasal sinuses 2nd ed, WB Saunders Co, Philadelphia (1991) 183-218

2. Bergara,A.R. and Itoiz,A.O. Arch Otolaryngol (1958) 61 271-283

3. Montgomery,W.W. and Pierce,L. Trans. Am. Acad. Ophthalmol.Otolaryngol. (1963) 67 46-54

4. McNeil,R.A. J. Laryngol. Otol. (1966) 80 953-961

5. Schenck,N.L. Laryngoscope (1975) 85 76-92

6. Failla,A. Laryngoscope (1968) 78 1833

7. Barton,R.T. Laryngoscope (1980) 90 47-52

8. Andersson,Ö.H., Liu,G., Karlsson,K.H., Niemi,L., Miettinen,J. and Juhanoja,J. J. Mater. Sci.: Materials in Medicine (1990) 1 219-227

Bioceramics, Volume 7 Edited by Ö.H. Andersson and A. Yli-Urpo
(*Proceedings of the 7th International Symposium on Ceramics in Medicine, Turku, Finland, July 1994*)

Clinical Applications of Bioglass Implants

J. Wilson,[1] A.E. Clark,[2] E. Douek,[3] J. Krieger,[4] W. King Smith[5] and J. Saville Zamet[6]

[1] Bioglass Research Center, University of Florida, USA
[2] Department of Prosthodontics, University of Florida, USA
[3] Department of ENT, Guy's Hospital, University of London, UK
[4] Department of Periodontology, University of Florida, USA
[5] East Broadway, Bel Air, Maryland, USA
[6] Department of Periodontology, Eastman Dental Hospital, University of London, UK

ABSTRACT

Bioglass® is a bioactive glass unique in its ability to bond to both hard and soft tissue. During the last decade it has been used clinically in a variety of situations in both medicine and dentistry. Devices are now available to treat conductive deafness, alveolar ridge resorption and bone loss due to periodontal disease. Where augmentation of bone is required to fill cystic and surgically created defects, the powdered form of Bioglass® has proved successful either alone or mixed with autologous or freeze-dried bone. Case histories from studies in several locations are provided to illustrate clinical applications.

INTRODUCTION

Bioglass® was first implanted in patients in March 1984 to start a clinical trial of middle ear devices. Since then its clinical use has expanded to use in many surgical and dental specialties. The material can be shaped at the time of surgery and a simple shape in bulk glass for use in the middle ear is now available. A similar shape, matched to a dental bur, is used for ridge maintenance after multiple extractions. The particulate, developed for the treatment of periodontal defects, has found use in various conditions where bone augmentation and bone fill are needed. As an extender of autologous bone the particulate has enabled defects to be filled with minimal morbidity at the host site. In addition it can substitute for freeze-dried bone, which is sometimes unacceptable to patients.

CLINICAL USES

Middle Ear Devices

The first clinical application of Bioglass® was as a middle ear prosthesis for the treatment of conductive hearing loss. Such hearing loss is commonly due to chronic otomastoiditis with and without cholesteatoma but may also be due to trauma and to congenital abnormality. Many materials previously used for treatment were unsuccessful in long-term use because of encapsulation by scar tissue, which dampens the sound before it reaches the oval window, but mainly because they did not bond to the ear drum. Extrusion is reported as reaching 80% in 3-5 years for other materials.[1] This is in spite of the fact that all of these materials required interposition of cartilage or bone pate between device and the ear drum. Devices implanted in two clinical trials, one at the University of Florida (Dr. G.E. Merwin) and a second at Guy's Hospital of the University of London (Mr. Ellis Douek) have now been *in situ* for up to 10 years without a single extrusion. The ability of Bioglass® to bond to the soft tissue of the ear drum without the need for bone pate or cartilage, thereby eliminating abrasion and inflammation at the interface, is believed to be the explanation for this success. The original design of the device (MEP) (Fig. 1A) was similar to those then in use, made from other materials. In a few cases the implants have begun to break up after 8-10 years in place. Preliminary examination of such an implant suggests that the junction between stem and platform is vulnerable, especially if an implant is cracked during implantation. Experience now dictates that any implant damaged during implantation must be rejected, since soft tissue will grow into the cracks and in time widen them sufficiently to fail. However, it is important to note that 8 or more years of successful clinical performance is superior to that of any other available device.

The latest design of this devices is truncated cone (Douek-MED®) which can be shaped at the time of surgery to accommodate any anatomical variation found in the middle ear. (Fig 1B) The implant can be contoured to ensure that it does not come in contact with tissues such as the facial nerve, where bonding must be avoided, but does fit well over the stapes or around the malleus where those ossicles remain. The vulnerable junction has been eliminated in this design.

Case History (E.D.)
A conical implant of Bioglass® was placed in a female patient under a fascial graft. Hearing results were satisfactory. The patient later developed a severe middle ear infection resulting in graft failure, which was treated abroad with appropriate antibiotics. Some time later the patient consulted the surgeon, asking for a replacement tympanic membrane graft so that she could resume swimming. Some hearing, surprisingly, was still present. On examination it was seen that, despite the large perforation, an area of the graft adherent to the surface of the cone had survived. She was re-operated on and provided with a second fascial graft and Bioglass® implant with excellent results.

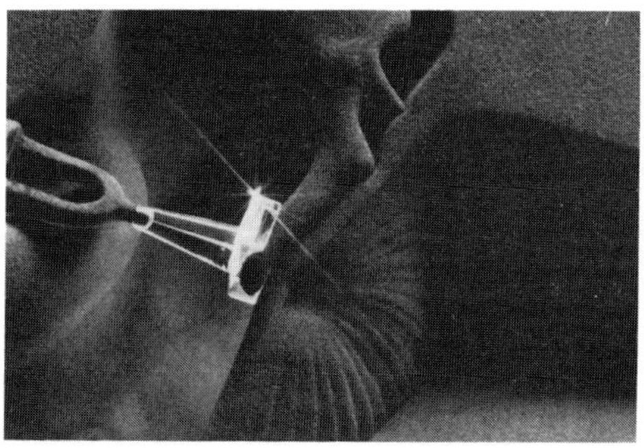

Fig 1A. MEP

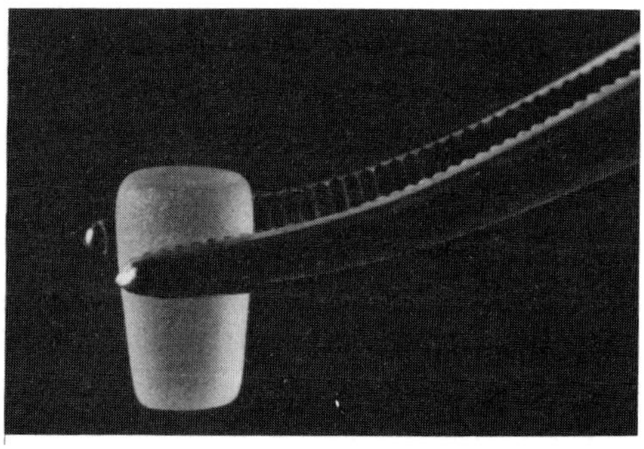

Fig. 1B. Douek-MED®

Cochlear Implants

The deafness due to inner ear disease cannot be treated in such a straight forward fashion. Intra and extracochlear implants must stimulate the nerve endings of the inner ear and the resultant signals must be converted electronically into sounds which the patient can interpret. The internal signals and the external electronics, meet at the surface of the mastoid where the transcutaneous connection must be made. Such a connection must prevent infection which could reach the meninges, with dangerous consequences. No metal or polymer can achieve a transcutaneous seal but electrodes coated with Bioglass® have been successfully used in this situation for several years. The electrode assembly is held in a connector made from alumina, coated with Bioglass®. The connector becomes bonded to the bone of the mastoid and thus immobilized, it passes through the skin and the soft tissue bonded to the coating ensures the integrity of the interface. Even though the interface is stressed daily when the patient attaches and removes the connection to the speech processor, these interfaces have remained stable for several years.

Endosseous Ridge Maintenance Implants

Conical implants of Bioglass® have been used for the maintenance of the endosseous ridge since clinical trials began in 1984, they have been commercially available since 1990. In clinical trial the loss and dehiscence rate compared favorably with results of hydroxyapatite implants.[2] These data have been described and the mechanisms responsible determined. It is now clear that soft tissue bonded to Bioglass® provides a compliant interface which closely resembles the natural one and at the same time, contributes to the long-term retention of these implants.[3]

Case History (W.K.S.)
C.E.H., 53 yr male with excessive maxillary bone loss and periodontal disease. Oral hygiene was not ideal, and this patient was not a candidate for endosseous titanium dental implants. With his history of alveolar bone loss, the treatment chosen was the removal of the remaining six maxillary teeth, placement of Bioglass® endosseous ridge maintenance implants (ERMI), and insertion of an immediate denture. On June 2, 1992 under light sedation and local anesthesia, the maxillary right first bicuspid through the left cuspid were extracted. Six implants were placed in the sockets of the extracted teeth after socket preparation with custom burs. All implants were placed at least 2mm below the crestal bone. Good tissue closure was obtained by maintaining the interdental papilla where possible, and suturing it across the open socket. By relieving the inside of the denture with an acrylic bur and placing a soft reline material, immediate placement of the denture is possible. The patient must be followed closely to relieve any denture irritation so that dehiscence over the implant does not occur. In my experience if an implant becomes exposed, careful reduction of the implant with a diamond bur will allow granulation of tissue over the implant. C.E.H. has an uneventful healing period with minimal discomfort. After six months, a permanent rebase of the denture was performed. At eighteen months a final panoramic radiograph was taken which showed the implants to be well integrated with the bone and very little change in the structure of the edentulous alveolar ridge. (Fig. 2A and B)

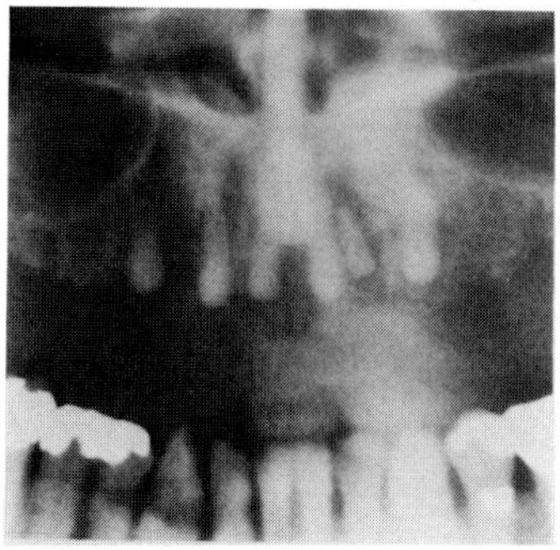

Fig. 2A. 2 months post-op.

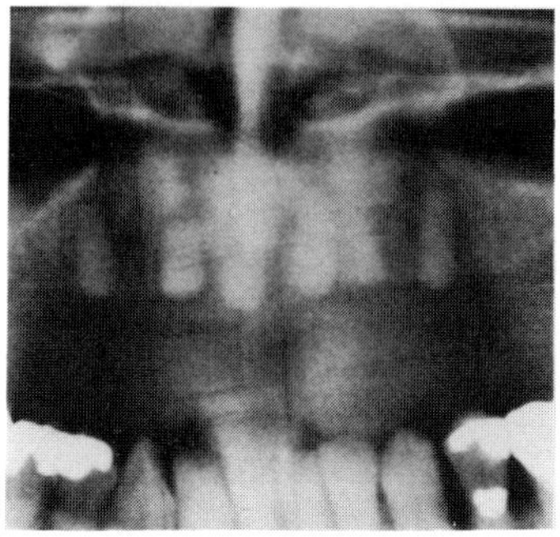

Fig. 2B. 18 months post-op

<u>Case History</u> *(A.E.C.)*

In July 1984, RW, a 35 year old white male, enrolled in the clinical trial, had three remaining maxillary teeth extracted and one ERMI placed in the upper left lateral incisor socket. Five mandibular teeth were extracted and two ERMI's placed, one in the lower right canine socket and one in the lower left first premolar socket. (Fig. 3A) Other mandibular sockets were too large for the available ERMIs. The mandibular central and lateral incisor natural teeth were not extracted. Six weeks later an upper, complete denture and a lower removable partial denture were made and delivered to the patient. At the most recent visit in August of 1993, (nine years post implantation) all three implants were in place (Fig. 3B) and the prostheses continue to function well.

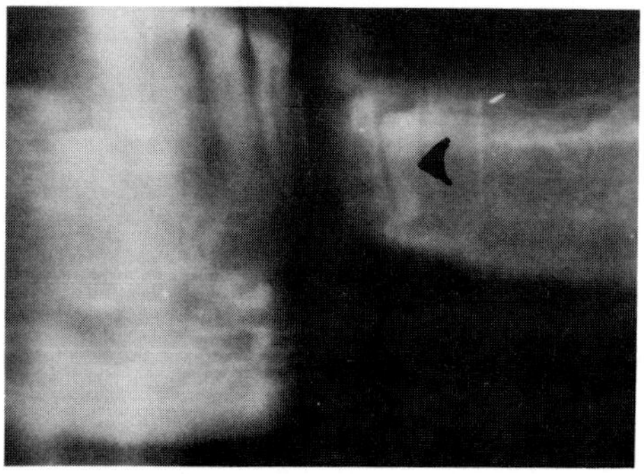

Fig. 3A. Implant immediately post-op.

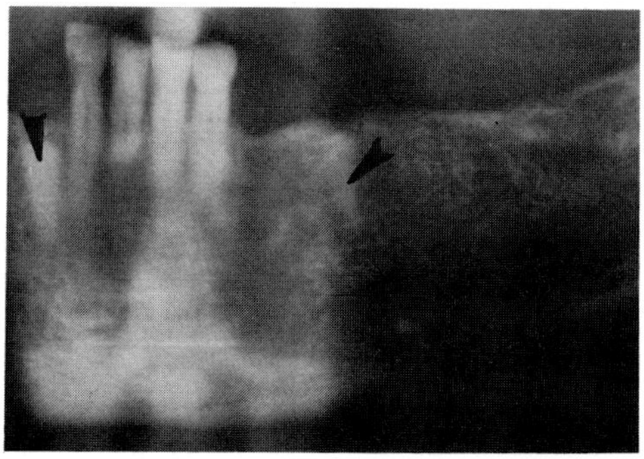

Fig. 3B. Mandibular implants 9 years post-op.

Periodontal Treatment (Perioglas®)

Particulate Bioglass® was developed initially for the treatment of the consequences of periodontal disease. In preliminary animal studies it was shown to be more effective than other commercially available materials. It was unique in allowing restoration of the fibrous attachment of the periodontal membrane quickly enough to prevent epithelial downgrowth.[4] These observations were confirmed in clinical trial from which five-year data will shortly be published.

Case History *(J.K.)*
A 45 year old white male (#006) enrolled in the clinical trial in 1989. An osseous defect distally on the lower left molar (18) was filled with Perioglas®. Bone fill measured radiographically after 25 months was 3.11 mm. (Fig. 4A and B)

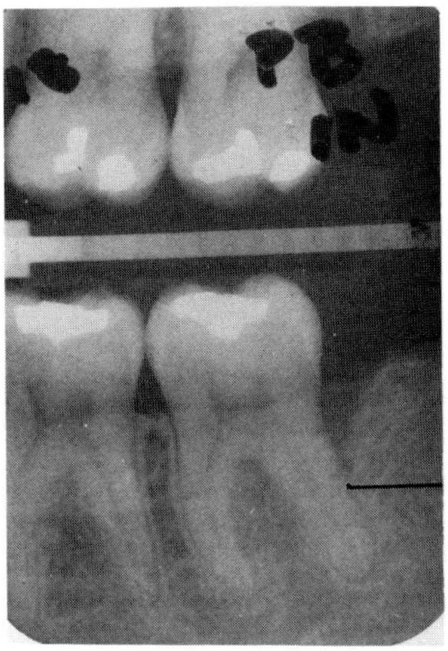

Fig. 4A. Pre-op.

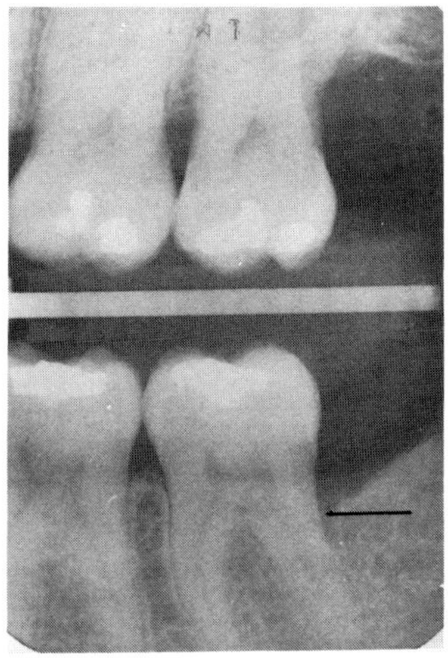

Fig. 4B. 25 months post-op. Bone level is marked.

The material is now commercially available and has been used successfully in other applications, such as is illustrated by the following case histories.

Case History *(W.K.S.)*
Identical teenage twins, AS and TS presented with identical malocclusions. An anterior open bite which did not respond to orthodontic treatment. Pre-surgical model surgery provided adequate occlusion by intruding the posterior segment 3mm and lowering the anterior segment by 3mm. The two posterior segments were particularly small and difficult to stabilize. Luhr-system miniplates were used. Surgery on AS was a routine LaForte 1 osteotomy with an acrylic splint wired to the maxillary teeth to position the segments. The mandibular teeth were wired to the maxillary teeth to establish their new relationship. The procedure was complicated by the smallness of the segments and a paper-thin sinus wall. It was expected that the splint would be needed for 4-6 weeks to compensate for deficiencies in stabilization and non-optimal bone-bone contact. Perioglas® was used to cover all

the Luhr mini-bone plates used and to fill areas of bony cuts. It did not migrate and stayed in place and maintained hemostasis during suturing. The intermaxillary fixation was released to allow free motion of the lower jaw. The same procedures were used on TS the following day. The ability of the particulate to augment autologous bone has been established in animal studies.[5] This property and the hemostatic effect of the material are proving valuable in maxillofacial surgery. Each patient received two units of blood and took only one dose of pain medication, that one was immediately post-op. After 4 days the maxillary segments were still mobile but by 9 days were very tight, and the splints were removed. After three weeks the patients were returned to an unrestricted diet. The use of Perioglas® in these procedures significantly affected the outcome and reduced the recovery period.

Case History (W.K.S.)
C.C.S. is a 46 yr male referred for the evaluation of an impacted mandibular right third molar surrounded by a cystic lesion. Removal of the tooth and cyst was indicated. Because of the proximity to the inferior alveolar nerve, it was explained to the patient that paresthesia was possible. Under light sedation with local anesthesia, the tooth was surgically removed as was the cystic tissue. The defect was filled with a mix of 1.0 cc demineralized bone and 1.0 cc Perioglas®. The incision was tightly closed with 3-0 black silk. The patient was placed on antibiotics for one week.
The post-operative healing period was uneventful. The pathology report was inflamed dentigerous cyst. There was no paresthesia involvement with the inferior alveolar nerve. A panoramic radiograph after 3 months showed excellent bone repair. (Figs. 5A, B and C).

Case History (J.S.Z.)
A 35-year old female presented complaining of an unsightly appearance related to her maxillary, partial denture, replacing 21 and 22 which had been lost through trauma some five years previously. Oral examination showed a considerable labial concavity in the alveolar ridge in 21, 22 area where the labial plate of bone had been lost. The patient also had a high lip line. She was anxious to replace the partial denture with a fixed bridge. In this case, membrane technology using Goretex membranes (W.L. Gore) was adapted for localized ridge augmentation. Remote incisions over healthy bone were used to reflect full thickness flaps, the bony defect was thoroughly debrided and fenestrations made through to the cancellous spaces. In order to "tent" the e-PTFE material, Orthosorb absorbable pins (Johnston & Johnson) were used. In previous cases collagen fleece (Pentopharm A.G.) or decalcified freeze-dried bone have been used to take up the underlying dead space and stabilize the blood clot. Perioglas® proved to be an excellent alternative with no danger of hyper-sensitivity reactions or transmission of disease. Perioglas® granules mixed with normal saline to a stiff consistency were packed beneath the membrane and around pins. Flaps were then sutured over the top of the membrane.
The membrane (the pins are absorbable in six months) was removed 6 months later and construction of definitive bridge-work was then completed, with satisfactory results.

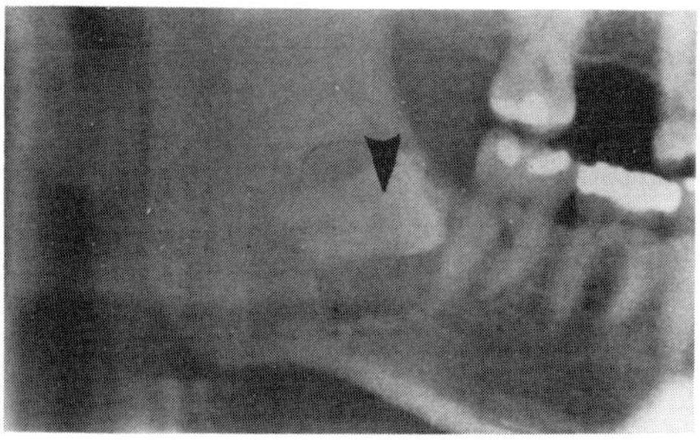

Fig. 5A. Showing impacted molar.

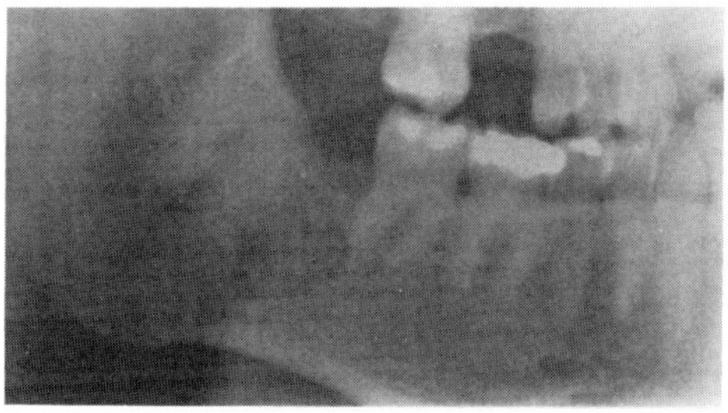

Fig. 5B. Post-op.

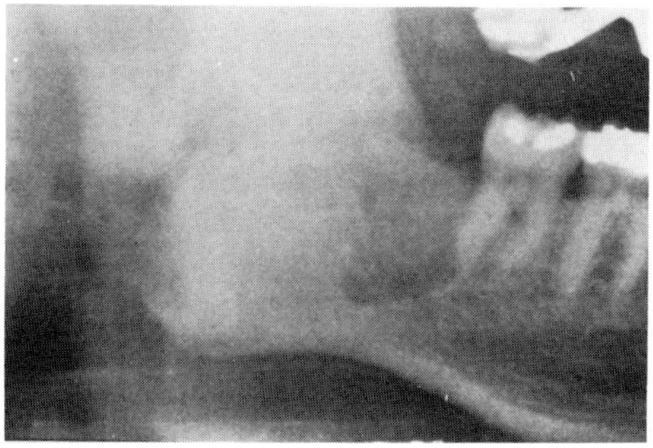

Fig. 5C. 3 months post-op.

CONCLUSION

Bioactive Bioglass® products, used as solid implants or as particulate Perioglas®, have proved successful in clinical practice. The particulate has proved to be valuable when mixed with autologous bone in promoting bone fill in defects, thus reducing the amount of bone needed and simultaneously reducing the morbidity associated with bone harvest. Used separately or in combination these implants can provide treatment for a wide variety of clinical conditions.

REFERENCES

1. Frootko, H. Biomaterials in Otology M. Nijhoff, Leiden, 1984, 169-176.

2. Hench, L., Stanley, H., Clark, A., Hall, M., and Wilson, J. In Bioceramics 4 Butterworth-Heinemann Ltd., Guildford, UK, 1991, 229-238.

3. Wilson, J., Clark, A., Hall; M., and Hench L. J. Oral Impl. 1993, XIX, 4, 295-302.

4. Wilson, J. and Low, S. J. App.Biom. 1992, 3, 123-129.

5. Wilson, J., Yu, L., and Beale, B. In Bioceramics 5 Kobunshi Kankokai, Japan, 1992 139-146.

ACKNOWLEDGEMENT

The authors gratefully acknowledge the support of U.S. Biomaterials Corporation and especially Dr. David Greenspan, Vice President of Research and Development.

Bioceramics, Volume 7 Edited by Ö.H. Andersson and A. Yli-Urpo
(*Proceedings of the 7th International Symposium on Ceramics in Medicine, Turku, Finland, July 1994*)

Preclinical and Clinical Application of Bioactive Bone Cement

T. Yamamuro, T. Nakamura, H. Iida, Y. Matsuda, K. Ido and Y. Senaha
Department of Orthopaedic Surgery, Faculty of Medicine, Kyoto University, 54, Shogoin Kawara-cho, Sakyo-ku, Kyoto, 606 Japan

ABSTRACT

A new bioactive bone cement, consisting of $CaO-SiO_2-P_2O_5-CaF_2$ glass powder and bisphenol-a-glycidyl methacrylate (Bis-GMA) resin, forms direct bonding with living bone tissue within a few weeks after implantation. Its compressive strength after hardening is 180MPa. The bioactive bone cement has been used to fix a hip prosthesis in 12 dogs, and the results of histological and mechanical examinations performed 4 to 25 weeks after implantation were compared with those obtained in another 12 dogs with PMMA bone cement. Femurs containing the prosthesis stem were transectioned into 5mm thick segments, and failure loads obtained in each group by push out test of the stem for each segment were compared to each other. The failure load with bioactive bone cement was 1.2 and 1.7 times greater than that with PMMA cement one month and three months after implantation respectively. The former further increased significantly as time goes thereafter, while the latter remained unchanged. The bioactive bone cement was used in a few aged patients with loosened hip prostheses for the purpose to substitute for a large bone defect and to fix a new prosthesis at revision operation. The longest follow up period is still 10 months, but the patients have been doing well with no adverse effect to date.

INTRODUCTION

In 1961, Charnley first introduced polymethylmethacrylate (PMMA) cement to fix hip prosthesis to bone, and it has been widely used in various prosthetic replacement operations, since then. PMMA bone cement, however, has several inherent problems such as nonadhesiveness with bone, high heat generation during its polymerization, and relatively low mechanical strength. For the purpose to overcome these problems, we developed a new bioactive bone cement which has a capability of bonding directly with living bone tissue in a few weeks, does not generate high heat when it hardens, and has significantly greater mechanical strength than PMMA cement after hardening[1]). In the present study, we investigated adhesiveness of bone cement with the femur and the acetabular bone of dogs which underwent prosthetic replacement of the hip by the use of either bioactive bone cement or PMMA bone cement.

The analyses were performed both histologically and mechanically. We also conducted clinical studies in a few high aged patients to confirm the adhesiveness of bioactive bone cement with the human bone by applying the bioactive bone cement to fill in a large bone defect remaining after extraction of a loosened hip prosthesis.

MATERIALS AND METHODS

Cement Preparation

Mother glass of AW glass-ceramic (CaO 46.7, SiO_2 35.6, P_2O_5 17.0, and CaF_2 0.7 weight ratio) was prepared. It was melted in a SiC furnace at 1500°C for 4 h. The melt was quenched between water-cooled steel rollers and formed into ribbons, then pulverized with an alumina ball mill [2]). The average diameter of the glass powder particle was 3 μm. The glass powder was silane-treated with γ-methacryloxy propyl trimethoxy silane. Benzol peroxide, 0.4% per unit weight ratio, was added. The Bis-GMA resin was prepared from equal weights of Bis-GMA and triethylene-glycol dimethacrylate. N,N-dimethyl-p-toluidine, 0.2% per unit weight of resin, was dissolved into the mixture [1]). Two types of bioactive bone cement were prepared; dough type and injection type. Their compositional weight ratios are shown in Table 1.

Table 1. Compositional Weight Ratio (wt %)

type of cement	bioactive cement powder	Bis-GMA	resin filler	Silica filler
dough type	73	17	5	5
injection type	61	29	5	5

Dough type was applied to the acetabular component with hand and injection type was applied to the femoral component with a syringe in animal experiment as well as in clinical application.

Mechanical Properties of Bioactive Cement

The cement paste before polymerization was placed in a steel mold (6mm in diameter, 12mm in length). After the composite has hardened, the specimen was removed from the mold and soaked in normal saline at 37°C. After one day of soaking, the cement was subjected to mechanical testings. Mechanical properties of bioactive bone cement of dough type and injection type are shown in Table 2.

Table 2. Mechanical Properties of Bioactive Cement

type of cement	compressive strength (MPa)	elastic modulus (GPa)	fracture toughness (MPa)
dough type	199	5.1	7.8
injection type	163	4.0	10.0

Animal Experiment

A hip prosthesis for dog was made of stainless steel for the femoral component and of UHMWPE for the acetabular component. The femoral component consisted of a head of 12mm in diameter and a stem of 50mm in length. The polyethylene liner had an outer diameter of 18mm and an inner diameter of 12mm. Twenty four female beagle dogs aging between 8 and 10 months were used. They were divided into two groups; bioactive cement group and PMMA cement group. All of them underwent prosthetic replacement of the left hip joint through a lateral approach by severing the greater trochanter. The femoral head was removed at the base of the femoral neck, the cancellous bone of the proximal half of the femur was removed using reamers and rasps, and a bone chip was inserted intramedullarly at a presumable level just distal to the tip of the prosthesis stem to function as a cement restrictor. Articular cartilage of the acetabulum was removed with a reamer to expose the subchondral bone, in which multiple anchoring holes about 2mm in diameter were made. In bioactive cement group, dough type cement was used to anchor the acetabular component and injection type cement was used to fix the femoral component. Similarly, in PMMA cement group, CMW 1 was used for the acetabulum and CMW 3 for the femur. After both components were set, the prosthesis head was reduced into the socket, and the greater trochanter was re-attached. The dogs were kept free after operation. They started walking on the operated hip in a week postoperatively.

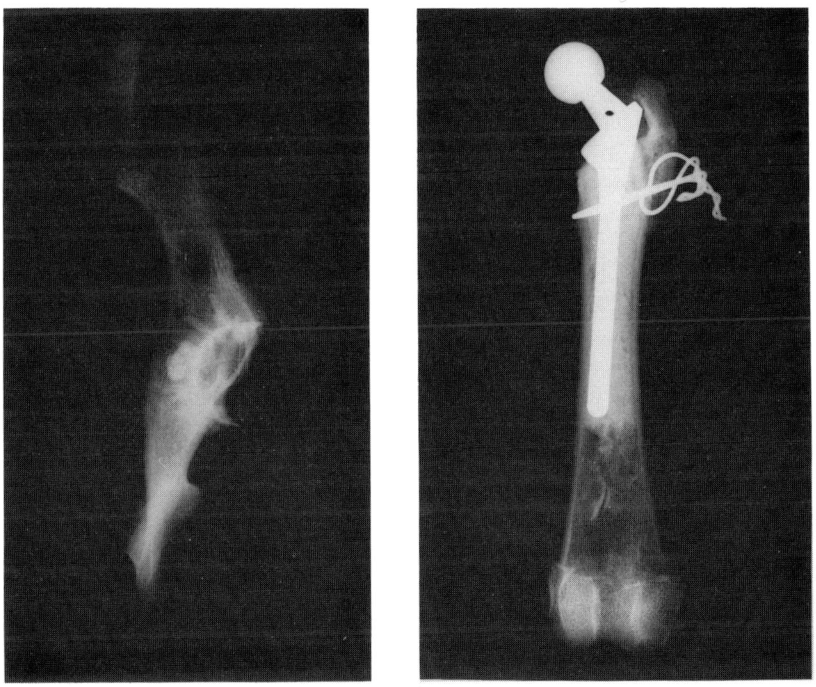

Figure 1. Radiograms of specimen harvested 3 months postop.

Mechanical Test

Animals were sacrificed one month, three months, and six months after operation, and the acetabular bones and femurs containing prosthesis component were harvested. The femurs were transectioned into eight segments with 5mm thickness each. Four segments were used for the mechanical test and another four were used for the histological examinations. For the mechanical test, bone cement in each segment was pushed out and its failure load was measured. An average value of failure loads of four segments was calculated, and it was compared between the bioactive cement group and the PMMA cement group at the corresponding postoperative stages.

Histological Examination

Four segments of the femur prepared as above and all specimens containing the acetabular component were subjected to undecalcified histological examinations such as contact microradiography and scanning electron micrography.

RESULTS

Results of Push Out Test

The failure load obtained by push out test between bone and bone cement at one month of implantation was 2.16±0.8 MPa for bioactive bone cement, and 1.75±0.6 MPa for PMMA bone cement. The failure load at three months of implantation was 3.10±0.7 MPa for bioactive bone cement and 1.83±0.8 MPa for PMMA cement. Thus, the failure load with bioactive bone cement was greater than that with PMMA cement about 1.2 times at one month and about 1.7 times at three months of implantation. The failure load with bioactive bone cement tended to further increase thereafter, while that with PMMA cement remained unchanged throughout the experimental period.

Histological Findings

In one month specimens, bioactive bone cement showed closer contact with surrounding bone tissue than PMMA cement, but evidence of direct bonding between bioactive bone cement and bone tissue was not clearly demonstrated. However, in three month specimens (Figure 2 and 3), it was demonstrated that bioactive bone cement around the stem as well as the socket made direct bonding with bone tissue. On the other hand, there was always a thin layer of connective tissue intervening between PMMA cement and bone tissue at any postoperative stage.

CLINICAL APPLICATION

Bioactive bone cement was used in a few clinical cases. The case illustrated in Figure 4 was a 84 year old female who had undergone cemented prosthetic replacement of the right hip in 1976 with a Charnley's prosthesis, and its revision operation in 1981. When the patient came to our clinic in 1993, the acetabular component was completely loose and the

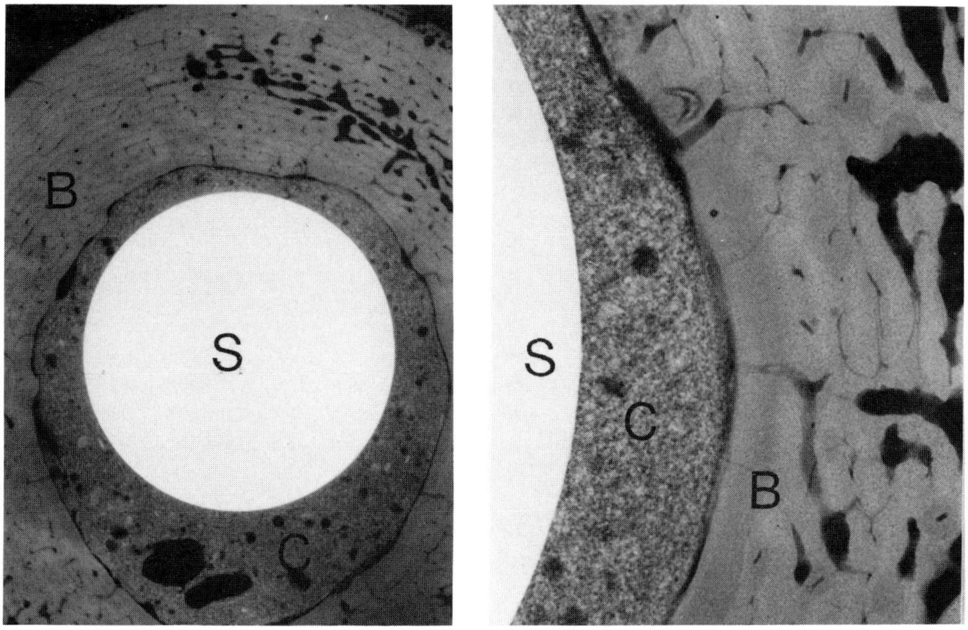

Figure 2. CMR of a transectioned segment of the femur harvested 3 months after operation, showing bone bonding with cement
S: stem C: bioactive bone cement B: bone

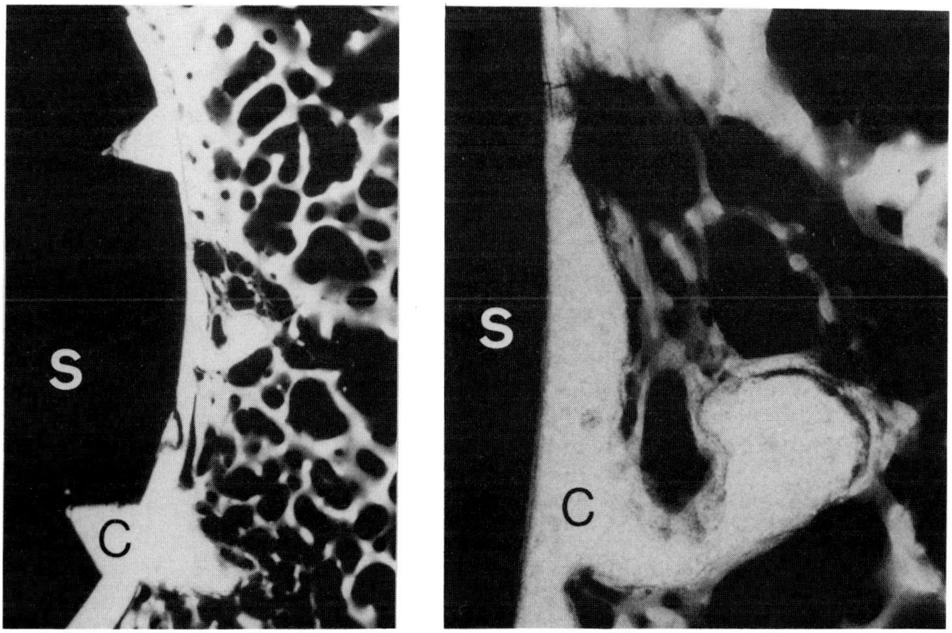

Figure 3. CMR of a cross sectioned acetabular component harvested 3 months after operation showing bone-cement interface
S: socket C: bioactive bone cement

head of the prosthesis was dislocated from the socket. As her bilateral iliac crests and right femoral head had been removed already at the previous operations, and in addition, as she was severely osteoporotic in general, it seemed impossible to harvest enough amount of autogenous bone to substitute for a large bone defect remaining after extraction of the socket. Therefore, bioactive bone cement was used to fix a new socket after thorough curettage of the connective tissue in the bone defect. In this case, bioactive bone cement of dough type was used. It hardened in about seven minutes in vivo. It generated heat slightly when it hardened so that surgeons felt with their hands slight warmness over its surface during the operation. As far as physical and laboratory examinations are concerned, no adverse effect has been observed. Partial weight bearing on the operated hip was commenced four weeks and full weight bearing eight weeks after operation. The longest postoperative follow up period is still only 10 months, but the patient has been doing well (Figure 4).

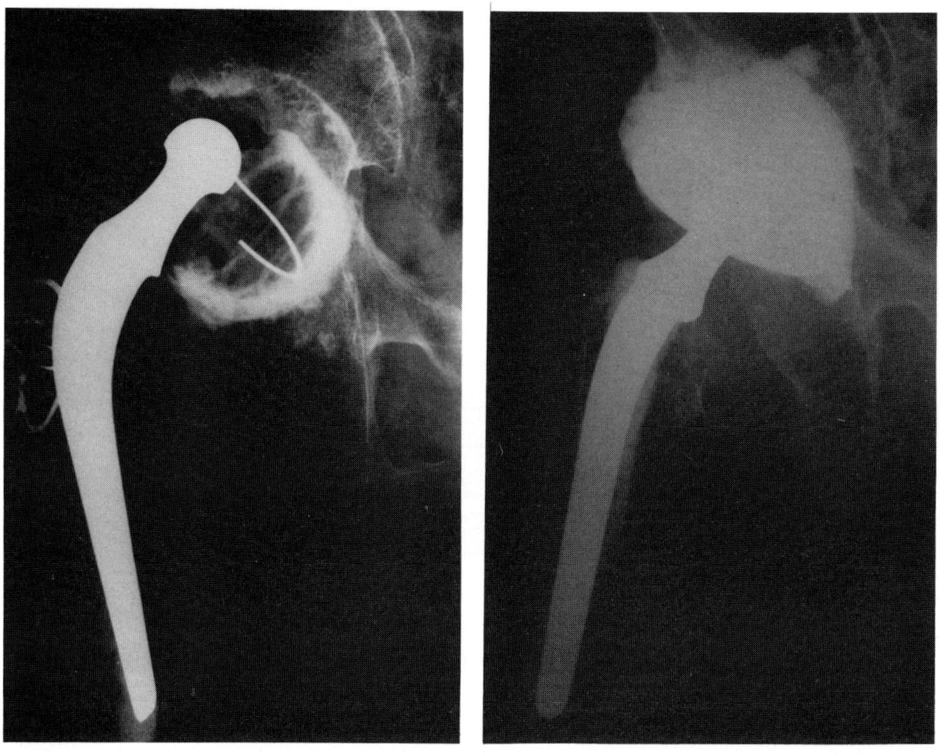

Figure 4. A 84 year old female developed severe loosening of the hip socket that was replaced with a new socket and bioactive bone cement. The right picture is 4 weeks postop.

REFERENCES

1. Kawanabe,K., Tamura,J., Yamamuro,T., Nakamura,T., et al. J. Appl. Biomater., 4: 135-141, 1993
2. Kokubo,T., Yoshihara,S., Nishimura,N., Yamamuro,T. et al. J. Am. Ceram. Soc., 74: 1739-1741, 1991

Bioceramics, Volume 7 Edited by Ö.H. Andersson and A. Yli-Urpo
(*Proceedings of the 7th International Symposium on Ceramics in Medicine, Turku, Finland, July 1994*)

The Application of HA-coated Dental Implants in Patients with Severe Bone Defects

G.L. de Lange, L. Kuiper and P.A. Blijdorp*
Academic Center for Dentistry (ACTA), Louwesweg 1, 1066 EA Amsterdam, N.L.
* Rijnstate Hospital, Arnhem, N.L.

ABSTRACT

Hydroxylapatite coated dental implants were selected for rehabilitation of patients with large jaw bone defects or severe bone atrophy. Bone reconstruction therapies were performed 3 months before implant surgery, in lower and/or upper jaw, including sinus floor elevations in the distal area of the maxilla with autogenous bone from the iliac crest, mixed with dense sintered HA. A total of 11 consecutive patients were evaluated radiographically, clinically and histologically. The results after 3-6 years (mean 4.3 years) include a total of 80 loaded fixtures and an implant survival rate of 97.5%. The failure rate (3.75%) observed is low and comparable with normal bone situations. Histological controls were performed and normal appearing bone tissues were observed with a high activity of osteoblast cells.

INTRODUCTION

One of the conditions for successful applications of dental implants is the necessity of withstanding high occlusal forces. This means that dental implants can only be used in sites where there is enough jaw bone of good quality or density. In these cases implants have shown to be very successful.[1] However, some patients have insufficient bone volume or large bone defects. When these patients need dental implants for their oral rehabilitation, augmentation procedures are necessary to increase the bone volume. After this procedure, fixtures can be placed in augmented bone to support a removable or fixed prosthesis. High failure rates, up to 45%, have been described for titanium implants placed in maxillary bone of poor quality.[2,3] Both the selection of the biomaterial and the clinical protocol are very critical under these circumstances. For these reason hydroxylapatite coated titanium have been used. Hydroxylapatite(HA) has shown to posses a high rate of biocompatibility for bone and periodontal tissues.[4,5,6,7] We have studied the efficacy of this ceramic material in bony sites with severe resorption, atrophy or defects. This was performed on patients who needed bone reconstruction therapies combined with dental implants.

All patients followed a precise protocol which included a thorough presurgical planning and complete prosthetic reconstructions. All patients were evaluated radiographically, clinically and histologically.

METHODS AND RESULTS

A total of 11 consecutive patients were investigated which were treated for partial or full edentulism between 1988 and 1991. All patients showed severe bone atrophy or severe bone defects (figures 1-3) and all had large difficulties with their oral function (speech, chewing, laughing etc). Bone reconstructions were performed under general anesthesia with autogenous bone, derived from the iliac crest and mixed with dense sintered hydroxylapatite granules with a ratio of 3:1. These procedures included a maxillary bone augmentation by a bone onlay technique (all patients), combined with sinusfloor elevation procedures (7 patients), combined with mandibular bone augmentations (2 patients). After a three months healing period, a total of 80 screw type implant fixtures were installed under local anesthesia. The fixtures had a diameter of 3.8 mm and were coated with a 80 - 100 μm thick layer of hydroxylapatite (Steri-Oss, Bausch and Lomp, Yoma Linda, CA, USA).
A three months healing period was used, without load, before the implant fixtures were uncovered and gradually loaded. Permucosal healing abutments were mounted into the fixtures and 3 weeks later the final abutments. Fixed bridges were made for occlusal rehabilitation in nearly all cases. All steps were carefully documented and much attention was paid to achieve optimal occlusion and maximal oral function and esthetics (Figures 4 and 6).
During implant surgery a biopsy was taken from the implant site. Thin ground sections and decalcified sections were made for histological evaluation.
All fixtures were checked every 3-6 months. Panoramic radiographs were made each year for radiographic evaluation. During the evaluation period, from 1988 on, three fixtures were lost (3.75%). Two were ailing after loading and another was lost after one year loading. These three fixtures showed an unfavorable peri-implant mucosa immediately after exposure from the abutment surgery. The remaining fixtures (96.25%) showed a healthy peri-implant mucosa, were firmly anchored in bone and could withstand occlusal loading.
Signs of bone resorption or peri-implant bone breakdown were not observed. The biopsies after three months showed normal bone structures including normal appearing bone marrow with bone trabeculae and a high activity of osteoblast cells (See figure 5).

DISCUSSION AND CONCLUSIONS

Patients with substantial loss of jaw bone always suffer from bad retention and stability of their removable prosthesis. Osseointegrated implant fixtures can support a fixed or removable prosthesis for reconstruction of completely or partially edentulous mandible or maxilla. These fixtures can be applied as long as there is sufficient bone volume left. Good clinical results have been described in mandibular bone but not in maxillary bone. Maxillary bone is often of poor quality and the maxillary sinus frequently is low. Titanium implants placed in acceptable maxillary bone situations show in general a failure rate up to 30% [2,8,9] and this increases up to 45% in compromised bone situations or extreme bone resorption. [2,10]
This study shows that hydroxylapatite (HA) coated dental implants can be very successful for application in patients with severe bone defects. The HA-coated implants used in augmented bone, could withstand loading and only 3.75% failed. The loss of fixtures we observed, is similar as described for routine cases where patients have selected for treatment in general practice because the ideal bone situation.
Bahat (1993) achieved a 5.5% failure rate in compromised situations of maxillary bone.[11] Although his patient cases still showed better bone conditions compared with our cases where there was no bone at all (see figures 1 and 3), he demonstrates clearly that with an appropriated presurgical planning, reconstruction and monitoring, a high success rate can be achieved.
Three main reasons can be identified for the succes rate in this patient series:

1. related to the biocompatibility of the biomaterials used,
2. related to mechanical loading
3. clinically related.

<u>1. The biocompatibility of the biomaterial used</u>

HA-coated implants in grafted bone sites offer certain advantages compared with implants with a titanium surface only. HA-coated titanium implant fixtures have shown a 2-3 times faster initial bone growth.[5] This means that the fixtures can be loaded earlier and more intensive. Another aspect is the high regenerative potential of the grafted bone as observed histologically. Vital bone cells and growth factors have been maintained and the use of HA granules apparently prevents bone resorption. This phenomenon has also been observed in previous studies from augmentation procedures of mandibular bone.[12]

<u>2. The loading of the implants in bone</u>

Mechanical load is an important factor in implant dentistry, especial in the distal area. The first and second molars

provide the area of greatest occlusal need and force. Much attention should be paid after inserting the implants in bone to avoid premature loading. For this reason a two step procedure has been used and the implants are kept unloaded below the mucosa, during the whole bone healing period. At the abutment surgery, the bone anchorage of the fixtures have been tested by placing the permucosal abutment into the implant fixture with controlled force.
It is known that osteoblast cells are very sensitive for mechanical loading. Gradual loading of the implant fixtures interacts with the high osteoblastic activity and can act as a physiological stimulus to the augmented bone by which osteoblasts and pre-osteoblast cells are stimulated in their phenotypic expression.

3. Clinical related aspects and the clinical protocol used

We have used a precise clinical protocol which includes thorough presurgical planning, simulation in reverse of the steps needed to reach the chosen goal, placement of the implants exactly according to the treatment plan and applying the highest possible biocompatibility of selected biomaterials. These point can be summarized by the so called **BIOM** principles of implant dentistry as described previously: maximum **B**iocompatibility of the biomaterials used, proper **I**mplant design, controlled **O**cclusal loading, and a high level of **M**aintenance and after care.[13]
Interfering in the occlusal anatomy can be avoided by omitting occlusal fixation screws. This improves the occlusal design and offers much more rigidity to the occlusal surfaces. Further aesthetic improvement can be achieved by a proper soft tissue management which as a rule includes gingival grafts, gaining attached gingiva around the fixtures and/or the reconstruction of interdental papillae.
This investigation demonstrates that despite discouraging statistics, implant dentistry can be successful in cases of poor bone quality or severe bone resorption.

The following conclusions are made:

1. HA- coated implants serve well in patients with severe bone resorption or bone defects.
2. There is a low risk for these implants in augmented bone
3. Implants show minimal loss, 3.75 % after 3-6 years.
4. Normal bone growth and bone turnover is observed in augmented bone area.
5. Composed grafts do not show signs of resorption.
6. HA-coated implants become well fixed in augmented bone. and can withstand occlusal loading.
7. Progressive loading interacts with bone deposition.
8. With an appropriated presurgical planning, reconstruction and monitoring a high success rate can be achieved.
9. The prosthetic result is dependent from restoring the original width of the upper arch and placement of the the implants in former tooth positions.
10. Final esthetics depends from soft tissue management.

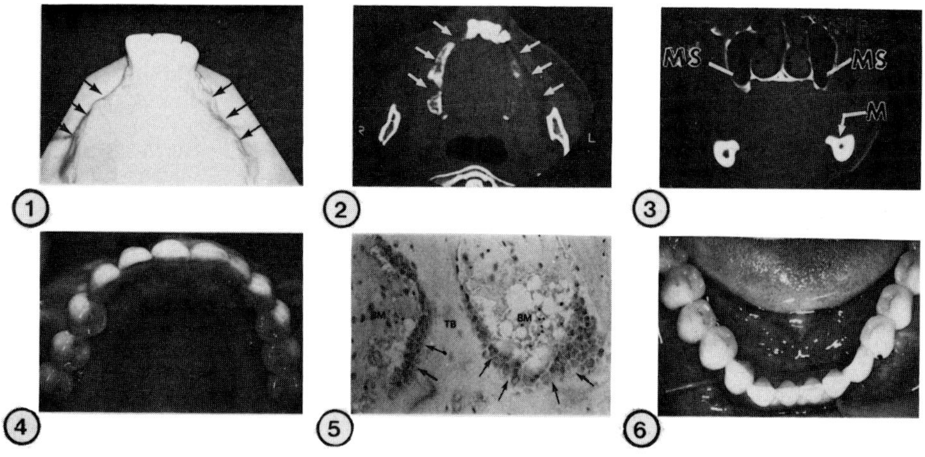

Figures 1-6. Clinical illustrations of a typical case of bone resorption defects. The figures demonstrate the problem before and the obtained results after bone reconstruction procedures and dental implants.

Figure 1. Occlusal view of upper jaw, plaster replica of the oral situation. Note the narrow width of the maxillary arch and the severe resorption defects of alveolar bone in the premolar and molar area at both sides (arrows) after premature loss of these teeth.

Figure 2. Computer tomograph of maxillary bone. Horizontal cross section at apical level of remaining teeth. Note the presence of only 3 anterior teeth and absence of canines, premolars and molars. There is severe bone loss and there is insufficient height and thickness of remaining alveolar bone for the placement of dental implants. Note the large bone defects of maxillary bone in the distal area at left and right (arrows) which corresponds with the edentulous molar area in fig 1.

Figure 3. Computer tomograph of upper and lower jaw. Frontal cross section made in the vertical plane in the molar area, the nasal cavity is in the middle. There is only a thin bone layer left of less than 1 mm thickness below the maxillary sinus (MS). Both left and right cross sections of lower jaw can be seen. The edentulous mandibular bone (M) shows severe resorption with only 3-5 mm bone height left, above the mandibular canal.

Figure 4. Occlusal view of the same patient, after bone reconstruction and rehabilitation with HA-coated dental implants and fixed bridges. The three natural teeth are maintained. Note the improved width of the maxillary arch (compare with fig.1) and the absence of occlusal screws.

Figure 5. Histology of augmented bone. Three months biopsy. There is normal appearing bone tissue with bone trabeculae (TB) and normal appearing bone marrow (BM). Many active osteoclast cells line the marrow spaces and have deposited new bone (arrows).

Figure 6. Occlusal view of the lower jaw after bone reconstruction procedures and placement of HA-coated implant screws. Fixed bridges have been made (porcelain fused to metal) in the edentulous distal parts. Note the improved width of the mandibular arch and the occlusal design of the fixed bridges in premolar and molar area.

REFERENCES

1 Adell, R., Lekholm, U., Rockler, B. and Brånemark, P.I. Int.J. Oral Surg. (1981) 10, 387-416

2 Engquist, B., Bergendal, T., Kallus, T. and Linden, U. Int. J. Oral Maxillofac. Implants (1988) 3, 129-134

3 Jaffin, R. A. and Bergman, C. L. J. Periodontol. (1991) 62, 2-4

4 De Lange, G. L. and De Putter, C. J. Oral Implantol. (1993) 19, 123-135

5 Cook, S. D., Kay, J. F., Thomas, K. and Jarcho, M. Int. J. Oral Maxillofac. Implants (1988) 3, 15-27

6 De Groot, K. In: Ceramics in Surgery, Elsevier, Amsterdam, NL (1983) 79-90

7 Yukna, R.A., Harrison, G, Caudill, R.F., Evans, G.H., Mayer, E.T. and Miller, S. J. Periodontol. (1985) 56, 540-547

8 Jemt, T., Lekholm, U. and Adell, R. Int. J. Oral Maxillofac. Implants (1989) 3, 211-217

9 Steenberghe, D., Lekholm, U., Bolender, Folmer, T.,C., Henry, P., Herrman, I., Higuchi, K., Laney, W., Linden, U. and Astrand, P. Int. J. Oral Maxillofac. Implants (1990) 5, 272-281

10 Nystrom, E., Kahnberg, K. E. and Gunne, J. Int. J. Oral Maxillofac. Implants (1993) 8, 45-52

11 Bahat, O. Int. J. Oral Maxillofac. Implants (1993) 8, 151-161

12 Blijdorp, P. A., Vanassche, B. J. E. and De Lange, G. L. J. Oral Maxillofac. Surg. (1988) 17, 354-357

13 De Lange, G.L., Dental Implantology, Bohn Stafleu Van Loghum, Houten, NL (1991) 63-75

Bioceramics, Volume 7 Edited by Ö.H. Andersson and A. Yli-Urpo
(*Proceedings of the 7th International Symposium on Ceramics in Medicine, Turku, Finland, July 1994*)

Efficiency of Hydroxyapatite-Tricalcium Phosphate-Composite (HAP-TCP) for Bone Defect of Tibia Fracture – Comparison Between HAP-TCP and Autogenous Iliac Bone

K. Suzuki* and H. Kurabayashi**

* Department of Orthopaedic Surgery, Fujita Health University School of Medicine, 1-98 Dengakugakubo, Kutsukake, Toyoake 470-11, Japan

** Department of Orthopaedic Surgery, Hiratsuka Municipal Hospital, 1-19-1 Minamihara, Hiratsuka 254, Japan

ABSTRACT

The efficiency of a hydroxyapatite-tricalcium phosphate-composite (HAP-TCP) filler for traumatic bone defects of the tibia was studied in 16 patients treated with HAP-TCP and 14 treated with autogenous ilium. The period of time until full-weight bearing was significantly shorter in the HAP-TCP group than in the autogenous bone group. However, there was no significant difference in the time required for recovery of at least 90% of the healthy side range of motion (ROM). Postoperative infections occurred in two patients from the HAP-TCP group and three from the autogenous bone group. In addition, pain at the bone removal area occurred in five patients from the autogenous bone group, but in none of the HAP-TCP group.

INTRODUCTION

Some type of filler is often necessary for bone defects in patients with comminuted and depressed fractures. Autogenous ilium is often used in such cases, but problems arise such as the limited amount of bone which can be obtained and residual symptoms at the donor site. In the present study, we packed traumatic bone defects of the tibia with a hydroxyapatite-tricalcium phosphate-composite (HAP-TCP) and compared the efficiency of this filler with that of autogenous bone.

SUBJECTS

There were 105 patients with tibia fractures who underwent surgery in our department from January 1991 to May 1993. Twenty-nine of these patients had bone defects which required filling. They were divided into a group receiving autogenous bone and a group receiving HAP-TCP as the filler.

Autogenous bone group: Autogenous bone was used from January 1991 to February 1992 in 14 patients with a median age of 52.6 years (22-80 years), composed of eight men and six women. The fracture was in the proximal tibia in seven cases, at the diaphysis in six cases, and in the distal tibia in one case. Five of the patients had an open fracture. A pseudoarthrosis was treated in three cases. One patient who had a stroke 2 months postoperatively and one who developed an infected pseudoarthrosis after the filling with autogenous bone and later treatment with HAP-TCP were excluded, so that 12 patients were available for assessment.

HAP-TCP group: HAP-TCP was used from March 1992 to May 1993 in 16 patients with a median age of 51.4 years (16-85 years): 12 males and four females. The fracture was in the proximal tibia in six cases, at the diaphysis in seven cases, and in the distal tibia in three cases. Five patients had an open fracture. A pseudoarthrosis was treated in two patients, one of whom had developed an infected pseudoarthrosis after packing with autogenous bone. All of these patients were eligible for assessment.

METHODS AND RESULTS

Weight-bearing recovery, ROM recovery, and complications were studied in both groups. Weight-bearing recovery was evaluated on the basis of the time required for walking with full weight-bearing after the operation, and ROM recovery was assessed as the time until recovery of at least 90% of the total active motion of the knee and ankle joints on the healthy side.

Weight-bearing recovery: The mean time until full weight-bearing walking was 5.2 ± 2.0 months ($M \pm SD$) in the autogenous bone group and 3.4 ± 1.3 months in the HAP-TCP group. The time was thus significantly shorter in the HAP-TCP group ($p<0.05$; Wilcoxon test).

ROM recovery: The mean ROM recovery time was 3.4 ± 2.6 months ($M \pm SD$) in the autogenous bone group and 1.8 ± 1.0 months in the HAP-TCP group. Thus no significant difference was seen between the two groups (ns; Wilcoxon test).

Complications: Postoperative infections occurred in three patients from the autogenous bone group and two from the HAP-TCP group. Extended resection of the infected focus was performed in one patient from the autogenous group, and the defect was packed with HAP-TCP after control of the infection. The other four patients were treated successfully by the administration of antibiotics. Pain at the site of bone removal occurred in five patients from the autogenous bone group, but no symptoms were noted in the HAP-TCP group.

CASE REPORTS

Cases 1 and 2 both had depressed fractures on the lateral articular surface of the tibia. Case 1 was a 54-year-old woman in the autogenous bone group whose injury was due to a traffic accident. The weight-bearing recovery time was 4 months and the ROM recovery time was 2 months (Figure 1-A,B).
Case 2 was a 56-year-old woman in the HAP-TCP group who was injured in a fall. Weight-bearing recovery took 2 months and ROM recovery occurred after one month (Figure 2-A,B).

Cases 3 and 4 both had comminuted fractures of the tibial condyle. Case 3 was a 53-year-old woman in the autogenous bone group who was injured in a traffic accident. The weight-bearing recovery time was 7 months and the ROM recovery time was 6 months (Figure 3-A,B). Case 4 was an 84-year-old woman in the HAP-TCP group who was injured in a fall. Weight-bearing recovery took 2 months and ROM recovery occurred after one month (Figure 4-A,B).

Cases 5 and 6 both had comminuted fractures of the diaphysis of the tibias with delayed union after Ender nailing. Case 5 was a 68-year-old man in the autogenous bone group with an open fracture who had been injured in a traffic accident. Bone grafting was performed 3 months after Ender nailing. The weight-bearing recovery

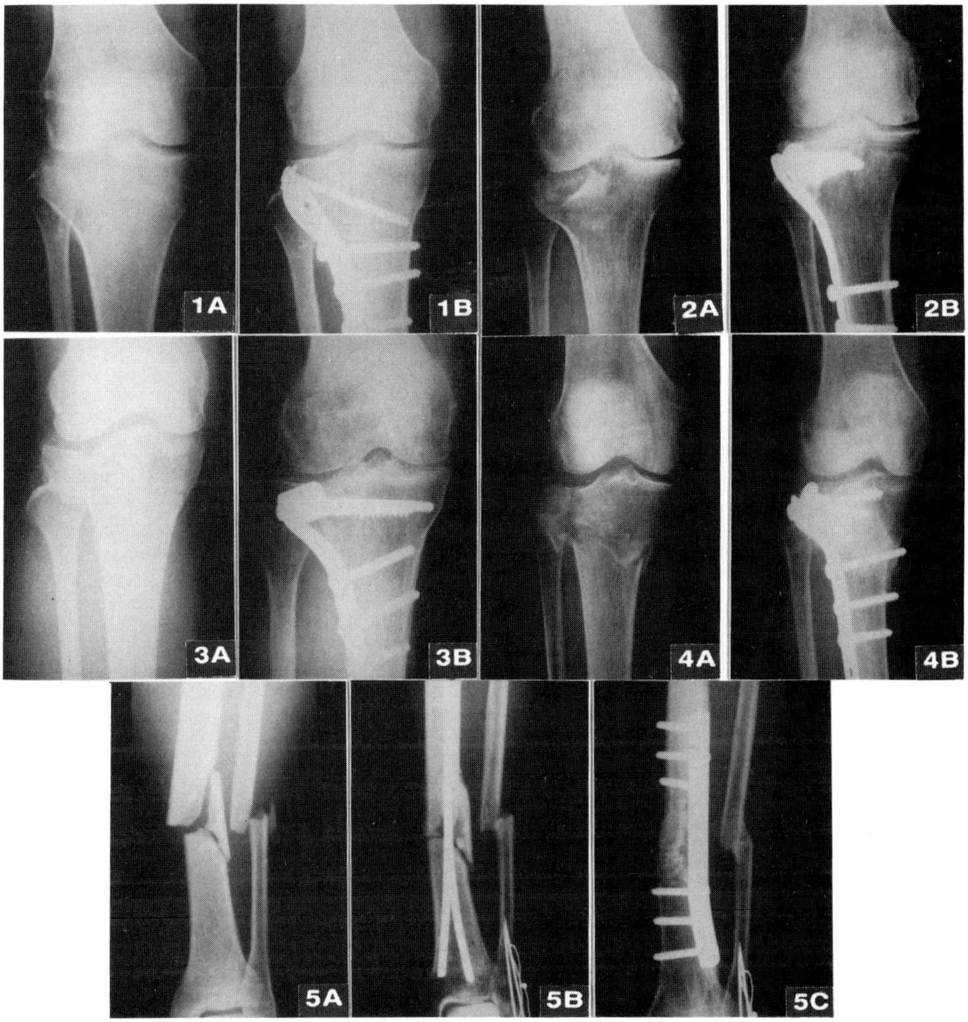

Figure 1 Case 1. A 54-year-old female (autogenous bone group).
1-A: Depressed fracture on the lateral; articular surface of the right tibia.
1-B: Four months after operation with full weight-bearing recovery.

Figure 2 Case 2. A 56-year-old female (HAP-TCP group).
2-A: Depressed fracture on the lateral articular surface of the right tibia.
2-B: Two months after operation with full weight-bearing recovery.

Figure 3 Case 3. A 53-year-old female (autogenous bone group).
3-A: Comminuted fracture of the right tibial condyle.
3-B: Seven months after operation with full weight-bearing recovery.

Figure 4 Case 4. An 85-year-old female (HAP-TCP group).
4-A: Comminuted fracture of the right tibial condyle.
4-B: Two months after operation with full weight-bearing recovery.

Figure 5 Case 5. A 68-year-old male (autogenous bone group).
5-A: Comminuted fracture of the diaphysis of the left tibia.
5-B: Three months after Ender nailing.
5-C: Eight months after the second operation with full weight-bearing recovery.

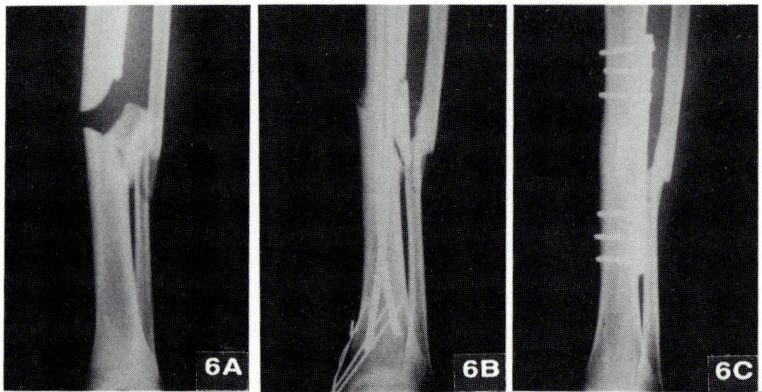

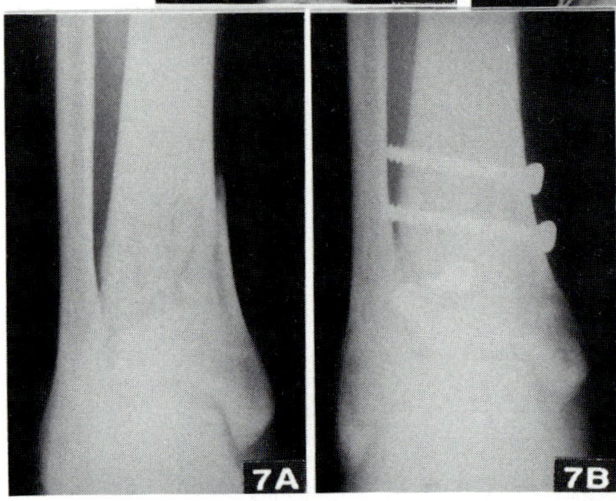

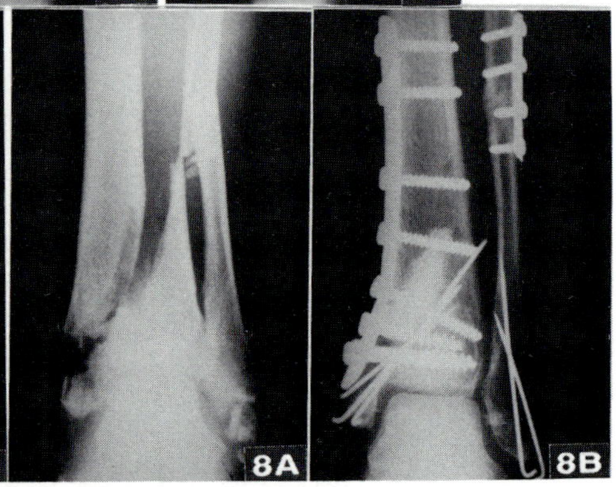

Figure 6 Case 6. A 22-year-old male (HAP-TCP group).
6-A: Comminuted fracture of the diaphysis of the left tibia.
6-B: Seven months after Ender nailing.
6-C: Four months after the second operation with full weight-bearing recovery.

Figure 7 Case 7. A 36-year-old male (autogenous bone group).
7-A: Comminuted fracture of the right distal tibia.
7-B: Four months after operation with full weight-bearing recovery.

Figure 8 Case 8. A 44-year-old male (HAP-TCP group).
8-A: Comminuted fracture of the left distal tibia.

time was 8 months and the ROM recovery time was 7 months after the second operation (Figure 5-A,B). Case 6 was a 22-year-old man in the HAP-TCP group who was also injured in a traffic accident. Filling with HAP-TCP was performed 7 months after Ender nailing. Weight-bearing recovery took 4 months and ROM recovery occurred one month after the second operation (Figure 6-A,B).

Cases 7 and 8 had comminuted fractures of the distal tibia. Case 7 was a 36-year-old man in the autogenous bone group who was injured in a fall. Weight-bearing recovery took 4 months and ROM recovery occurred after 3 months (Figure 7-A,B). Case 8 was a 38-year-old man in the HAP-TCP group who was injured in a traffic accident. Both weight-bearing recovery and ROM recovery took 2 months (Figure 8-A,B).

Case 9 was a 38 year-old man with an open comminuted fracture of the tibial diaphysis (Figure 9-A). Bone union was achieved by filling the defect with autogenous bone, but

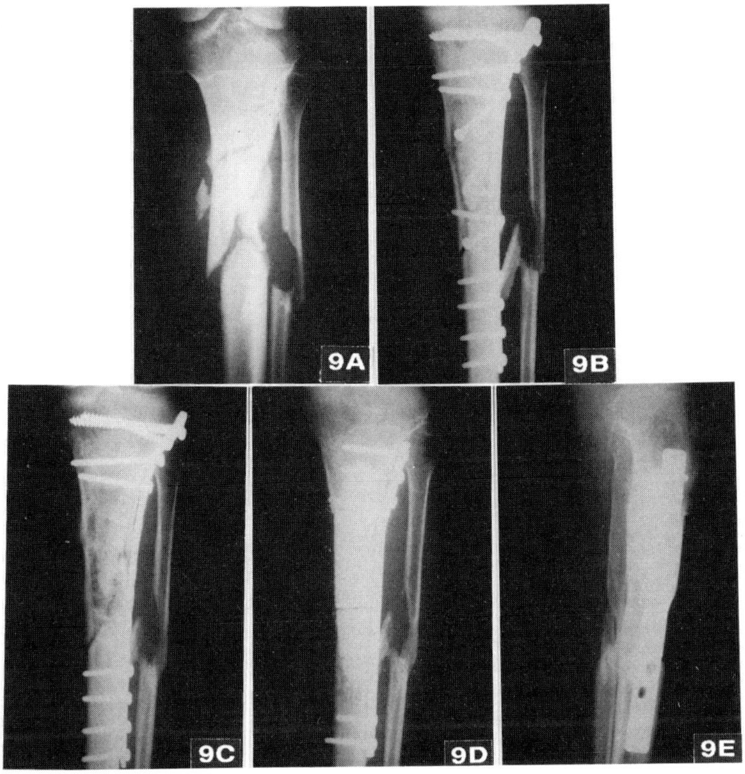

Figure 9 Case 9. A 38-year-old male.
9-A: Open comminuted fracture of the left tibial diaphysis.
9-B: Four months after the first operation filling by autogenous bone, infection had developed.
9-C: Six months after autogenous bone grafting, the infected focus was excised twice, and soon thereafter the second internal fixation packing of the defect with HAP-TCP was performed.
9-D: Six months after the second operation with full weight-bearing recovery.
9-E: Lateral view at the same time as Figure 9-D.

infection developed during partial weight bearing after 4 months (Figure 9-B). The original organism was *Staphylococcus epidermidis* and this was substituted by methicillin-resistant *S.aureus* (MRSA). Appropriate antibiotics were administered and the infected focus was excised twice (Figure 9-C). Six months after autogenous bone grafting, extended local excision, packing of the defect with HAP-TCP, fixation with a plate, and pedicle skin grafting were performed. The infection was brought under control , with weight-bearing recovery occurring 6 months and ROM recovery one month after HAP-TCP filling (Figure 9-D,E).

DISCUSSION AND CONCLUSIONS

HAP-TCP is a composite of hydroxyapatite (HAP) and tricalcium phosphate (TCP) which shows excellent biocompatability and bone conductivity (1). TCP also fuses strongly with the surrounding bone because it acts as a bone substitute.

Nakamura (2) inserted fresh autogenous bone grafts into rats and found that the compression strength of 67.3 MPa before grafting was reduced by about 25% 8 weeks after grafting. When autogenous bone was used as the filler, the graft was gradually replaced by new bone and its mechanical strength was reduced during this process.

Hori et al. (3) found in experiments using rabbits that the compression strength of cancellous bone was 233.8 kg/cm^2 and that of HAP was 25.5 kg/cm^2 , but the compression strength of HAP increased to 165.6 kg/cm^2 when it was used to pack a bone defect and was then removed after 4 weeks. When HAP-TCP was used as a filler, it fused directly with the surrounding bone from an early stage. TCP formed a strong union with the surrounding bone by creeping substitution, and HAP gradually increased in strength after grafting because of its excellent bone conductivity, so the final strength approached that of normal cancellous bone.

These findings explain why weight-bearing recovery was significantly faster in the HAP-TCP group than in the autogenous bone group in the present study.

REFERENCES

1. Takado, T., Harii, K., Komuro, Y. and Yonehara, Y., J. Japan Soc. Plastic Reconstr. Surg. 1992, 12, 660-667

2. Nakamura, S., Kitazato Igaku 1988, 18, 406-419

3. Hori, M., Munemiya, M., Takahashi, S., Sawai, K., Niwa, S., Tagai, H., Kobayashi, M., Ono, M. and Takeuchi, K., J. Chubu Orthop. Traumatol. 1984, 27, 2133-2135

Bioceramics, Volume 7 Edited by Ö.H. Andersson and A. Yli-Urpo
(*Proceedings of the 7th International Symposium on Ceramics in Medicine, Turku, Finland, July 1994*)

Comparison of Hydroxyapatite-Coated with Harris/Galante Porous Uncemented Total Hip Replacement – A Retrospective Matched Study and Experimental Study for the Improvement of HA-Coating

K. Hayashi, Y. Nakashima, I. Inadome, Y. Sugioka and I. Noda*
Department of Orthopaedic Surgery, Faculty of Medicine, Kyushu University, Fukuoka, Japan
* Research and Development Section, Bioceram Division, Kyocera Corp., Japan

Abstract

We compared the clinical results of Harris/Galante Porous (HGP) uncemented total hip replacement (THR) with that of hydroxyapatite (HA)-coated THR (HA-coating onto flat titanium) using a matched study (sex, age, weight, duration of follow-up, and diagnosis). Incidence of postoperative thigh pain was 31% in HGP and 6% in HA ($p<0.005$). Incidence of clear zone of zone 1 was 94% in HGP and 22% in HA ($p<0.001$), and that of zone 4 was 91% in HGP and 44% in HA (N.S.).
HA-coating onto the porous titanium created by the Shielded Arc spray technique did not obstruct the pores of the surface and the sprayed layer was more tightly attached to the titanium substrate because of the lack of cracks or voids of the sprayed titanium layer commonly created by the plasma spray technique.

Introduction

We had high incidence of postoperative thigh pain of Harris/Galante Porous (HGP) type uncemented total hip replacement (THR) (1). Therefore we tried to perform hydroxyapatite (HA)-coated THR to improve the fixation of the uncemented THR. HA-coated titanium implant was created, based on our previous experimental study (2-10). Recent clinical trials on HA-coated THR have shown encouraging short-term results, however, almost few reports have performed a matched study (sex, age, weight, duration of follow-up, and diagnosis) with other uncemented THR. We report a comparison study of HA-coated THR with HGP type THR, using a matched study. HA-coating onto flat titanium substrate used in our clinical trial has the possibility to fracture under shear loading (7). Therefore we developed HA-coating onto porous titanium by Shielded Arc spray technique. In this implant HA-coating can be applied to the porous titanium surface without obstruction of the pores of the surface and bone-implant attachment strength was greater than that of bead-coated titanium and HA-coating onto flat titanium (8). This time we compared the attachment strength of sprayed titanium layer to the titanium substrate created by the Shielded Arc spray technique with that by the plasma spray technique.

[Clinical Study]

1. Materials and Methods

All primary THR that were performed at Kyushu University were retrospectively reviewed. There were 150 HGP THR (mean age ; 58 years, mean duration of follow-up ; 5 years) and 35 HA THR (mean age ; 52 years, mean duration of follow-up ; 1.5 years) (Kyocera Corp). Sixteen cases were selected for matched study out of both groups. Sex, age, and body weight were matched and duration of follow-up was 1.5-2.0 years. We investigated these cases regarding to postoperative thigh pain and radiographic findings. All these cases had osteoarthritis.

2. Results

In HGP THR, 31% of cases had thigh pain during 6 months to 1.5 years after operation and over 90% of case showed radiolucent zone at zone 1 and 4 at 5 years after operation. In HA THR , 6% of case had thigh pain and 10%-15% of cases showed radiolucent zone at zone 1 and 4.

In matched study, 31% of HGP THR and 6% of HA THR had postoperative thigh pain ($p<0.005$). Regarding to radiographic findings, 94% of HGP THR and 22% HA THR showed clear zone at zone 1 ($p<0.001$), and 91% of HGP THR and 44% THR showed clear zone at zone 4 (N.S.) (Table 1, Fig. 1).

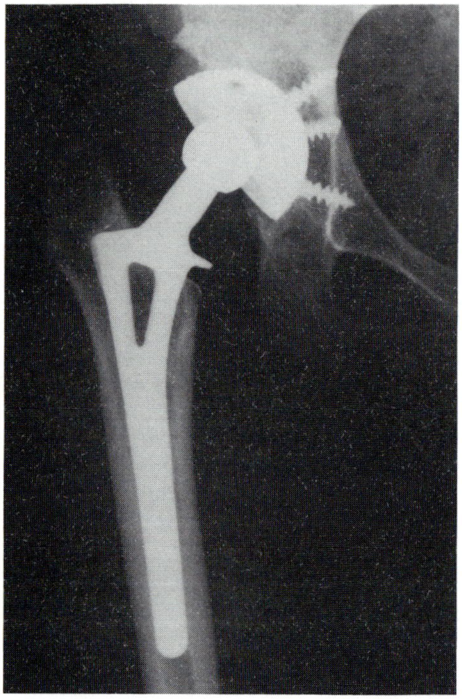

Fig. 1 : HA THR (2 years postoperation)

Table 1 : Matched study of HGP and HA THR

	HGP	HA
Thigh pain	31%	6% ($p<0.005$)
Clear zone		
zone 1	94%	22% ($p<0.001$)
zone 2	91%	44% (N. S.)

[Experimental study for the improvement of HA-coating]

1. Materials and Methods

We compared the cut surface perpendicular to the implant surface of the sprayed titanium layer created by the Shielded Arc spray technique with that by the plasma spray technique using scanning electron microscopy (SEM). Then we also compared the attachment strength of the sprayed titanium created by the Shielded Arc technique to the titanium substrate with that created by the plasma spray technique using ACT-JP test (11, 12). In ACT-JP test, we sprayed small alumina particles against the surface of the sprayed titanium layer, then we measured the volume of the worn titanium (Fig. 2). Therefore, the greater the volume of worn titanium, the weaker the attachment of the sprayed titanium layer to the titanium substrate in this test.

2. Results

In the sprayed titanium layer created by the plasma spray technique, voids or cracks were observed, however, those were not observed in the sprayed titanium layer created by the Shielded Arc spray technique (Fig. 3, 4). The volume of the worn titanium of the Shielded Arc spray technique was six times lower than that of the plasma spray technique. The attachment strength of the sprayed titanium layer created by the Shielded Arc spray technique was six times greater than that created by the plasma spray technique (Fig. 5).

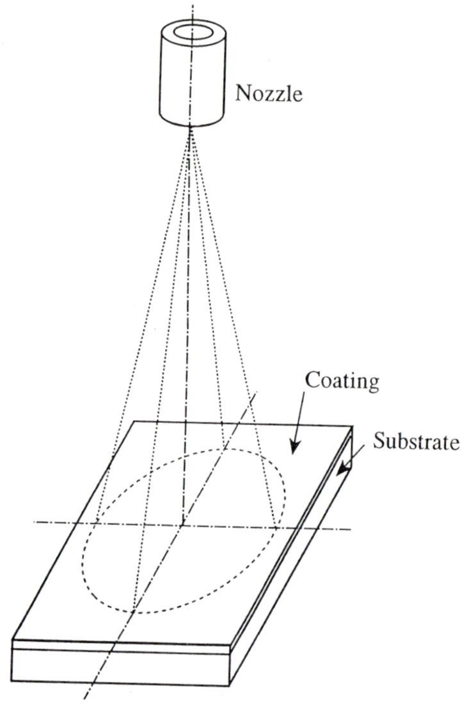

Fig. 2 : ACT-JP test

Fig. 3 : Cut surface of Low pressure plasma spray (SEM x25)

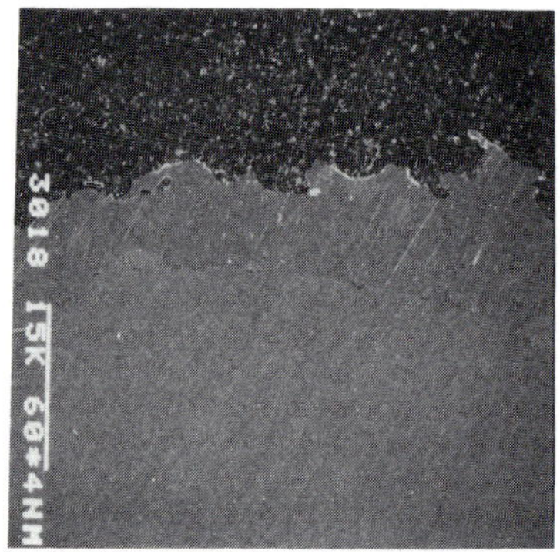

Fig. 4 : Cut surface of Shielded Arc spray (SEM x25)

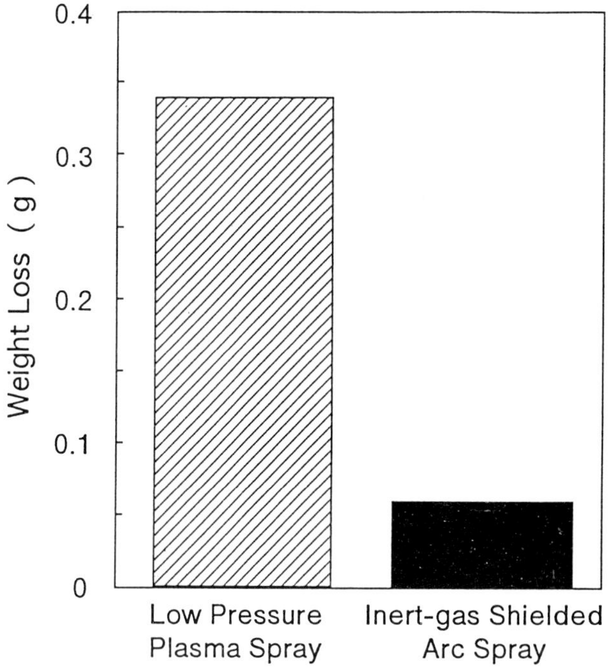

Fig. 5 : The volume of the worn titanium in ACT-JP test

Discussion

HA-coated titanium femoral prostheses showed a superior short-term clinical results to HGP femoral prostheses even though HA was coated onto flat titanium substrate. However, we found that HA-coating should be protected from shear loading because mechanical failure occurred at coating-titanium substrate interface in push-out test of transcortical dog model (7). In the implants with HA-coating onto porous sprayed titanium layer created by the Shielded Arc spray technique, the failure site was not coating-titanium, but bone-coating interface. HA can be applied to the surface of titanium without obstruction of the pores of the surface and the bone-implant interface shear strength of this implant was significantly greater than that of bead-coated porous titanium and HA-coating onto flat titanium implants (8). The attachment strength of sprayed titanium created by the Shielded Arc technique to titanium substrate was six times greater than that by the plasma spray technique. HA-coating onto porous sprayed titanium created by the Shilded Arc spray technique may be a promising composite to make the clinical results of HA-coating THR better.

Reference

1. Hayashi K. and Sugioka Y. J. Jpn. Orthop. Assoc. (in Japanese), 1992, 66, 681-682

2. Hayashi K., Matsuguchi N., Noda I. et al. In Transaction of the Third World Biomaterials Congress, Business Congress, Business Center for Academic Societies, Tokyo, Japan, 1988, 314

3. Hayashi K., Matsuguchi N., Uenoyama K., Kanemaru T., and Sugioka Y. J. Biomed. Mater. Res., 1989, 23, 1247-1259

4. Hayashi K., Uenoyama K., Matsuguchi N., and Sugioka Y., In Handbook of Bioactive Ceramics, CRC Press, Boca Raton, 1990, 149-153

5. Hayashi K., Noda I., Uenoyama K., and Sugioka Y. J. Biomed. Mater. Res., 1990, 24, 1111-1113

6. Hayashi K., Uenoyama K., Matsuguchi N., and Sugioka Y. J. Biomed. Mater. Res., 1991, 25, 515-523

7. Hayashi K., Inadome T., Mashima T., and Sugioka Y. J. Biomed. Mater. Res., 1993, 27, 557-563

8. Nakashima Y., Hayashi K., Inadome T., and Sugioka Y. In Bioceramics, Butterworth-Heinemann Ltd, Philadelphia, USA, 1993, 6,449-453

9. Hayashi K., Nakashima T., Inadome T., and Sugioka Y. In Bioceramics, Butterworth-Heinemann Ltd, Philadelphia, USA, 1993, 6, 454-459

10. Takamura K., Hayashi K., Ishinishi N., Yamada T., and Sugioka Y. J. Biomed Mater. Res., 1994 (in press)

11. Noda I., Okimatsu H., Nishio Y., and Fujisawa A. In Transaction of Fifteen Japanese Society for Biomaterials. Business Center for Academic Societies, Tokyo, Japan, 1993, p63

12. Arata T., Ohmori A., and Li C. J. In Trans. JWRI, 1987, 15, 44-54

Author Index

Subject Index